The Gaṇitatilaka and its Commentary

The Gaṇitatilaka and its Commentary: Two Medieval Sanskrit Mathematical Texts presents the first English annotated translation and analysis of the *Gaṇitatilaka* by Śrīpati and its Sanskrit commentary by the Jaina monk Siṃhatilakasūri (13th century CE). Siṃhatilakasūri's commentary upon the *Gaṇitatilaka* is a key text for the study of Sanskrit mathematical jargon and a precious source of information on mathematical practices of medieval India; this is, in fact, the first known Sanskrit mathematical commentary written by a Jaina monk, about whom we have substantial information, to survive to the present day.

In presenting the first annotated translation of these two Sanskrit mathematical texts, this volume focusses on language in mathematics and puts forward a novel, fresh approach to Sanskrit mathematical literature which favours linguistic, literary features and textual data. This key resource makes these important texts available in English for the first time for students of Sanskrit, ancient and medieval mathematics, South Asian history, and philology.

Alessandra Petrocchi completed a PhD in Sanskrit and Classical Indian Studies at the University of Cambridge, UK, and is currently a Leverhulme Early Career Fellow in the Faculty of Linguistics, Philology and Phonetics at the University of Oxford, UK, undertaking research into the early vernaculars of Italy and the history of numerals in the Renaissance. She has published several papers on Sanskrit sources on mathematics, comparative and historical Indo-European linguistics, and Italian Renaissance literature. Her research interests include textual criticism, manuscript studies, textual traditions and networks of knowledge across the Mediterranean, and the language-culture-literature interface.

Scientific Writings from the Ancient and Medieval World

Series editor: John Steele, Brown University, USA

Scientific texts provide our main source for understanding the history of science in the ancient and medieval world. The aim of this series is to provide clear and accurate English translations of key scientific texts accompanied by up-to-date commentaries dealing with both textual and scientific aspects of the works and accessible contextual introductions setting the works within the broader history of ancient science. In doing so, the series makes these works accessible to scholars and students in a variety of disciplines including history of science, the sciences, and history (including Classics, Assyriology, East Asian Studies, Near Eastern Studies, and Indology).

Texts will be included from all branches of early science including astronomy, mathematics, medicine, biology, and physics, and which are written in a range of languages including Akkadian, Arabic, Chinese, Greek, Latin, and Sanskrit.

The Foundations of Celestial Reckoning
Three Ancient Chinese Astronomical Systems
Christopher Cullen

The Babylonian Astronomical Compendium MUL.APIN
Hermann Hunger and John Steele

The Gaṇitatilaka and its Commentary
Two Medieval Sanskrit Mathematical Texts
Alessandra Petrocchi

www.routledge.com/classicalstudies/series/SWAMW

The Gaṇitatilaka and its Commentary

Two Medieval Sanskrit Mathematical Texts

Critically Revised with Introduction, Annotated Translation, and Explanatory Analysis

Alessandra Petrocchi

Routledge
Taylor & Francis Group
LONDON AND NEW YORK

First published 2019 by Routledge

2 Park Square, Milton Park, Abingdon, Oxon OX14 4RN
605 Third Avenue, New York, NY 10017

Routledge is an imprint of the Taylor & Francis Group, an informa business

First issued in paperback 2022

Publisher's Note

The publisher has gone to great lengths to ensure the quality of this reprint but points out that some imperfections in the original copies may be apparent.

British Library Cataloguing-in-Publication Data
A catalogue record for this book is available from the British Library

Library of Congress Cataloging-in-Publication Data
Names: Petrocchi, Alessandra, author. | âSrâipati (Son of Nâagadeva)
Title: The Gaṇitatilaka and its commentary : two Medieval Sanskrit mathematical texts : critically revised with introduction, annotated translation, and explanatory analysis / Alessandra Petrocchi.
Description: Abingdon, Oxon ; New York, NY : Routledge, 2019. |
Series: Scientific writings from the ancient and Medieval world |
Includes bibliographical references and index.
Identifiers: LCCN 2018042798 (print) | LCCN 2018049458 (ebook) |
ISBN 9781351022262 (ebook) | ISBN 9781351022255 (web pdf) |
ISBN 9781351022248 (epub) | ISBN 9781351022231 (mobi/kindle) |
ISBN 9781138496217 (hardback : alk. paper)
Subjects: LCSH: Mathematics–India–HIstory. |
Mathematics–India–Early works to 1800. |
Mathematics–History–11th century. |
Mathematics, Medieval. | Hindu mathematics.
Classification: LCC QA27.I4 (ebook) | LCC QA27.I4 P48 2019 (print) |
DDC 510.954–dc23
LC record available at https://lccn.loc.gov/2018042798

ISBN: 978-1-138-49621-7 (hbk)
ISBN: 978-1-03-233853-8 (pbk)
DOI: 10.4324/9781351022262

Typeset in Times New Roman
by Newgen Publishing UK

To Irisha

Contents

List of tables x
Preface xi
List of abbreviations xiii

PART I
Introduction 1

1 Introducing the *Gaṇitatilaka* and its commentary 3

1.1 Sanskrit mathematical writings in medieval India 3
1.2 Authorship and content of the Gaṇitatilaka *5*
1.3 The Gaṇitatilaka, *a work on* pāṭīgaṇita *7*
1.4 Siṃhatilakasūri, the commentator of the Gaṇitatilaka *12*
1.5 The genre of mathematical commentaries 16
1.6 Aspects of style and language in Siṃhatilakasūri's commentary 19
1.7 Reading the Gaṇitatilaka *and its commentary 23*

2 On the edition of the *Gaṇitatilaka* and some methodological notes 25

2.1 The edition 25
2.2 The Sanskrit manuscript 26
2.3 Linguistic phenomena and oddities: Between the manuscript and the edition 28
2.4 Lines of transmission and internal evidence 34
2.5 Methodological notes: A philological perspective 36
2.6 Theory and praxis of textual criticism 38
2.7 A new verse numbering system 39

PART II
Translation 43

Some notes on the English translation 43
Editorial conventions 46

3 Translation of the *Gaṇitatilaka* and its commentary: The *Gaṇitatilaka* composed by Śrīpati together with the commentary written by Siṃhatilakasūri 47

3.1 Homage to the Jina 47
3.2 Benedictory verse (maṅgalācaraṇa) *[GT 1] 47*
3.3 Section on technical terms (paribhāṣā) *[GT 2–12] 48*
3.4 Arithmetical operations with integers [GT 13–34] 51
3.5 Arithmetical operations with fractions [GT 35–51] 82
3.6 Arithmetic of zero [GT 52] 120
3.7 Classes of simplification of fractions [GT 53–63] 125
3.8 Type-problems of fractions [GT 64–92] 156
3.9 Inverse operation [GT 93–94] 221
3.10 Rules on proportion [GT 95–117] 226
3.11 Rule of five (pañcarāśika) *[GT 107–117] 240*
3.12 Barter (bhāṇḍapratibhāṇḍa) *[GT 112–117] 253*
3.13 The sale of living beings (jīvavikraya) *[GT115–117] 255*
3.14 Practices [GT 118–133] 257
3.15 Commission to the moneylender (vyājopajīvivṛtti) *[GT 120] 259*
3.16 Rule on interest [GT 122] 263
3.17 Rule on time and double capital [GT 124] 263
3.18 Conversion of several bonds into one (ekapatrakaraṇa) *[GT 127] 265*
3.19 Equating instalments of capital (samīkaraṇa) *[GT 131] 275*

PART III
Text analysis 283

4 Text analysis 285

4.1 Preliminaries 285
4.2 Benedictory section (maṅgalācaraṇa) *[GT 1] 285*
4.3 Section on technical terms (paribhāṣā) *[GT 2–12] 286*
4.4 The eight arithmetical operations with integers [GT 13–34] 287
4.5 The eight arithmetical operations with fractions [GT 35–51] 313
4.6 Classes of simplification of fractions [GT 53–63] 332
4.7 Type-problems of fractions [GT 64–92] 348
4.8 Inverse operation [GT 93–94] 376
4.9 Rules on proportion [GT 95–117] 378
4.10 The rule of five (pañcarāśika) *[GT 107] 389*
4.11 Barter (bhāṇḍapratibhāṇḍa) *[GT 112] 397*
4.12 The rule regarding the sale of living beings (jīvavikraya) *[GT 115] 399*
4.13 Practices [GT 118–133] 400
4.14 Commission to the moneylender (vyājopajīvivṛtti) *[GT 120] 402*

4.15 Rule on interest [GT 122] 405
4.16 Rule on time and double capital [GT 124] 405
4.17 Conversion of several bonds into one (ekapatrakaraṇa) *[GT 127] 407*
4.18 Equating instalments of capital (samīkaraṇa) *[GT 131] 414*

Appendix 1: List of mathematical rules and sample problems supplied by the SGT and found in other works 418
Appendix 2: Rules and sample problems of the GT occurring in other works 419
Appendix 3: Glossaries 420
3.1 Glossary of mathematical terms (Sanskrit–English) 420
3.2 Glossary of mathematical terms (English–Sanskrit) 423
3.3 Glossary of measuring units 425

Bibliography 427
Index 434

Tables

1.1 List of the mathematical topics treated in the *Triśatī* and the *Līlāvatī* 10
2.1 List of the mathematical topics treated in the GT together with new verse numbers 40

Preface

The Gaṇitatilaka and Its Commentary: Two Medieval Sanskrit Mathematical Texts presents the first English annotated translation and analysis of the *Gaṇitatilaka* by Śrīpati and its Sanskrit commentary by the Jaina monk Siṃhatilakasūri (13th century CE). The *Gaṇitatilaka* is a Sanskrit mathematical text written by Śrīpati, an astronomer-mathematician from 11th-century CE Maharashtra. The *Gaṇitatilaka* has come down to us together with Siṃhatilakasūri's commentary in an incomplete manuscript. The only edition available of both Sanskrit texts is by Kāpaḍīā (1937). Siṃhatilakasūri's commentary upon the *Gaṇitatilaka* is a key text for the study of Sanskrit mathematical jargon and an important source of information on mathematical practices of Medieval India. To my knowledge, this is, in fact, the first known Sanskrit mathematical commentary written by a Jaina monk, about whom we have substantial information, to survive to the present day. Siṃhatilakasūri's commentary has never before been studied or translated into English. The *Gaṇitatilaka* has been translated by Sinha and published in 1982 as a journal article; this translation is, however, clumsy, not annotated, and there has been a real need to fully revise it.

In presenting the first annotated translation of these two Sanskrit mathematical texts, this volume focusses on *language in mathematics* and *language and mathematics*; as a philologist, my work on ancient sources involves, *apertis verbis*, text-critical analysis and exegetical methods rather than modern mathematical formalisation. This book put forward a novel, fresh approach to Sanskrit mathematical literature which favours linguistic, literary features and textual data.[1]

I am grateful to the following colleagues who have helped me at various stages of my research: John Cort, Daniele Cuneo, Hugo David, Phyllis Granoff, Eivind Kahrs, Vincenzo Vergiani, and Steven Vose. I am particularly

1 This book is partially based on the research I carried out for my doctoral dissertation in Classical Sanskrit, which was submitted in 2017 to the Faculty of Asian and Middle Eastern Studies, University of Cambridge (UK), and successfully defended in the same year. PhD thesis (2017): "The *Gaṇitatilaka* and its commentary by Siṃhatilakasūri: An annotated translation and study", Faculty of Asian and Middle Eastern Studies, University of Cambridge.

indebted to Paul Dundas and Takao Hayashi with whom I had long, fruitful discussions concerning medieval Jainism and Sanskrit mathematical texts. I also would like to express my gratitude to Kim Plofker and Clemency Montelle for encouraging the publication of this work. A special word of thanks goes to John Steele, the editor of the series *Scientific Writings from the Ancient and Medieval World,* and to the team at Routledge, in particular Elizabeth Risch, Laura Lawrie, and Emma-Leigh Craig. I greatly appreciate their professional and kind help during the publication process. It goes without saying that any mistakes and inaccuracies in this book are my sole responsibility.

The current volume is divided into three parts: in Part I, Chapter 1 introduces the *Gaṇitatilaka*, describes aspects of style and language of the mathematical commentary, and presents new evidence to contextualise the commentator Siṃhatilakasūri. Chapter 2 investigates the peculiarities characterising the edition by Kāpadīā, explains linguistic phenomena and oddities, and attempts to make observations about the textual history and transmission of both mathematical works; this chapter also includes the author's theoretical perspective and methodological approach. Part II contains a complete annotated English translation of both the root-text by Śrīpati and the commentary by Siṃhatilakasūri. I have also critically revised the Sanskrit edition by Kāpadīā; literary analysis and philological discussions are examined in the footnotes. Part III elaborates the rules formulated in the root-text and explained by the commentary while also elucidating aspects related to the vocabulary used; it also examines the commentator's execution of the sample problems.

Abbreviations

BSS	*Brāhmasphuṭasiddhānta*
GSK	*Gaṇitasārakaumudī*
GSS	*Gaṇitasārasaṃgraha*
K	Kāpadīā's edition
L	*Līlāvatī*
PG	*Pāṭīgaṇita*
SGT	Siṃhatilakasūri's *Gaṇitatilaka*
SŚ	*Siddhāntaśekhara*
TŚ	*Triśatī*

Part I

Introduction

1 Introducing the *Gaṇitatilaka* and its commentary

1.1 Sanskrit mathematical writings in medieval India

In early Sanskrit literature,[1] works on astronomy, astrology, and mathematics were all part of the specialised branch of learning known as *jyotiḥśāstra* or "the science (*śāstra*) of heavenly bodies".[2] As far as can be discerned from extant sources, in India the development of "mathematics" (*gaṇita*) as an independent literary subject may be traced back to the end of the first millennium CE;[3] until then, mathematics was incorporated into astronomical works.[4] By the 5th century CE, astronomers had developed the medieval *siddhānta* genre as a standard textual tradition and organized astronomical schools based on the authority of different authors and their works.[5] Some chapters in medieval *siddhāntas* were explicitly devoted to mathematics independently from astronomy. The first of the fully preserved astronomical texts that integrated both mathematical methods with astronomical

1 A complete presentation of the history of mathematics in India is beyond the scope of this book; here I present only the basics in order to help readers contextualise the Sanskrit texts translated in the present book. The most comprehensive work on the history of mathematics in India is by Plofker (2009), to which readers are referred to for further information, with references, on most of the points discussed here. Classic and more recent contributions include: Bag (1979), Colebrooke ([1817] 1973), Datta and Singh 1962 [1935]), Delire (2016 and 2016), Filliozat (2004), Hayashi (1994, 1995a, 2003, 2013), Joseph (2011), Kaye (1908, 1915, 1927–1933), Keller (2006 and 2010), Kusuba (1993), Kusuba and Plofker (2013), Patte (2004), Pingree (1981), Plofker (2007), Plofker et al. (2017), SaKHYa (2009), Sarasvati Amma (1979), Sarma (2003 and 2009), Sen and Bag (1983), Shukla (1959), Srinivasiengar (1967), Staal (1999, 2006), and Thibaut (1875). The reader interested in the history of Indian mathematical astronomy in the Indian subcontinent can refer to the numerous publications by Montelle and Plofker.

2 The term *jyotiṣa* refers to a mix of astronomy, astrology, and calendrics; a thorough history of *jyotiḥśāstra* literature is found in Pingree (1981).

3 The word *gaṇita* was also used to distinguish quantitative astronomy from the more discursive discipline of astrology and to denote any type of computational practice.

4 As Plofker (2009, 122) observes: "[...] perhaps the first independent texts on *gaṇita* in its broader sense developed in parallel with early works on mathematical astronomy, or even before them, but simply failed to survive."

5 The development of the different schools or *pakṣa*s is discussed in Plofker (2009, 66–72) and in Montelle (2011, 156–284).

explanations is the *Āryabhaṭīya* by Āryabhaṭa (476 CE),[6] where the earliest surviving *siddhānta* chapter on mathematics is found.[7] Medieval *siddhānta*s were part of an orally transmitted textual corpus; authors employed concise, densely written verses to serve a mnemonic function. As it was common in other Sanskrit disciplines, long explanatory prose commentaries composed by later writers accompanied verse treatises. The first surviving commentary on the *Āryabhaṭīya* was composed by Bhāskara in 629;[8] Brahmagupta was Bhāskara's contemporary and the author of the astronomical treatise *Brāhmasphuṭasiddhānta*, where two separate chapters on mathematics are found. In this text, Brahmagupta criticises earlier works – particularly Āryabhaṭa's – and establishes new astronomical parameters; his contribution marks the beginning of the astronomical school called *Brāhmapakṣa*.[9] The 12th chapter of the *Brāhmasphuṭasiddhānta* contains 66 verses dealing with arithmetic, while the 18th chapter has 101 verses; the author expounds rules for finding unknown quantities, including methods for second-degree indeterminate equations and methods for the pulverizer. As opposed to later texts in the Sanskrit genre of arithmetic calculation (*pāṭīgaṇita*),[10] this work does not bother explaining the first six arithmetical operations with integers,[11] and begins with the application of methods for fractions (12.2–5), followed by the procedure for finding out the cube and cube-root of integers (12.6–7).

Mathematical chapters of *siddhānta* texts were shortly followed by works fully dedicated to mathematics independently of astronomy. Śrīdhara (ca. 8th–9th century CE), Mahāvīrācārya (9th century CE), Śrīpati (11th century CE), and Bhāskarācārya (12th century CE) wrote seminal works which gave a broader and clearer shape to the *śāstric* didactic genre of mathematical literature; the expression *gaṇitaśāstra* denotes the specialised body of knowledge and Sanskrit literature on mathematics which appear to have fully emerged after the time of Brahmagupta. Mathematics came to be subdivided into *pāṭīgaṇita* and *bījagaṇita* (literally "mathematics by means of seeds"), which roughly correspond to arithmetic including geometry and algebra, respectively.[12] The most well-known Sanskrit texts (8th–14th centuries CE) on *pāṭīgaṇita* which have survived are:

6 Clark (1930) as well as Shukla and Sarma (1976) provide an English translation of the work.

7 This work begins with rules for finding out the square and square-root of integers (verses 2.3–4).

8 Keller (2006) gives a translation and analysis of Bhāskara's commentary on the mathematical chapter of the *Āryabhaṭīya*. Bhāskara is often referred to as "Bhāskara the elder" or "Bhāskara I" to distinguish him from the 12th-century astronomer-mathematician bearing the same name and known as Bhāskara II or Bhāskarācārya. This latter is the author of the *Līlāvatī*, the most famous Sanskrit mathematical work ever composed.

9 Śrīpati and Bhāskarācārya were both followers of this school.

10 On this expression, see more in § 1.3.

11 The six operations are: addition, subtraction, multiplication, division, square, and square-root.

12 These are also known as *vyaktagaṇita* or "mathematics of visible (or known) numbers" and *avyaktagaṇita* or "mathematics of invisible (or unknown) numbers", respectively.

- the Bakshālī Manuscript (3rd–12th century CE?)[13]
- the *Pāṭīgaṇita*, *Triśatī*, and the *Gaṇitapañcaviṃśi* by Śrīdhara[14]
- the *Gaṇitasārasaṃgraha* by Mahāvīrācārya[15]
- the *Gaṇitatilaka* by Śrīpati
- the *Līlāvatī* by Bhāskarācārya[16]
- the *Pañcaviṃśatikā*[17]

Commentaries on these works are:

- the anonymous and undated Sanskrit commentary on the *Pāṭīgaṇita*[18]
- Kannaḍa and Sanskrit commentaries on the *Gaṇitasārasaṃgraha*[19]
- Siṃhatilakasūri's commentary (13th century CE) on the *Gaṇitatilaka*
- numerous commentaries on the *Līlāvatī*
- two Old-Gujarātī commentaries[20] on the *Pañcaviṃśatikā*

It seems safe to conclude that Siṃhatilakasūri's mathematical commentary is, so far, the first known Sanskrit commentary upon a work on *pāṭīgaṇita* written by a Jaina monk, about whom we have substantial information, to survive to the present day.

1.2 Authorship and content of the *Gaṇitatilaka*

The *Gaṇitatilaka* (henceforth GT) is a Sanskrit mathematical text written by Śrīpati, an astronomer-mathematician who hailed from 11th-century CE

13 See the English translation and commentary by Hayashi (1995a). The date of the Bakshālī Manuscript is still a matter of debate; this work has recently received popular attention due to the results of the Bodleian Library's 2017 project of radiocarbon dating portions of the birch-bark fragments of the manuscript. Plofker et al. (2017) discuss in detail some of the problems concerning the findings suggested by the palaeographic investigation.

14 Shukla (1959) provides an edition and English translation of the *Pāṭīgaṇita*; the *Triśatī* was first edited by Dvivedī (1899) and then partially translated by Ramanujacharia and Kaye (1912–1913). Some of Śrīdhara's mathematical methods are worked out in Datta and Singh ([1935], 1962). The *Gaṇitapañcaviṃśi* is discussed in Hayashi (1995b); on the relations between this text and the *Pañcaviṃśatikā*, see Hayashi (1991).

15 English translation and Sanskrit text by Raṅgācārya (1912). Morice-Singh (2015) and Petrocchi (2015) are two recent studies on some aspects of the *Gaṇitasārasaṃgraha*.

16 English translation by Colebrooke ([1817] 1973). Edited by Āpaṭe (1937) with the commentaries *Buddhivilāsinī* by Gaṇeśa Daivajña and *Līlāvatīvivaraṇa* by Mahīdhara, and by Sarma (1975) with the commentary *Kriyākramakarī* by Śaṅkara and Nārāyaṇa. See Pingree (1981, f32) for other published editions of the work.

17 Hayashi (2017) is an edition and English translation of the root-text (written before CE 1429) and its Old-Gujarātī commentary by Śambhudāsa.

18 The Sanskrit commentary on the *Pāṭīgaṇita* is anonymous and undated; according to Shukla (1959), it may belong to the early medieval period.

19 See Pingree (1981, 60).

20 These are dated by Hayashi (2017) to between 1428 and 1730 CE.

Maharashtra.[21] It has been handed down to us together with Siṃhatilakasūri's commentary in an incomplete manuscript. Siṃhatilakasūri's commentary upon the GT (henceforth SGT) is a seminal work for the study of Sanskrit mathematical jargon and Indian mathematical practices. The SGT has never before been fully studied or translated into English; the GT has been translated by Sinha (1982) as a journal article.[22] The Sanskrit edition of these works is by Kāpadīā (1937, henceforth K).[23]

The GT is a versified work on *pāṭīgaṇita*; it is a fragmentary text consisting of 133 metrical stanzas containing procedural rules and sample problems which begin with basic arithmetic and end with investment computations. The mathematical topics treated are the following:

- the eight arithmetical "operations" (*parikarman*) with integers: addition, subtraction, multiplication, division, square, square-root, cube, and cube-root
- the eight arithmetical operations with fractions: addition, subtraction, multiplication, division, square, square-root, cube, and cube-root of fractions
- arithmetic of zero
- five "classes" (*jāti*) of "simplification" (*savarṇana*) of fractions: the class of simple fractions, the multi-part class, the class of fractional increase, the class of fractional decrease, and the chain-simplification class
- nine "type-problems" (*jāti*) of fractions:[24] the visible quantity type-problem, the remainder type-problem, the difference type-problem, the remainder-root type-problem, the type-problem of the fraction executed with the root and the visible quantity, the type-problem of the visible quantity at both tips, the type-problem of the fractions and the visible quantity which is a fraction, the type-problem of the coefficient of the square-root of the fraction, and the type-problem of the subtractive square
- the procedure known as the "inverse operation"
- the "rule of three" (*trairāśika*) and its derivatives: rule of three, rule of five, barter, and rule of the sale of living beings
- rules on "practices" (*vyavahāra*): practice of mixture, and relating rules such as commission to the moneylender, rule on interest, rule on time and

21 The Sanskrit compound *gaṇitatilaka* is lit. "ornament of mathematics". Śrīpati was a prolific writer on *jyotiḥśāstra*, and numerous references to his name and works in primary sources attest that Śrīpati was considered an authority in his field; he wrote texts on astronomy, astrology, and mathematics. According to Pingree (1981, 25), Śrīpati wrote some of his works at Rohiṇīkhaṇḍa (about 150 miles south of Ujjainī) between 1039 and 1056 CE.

22 Sinha's translation presents several problems, and it is not annotated.

23 Note that, in Part I, for the convenience of the reader the referencing system adopted to discuss aspects of both the GT and the SGT relates only to the edition (page numbers and line) by Kāpadīā (K); for instance, K 48, 23 means "Kāpadīā's edition p. 48 line 23". The English translation and text analysis parts (Parts II and III) employ the verse numbering system proposed by Hayashi and reproduced to some extent in § 2.7.

24 I discuss the use of the GT of term *jāti* to denote both the "classes" and the "type-problems" of fractions in Part III.

double capital, conversion of several bonds into one, equating instalments of capital.

Śrīpati provides numerous sample problems on the rules enunciated which share common features with those found in other Sanskrit works on arithmetic. In the GT, several creative riddles and recreational problems describe a variety of situations drawn from the animal world, everyday life, and the marketplace. A distinctive feature is their emphasis on the practical aspect of learning. Finally, they demonstrate that mathematical works – both root-texts and commentaries – served as textbooks also for teaching commercial mathematics and mercantile skills to young pupils.

1.3 The *Gaṇitatilaka*, a work on *pāṭīgaṇita*

In the first verse (GT 1), Śrīpati styles his text as a work on *pāṭīgaṇita*.[25] Although the origin of the term *pāṭī* remains unclear, *pāṭīgaṇita* came to denote the branch of mathematics concerning arithmetic and geometry.[26] Some scholars claim that in the compound *pāṭīgaṇita*, the term *pāṭī* derives from *paṭṭa* or *paṭa* meaning the calculating board;[27] Hayashi (2014) observes that one should understand *pāṭīgaṇita* as "mathematics (*gaṇita*) by means of algorithms (*pāṭī*)" – thus "procedural mathematics."[28] In Śrīdhara's *Triśatī*, one finds the earliest instances of the technical term *pāṭī* (Pingree 1981, 58).[29] A work on *pāṭīgaṇita* consists of rules and sample problems on "operations" (*parikarman*) and "practices" (*vyavahāra*). Brahmagupta is the first known author to classify topics in the following way:[30]

i) 20 "operations" (*parikarman*): addition, subtraction, multiplication, division, square, square-root, cube, cube-root, five *jāti*s or "classes" for the simplification of compound fractions, rule of three, the inverse rule of three, rule of five, seven, nine, and eleven, and barter and
ii) eight "practices" (*vyavahāra*):[31] mixture, series, figures, excavations, piles, sawing, mounds of grain, and shadows of gnomons.

25 The term *pāṭī* is used by Śrīpati also, for instance, in K 6, 21, K 18, 3, K 23, 3, K 25, 21, and K 53, 20.

26 Note that when the expression *pāṭīgaṇita* is translated as "arithmetic", it also implies arithmetical computations on geometry.

27 The origin of the term *pāṭī* is still open to question. Datta and Singh ([1935] 1962, 8) emphasise that calculations were probably performed on sand spread on the ground.

28 Hayashi (2014, 1–4); see also Colebrooke ([1817] 1973, 1) and SaKHYa (2009, XXXII). In this context, "algorithm" denotes the series of mathematical steps meant to carry out the procedure formulated in each *karaṇasūtra* or "procedural rule".

29 *Triśatī* verse1.

30 See *Brāhmasphuṭasiddhānta* verses 12.8–9 and 12.14–54.

31 Colebrooke ([1817] 1973) and Pingree (1981, 57) translate *vyavahāra* as "determination", Keller (2014) as "practice", and Plofker (2009) and SaKHYa (2009) as "procedure".

In the astronomical treatise *Siddhāntaśekhara,*[32] Śrīpati dedicates two separate chapters to mathematics: these are the *adhyāya* 13th (*vyaktagaṇita*), which deals with arithmetic and comprises 55 verses, and *adhyāya* 14th (*avyaktagaṇita*), which deals with algebra and comprises 37 verses. Probably following Brahmagupta,[33] in the *Siddhāntaśekhara Śrīpati* mentions 20 *parikarmāṇi* and specifies the classical separation between *parikarman* and *vyavahāra.*[34] In the GT, these two terms do not occur but are instead found in the SGT; it is worth noting that the commentator considers the "procedural rule" (*karaṇasūtra*) on inverse operation to be the 31st *parikarman,* as the following sentence demonstrates:[35]

> Thus far, thirty-one operations have been accomplished.[36]

Here Siṃhatilakasūri observes that, up to the rule of "inverse operation", 31 *parikarmāṇi* have been explained;[37] these are: the eight operations with integers, the eight operations with fractions, five classes of fractions, nine type-problems of fractions, and the inverse operation.[38] As also noted by SaKHYa (2009, xlv), Siṃhatilakasūri's remark may suggest that the SGT presents a subdivision of *parikarman* and *vyavahāra* in which the *parikarmāṇi* are 36 and run up to the *jīvavikrayavidhi* or the "rule of the sale of living beings"; after two sample problems, the GT begins the treatment of the *vyavahāras*.

The *Pāṭīgaṇita* and the *Triśatī* by Śrīdhara, and the *Līlāvatī* by Bhāskarācārya are the mathematical texts more frequently quoted by Siṃhatilakasūri in his commentary.[39] Śrīdhara seems to have been the first mathematician to devote a separate treatment to different branches of mathematics; Bhāskarācārya is the author of the most famous Sanskrit mathematical work ever composed: the *Līlāvatī*, a text on *pāṭīgaṇita* that soon

32 Edited by Miśra (1932–1947). It is beyond the scope of the present book to compare the mathematical topics expounded by Śrīpati in the *Siddhāntaśekhara* and in the GT; on this topic, see Pingree (1981, f30).

33 I find it interesting that Śrīpati makes this distinction in his astronomical work (namely in chapter 13, 1) and not in the GT, which is a work fully dedicated to mathematics. Is this possibly a clue that the *Siddhāntaśekhara* was written before the GT? The relative chronology of the two texts is uncertain.

34 SaKHYa (2009, xl–xli) provides a table in which 11 works are compared on the way the subdivision into *parikarman*/*vyavahāra* occurs; this classification is, however, mainly based on SaKHYa's understanding and not always on authors' explicit information.

35 K 68, 2.

36 See also the commentator's introductory line to *trairāśika* or the "rule of three" (K 68,3): "Now the rule of three, which is the thirty-second [operation treated in this work], is undertaken".

37 In this way, the commentator does not include rules quoted by him from other sources.

38 It appears that the rule of zero is not counted by the SGT as a separate *parikarman*.

39 The commentator must have had access to different recensions of the *Triśatī* and the *Līlāvatī* as alternative readings between the verses quoted in the edition by Kāpadīā and the editions of the *Triśatī* by Dvivedī and *Līlāvatī* by Sarma and Āpaṭe attest.

became the standard Sanskrit textbook.[40] The *Līlāvatī* is also the Sanskrit mathematical text more commented upon; the poetic structure of his mathematical texts, together with the fact that the author provides his own – although brief – commentary, with sample problems and the solutions, may have contributed to the lasting popularity of Bhāskarācārya's works. He is the author of the astronomical text *Siddhāntaśiromaṇi* (1150 CE), with his own commentary *Vāsanābhāṣya*; Bhāskarācārya also wrote the *Bījagaṇita*, a mathematical text on algebra. Although less popular than the *Līlāvatī*, presumably because it deals with more complex topics, it was similarly considered the standard textbook on algebra. His works became so popular that intellectuals were probably committed to preserve and transmit these texts from generation to generation. Bhāskarācārya's texts have dominated the scholarly field of Sanskrit mathematics ever since; several hundred manuscripts of his texts are available in India and abroad. Even today, his works are considered the standard Sanskrit texts on their subjects. Notably, medieval Sanskrit mathematical texts survive in a relatively small number of manuscripts, except the *Līlāvatī*.

Kāpadīā (1937, LXI) points out that the GT follows Śrīdhara's *Triśatī*[41] and that Śrīpati has borrowed from this work three examples; a close analysis shows, however, that the relationship between the two texts is more elaborate and that the GT displays some specific characteristics. I list here the topics of the two main textual sources quoted in Siṃhatilakasūri' s mathematical commentary, namely the *Triśatī*, and the *Līlāvatī*. The *Triśatī*, has 180 verses, of which 107 are sample problems;[42] the *Līlāvatī*, contains 277 verses including sample problems:

Four sample problems occurring in the *Triśatī* are also found in the GT;[43] it is worth noting that these are not acknowledged by the commentator as quotations.[44] The rule quoted by the GT from the *Brāhmasphuṭasiddhānta* is recognised as such by the commentator in the introductory line.[45] The

40 See Plofker (2009, 182–196) for an overview of the contents of Bhāskarācārya's mathematical texts; his rules are also analysed in Bag (1979), Srinivasiengar (1967, 79–94), and Datta and Singh ([1935] 1962).

41 This seems also to be Shukla's view (1959, XXVIII).

42 Dvivedī (1899) does not provide verse numbers for the first eight verses listing measuring units and notational places.

43 K 41, 15–16 corresponds to *Triśatī* example 25, K 70, 11–12 is the same as *Triśatī* 37, and K 74, 8–9 is similar to *Triśatī* 38 (the formulation of the sample problem is, however, slightly different). Kāpadiā overlooks that K 42, 9–12 corresponds to *Triśatī* 27.

44 Mathematicians do not always specify their sources, and each other's rules were sometimes silently embedded in one's own text. Borrowing was a common practice in Sanskrit literature; the notions of authoriality, textual authority, and textual reuse characterising the Sanskrit landscape make it difficult to elucidate the relation of Indian authors to other authors and habits of reading and composing texts. The same is true for attributions of a particular rule to a particular author.

45 K 85, 19 is from *Brāhmasphuṭasiddhānta* verse 12.14.

Table 1.1 List of the mathematical topics treated in the *Triśatī* and the *Līlāvatī*

Triśatī	*Līlāvatī*
List of notational places	List of measuring units
List of measuring units	List of notational places
Arithmetical operations with integers (starting with the addition and subtraction of series)	Arithmetical operations with integers
Arithmetical operations with fractions	Classes for the simplification of fractions
Classes for the simplification of fractions	Arithmetical operations with fractions
Rules of three, five, seven, and nine	Arithmetic of zero
Barter	Inverse operation
The sale of living beings	Miscellaneous rules (concurrence, supposition, etc.)
Practice of mixture	Rules of three, five, seven, nine, and eleven
Commission to the moneylender	Practice of mixture
Bonds	Mathematics of gold
Mathematics of gold[a]	Practice of series
Equal interests	Combinations of syllables in metres
Practice of series	Practice of figures
Progressions	Practice of excavations
Practice of figures	Practice of piles
Practice of excavations	Practice of sawing
Practice of piles	Practice of mounds of grains
Practice of sawing	Practice of shadows
Practice of mounds of grains[b]	Pulverizer
Practice of shadows[c]	Permutations

a By the expression "mathematics of gold," I mean rules and procedures dealing with calculations related to gold; these calculations involve, for instance, how to exchange a certain quantity of gold of purity x against another piece of gold of purity y.

b Excavations and computations on mounds of grains denote calculations to measure volume of solids and heaps of grains.

c These are rules to ascertain heights and distances of light sources from their shadows.

verse quoted by the GT in K 85, 8–9 and numbered by Hayashi as GT 122 remains unfortunately unidentified; this is a sample problem on interest. Siṃhatilakasūri quotes from the *Triśatī* and the *Līlāvatī* in order to supply alternative methods, clarify some passages, and provide sample problems; quotations from authoritative sources also help the commentator to add authority to his own work. When the commentator mentions the sources of his quotations, one often finds the expressions *tryśatyām* ("In the *Triśatī*") and *līlāvatyām* ("In the *Līlāvatī*").[46] Nevertheless, Siṃhatilakasūri does not always name the sources used; for instance, he provides sample problems on the rule of seven and on the rule of nine which correspond to *Triśatī*

46 See K 9, 2 and 8.

51 and 52, respectively, but these are not stated as such. The commentator provides a sample problem on the rule of nine by providing a verse which is the same as *Triśatī* 52, but without acknowledging it. The same pattern occurs when the commentator supplies a sample problem on the rule of nine which corresponds to *Līlāvatī* verse 86. The quotation from the rule of five, seven, nine, and eleven from *Līlāvatī* verse 82 is instead acknowledged in the commentator's introductory line; a passage from *Līlāvatī* verse 75 cited by the SGT (see K 82, 9–10) is introduced by *līlvātyām.*[47] Siṃhatilaksūri often quotes brief passages from the *Triśatī* in the middle of an explanation to indicate that that rule from the *Triśatī* should be applied; for instance, *Triśatī* verses 11 and 31 are quoted by the SGT in K 9, 8 and K 75, 24–25, respectively, and in both cases, the commentator mentions his source. Similarly, while explaining the procedure of cube, he quotes a passage from *Triśatī* verse 15, and states it explicitly.[48]

This analysis indicates that the commentator specifies his mathematical sources only when he supplies rules, and not when he supplies sample problems which also occur in the *Triśatī* and the *Līlāvatī*; these, in fact, are not clearly designated by the commentator as verses of those texts. Furthermore, the commentator usually refers to Śrīpati by using a verb in third person or by the expression *sūtrakāra* or the "author of the *sūtras*" (see, for instance, K 1, 10), whereas while explaining a sample problem he sometimes alludes to the author of the sample problems by the tem *pṛcchaka* (see, for instance, K 65, 24). The word *pṛcchaka* (lit. "inquirer") may not necessarily denote Śrīpati but rather the unknown author of the sample problem; sample problems were part of oral traditions and it is thus difficult to establish whether the authors of mathematical treatises are also the authors of the sample problems they present.[49] In the SGT, mathematical quotations have an immediate, practical purpose; those on grammar are erudite and ornamental.[50] The commentator rarely quotes or reuses passages from his own work; an exception is the

47 Hayashi (2013, 70) has overlooked this quotation.

48 In K 31, 25–26, the SGT quotes *Pāṭīgaṇita* verse 37; according to Hayashi (2013, 59) this verse testifies that Siṃhatilakasūri had both texts by Śrīdhara available while commenting upon the GT, since this rule does not occur in the *Triśatī*. It is mistakenly treated in Sinha's translation (1982) as a genuine GT's verse and numbered as GT 47.

49 The commentator's introductory line to a sample problem on the "rule of five" (K 75, 5) adduces further evidence to this argument: "In this regard, in the illustrative verse [which follows] the author enunciates a well-known sample problem". The source is, however, not traced.

50 The commentator does not mention the sources of quotations which are not mathematical; for instance, in K 39, 26 the commentary provides the following passage: *sarvannāmnā 'nusandhir vṛtticchannasya* or "[…] the investigation of that which is implied in a compound by means of a pronoun […]". Here the commentator uses a passage (5.1.11; see the edition by Acharya, 1953) which occurs in the *Kāvyālaṅkārasūtra* by Vāmana (7th century CE), a text concerning figures of speech and poetry, without acknowledging it.

passage *viṣamasama* ("*odd* and even"), which is from the explanation of the square-root of integers.[51]

1.4 Siṃhatilakasūri, the commentator of the *Gaṇitatilaka*

According to Mehta and Malvania (1969, 166), Siṃhatilakasūri the author of the mathematical commentary on the GT is also the author of the following Sanskrit works:

a commentary on the *Bhuvanadīpaka*
the *Vardhamānavidyākalpa*
the *Mantrarājarahasya*
the *Parameṣṭividyāyantrastotra*
the *Laghunamaskāracakra*
the *Ṛṣimaṇḍalayantrastotra*.[52]

Pingree (1981, 61) observes: "A commentary on the Gaṇitatilaka was written by Siṃhatilakasūri Sūri, the pupil of Vibudhacandra; this Siṃhatilakasūri wrote a *vṛtti* on the Bhuvanadīpaka of Padmaprabha Sūri at Vijāpura in 1269".[53] Mehta and Malvania (1969, 166) claim that the mathematical commentary by Siṃhatilakasūri was written in 1330 (*vikrama* era);[54] this information is not given by Kāpadīā who only specifies that the extant manuscript of the GT and its commentary could be dated between the 15th and the 16th century.[55] In the colophon of the *Mantrarājarahasya*, Siṃhatilakasūri himself states that he composed the work in 1327 (*vikrama* era).[56] In the introduction (1937, LXIV), Kāpadīā states: "There is a work styled as *Vardhamānakalpa* whose author is named as Siṃhatilakasūri S. There is another work viz. *Bhuvanadīpikāvṛtti* of which the author bears the same name. But it remains to be ascertained whether these Sūris are identical with *Gaṇitatilakavṛttikāra.*"[57] The analysis which I have carried out provides unquestionable evidence

51 This passage is part of the multifaceted system of internal cross-references used by the commentator (see more in § 1.6). It first occurs in K 9, 29; it is also found, for instance, in K 25, 3 (here *samaviṣamam*), K 60, 23, K 61, 10, and K 65, 5.

52 The last three works mentioned earlier are short compositions included in the edition of the *Mantrarājarahasya* by Jinavijaya (1980).

53 Vijāpura could denote Bījāpūr of Rajasthan, at that time a popular Jaina centre; see K.C. Jain (1972). The *Bhuvanadīpaka* is a work on *jyotiṣa* composed in the 12th century CE by Padmaprabhasūri, a monk of the Nāgapurīya Tapāgaccha. The *Bhuvanadīpaka* deals with the branch of astrology called *praśna* or "interrogations", and it was extremely popular in medieval India. Pingree's Census (vol. 4 p. 173) lists a considerable number of manuscripts of this text.

54 The *vikrama* era is 56–57 ahead of the solar Gregorian year.

55 Sinha (1982, 114) tentatively suggests 1275 AD as the date of Siṃhatilakasūri's commentary.

56 It corresponds to approximately 1270 CE.

57 Notably, Kāpādīā points out that in the Añcalagaccha, another early medieval Śvetāmbara worshiping-images sect, a *sūri* bearing the name Siṃhatilaka was known.

that Siṃhatilakasūri the commentator of the GT is also the author of the *Mantrarājarahasya*; the opening and closing verses occurring in these works tell us something about the author's lineage: he mentions his teacher Vibudhacandrasūri, whose teacher was Yaśodevasūri.[58] Also, in both works the author worships the *devī* Kuṇḍalinī, he uses the expression *sāhlādadevatā,*[59] and the peculiar spelling *adhi* (for *adhika*). In the dedicatory colophon at the end of the *Mantrarājarahasya*, the author specifies his own name (Siṃhatilakasūri) and the date (1327), and that he himself also wrote a commentary upon it, which is entitled *Līlāvatī.*

According to Dundas (1998),[60] Siṃhatilakasūri the author of the *Mantrarājarahasya* and of the mathematical commentary upon the GT, lived between the second half of the 13th century and the first decades of the 14th century in the area of Gujarat and Rajasthan. The term *sūri* denotes that he was the leader of a group of religious mendicants. As Dundas (1998, 38) explains, Siṃhatilakasūri was a Śvetāmbara Mūrtipūjaka monk[61] affiliated to the Kharataragaccha, one of the Jaina monastic orders in early medieval Gujarat and Rajasthan. Between the 12th and the 16th century, Śvetāmbara Mūrtipūjaka *gaccha*s arose in Western India with the intent to correct lax conduct, particularly of the temple-dwelling monks. Dundas (2002, 139), points out that "from the eleventh-century, in Western India there appeared a variety of image-worshipping Śvetāmbara *gaccha*s, each claiming to represent a true Jainism as near to that stipulated in the scriptures as possible and headed by charismatic teachers, invariably entitled *sūri,* [...]."

Siṃhatilakasūri was an author-monk involved in the codification of a Jain ritual that can be styled as tantric. Sanderson (2009) shows that from

58 One Yaśodevasūri is mentioned in Balcerowicz and Potter (2013, 284); the date assigned to him is 1118 CE.

59 In the SGT (see K 1,5), it is part of a long compound where it means "joyfully [worshipped] by deities".

60 To my knowledge, Dundas is the only scholar to have investigated Siṃhatilakasūri's *Mantrarājarahasya*. In his studies, Dundas claims that Siṃhatilakasūri belonged to the Kharataragaccha. Nevertheless, in a private conversation which I had with Dundas and Gough (2012), who has partly studied Siṃhatilakasūri's works on rituals and *maṇḍala*s, has emerged that regarding Siṃhatilakasūri's sectarian affiliation there is no hard evidence for him in Kharataragaccha sources. As no other evidence has so far been found either by me or by others, here I rely on Dundas (1998). The *Kharataragaccha pratiṣṭhā lekha saṃgraha* (vol. 1, Vinyasāgara 2005a) has one entry for Siṃhatilakasūri (entry no. 108); the entry is for an image consecrated in VS 1439 (1382/3 CE, *pauṣa vadi* 8), which is a date too late for Siṃhatilakasūri the author of the SGT; so it might be a case of homonymy. Vinyasāgara (2005b and 2006), does not mention any Siṃhatilakasūri and neither does the *Kharataragacchagurvāvali* by Jinapāla (13th century), which describes the activities of Kharataragaccha leaders from the beginning of the 11th century (edited by Jinavijaya, 1956). It is possible that Siṃhatilakasūri may have belonged to an early branch of the Kharataragaccha; in a private conversation, Vose (2013) has suggested the Rudrapallīyagaccha to me, while Dundas mentioned the Bṛhat Kharataragaccha. On the origin of the Kharataragaccha, see the introduction in Dundas (2002, 140–142); other aspects are investigated in Balbir (2012) and Cort (2009b).

61 A study on Śvetāmbara Mūrtipūjakas monks is found in Cort (2001).

the 5th to the 13th century, all major Indian religious traditions were either absorbed by Śaivism or remodelled along Śaiva lines. In early medieval India, tantric practices, such as the use of *mantra*s and diagrams for esoteric rituals, were widely spread and Jainism was among the traditions affected and reshaped by these developments; however, the use of *mantra*s was not entirely new for Jainas.[62] Among Siṃhatilakasūri's works on Jaina mantras and tantric rituals, the *Vardhamānavidyākalpa* deals with a ritualistic formula called *vardhamānavidyā*, which was used in the ceremony during which a monk was ordained to a title lower than that of *sūri*.[63] For the understanding of the development of medieval Jainism, Siṃhatilakasūri's *Mantrarājarahasya* is of special interest; it deals with the *sūrimantra*, a Jaina ritualistic formula "from whose power the very continuity of Jainism was regarded as deriving." (Dundas 2000, 232). Among the medieval Śvetāmbara textual sources on the *sūrimantra*,[64] the *Mantrarājarahasya* is the longest and probably the earliest available (Dundas 1998, 38); the performance of this *mantra*, which was transmitted privately and secretly, was meant for a special ritual: the ceremony during which a monk was ordained by a *sūri* to the rank of *sūri*, the title at the top of the monastic hierarchy.

Siṃhatilakasūri's *Parameṣṭividyāyantrastotra*, *Laghunamaskāracakra*, and *Ṛṣimaṇḍalayantrastotra* deal with the drawing of mystical diagrams used in Jaina rituals and meditation. The *Ṛṣimaṇḍalayantrastotra* concerns the representation of the *ṛṣimaṇḍala*, a ritual diagram which was developed in the medieval period and is still one of the most popular *maṇḍala*s for both Śvetāmbaras and Digambaras. According to Gough (2012, 19), Siṃhatilakasūri's text is the earliest datable text on the diagram.[65]

By the time of Siṃhatilakasūri, the Kharataragaccha had grown to become one of the pre-eminent Śvetāmbara mendicant lineages in Western India. "Monks have been the primary bearers of the intellectual tradition within the Śvetāmbara Jains tradition" (Cort 2001, 327). It is likely that at the earlier stages of his monastic life Siṃhatilakasūri roamed around with his guru Vibudhacandrasūri and studied a wide range of religious and secular texts and literary skills with preceptors and senior monks. It was a standard practice that, once initiated as *sūri*, the junior monk changed his name; we do not know what Siṃhatilakasūri's name originally was. He must have had young pupils or his own group of mendicants travelling and studying with him, as was common among senior monks of his monastic order of the

62 Dundas (1998) provides a sophisticated analysis of Śvetāmbara Jaina *mantraśāstra*; see also Jhaveri (1944) and Roth (1974). On Jaina ritual diagrams, see Caillat and Kumar (1981), Cort (2009a), and Gough (2017).

63 See Dundas (1998, 36 and 2009, f60).

64 According to Dundas (1998, 38), Siṃhatilakasūri's *Mantrarājarahasya* encodes several versions of the *mantra*.

65 "The only known pre-modern representations of the Ṛṣimaṇḍala are Śvetāmbara, and they seem to be modeled on Siṃhatilakasūri's thirteenth-century" (Gough 2012, 28).

time.[66] In his article on the *Līlāvatīsāra* (1285) by Siṃhatilakasūri's contemporary Jinaratnasūri, a monk of the Kharataragaccha, Cort (2009b) provides a sophisticated analysis of the medieval Śvetāmbara literary culture. To some extent, his findings can be applied to the context of which Siṃhatilakasūri was part, since Jinaratnasūri lived in late 13th century North-West India.[67] The production of literature was an important activity in the medieval Kharataragaccha; Dundas (2007) demonstrates that this was equally true in the medieval Tapāgaccha. The extent to which Jaina scholars received reward for their scientific activities, encouraged and assisted by their patrons, must have varied from individual to individual.[68] The literary culture was composite, all the branches of Sanskrit traditional knowledge were cultivated as Jaina monks were extremely prolific and versatile writers. Reflecting on the public role of the literary production among Jainas at that time, it can be understood that narratives and plays were for instance clearly intended for an audience that was larger than just a community of monks. Cort (2009b underlines that among the audience of Jaina scholars were well-educated lay people, Jaina merchants, and also non-Jaina intellectuals and members of the local social elite. Although the Jaina literary culture has tended to be multilingual, Cort observes that in the 13th century, Kharataragaccha monks focused on the productions of texts in Sanskrit. "The instruction of Sanskrit at both the elementary and advanced levels must have been thoroughly institutionalized in the Kharataragaccha at the time" (Cort 2009b, 16).

Given this background, what can be said about Siṃhatilakasūri's readers? Why did he write his mathematical commentary? Was this text commissioned to him? The fact that the SGT is written in Sanskrit means that it was used for educational purposes and addressed to an audience having a high level of literacy. It is difficult, however, to give a full picture of the social and cultural environment in which mathematicians were working and in which mathematics was taught and practiced in pre-modern India. In discussing aspects of intertextuality and reception of the SGT, its literary context and historical setting, it is possible to infer that among his readers were young pupils of the lay community (Jaina and not) as well as young Jaina monks receiving

66 Cort (2001) is a study on the curriculum of a Śvetambāra monk; see also the PhD dissertation by Vose (2013).

67 An investigation of the literary works of Kharataragaccha Jaina scholars active during the 13th–15th centuries sheds light on the monastic education and the curriculum expected of a monk of Siṃhatilakasūri's time.

68 An equally important predecessor for the literary culture in which Siṃhatilakasūri participated was, only one or two generations before, the literary circle that developed around the brothers Vastupāla and Tejaḥpāla in the late 12th and first half of the 13th century. Vastupāla and Tejaḥpāla served as ministers under the local Vāghelā kings and under their political influence, Sandesara (1953) shows that at that time in Gujarat a period of extraordinary cultural and literary activity took place. Sheikh (2010) provides an in-depth political, social, and cultural story of Gujarat between 1200 and 1500 CE.

a mathematical education.[69] Also, his work might have served as a reference book which teachers could draw upon for their lessons. The wide range of sample problems provided, the meticulous explanations, the emphasis on practical situations – all these would have helped a young student, a merchant, and a teacher. Textual authority has to do with cultural legitimation; historical and socio-cultural circumstances affect the reception and use of the texts and one wonders whether sectarian divisions might have prevented a Hindu audience to fully appreciate Siṃhatilakasūri's work. Notably, the Jaina Siṃhatilakasūri comments upon a text written by the Hindu Śrīpati and quotes Hindu authors, such as Śrīdhara, Brahmagupta, and Bhāskarācārya,[70] while he does not mention the *Gaṇitasārasaṃgraha* by the Digambara Mahāvīrācārya (who hailed from South India). This corroborates the idea that the transmission and reception of mathematical manuscripts was not as uniform as we may assume.

1.5 The genre of mathematical commentaries

The most typical pattern of exposition of the GT and its accompanying commentary is the following:

GT:
- procedural rule (*karaṇasūtra*)
- sample problem (*udāharaṇa*)

SGT:
- explanation (*vyākhyā*) of the terms of the procedural rule[71]
- presentation (*nyāsa*) of the numerical data given in the sample problem
- computation (*karaṇa*)
- verification (*ghaṭaṇā*)[72]

The GT consists of "procedural rules" (*karaṇasūtra*) and stanzas presenting "sample problems" (*udāharaṇa*); the SGT provides prose commentaries on the rules and shows the execution of the sample problems. Siṃhatilakasūri

69 It appears that no primary sources give us a clue on mathematical education among Jainas. In Petrocchi (2017) I provide an analytical framework to understand the role played by mathematical enquiry within Jaina philosophy.

70 An investigation of the 2006 *Catalogue of the Jain manuscripts of the British Library* by Balbir et al. shows that Jainas were regularly commenting upon texts written by Hindus.

71 With respect to the sample problems, in the SGT the term *vyākhyā* is sometimes missing; when there is no need to clarify technical terms, the commentator begins his prose straight away by the presentation of the numerical data to execute thereafter the sample problem (see, for instance, K 25, 23). However, it is also true that *vyākhyā* is at times wanting due to manuscript problems (see, for instance, K 27, 8 and K 34, 28).

72 In the SGT, "verifications" mainly occur at the end of the execution of sample problems on type-problems of fractions.

introduces each Śrīpati's rule with a brief line mentioning the topic and the number of verses or metrical stanzas, which conveys crucial information about the mathematical subject expounded in the rule.[73]

In order to fully appreciate aspects of Siṃhatilakasūri's work, it is important to comprehend the role played by commentaries in *gaṇitaśāstra,* the specialised branch of Sanskrit mathematical literature.[74] A key feature of Sanskrit mathematical literature concerns the dichotomy between orality and literacy.[75] Like other genres of Sanskrit literature, Sanskrit mathematical texts in verse were part of an orally transmitted textual corpus; they were meant to be memorized by heart and orally transmitted. Mathematical texts written in verse probably served as a memory aid for teachers of mathematics, while commentaries might have provided a teacher with a kind of guideline for his lessons. Conciseness, emphasis on essential points, and links from formula to formula are some of the salient features of mathematical literature. Filliozat (2004) highlights that the Sanskrit mathematical text is a literary text, because it imitates the form and spirit of a poetic text. The use of synonymous words and metaphorical expressions represent an interesting feature of Sanskrit mathematical literature. I have argued elsewhere (see Petrocchi 2016b) that the various systems for writing numerical expressions found in Sanskrit texts on *gaṇita* represent literary devices purely functional to the specific features of this literature. The linguistic practices adopted by authors reflect the tremendous influence of orality in the Sanskrit tradition; mathematical verse employ, in fact, a refined writing style while remaining mathematically meaningful.

The circulation of two different genres of Sanskrit mathematical writings – verse-treatises and prose-commentaries– reflects two distinct modes of mathematical discourse by which the *śāstric* tradition of mathematics reveals itself as a culturally shaped discipline. Scholars have often seen the contrast between treatises and commentaries as the material manifestation of what was to be known by heart (the treatise) and what was intended to present the explanations and visual representations that were given separately (the commentary). I argue that, with regard to Sanskrit mathematical

73 The commentator also specifies the number, and sometimes the topic, of sample problems; furthermore, he clarifies the number of "verses" (*śloka*) or "metrical stanzas" (*vṛtta*) enunciated by the author. The expressions *uddeśakaśloka* ("exemplifying verse") and *uddeśakavṛtta* ("exemplifying metrical stanza") refer to a verse and a metrical stanza, respectively, which enunciate a sample problem. In the introductory line, the commentator uses the term *uddeśaka* as a noun or as an adjective, in the latter case it qualifies *śloka* ("verse") or *vṛtta* ("metrical stanza"). Note that the GT is a versified treatise written in variegated metrical stanzas, this is specified by Śrīpati himself in the incipit of his work (see GT 1). The distinction between "verse" and "metrical stanza" does not concern the content of a rule/sample problem but it is purely a matter of prosody, the commentator always differentiates between the two.

74 An investigation of the main characteristic of Sanskrit commentarial literature of mathematics is found in Plofker (2009, 210–216).

75 The relationship between orality and literacy in mathematical literature is discussed by Filliozat (2004) and Yano (2006).

literature, versified treatises and explanatory prose commentaries represent two different textual forms complementing each other and responding to specific pedagogical, mathematical, and literary needs. Without a commentary, the algorithms explained in the terse, aphoristic style of *mūla*-texts (i.e., the "root" text) are very hard to understand; in mathematical commentaries one finds the graphic figures that orally presented verse texts cannot present. "Thus they served as a bridge between the oral pronouncements of the base-texts and the actual manipulations that a mathematician needed to perform in order to solve problems" (Plofker 2009, 213).

Since early times, Sanskrit developed a powerful tradition of exegetical techniques; the early commentaries on Pāṇini's grammar embodied the prototypes of what commentaries should look like. As highlighted by Bronkhorst (2006, 773), in Sanskrit classical literature commentaries go far beyond the root-text they comment upon, although they claim they do not, as it is said that the root-text "contains in condensed form what the commentaries bring out".[76] Despite the fact that mathematical commentaries share some features with other kinds of Sanskrit commentaries, they also reflect and embody multifaceted practices. Bronkhorst (2007, 773) wonders whether and to what extent the commentarial way of presentation can influence the contents of the science concerned: "Could it be that the history of science is to some extent determined by the form of expression chosen by its representative?" I believe that the Sanskrit commentarial literature on mathematics represents a well-defined literary paradigm whose function exceeds the common services provided by a commentary in other traditional disciplines.

It is true that Sanskrit texts in verse often require commentaries to be fully understood, independently of the subject treated; this is even more appropriate in the case of a highly technical and specialised field of inquiry such as mathematics. Once learned by heart, pupils had to understand how to apply the rules enunciated in the root-text, how to carry out procedures and work with numbers, and hence "how to practice mathematics." Although the commentarial style varies from author to author and the level of details supplied makes a mathematical commentary a more or less meticulous and comprehensive pedagogical medium, commentaries do not simply interpret and expand the *mūla*-text repeating rules in a more articulated prose-form.[77] Mathematical commentaries establish relations, unveil what is implied in a verse, and construct an interwoven network of textual features. In brief, a *mūla*-text by itself is not only difficult to comprehend but it does not fully serve the purpose of both teaching and learning mathematics. A root-text, whose format repeats the pattern "rule and sample problem", does not show how to carry out the procedures described by the rules enunciated.

76 Bronkhorst (2006) examines the relationship between Sanskrit commentaries and the history of science in India.

77 Keller (2010) argues that some commentaries emphasise mathematical ideas and others emphasise mathematical practices.

As mathematical treatises and mathematical commentaries connote two distinct kinds of texts, I contend that they also denote two different educational aids and that the design of these texts is an indicator of their educational landscape.

Mathematics is a practical discipline and its technical nature influences its mode of discourse as a literary genre. The production, manipulation, and use of Sanskrit mathematical treatises in verse poetry, on the one hand, and explanatory prose commentaries, on the other hand, testify specific textual habits developed within professional settings.[78] At *prima facie*, one can denote *mūla*-texts as "primary texts", meaning that these are texts that stand alone, and commentaries as "secondary texts", texts that need another to exist. While this classification may be useful in a general sense, it does not bring out the relationship between the two. In mathematical literature, a root-text and its commentary typify two diverse works that complement each other by responding to different literary and pedagogical purposes.

1.6 Aspects of style and language in Siṃhatilakasūri's commentary

As much as the mathematical jargon and hermeneutical strategy of the SGT mark significant ruptures to the same degree they testify continuity of traditional Sanskrit commentarial strategies; the weight of authoritative sources and the influence of local usages emerge at length from Siṃhatilakasūri's writing practices. Literary and linguistic aspects of the SGT are treated in more detail in both the English translation and analysis; here I limit myself to mention some of the most recurring features characterising the commentary.

The SGT begins the explanation of a rule by the term *vyākhyā* ("explanation, commentary"); the rule is then analysed, grammatical cases are shown, synonyms provided, and most importantly the algorithm formulated in the rule is almost always connected to the sample problem given by Śrīpati after the rule. In this manner, the commentator links theory to practice; one often finds, in fact, the expression *vakṣyamānodāharaṇayukytā* or "by means of the reasoning (*yuktyā*) of the sample problem (*udāharaṇa*) that will be formulated next".[79] In order to elucidate the steps enunciated in the rule, the commentator relates the concepts mentioned by Śrīpati to the actual mathematical scenario of the sample problem. This is a most ingenious stratagem employed by the SGT to promote clarity and learning; his elaboration serves as a bridge between the rule and its sample problem, and technical terms and their application become more intelligible to students.

78 As Chemla (2004, VIII) points out: "In exactly the same way as for the other outcomes of scientific activities, all kinds of factors, cognitive as well as cultural, technical, social or institutional enter the shaping of texts [...] scientific activity goes along with the production, the manipulation, and the use of texts".

79 K 3, 16 and K 41, 10.

The attention to the actual calculations is a major characteristic of Siṃhatilakasūri' style. His commentary is uniquely precious as a source of direct information about how medieval Sanskrit mathematics was actually written; the commentator shows in great detail the way numbers were displayed in the notational layouts and manipulated to perform a computation algorithm. The word *nyāsa* denotes the "setting down, presentation" of the numerical data, and it occurs throughout the various stages of the execution of an algorithm;[80] *nyāsa*s are extremely valuable to understand the computational practices they embody. Siṃhatilakasūri's mathematical language reminds readers that the arrangement of quantities in the calculating board creates a set of space-relations which makes the layout in continuous transformation, as it evolves and changes during each phase of the execution of an algorithm. On account of this constant reshaping, the commentator adapts his own vocabulary. The lexicon includes verbs such as *sthā* "to place", *likh* "to write", *cal* "to move" and the prepositions/adverbs *paścāt* or "behind" and *agra* or "in front of"; the narrative movement reflects a major concern with the way numbers are to be reproduced in the layouts. The adjectives *pūrva* ("first") and its synonyms *ādya, prathama, prācya,* and *prāk*, or *agretana* ("next") and its synonym *para,* or *dvitīya* ("second") and *tṛtīya* ("third") qualifying *aṅka* ("number") or *pakṣa* ("side") identify in a clear manner the quantity which is computed by the commentator each time.

When compared, for instance, to the anonymous early medieval commentary on the *Pāṭīgaṇita*, linguistic analysis indicates that Siṃhatilakasūri develops more comprehensive expository and pedagogical strategies. His explanatory style is meticulous, accurate, and sometimes even verbose in order for students to avoid ambiguities; rhetorical features constantly rotate around the configuration of the notational layout. Visual reasoning plays a significant role in the commentator's communicative intentions; students are, indirectly,[81] urged to pay attention to the representations of the algorithm being performed.

Also, it should be noted that in the SGT while, on the one hand, the persistent use of internal cross-references creates a versatile textuality, an interwoven network of interpretative practices, on the other hand, references to the *Triśatī* and the *Līlāvatī* validate the commentator's arguments on the basis of their textual authority. To the linear exposition and systematic clarity which characterise the commentator's way of producing knowledge corresponds the making of a structured "inner text"; there is, in fact, a repetitive texture underlying the composition that produces intertextual elements consisting of a system of markers. The component "parts" are sequences of details, passages, quotations, references,

80 Plofker (2015) investigates the use of layout and notation in Sanskrit scientific manuscripts. In Sanskrit mathematical manuscripts *nyāsa*s are given in square boxes such as these: | |; see, for instance, in Hayashi (1995a) and Shukla (1959).

81 The commentary employs what we may call "zero person" constructions; in the "zero person" construction, there is no overt subject, and the verb is in the third person singular form. Passive constructions are also widely found.

and keywords, which serve the commentator's communication plan; they craft a multi-layered text that only a close text critical analysis can grasp in its full scope.

Other linguistic and stylistic features of the SGT are:[82]

- a normative language mainly expressed by the optative mood with a "quasi-imperative" connotation and sometimes by constructions with infinitive[83]
- un-Pāṇinian types of compounds[84]
- a repertoire of terms which reflect local usages and are part of the commentator's own idiom[85]
- forms of *yat* used as determinative pronouns to call attention to a substantive mentioned in the root-text[86]
- *kṛtvā, etena,* and *yuktyā* followed by a quotation or a reworded step from the root-text[87]
- linguistic formulae typical of the Sanskrit commentarial tradition[88]
- the use of compounds ending with *ādi* and *prabhṛti,* sometimes with the meaning of "[the number] beginning with"[89] and sometimes with meaning of "such as"[90]
- *rūpa* and *lakṣaṇa* at the end of a compound with the meaning of "consisting of, namely"[91]
- subjective genitives of reference or specification[92]
- the use of causal ablative and instrumental abstract forms[93]
- numerous instances of locative constructions expressing what Whitney (1924, 101) refers to as "locative of situation", and no copula occurs[94]
- terms occurring only once[95]
- forms of the verbal adjective *jāta* ("become") qualifying an obtained result[96]

82 I have given only a limited set of examples.

83 K 12, 25 and K14, 14. See also the frequent use of adverbial phrases, such as the formula with *artha* ("in order to"), and the dative case to express purpose.

84 Members of some compounds are arranged differently from what would be expected in Pāṇinian usage (see, also, the use of *dvaya* in *aṅkadvaya* and *pakṣadvaya*; K 5, 22 and K 77, 8).

85 See the term *bhāgadāyin* (K 65, 25), the verbal root *maṇḍ-* "to arrange" (lit. "to decorate, distribute"), and the use of *madhye* and *madhyāt* when a sum and a subtraction occur, respectively.

86 K 10, 13–14, K 37, 7–8, and K 86, 16.

87 K 57, 3–4 and 8, K 19, 9–10, and K 75, 21.

88 See the expressions *iti tattvam* (K 7, 15,) *iti sambandha* (K 15, 25), and *śeṣam spaṣṭam* (K 18, 9).

89 K 3, 16 and K 5, 13.

90 K 5, 4 and 11.

91 K 4, 11 and 20.

92 Note that this type of grammatical formula borders on locative constructions; see K 20, 25 and K 56, 15.

93 K 5, 20–21 and K 18, 24.

94 K 77, 17 (*bhāge*) and K 78, 17 (*apavarte*).

95 See *udāhṛti* (K 13, 28), *kalpanā* (K 43, 6), *prapañca* (K 30, 10), *yantraka* (K 11, 20), and *saṃvādana* (K 90, 5).

96 K 57, 6 and K 62, 15.

- *yathā* ("thus") preceding a layout[97]
- the verbal adjectives *bhagna, gata,* and *nivṛtta* denoting numbers that are "erased"[98]
- a distinct language to describe the computation carried out[99]
- *etena* ("by this" or "therefore") and *tatra* ("therefore") often used as conjunctive adverbs connecting two mathematical steps[100]
- *pūrvavat* underlines that a procedure has to be performed "as before", meaning as previously shown or as formulated in a previous procedure[101]
- there is a progressive texture which reflects the carrying out of the computation step by step; this movement in the discourse is created by the temporal adverbs *prāk, paścāt, tatas, param, prathamam,* and *pūrvam*[102]
- numerous adverbs and prepositions indicate the position of a quantity or the relationship between quantities in the layout, such as *adhastana, upari, urdhvā*, and *adhas* ("underneath", "above", "at the top", or "below")[103]
- expressions such as *sthāne sthāpya* or *anyatra sthāpayitvā* and *anyatra lekhya* ("having placed, written elsewhere") specify when intermediate results are placed aside[104]
- prepositional phrases are sometimes introduced by *ekatra/paratra* or *ekatas/ekatas* or "on the one hand, on the other hand"[105]
- fractions are often expressed by *bahuvrīhi* compounds where the denominator is in apposition to the numerator[106]
- when the explanation of a topic is concluded, the name of the procedure is followed by s*amāpta* ("concluded, accomplished")[107]
- *vāra* and *velā* are usually followed by numerals and accompanied by verbs of action ("to write, to multiply" and so forth)[108]
- the explanations sometimes begin with a clear statement concerning the quantities involved in the procedure[109]
- the sample problems are always carried out in detail
- the commentator often provides technical definitions.[110]

97 K 31, 1 and K 70, 3.
98 K 14, 7, K 17, 9, and K 31, 2.
99 See the frequent use of *prāglikhita* ("previously written"), *prāgukta* ("aforementioned"), and *prāgdṛṣṭa* ("previously obtained"); K 8, 3, 13, and 14.
100 K 14, 11, and K 65, 25.
101 K 14, 24, and K 80, 20.
102 K 10, 7, K 15, 5, and K 80, 12.
103 K 4, 13, K 15, 24, and K 25, 14.
104 K 12, 2, and K 36, 10.
105 K 9, 4, and K 24, 20–21.
106 See *ekaḥ saptaviṃśaticchedaḥ* (K 26, 23) and *dvau tricchedau* (K 28, 14).
107 K 30, 11, and K 34, 13.
108 On p. 5 line 27 *vibhajya ekavelaṃ* is "having divided once" and on p. 44 line 12 *vāradvayaṃ sthāpya* is "it should be placed twice".
109 On p. 39 line 11 *aṅkadvayāpekṣā prakriyā* is "this procedure concerns a pair of quantities [each time]", and on p. 90 lines 15–16 explain that rule enunciated (GT 131) concerns four quantities.
110 K 35, 15 and K 80, 14–15.

1.7 Reading the *Gaṇitatilaka* and its commentary

As a type of literature, mathematics partakes to the values and customs of the society that surrounds it; the translation of Śrīpati and Siṃhatilakasūri's works provide, in fact, insight into the history of mathematical literature and education in Sanskrit sources and bring to light features related to social life and economic activities, since aspects of everyday life and business practices are often mentioned. Sample problems are a valuable source for the history of medieval India as they provide insight into local cultures, the mathematical ability and mind-set that students had to develop, and they give us a window into economic information pertaining to the period (prices, products, and various measuring units). The stress on commercial problems in texts on *pāṭīgaṇita* suggests that some students were members of mercantile castes, and that those who were not had in any case to understand the basics of financial transactions. Accountants and calculators seem to have been regularly needed for all sorts of administrations; one would imagine that they must have had a solid mathematical education. Also, the GT and its commentary are unique sources to explore the relationship between language and mathematics in Sanskrit literature. The linguistic attainment in this area is fascinating, as it shows a wide range of technical terms to denote concepts and their explanations.[111] The language of mathematics speaks of objects which are distinctively mathematical in nature; nevertheless, each author provides a specific narrative argument to his reasoning. Furthermore, it is common for works on *pāṭīgaṇita* to follow a distinct organisation of the content; there was neither a standard paradigm nor a single tradition, and notwithstanding authors often seem to be aware of previous contributions and partly rely on them, multiple views were an acceptable reality. Mathematicians, like other thinkers, have their own ideas, literary styles, and a unique way of dealing with the interpretation and production of mathematical knowledge. In the GT, Śrīpati expounds rules in his own way and presents the procedures of type-problems of fractions which occur neither in the *Triśatī* nor in the *Līlāvatī*.

I have emphasised that the SGT preserves many particular aspects of the actual arithmetical calculations carried out on the writing board; this feature is not found in works written in verse and in the other Sanskrit commentaries available to us is not as developed. There is is no doubt that Siṃhatilakasūri's works represent invaluable sources of study for the history of Jainism and the history of Indian mathematics; in this regard, it is interesting to observe that while, on the one hand, the monk Siṃhatilakasūri is the author of a text on mathematics, which is a subject appreciated by a large audience that includes the lay community, on the other hand, his role as a literate and committed

111 I have investigated elsewhere (Petrocchi 2016b) the way in which mathematical representations and the Sanskrit language, two well-defined meaning-making systems, have interacted in construing mathematical knowledge in the medieval Indian context.

monk is revealed by his religious writings. He is, in fact, the author of works meant to circulate exclusively within the Jaina Śvetāmbara monastic order; these testify Siṃhatilakasūri's role as interpreter of the *sūrimantra* and his effort in consolidating the sectarian identity of his *gaccha*. His works embody the development of a variety of literary genres among Jainas and contribute significantly to our understanding of the history of medieval Jainism.

As I have earlier mentioned, Bhāskarācārya II is the author of the most famous Sanskrit mathematical work ever composed: the *Līlāvatī*, which is also the mathematical text more frequently commented upon. The popularity of Bhāskara's treatises and the zeal with which these were copied in manuscripts has, presumably, contributed in some measure to the oblivion and disappearance of earlier and later mathematical works; no longer being reproduced and thus transmitted, they finally stopped. A similar fate, although perhaps for different reasons, has unfortunately characterised the GT and the SGT; I hope that the present book justifies the attention they deserve.

2 On the edition of the *Gaṇitatilaka* and some methodological notes

2.1 The edition

The texts of the GT and its commentary by Siṃhatilakasūri were published in 1937 in Gaekwad's Oriental Series with an introduction and appendices by Hirālāl Rasikdās Kāpadīā. His remains, so far, the only edition available. Kāpadīā was a prolific author, translator, editor of various works on Jainism, and a Jaina himself. The front page of the edition of the GT highlights that Kāpadīā served as an Assistant Professor of Mathematics at Wilson College in Bombay.[1]

In his edition of Śrīpati and Siṃhatilakasūri's works, Kāpadīā devotes a major part of the 65-page English introduction to the mathematics found in early Jaina literature, and then presents the GT and the SGT in only 20 pages. He was clearly aware of the contributions to the history of mathematics by scholars of his time; he names, among others, Datta, Kaye, Thibaut, and Weber.[2] Surprisingly, Kāpadīā pays more attention to the *mūla*-text by Śrīpati, although this is not investigated in detail, than to the commentary; he briefly reconstructs the main steps of some of the rules given by Śrīpati and lists some of the mathematical terms included. No textual or mathematical analyses are given, which appears odd after such a large introduction on Jaina mathematics. Also, his outline of the topics of the GT is concise and sometimes inaccurate. It is clear that Kāpadīā's intention was nothing but to produce an edition of a mathematical commentary by a Jaina mathematician, for which modern scholarship remains grateful.

His English introduction is followed by a Sanskrit introduction (*prastāvanā*) of eight pages, where Kāpadīā mentions the mathematical topics covered by the GT, the works quoted by the commentator, and a list of both Śrīpati and Siṃhatilakasūri's works. On pages 79–81, Kāpadīā provides a table of contents of the GT where, however, he missed to mention the *sūtra*s on barter, mixture, interest, and the rule on time and double capital which Siṃhatilakasūri specifies to be a rule from the *Brāhmasphuṭasiddhānta*. At the end of the Sanskrit

1 Kāpadīā (1934) wrote an article on Jaina hymns and magic squares.
2 Datta seems to have contributed to Kāpadīā's English introduction; see Kāpadīā (1937, LXIX).

text, the editor gives three appendices: appendix 1 gives the translation of the sample problems by Śrīpati together with the solutions; appendix 2 presents tables of the measuring units mentioned in the GT, which are surprisingly compared with those found in the *Śatapatha Brāhmaṇa*, in the *Lalitavistara*, and in the *Arthaśāstra*; appendix 3 mentions Sanskrit words with numerical significations according to the *bhūtasaṃkhyā* notation.[3] The edition ends with the section *Addenda et Corrigenda*.[4]

Kāpadīā's edition can by no means be called critical, although it is styled as such, and could be better described as a "transcribed copy" from the manuscript. The critical apparatus is almost non-existent, and nonsensical readings, gaps, and other textual irregularities have been fully ignored. Fidelity to the original have been at times compromised;[5] several ambiguities make it a challenging task to translate and analyse the GT and the SGT by simply relying on the edition, specially given the fact that the only extant manuscript is not accessible. Also, the edition lacks verse numbers for about 45 per cent of the extant verses;[6] as pointed out by Hayashi (2013), the criteria for numbering followed by Kāpadīā and Sinha (1982) are inconvenient and incorrect since: (i) they include verses quoted by the commentator from other works, (ii) they exclude several genuine verses, and (iii) they ignore the metrical construction of the Sanskrit text. For these reasons, Hayashi has provided a new verse numbering system, which is followed in the present volume.[7]

2.2 The Sanskrit manuscript

In the English introduction of the edition, Kāpadīā (1937, LXVII f1) states: "I am led to believe that there is no other Ms. of this work available elsewhere, for, I have received up till now no reply in connection with inquiry made by me in my article "A note on Śrīpati and his Gaṇitatilaka" published in "The

3 This is a method of expressing numbers by means of symbolic words. The *bhūtasaṃkhyā* notation is an interesting example of a particular system of notation and literary practice widely used in mathematical and astronomical literature; an investigation of this system is given in Petrocchi (2016b), see also Sarma (2009).

4 This section presents some problems as it often provides the same anomalous expression instead of the correction. Also, in the *Addenda et Corrigenda*, on the top of some words small footnotes occur, but no references are given.

5 See the use of commas, em dashes, and corrections in the running text. One wonders whether the editor has sometimes silently emended the form *adhi,* which is part of the commentator's own idiom (a feature not mentioned by the editor in the introduction), to *adhika*. The reason to raise this question is due to the fact that, in a like manner, instances of the irregular *saptāviṃśati* have been emended by Kāpadīā while preparing the edition.

6 In the introduction, the editor observes that the verses of the GT were not numbered in the manuscript and that the manuscript was not divided into chapters or sections which could have helped the reader to understand its contents. It is noteworthy that from page 50 onwards, the headings which in the Sanskrit text have preceded each procedure up to the *mūlāgrabhāgajāti* no longer occur.

7 See Table 2.

Indian Historical Quarterly" (VOL. VIII, no. 2. p. 351)." The manuscript was privately owned as the editor discloses that it was lent to him by a certain Sheth Mohanlal Hemacand Jhaveri. Despite my efforts, I have unfortunately not yet been able either to locate the manuscript or to discover other extant witnesses of the GT and the SGT.

Kāpadīā's description of the manuscript is, unfortunately, brief and vague. He observes that the manuscript consists of 173 palm leaves written in Jaina *devanāgarī* characters and suggests its date may be the 15th century as the earliest limit and the 16th century as the latest.[8] The manuscript is said to be fairly correct and in a good state of preservation, with only leaves 60*b* and 61*a* being illegible and leaves 37th and 64th missing.[9] Text analysis has demonstrated that the missing 37th and 64th folios correspond, in the published edition, to p. 19 line 16 and p. 32 line 28, respectively, and occur while the SGT is explaining the "subtraction of fractions" (*bhinnavyavakalita*) and the "multi-part class" of fractions (*prabhāgajāti*). Kāpadīā points out that, in the manuscript, the work is not divided into chapters or sections and observes that: "At times we come across a few corrections made by the scribe" and that "[...] the *nyāsa*s are given at times in vertical lines and at times in horizontal ones" (Kāpadīā 1937, LXVIII). Kāpadīā states that there were "some slightly mistakes here and there" and the MS is "fairly correct".

The editor does not go into detail about visual and para-textual features, such as the use of different ink colours, scribal marks, and marginal notes; neither does he specify whether for editorial reasons had to make readjustments to the *nyāsa*s. Evaluation of the scribe's writing techniques, space-intervals, and correction events is absent; no information is given on folios which may have disclosed characteristics of palimpsests, and it would not be surprising if this was, in fact, the case of folios 60*b* and 61*a*.[10] Seemingly, other crucial elements which would have assisted scholarship to reconstruct the composition history of the witness and propose more conclusive emendations are: scripts and hands, the use of punctuation, distribution of the space on the page, location of the annotations on the page, and typology of the link between main text and annotations.[11] In addition, did corrections made *in scribendo* occur? Were annotations written with the same ink used from the

8 Kāpadīā has not provided photos of the manuscript.

9 While folio 64 is clearly designated as missing in a footnote by Kāpadīā (1937, 32), the edition does not give any details about the missing folio 37; I would like to thank Hayashi who has helped me to locate it. Folios 60b and 61a, which are illegible, in the edition correspond to the last lines of the SGT passage on GT 54 (see, in the edition, p. 31).

10 Palimpsests are manuscripts from which the primary text has been effaced to make room for fresh writing; manuscripts that for a variety of reasons were not being used were subject to a special abrading process to erase the existing text and re-use the obtained blank surface. A number of ancient works in different languages have survived only as palimpsests.

11 The collection of essays edited by Vergiani, Cuneo, and Formigatti (2017) cast light on codicological features and the wealth of information which can be drawn from some often neglected aspects of Indic manuscripts.

main text?[12] Were direct quotations somehow marked? What about the use of word-dividers and reference signs? This type of information would have helped us to understand whether marginal annotations were written together with the main text or added later by a different hand.[13]

In order to be able to give a complete evaluation of the textual history of classical works, it is vital to investigate the material aspects relating to their transmission; this consideration holds all the more for texts surviving in one only manuscript. As a consequence of a scarce attention to codicological issues, valuable features of the uniquely extant witness of the GT and SGT remain unexplored.

2.3 Linguistic phenomena and oddities: Between the manuscript and the edition

Although the publication by Kāpadīā is poorly edited, much of the difficulty in working on the two Sanskrit texts is related to the absence of witnesses; we cannot inspect the original manuscript and compare it to the printed edition in order to solve ambiguities. It is challenging to draw firm conclusions about the linguistic peculiarities which characterise the commentary, for without such knowledge here some of the discussions are likely to remain unresolved, especially those concerning "transcriptional probability". Nevertheless, there are a number of linguistic phenomena and oddities which can be undoubtedly ascribed to the manuscript; some are to be attributed to editorial inaccuracies. Bad Sanskrit and problems with figures in the layouts are partly to be blamed on the transmission.

Below I list the most relevant syntactic, orthographic, phonetic, and morphological irregularities occurring in the Sanskrit commentary:[14]

- lacunae of various kinds[15]
- transposition[16]

12 Royse (2008) provides a thorough investigation of scribal habits in early Greek manuscripts, most of his arguments can be extended to the study of Indic sources.

13 For instance, on page 35, footnote 1 supplies a layout which should have occurred in line 11, and one wonders whether this kind of intervention was based on a scribal note.

14 These include both problems of Kāpadīā's edition and what I believe to represent scribal features (note that it is sometimes difficult to distinguish between the two). I have presented only a limited set of examples, the references given in the footnotes below are by no means exhaustive; see more in the critical apparatus of the English translation.

15 Gaps are noted, for instance, on page 15 lines 2–10 and page 72 line 1. In the SGT, scribal leaps could explain some of the missing passages. Also, the term *nyāsa* is often wanting and one wonders whether it may due to the scribal choice to save space on the manuscript (a common practice in works in Classical Greek and Latin too); the same holds true for some verbal numerical expressions which are missing and are replaced, instead, only by numbers in figures.

16 See K 5, 5 where the passage *āhośabdo vikalpārthe* should occur at the end of line 8, on p. 16 line 23 *saṅkalite* is mistakenly found as it should occur in the line above, and on p. 92 lines 15–16 what refers to the first *khaṇḍa* ("instalment") is found instead in the line below describing the second *khaṇḍa*.

- anacoluthon[17]
- homoeoteleuton[18]
- harmonization to parallel passages[19]
- haplography[20]
- dittography[21]
- scribal abbreviations[22]
- metathesis[23]
- *constructio ad sensum*[24]
- nonsense readings[25]
- interpolations[26]
- confusion of similar letters[27]
- drop of initial or final consonants[28]
- irregular use of the *anusvāra* at the end of a verse or at the end of the first half of a verse[29]
- sandhi irregularities absence of consonantal *sandhi*[30]

17 K 24, 28 (*yadā*/*tadā*), K 37, 6 (*yatra*/*tatra*), and K 41, 7 (*yadā*/*tadā*).

18 See p. 46 line 4 *śeṣeṣu pañcadasu pañcabhakteṣu.*

19 K 52, 20 (*bhāgaikya*), K 53, 20 (the abbreviation *yū*), and K 54, 13 (*agra*).

20 K 26, 13 (*ghaneti* instead of *ghaneneti*) and K 27, 26 (*śeṣe caika* instead of *śeṣāṅkaḥ*), which might be blamed on homoeoarcton.

21 K 43, 27 (see the syllable *tya*), K 64, 18 (*aṣṭau*) and K 66, 11 (*guṇaḥ*).

22 K 15, 25 (*udā*). From SGT 99 to SGT 106, a system of shortening a passage quoted by the commentary from GT 42 has been employed by the scribe; the type of abbreviation used is "suspension". This idiosyncratic practice exemplifies the orthographic and morphological features of the surviving witness and it is closely related to the scribe who copied the manuscript; it saves the time of the scribe and the space available on a manuscript. Also, the layouts often present anomalous abbreviations and it would not be surprising if these testify deliberate scribal modifications.

23 On p. 27 line 2 *ghanalavamūle* should be emended to *lavaghanamūle,* and on p. 64 line 4 *triṃśanmadhye ṣaṭkṣepe* should be emended to *ṣaṭtriṃṣanmadhye pañcakṣepe*.

24 See *jātau dviccheda eka* on p. 59 line 9, *bhāganubandheṣv aṅke* on p. 20 line 10 and *saṃyateṣv aṅke* line 11. The scribe has produced nonsense plurals and verbal forms via grammatical attraction.

25 K 27, 3 (*trir gamayet*), K 58, 9 (*samacchedena samam*), and K 74, 23–24 (*trai śrityeya syāt*).

26 K 27, 23–24 (the passage from *etat* to *ghanamūlam*), K 28, 27 (*śunya*°), K 43, 28 (*yāvat tathaiva*), and K 55, 24 (in the layout, see the final abbreviation *mū*). It is not possible to establish with certainty all instances of interpolations as it might be that some represent instead scribal erroneous readings. Presumably, in Kāpadīā's edition, other interpolations indicate the occurrence of two phenomena that in textual criticism are known as "shift" and "replacement".

27 See K 14, 14 where *jātaḥ* occurs instead of *yathā*, on p. 14 line 16 one should find *trinighnayā* instead of *vinighnayā*, and on p. 42 lines 15 and 18 *svāṃśa* occurs instead of *sāṃśa*.

28 K 32, 3 (*nikṣipet* instead of *vinikṣipet*), K 40, 13 (*dadhyāt* instead of *vidadhyāt*), and K 60, 7. These might represent scribal trend towards brevity.

29 K 9, 25, and K 13, 24.

30 Hiatus occurs mainly between two *pādas*; in the SGT, there are numerous instances of *sandhi* irregularities. See, for instance, K 35, 2 where it should be *ekacatvāriṃśac catu*° rather than *ekacatvāriṃśat catu*°, and K 35, 9 where it should be °*catvāriṃśaj* instead of °*catvāriṃśat*. Another example is the frequent missing sandhi between the term *ṣaṭ* ("six") and the following term; see K 42, 2 and 24.

  - vowel hiatus[31]
  - problems concerning the final *visarga*[32]
- awkward syntax
  - lack of concord between adjective and substantive[33]
  - clumsy use of numerals[34]
  - occasional pleonasms[35]
  - incorrect verb-forms[36]
  - syntax of cases[37]
  - un-Pāṇinian grammar[38]
  - a predilection for abstract forms[39]
  - absolutives used as main verbs[40]
  - causative with non-causative meaning[41]
  - overuse of relative pronouns[42]

31 K 22, 19 (*tri-eka*°), K 59, 25 (°*sya-eka*°), and K 78, 1 (*sarvatra eka*°). There are various instances of *sandhi* irregularities concerning *iti* and a preceding or following word beginning with a vowel (see, for instance, K 89, 22).

32 Throughout the SGT, the omission of a *visarga* – or its *sandhi* equivalent – occurs regularly. For instance, see K 11, 21, (*ghanaḥ antyasya*), K 16, 16 (°*viṃśatiḥ evaṃ sarvatra*), and K 49, 18 (*trayaḥ adhaś*). In the present book, omissions of this kind will not be indicated in the footnotes. The editor tries to solve the problem with the *sandhi* concerning the final *visarga* by inserting a comma. In the SGT, the *visarga* is, however, sometimes dropped and the correct *sandhi* occurs (see, for instance, K 11, 13 and K 12, 5). It may be the case that a *daṇḍa* occasionally occurred when *yathā* is followed by a layout.

33 This often occurs between the adjective *jāta* ("obtained") and the verbal numerical expression which it should qualify; see p. 32 line 20, p. 70 line 19, and p. 80 line 8.

34 Methods of forming cardinals are standard, such as i) the additive method (see the use of *adhika*, *uttara*, and *ca*) the subtractive method (mainly by means of *ūna*), and multiplicative expressions ("times"). However, in the SGT ordinals sometimes occur instead of cardinals. See K 10, 11 where *eka* is the "first", in K 77, 8 the SGT uses *pakṣadvaye* to mean "in the second side", and in K 81, 12 one finds *dvipakṣe* instead of *dvitīyapakṣa*. In K 62, 21 *traya* is "three times, a triad" while it frequently occurs (see K 61, 26 and K 62, 9) denote the number "three".

35 On p. 77 line 1 *ekas tribhāgas* occurs, and on p. 60 line 14 one finds *pañcabhāga ekaḥ*.

36 K 6, 26 (*hṛyate* instead of *hṛyete*).

37 Some of the instances of awkward syntax may represent scribal shifts in construction. On p. 10 line 11 *dvādaśānāṃ* should be governed by a noun in the locative case (*pāte*), on p. 10 line 16 it should be *evaṃ sati kṛte* rather than *evaṃ sati kṛtiḥ*, and on p. 24 lines 23–24 *ekonaviṃśatir ekonaviṃśatitame* should be emended to *ekonaviṃśater ekonaviṃśatyā bhāge*.

38 In the SGT, instances of un-Pāṇinian grammar occur. I suggest that some of these features may be scribal and may have been created by analogy with existing forms, while others may be dialectical. An example of "vernacular Sanskrit" is a certain usage of numerals, which I have mentioned earlier.

39 These are usually constructed with a subjective genitive but other usages occur, too; see K 5, 23 (°*adhikatayā*), K 69, 8 (°*chedatayā*), and K 73, 9 (°*aṅkasya haratvāt*).

40 The commentator's explanation of the rule of multiplication (p. 5) widely employs absolutives as main verbs, see also *kṛtvā* in K 24, 20 and *maṇḍayitvā* in K 64, 1.

41 K 4, 56 (*melitā*) and K 10, 16 (*prāpyate*, after emendation).

42 I refer particularly to the recurrent use of forms of the relative pronoun *ya*. Also, it appears that the omission of correlatives in the SGT is much more frequent that is normal (Speyer, 1880: 349–350). See K 13, 28 (*yatra*), K 14, 1 (*yasya*), and K 15, 24 (*ye*).

- inexact results expressed by verbal numerical expressions[43]
- *daṇḍa*s missing or wrongly placed[44]
- unusual manner of marking quotations[45]
- "split" or incomplete compounds[46]
- deviant orthography[47]
- morphological peculiarities[48]
- oddities in the layouts
    - inconsistency in the way digits are placed in the layouts[49]
    - inconsistent layouts[50]
    - mistaken layouts[51]
    - missing layouts[52]

For scholars working on texts preserved in one only manuscript there are complications in terms of textual criticism, even more when the original manuscript is missing. For one thing, when there is only one manuscript, one is entirely dependent upon that manuscript; there is nothing to fall back on if

43 In the SGT, while the commentator is carrying a procedure, one finds verbal numerical expressions which are correct from a syntactical point of view but which do not correspond to the result of the calculations performed. See K 27, 19 where one would expect *saptacatvārīṃśat* ("forty-seven") rather than *pañcacatvāriṃśat* ("forty-five"), it is noteworthy that the layout gives the exact number 47), and on p. 70 line 8 *ekonacatvāriṃśat* ("thirty-nine") should be emended to *ekonapañcaśat* ("forty-nine").

44 K 41, 26 (before and after the layout), K 45, 3 (after the layout), and K 49, 14 (before *atra*). There are other numerous instances of missing *daṇḍa*s.

45 Quotations are placed between quotation marks; there are few exceptions (see, for instance, p. 41 line 24 *aṃśacchedāv ityādinā*). Unusual ways of separating quotations frequently occur; see page 17 line 7, page 24 line 19, and page 35 line 4.

46 An example of split-compound concerns *prāk,* which often is not compounded when it should be: see, for instance, K 7, 9, and K 8, 3 (cf. *prāgrāśir* in K 17, 19 and *prāg rāśis* in line 26). Split-compounds sometimes concern numerals: K 60, 10, and 12.

47 The irregular *saptāviṃśati* occurring in the manuscript have been emended by Kāpadīā while preparing the edition (see his English introduction).

48 An example is the reading *adhi* for *adhika* which often occurs in the SGT and which, as I will mention further in this section, is part of Siṃhatilakasūri's idiom.

49 In the layouts, the numbers sometimes follow a positional system (units below units, and so forth) but other times they do not; the criterion is, however, unclear and inconsistent. Because the only surviving witness of the GT and its commentary is not accessible, it is not possible to determine whether in the manuscript digits were written below each other exactly in the same way they are reproduced in the edition. See, for instance, K 63, 25 and K 92, 3. Also, cf. the following layouts: K 28, 7, 13, 26, and 29; K 44, 1, K 47, 6, K 75, 16, and K 91, 19.

50 In the section regarding the rules of five, seven, nine, and eleven, the layouts do not follow the same pattern when one would expect so: the denominator 1 do not always occur, as well as a 0, in the second side where the second "fruit" is to be ascertained. This feature might reflect either scribal trend towards brevity or instances of interpolations. Cf. K 78, 1 and 12, K 79, 1, 2, 11, and 20, K 80, 6, K 81, 10 and 25. Another example concern inconsistency in the abbreviations occurring in the layouts: cf. K 59, 8 and 25; K 51, 15 and K 52, 20.

51 Erroneous layouts at times occur; for instance, on page 19 line 6, page 31 line 23, and page 54 line 13.

52 Layouts are wanting, for instance, in K 16,12, and K, 36, 19.

the manuscript is illegible, and this can be a severe problem. Notwithstanding it is not possible to trace out the history of transmission of the manuscripts of the GT and SGT as no textual witnesses are currently accessible, text-critical analysis provides significant data to identify linguistic features of both works. A close investigation shows that the Sanskrit text of the GT is almost intact,[53] while that of the commentary is at times corrupt and presents distinctive aspects. This is not a marginal matter since it says us something about the descendants in the family lines of the two works. One of the reasons could be down, in fact, to separate manuscript families, where the text of the GT has been preserved in a better state, or to different handwritings used in the exemplar(s) from which the last scribe was copying.[54] In the SGT there is a sort of textual consistency in terms of language and style, and I would not argue for different literary strata; nevertheless, the fact that no other sources survive and that the only extant manuscript is not accessible makes it difficult to distinguish graphic errors from phonetic irregularities. Without references, instances of scribal explanatory supplements on the one hand and omitted material on the other hand are problematic to ascertain. It is arduous in our case to determine whether errors were created by unintentional, transcriptional changes, or deliberately, when the copyist attempted to emend a passage which he found corrupt. Scribal mistakes such as haplography, parablepsis, dittography, and harmonizations are common to all scribes of the classical and medieval world.[55]

In the commentary, there is a particular type of textual problem that does not allow a smooth running of the text from a mathematical point of view. The occurrence of terms which still "make sense" for the text at the grammatical and semantical level but are not consistent with the mathematics explained is significant. These sometimes exemplify cases of scribal "substitution" and "harmonization" both to the context and to parallels; for instance, on p. 47 line 23 *catvāriṃśadadhi(ka)śatam*[56] (140) should be *catvāriṃśadadhi(ka) dviśatī* (240), and on p. 87 line16 *pañcaśatānāṃ* (500) should be emended to *catuḥśatānāṃ* (400).[57] These uncertainties, which cannot be ascribed to the commentator, can be blamed on the transmission, and also point out the scribe's, or perhaps earlier scribes', lack of knowledge of mathematics. It is possible that there were already corrupted passages in the earlier exemplar(s),

53 Among the exceptions are K 16, 30 (see *anvitāt*), K 49, 9 (see *tryaṃśau*), and K 50, 28 (*kariṇī*).

54 Length is not a significant factor to justify differences in the transmitted corpora.

55 With respect to Old Testament writings, the active role played by scribes in the creative literary process is thoroughly investigated by Tov (2012); most of his arguments on scribal practices can be extended to Classical Sanskrit literature.

56 The morphological peculiarity *adhi* has been ignored by the editor in his introduction. Note that in the edition, throughout the SGT the missing syllable *ka* of *adhika* occurs between round brackets.

57 On page 14 line 24 *sapta* should occur instead of *nava,* on page 17 line 5 it should be *dvika* ("two") rather than *trika* ("three"), and on page 47 line 23 *dviśatī* ("two hundred") should occur instead of *śata* ("one hundred").

and one cannot rule out the hypothesis that the last scribe used more than one copy of the work to revise corrupt readings; the scribe might had added or substituted numbers in layouts and changed technical words with those he knew and understood.[58] In the SGT, instances of "vernacular Sanskrit" occur; some of the uncommon grammatical and stylistic forms found in Siṃhatilakasūri's work reflect, in fact, a local mathematical language. The existence of such a jargon is borne out by the rich technical vocabulary employed by the commentator. The reading *adhi,* which occurs throughout the whole commentary instead of the adjective *adhika* ("additional to, plus"), can be taken as an example; however, it is worthy to note that *adhika* rather than the form *adhi* is sometimes found.[59] Notably, both forms *adhi*/*adhika* in Siṃhatilakasūri's *Mantrarājarahasya* occur too. The use of *madhye* to denote a sum, *madhyāt* to denote a subtraction, and the terms *saṃmīlana* and *mīlana* for *saṃmelana* and *melana* represent instances of local expressions.[60]

The edition contains printing errors, and one should take into account that the editor might have encountered difficulties with the orthography of the manuscript and thus inadvertently caused morphological oddities and some of the other phenomena aforementioned that were not in the witness.[61] One of the main problems of Kāpadīā's edition concerns its unclear editorial techniques.[62] An example of ambiguous editorial methodology concerns the use of a long em dash, which is normally employed to denote a gloss

58 I wonder whether this is the reason for the unusual occurrence of *pratiloma* ("reverse order") instead of *anuloma* ("regular order") on page 9 line 29, page 13 line 26, and page 66 line 7.

59 K 4, 20, K 8, 7, and K 14, 22. I have earlier mentioned that both forms *adhi*/*adhika* in Siṃhatilakasūri's *Mantrarājarahasya* occur too.

60 K 3, 17 and 18, K 3, 28, K 14, 7, and K 27, 15. A lexical observation worth making concerns the fact that such a usage of *madhye* and *madhyāt* is not attested either by Sanskrit dictionaries, such as Apte or Monier Williams, nor in other Sanskrit mathematical texts mentioned in the present work.

61 Manuscripts can be difficult to decipher, and it is not surprising that they are sometimes read in different ways by modern editors.

62 In the introduction Kāpadīā (1937, LXVIII) mentions the editorial conventions he has used: that which appeared to him "superfluous" has been placed within square brackets []; an "alternative" or a "supplement" has been indicated in parenthesis (). This practice is grounded on apparently equivocal, arbitrary basis, and the outcome is often misleading: for instance, in K 4, 4 in the running text, the number 319 is given as (319). Is this, according to the editor's language, an "alternative" or a "supplement"? One would agree that it cannot be an "alternative", since it is preceded by the verbal expression of the same number given by the commentator; for the same reason, why should the editor provide the number in figures as a "supplement"? One is tempted to believe that it indicates a scribal note not designated as such by the editor. In K 9, 1 the commentator's passage is abruptly interrupted by the following: (*parikarmāṣṭake ślo.* 8). Was this found in the original manuscript or it is rather an (intrusive) editorial addition? The same holds true in the line below, where one finds (*pari*° *ślo.* 9). By contrast, it is noteworthy that on p. 92 Kāpadīā provides a reference to a rule quoted by the commentator in a footnote; furthermore, K 60, 9 emends *dva* to *ha* in *ha*(*dva*), where *dva* is instead the correct reading. Although ambivalent, suggestions given by the editor are sometimes correct and helpful (see, for instance, K 41, 3).

by the commentator; this practice, however, is not consistent and em dashes are at times wrongly placed.[63] The editor uses an em dash also: (i) to divide an expression for reasons of metre,[64] and (ii) to avoid an occurring hiatus.[65] Anomalous features are almost never underlined by the editor, and footnotes are instead used to mention the meters of Śrīpati's verses.[66] It is not clear whether by *iti pratibhāti* (lit. "so it seems"), which is sometimes found in the footnotes, the editor denotes an expression improved by himself or by the scribe.[67] The editor sometimes inserts a question mark in the running text to underline an unusual word but this usage is erratic and intrusive;[68] punctuation is not uniform and occasionally mistakenly used.[69]

2.4 Lines of transmission and internal evidence

As it is common for South Asian manuscripts, the texts of the GT and the SGT are likely to have been copied a number of times in their transmission and therefore to contain a relatively abundant distribution of errors. Texts usually become more corrupt as they descend from successive transmissions, this is even more true considering the significant time gap between the dates of Śrīpati and Siṃhatilakasūri (11th and 13th centuries CE, respectively) and the date ascribed by Kāpadīā to the manuscript (15th–16th century). Both texts have thus been copied for use over centuries and must have been subject to changes. A direct consequence of the fact that the manuscript used for the edition represents, so far, the only and late witness to the tradition is that the existence of a common ancestor remain an open issue until other manuscripts will be discovered. Nevertheless, the transmitted manuscript represents a textual tradition – as there may have existed not only one – by which the GT and SGT were handed down. The witness has come down to us incomplete and one can only conjecture about the reasons: was the exemplar before

63 K 21, 13 and K 39, 12. Note that in the English translation, I have not pointed out this problem all the times it occurs. One would like to know whether a gloss was so marked in the witness, in which case the editor has just reproduced the practice.

64 K 7, 25 and K 44, 11.

65 K 3, 16. Hiatus is more frequently eluded by the editor by placing a comma.

66 Few exceptions are page 13 f2, where a correction to a result is suggested, and page 29 f2–5, where a reference from a rule quoted by the commentator is given. In the edition, footnotes present curious features, such as the fact that numbers are often unreasonably repeated, and the footnote does not have its correspondent in the text (for instance, K 86 footnote(s) 5).

67 For instance, on page 22 line 8, footnote 1 reads '*harāṃśayoḥ' iti pratibhāti* as in the text one finds, in fact, *hararāśayoḥ* instead of *harāṃśayoḥ*, and on page 76 footnote 1 rightly emends the expression *navādhi(ka)dvādaśaśatī* to *dvādaśādhikānavaśatī*. See also K 22, 10 and K 85 footnote 5. Footnote 3–4 on page 86 appears to refer to a handwritten note in the manuscript. Sometimes the editor provides a footnote with an emendation without clarifying whether this is his own correction, an example is on page 77 line 4 where the peculiar form *dvipakṣe* stands for *dvitīyapakṣe* and it is so given in the footnote.

68 K 19, 22 and K 23, 8.

69 K 39, 17 and K 44, 26 (between *dvau* and *tathaiva*).

the last scribe already incomplete? Or did something occur unexpectedly to the last scribe that forced him to interrupt the copying the manuscript? Is it conceivable that Śrīpati himself left unfinished his work? Did someone, presumably from a Jaina milieu, commissioned the copying of a fragmentary work – literature and learning that would otherwise have perished? If this is the case, the incompleteness might represent one of the reasons why the manuscript did not circulate too much.[70] Are we dealing with an apograph presenting an already "incomplete" work or are the lost folios which make fragmentary the manuscript a damage which it may have happened over the years? Unfortunately, there is no evidence to support any of these hypotheses; text-critical analysis of Śrīpati's works and Siṃhatilakasūri's commentary cannot provide basis to draw any definitive conclusion about the literary growth, textual history, and transmission of both mathematical texts.

As far as we know, both the root-text written by the Hindu Śrīpati and its commentary by the Jaina monk Siṃhatilakasūri have come down to us in one only extant manuscript written in Jaina Devanāgarī script, this means that they have been preserved through a Jaina line of transmission.[71] I have argued that syntax and grammar indicates that text of the GT presents far less corruptions than the SGT; this points to the fact that works which have survived in the source edited by Kāpadīā might have independently descended from different lines of transmission. The same consideration makes it seem likely that last scribe or the copyists of earlier manuscripts had more than one exemplar before them. It is also reasonable to consider that the last scribe was copying from manuscripts written in different scripts and handwritings, with the copy of the GT being more regular and precise. The earlier GT by the Hindu Śrīpati must have been transmitted, in fact, in a different script before the commentary by Siṃhatilakasūri was composed, and therefore not in the Jaina Devanāgarī script, which is clearly related to the Jaina line of transmission of the commentary. Notwithstanding the only surviving witness exemplify a distinct line of tradition, it is safe to assume that there were others, which may had remained more or less independent of each other. We can infer that the GT had circulated in a separate family of textual tradition before the appearance of the SGT.[72] Although this scenario is the most likely one, a caveat is in order; this remains a unique manuscript as each scribe also maintained a certain level of independence, and the idiosyncratic spellings and forms recorded in a manuscript do not appear in every other witness of the original text. The transmitted manuscript almost certainly incorporates changes made by agents (scribes, redactors, readers) other than the authors, it definitely includes material and other elements which are non-authorial.

70 This appears to have badly affected the GT's transmission, reception, study, and popularity.

71 This is also confirmed by the initial invocation *namo vitarāgāya* and the use of the phoneme *cha* in the final page.

72 The hypothesis that other commentaries upon the GT have been written but failed to survive cannot be ruled out.

Contaminatio, resulting from the use of now one and now another manuscript, and *conflatio,* resulting from the combining of elements from two or more manuscripts, exemplify scribal practices which are common in classical sources.[73] In the absence of *contaminatio*, the manuscript used by Kāpadīā then contains the same errors that were in the exemplar(s), minus those that the scribe has corrected and plus some additional one created by him.

Also, what is more is that according to Kāpadīā the manuscript ends with two symbols, the *devanāgarī* phoneme *cha* and the number 7. It is worthy to note that the phoneme *cha* is a symbol commonly occurring in the final rubric of Jaina manuscripts of the Śvetāmbara tradition to mark the end of a chapter, a section, and/or the manuscript itself; the number following the symbol *cha* represents the number of chapters, which in the GT can be counted to seven if one includes the section on technical terms.[74] It must be ascertained, in fact, whether it is the same hand writing, the same scribe to have placed *cha 7* to mark the end of the manuscript; it could be interpreted not only as an indication of the religious affiliation of last scribe but also as a potential meaningful sign of the incomplete manuscript from which the last scribe was copying. Further inspection could reveal whether these signs represent instead a later addition made by an external intermediary, such as an inspector, who marked the manuscript as "finished". Naturally, there is a fair measure of tentativeness in all of this, but that – by its very nature – "it goes with the territory".

2.5 Methodological notes: A philological perspective

Given the technical nature of the subject, scholars working on the history of mathematics in India have been, so far, almost exclusively mathematicians. Notwithstanding I certainly appreciate such contributions, I believe that a text-critic can also offer a comprehensive investigation of Sanskrit mathematical sources and provide a perspective which differs from and can be integrated with what has been done so far. Although most driving factors in mathematical activity are epistemic in nature, mathematical literature is, in fact, first and foremost a species of discourse; this consideration holds all the more in Sanskrit literature where the requirements of the *śāstric* tradition, the literary format, and the characteristics of the Sanskrit language have played a major

73 Tanselle (1990) and Royse (2008) provides a thorough analysis of scribal practices in classical sources. Confluence of readings is a common pattern in manuscript transmission, when a variant was noticed, it might be introduced into the new copy by correction or it might be noted in the margin or between the lines. On the different aspects of this phenomenon, see West (1973).

74 These are the following: the section on technical terms, the eight fundamental arithmetical operations with integers, the eight fundamental arithmetical operations with fractions, the classes of simplification of fractions, the type-problems of fractions, the rules on proportion, and the *vyavahāras* or "practices".

role in authors' expository strategies. The move towards a better understanding of how Sanskrit mathematical works expound their subject is the first benefit of the text-critical approach employed in the present study. The key to using textual criticism in order to investigate the works here translated is, on the one hand, to indicate why the wording of these works look the way they do and took on their present shapes as a result of the process of transmission, and, on the other hand, to tackle the difficult passages by means of philological and historical-critical analyses. This volume is engaged with a different approach to mathematical literature where the primary aim is to examine Sanskrit mathematical writings as "literary writings"; my focus is on *language in mathematics* and *language and mathematics,* by which I mean the lexicon and the explanatory style used by the authors to expound mathematics, and how these relate to the treated subject. The attention to the wordings of both the GT and the SGT could foster, *inter alia*, a greater appreciation of Sanskrit literature on mathematics and a reconsideration of the extraordinary ways the Sanskrit language has been adapted to the needs of scientific expression.

Although Sanskrit mathematics has enjoyed a resurgence of interest in recent years, there are yet areas of research that deserve more attention. In order to uncover aspects that have been largely obscured by disciplinary schisms, this book uses interdisciplinary methods to explore the mathematical content as well as the linguistic peculiarities which characterise the GT and its commentary. As a philologist, I intend to reorient the attention of readers to crucial but under-studied characteristics of late medieval literature on mathematics, such as textual features concerning language and expository techniques. In my analysis of the algorithms formulated by Śrīpati and explained by Siṃhatilakasūri, I purposely keep mathematical formalisation to a minimum in order to faithfully reproduce the arguments put forward by the authors and avoid the danger of tinging technical terms with some bias.[75] This book does not offer modern mathematical parallels, nor does it transpose every algorithm into modern terms.

There has been a real need not only to provide an accurate English translation and thorough analysis of Śrīpati's root-text and its accompanying commentary by Siṃhatilaksūri but also to apply text-critical analysis to the edition by Kāpadīā. My emphasis on exegetical and linguistic aspects intends to bring to light the commentator's versatile textuality as well as the richness of forms defining, in both the *mūla*-text and its commentary, the interplay between

75 It is true that modern mathematical formulae tell us, in some ways, what technical terms stand for, but I agree with Staal (1995, 90) who, while discussing the use of modern notation to analyse the rules by the astronomer-mathematician Brahmagupta, observes: "[…] 'mathematical equivalence' is not a simple concept and it is equally clear that those very formulas are our constructs and were not in his mind. If they were, he would not have expressed them in words, used two rules and two terms instead of one. In sum, we must apply what Truesdell wrote about 'Newton's equations' (p. 7) and say: it is true that we, today, can easily read such formulas into Brahmagupta's words, but we do so by hindsight" (Staal 1995, 90).

knowledge and mathematical discourse, between numbers and words. It is against the background of the assessment of the language and textual exposition of the SGT that I focus primarily on the commentator's way of communicating mathematical ideas and on his skilful strategies of interpretation and representation, which shape the narrative structure of the text and contribute to establishing its originality. It is the contention of this study that the analytical functions of textual criticism are well suited for such aims.

2.6 Theory and praxis of textual criticism

The primary goal of textual criticism has traditionally been to establish, so far as this is possible, the actual text that the author wrote. Text-critics employ procedures to accomplish this goal which consist of a number of phases, such as *collatio*, *recensio*, *examinatio*, and *emendatio*; if there is only one witness to a text, *collatio* and *recensio* are synonymous. In textual criticism, special complications arise when texts are preserved in one only extant witness; this is, however, by no means a rare case either in Classical Sanskrit studies or in other cognate disciplines. When working on texts preserved in a single copy, text-critics have no recourse in the event of an error; if a Biblical manuscript has lost a line, scholars can determine its reading from another copy, but if the copy of the SGT has lost a line (and we can tell that it is missing because the surrounding lines make nonsense), how can we correct it? In the case of multiple manuscripts, the nature of the mistakes may tell us something about the original; not so when there is only one copy. The task of editing a book preserved in only one manuscript is arguably the most complex and difficult in textual criticism, for the scholar must reconstruct completely wherever the scribe has failed. This consideration holds all the more for scholars working on the GT and SGT: the surviving source is not accessible and the only edition available exhibits several anomalies. Such a challenging situation poses major problems for the philologist. One of the main difficulties on interpreting the SGT has been to rely solely on the edition by Kāpadīā; the task of translating and examining the two works has unavoidably involved also that of critically editing. The GT and SGT are not only incomplete but also full of textual oddities which may only (if at all) be finally resolved by *examinatio*, and more specifically by examining the only witness available or, ideally, multiple manuscripts.

Textual criticism often involves the attempt to discover the original form of details in a composition, although what constitute an "original text" is subject to much debate. In the present book, the evaluation of the readings contained in the edition will not aim at the establishment of an "original text", a "lost archetype", whether it is designed as the textual representative upon which Kāpadīā has based his edition, the scribe's exemplar(s), or the text that stood at the beginning of the transmission process. My goal is not to recover an earlier, more authentic (and, therefore, superior) form of the text; neither will the evaluation attempt to purge the wordings of the edition of errors and corruptions so as to establish a more pristine form of the works.

We do not know whether the exemplar(s) of the last scribe developed from a common ancestor that is no longer preserved in the extant textual witness; we do not know the way the manuscript represents two texts that are closer to the originals than other textual representatives. My emphasis is, therefore, on the "real" text as it has been preserved and made known in the edition by Kāpadīā. The manuscript used by the editor represents the textual evidence of a source which, as it stands, is a witness in its own right of a particular textual tradition in a particular place at a particular time.

The procedure of emending classical texts is one of the most subjective aspects of textual criticism; although philologists handle differently errors and alterations, I have followed principles and criteria which are usually employed by textual critics in Sanskrit studies, Classics, and by New Testament scholars.[76] I have not normalised all the spellings or regularised features which I believe were part of the commentator's language and style. My attempts to restore the text have been guided by Kane's (1969, 155–169) persuasive defence of conjectural emendation, in which he observes that an editor's not challenging bad manuscript readings can do more damage to the text than offering unsatisfactory conjectures. I suggest only conjectures which I consider necessary to solve textual problems; I believe that if I have erred in this, it has been on the side of caution. When suggesting a conjecture emendation, the critical methods I have followed are based on a textual approach: author's jargon, style, coherence of the text, consistency, and syntax; this means, however, that my conjectures may not correspond to neither the original nor to the manuscript on which Kāpadīā has based his edition. The readings proposed are inferred by internal evidence; a corrupted or suspect passage has been replaced with a reading that appear reasonable, linguistically plausible, and in accordance with the authors' vocabulary and style. I have proposed emendations taking into consideration author's intentions and on the basis of principles of what in textual criticism is known as "intrinsic probability".

2.7 A new verse numbering system

Below I list the rules expounded in the GT[77] followed by the new verse numbers proposed by Hayashi (2013) and employed in the present work.[78] I have named

76 Goodall (2004), Katre (1954), Murthy (1996), Rath (2012), Royse (2008), Tov (2012), and West (1973).

77 Appendix 1 provides a list of rules and sample problems supplied by the SGT.

78 Hayashi (2013, 62) explains: "I give one number to a stanza comprising two verse lines of two quarters each but, when a half stanza of one verse line stands alone, and is commented upon by the commentator independently of the next verse lines, I give it one number (with a bar above it in the table) and refer to its metrical counterpart in parentheses if it exists." Each of the five pairs of half stanzas "must have coupled together into one stanza having one number without changing their order because it was customary in Sanskrit mathematical works to insert stanza(s) for example(s) between two verse lines of stanza for rules." Hayashi notes that the pair 40+44 is irregular because not only two stanzas for examples (verse 41 and 43) but also one stanza for a rule (verse 42) intervene between them.

Table 2.1 List of the mathematical topics treated in the GT together with new verse numbers

Benediction, 1
Section on technical terms: list of notational places and measuring units, 2–12
Addition, $\overline{13}$ (+$\overline{15}$). Sample problem, 14
Subtraction, $\overline{15}$ (+$\overline{13}$). Sample problem, 16
Multiplication, 17–$\overline{18}$. Sample problems, 19–20
Division, 21. Sample problem, 22
Square, 23–$\overline{24}$. Sample problem, 25
Square-root, 26. Sample problem, 27
Cube, 28–30. Sample problem, 31
Cube-root, 32–33. Sample problem, 34
Addition of fractions, $\overline{35}$ (+$\overline{38}$). Sample problem, $\overline{36}$ (+$\overline{37}$). Sample problem, $\overline{37}$ (+$\overline{36}$)
Subtraction of fractions $\overline{38}$ (+$\overline{35}$). Sample problem, 39
Multiplication of fractions, $\overline{40}$ (+$\overline{44}$). Sample problem, 41
Division of fractions, 42. Sample problem, 43
Square of fractions, $\overline{44}$ (+$\overline{40}$). Sample problem, 45
Square-root of fractions, $\overline{46}$ (+$\overline{48}$). Sample problem, 47
Cube of fractions, $\overline{48}$ (+$\overline{46}$). Sample problem, 49
Cube-root of fractions, $\overline{50}$. Sample problem, 51
Arithmetic of zero, 52
Class of simple fractions, $\overline{53}$ (+$\overline{55}$). Sample problem, 54
Multi-part class, $\overline{55}$ (+$\overline{53}$). Sample problem, 56
Class of fractional increase, 57. Sample problems, 58–59
Class of fractional decrease, 60. Sample problem, 61
Chain-simplification class, 62. Sample problem, 63
Visible quantity type-problem, $\overline{64}$ (+$\overline{67}$). Sample problem, 65 (=*Triśatī* 25), sample problem, 66 (=*Triśatī* 27)
Remainder type-problem, $\overline{67}$ (+$\overline{64}$). Sample problems,[a] 68–69
Difference type-problem, 70. Sample problems, 71–72.
Remainder-root type-problem, 73. Sample problems, 74–75.
Type-problem of the fraction executed with the root and the visible quantity, 76. Sample problems, 77–79
Type-problem of the visible quantity at both tips, 80. Sample problems, 81–83
Type-problem of the fractions and the visible quantity which is a fraction, 84. Sample problems, 85–86
Type-problem of the coefficient of the square-root of the fraction, 87. Sample problems, 88–89

Table 2.1 (Cont.)

Type-problem of the subtractive square, 90. Sample problems, 91–92
Inverse operation, 93. Sample problem, 94
Rule of three quantities and its inverse, 95. Sample problems, 96–$\overline{106}$
Rule of five, 107. Sample problems, 108–111
Barter, $\overline{112}$ (+$\overline{115}$). Sample problems, 113–114
Rule of the sale of living beings,$\overline{115}$ (+$\overline{112}$). Sample problems, 116–117.
Practice of mixture, 118. Sample problem, 119
Commission to the moneylender, 120. Sample problem, 121
Rule on interest, 122. Sample problem, 123
Rule on time and double capital, 124. Rule from *Brāhmasphuṭasiddhānta* 12.14. Sample problems, 125–126
Conversion of several bonds into one, 127. Sample problems, 128–130
Equating instalments of capital, 131. Sample problems,132–133

a There is a problem in the table provided by Hayashi (2013, 66): the sample problem numbered as GT 69 concerns, in fact, the remainder type-problem.

all the procedures after Siṃhatilakasūri's usage;[79] in the line preceding each *karaṇasūtra* ("procedural rule") the commentator, in fact, provides the title of the rule.[80] I have also specified passages from other sources occurring in the GT, such as Brahmagupta's *Brāhmasphuṭasiddhānta*, and Śrīdhara's *Triśatī*.[81] According to the edition by Kāpadīā, the only extant manuscript of the GT and its commentary ends after the commentator's explanation of *Līlāvatī* verse 92; the GT is, in fact, a fragmentary text. At a speculative level, one can assume that the lost part of the GT would have included the other seven *vyavahāra*s expounded in the mathematical chapter of Śrīpati's *Siddhāntaśekhara*. These are, respectively: *śreḍīvyavahāra* ("practice of series"), *kṣetravyavahāra* ("practice of figures"), *khātavyavahāra* ("practice of excavations"), *citivyavahāra* ("practice of piles"), *kracacavyavahāra* ("practice of sawing"), *rāśivyavahāra* ("practice of mounds of grain"), and *cāyavyavahāra* ("practice of shadow").[82]

79 Regarding the translation of the names of the rules, I do not follow Hayashi.

80 Śrīpati too sometimes mentions the name of the procedures either in the *karaṇasūtra* or in the related sample problem; see, for instance, K 30, 24 and K 46, 10.

81 See also Appendix 2.

82 Rules on these *vyavahāra*s also occur in the *Triśatī*, *Gaṇitasārasaṃgraha*, and *Līlāvatī*.

Part II

Translation

Some notes on the English translation

Kāpadīā (1937, LXVII) observes that in the manuscript the entire work runs as a continuum; both my English translation and textual analysis (Part III) respect the arrangement of the topics found in the edition. This is a philological translation which tries to be accurate, literal, and as faithful as possible to the grammatical structure of the original; I have aimed at cultivating a literary and aesthetic flavor, too.[1] I have attempted to present a consistent translation of both works which, while reflecting the original ideas, would be accessible not only to Sanskrit scholars and philologists but also to those interested in the history of mathematics. Although my translation aims at a readable, modern presentation of the text, it retains as much of the original rhetoric and style as possible without, however, trying to be "archaic".[2] Whenever possible and relevant, I have adopted the same English word for the same Sanskrit term but different English words for different Sanskrit terms to emphasise the commentator's use of synonyms. Moreover, I have provided the translation of every Sanskrit technical term. In my analysis, I reconstruct passages, layouts, and algorithms always taking into consideration the commentator's explanation.

Other remarks on the English translation are the following:

- Although this present work uses the new verse numbers proposed by Hayashi (2013) and reproduced in § 2.7, for the convenience of the reader the referencing system relates to the edition (page number and

1 For instance, for the sake of a smooth translation, I have not translated the numerous *yathā* ("thus") occurring before a layout and I have instead used a colon; similarly, I have not always rendered the adverb *tatas* ("afterwards, then"), by which the commentary often begins a sentence and a new mathematical step, and the numerous occurrences of compounds ending with *ādi* and *lakṣaṇa* or "such as, namely".

2 One of my aims has been to reflect the linguistic usages employed by the commentator, which may result sometimes in unusual expressions (see, for instance, the numerous instances of "and so forth" when the commentator does not formulate full verbal numerical expressions or quotations).

line) by Kāpadīā (K) too. In the English translation, beginning with the benedictory verse and the section on technical terms, each procedural rule by the *mūla*-text is accompanied by a footnote giving page number and lines

- Rules of the root-text are indented and preceded by their verse number
- I have adopted the following referencing system: "GT 5" means "*Gaṇitatilaka* verse 5", while "SGT 5, 2" means "Siṃhatilakasūri's commentary on the GT, verse 5 line 2"
- When terms from the *karaṇasūtra*s are reproduced exactly in the commentary, they are printed in bold letters
- The annotated English translation does not mention the numerous inaccuracies found in Sinha's translation of the *mūla*-text by Śrīpati and Kāpadīā's translation of the sample problems
- When the commentator uses the same terms from the *karaṇasūtra* but in different grammatical cases, they are placed within quotation marks such as " "
- As far as possible, when the commentary provides a gloss, I give synonyms in order to reflect the literary practices adopted by the commentator (the root-text is in bold), for instance: **having subtracted** [also termed] "having deducted". When it is not possible to use two different English terms, I use the following system: the "root" (*pada*>*mūla*)
- An effort has been made to limit citation of Sanskrit terms in parentheses; however, in order to make the basic mathematical vocabulary visible to the reader and to take note of the polysemy of certain crucial Sanskrit terms, a Sanskrit word or expression is instanced when it designates a key notion, at its first occurrence, when it is required to grasp the sense of a gloss, and when it is part of the commentator's repertoire
- Sanskrit terms of technical usage have been treated in the footnotes
- In the translation, the exposition of literary techniques employed by the commentary, such as analysis of complex grammatical formations and specification of word-division, are unavoidably fully clear only to Sanskritists
- When the commentator quotes passages from the same rule of the *mūla*-text is commenting upon, in the translation these are printed in bold letters and no references are given in the footnotes. When he quotes passages from other rules of the GT, these are written between quotation marks, I give then the full reference in a footnote
- When the SGT quotes an entire rule or sample problem from other sources, this is given in a smaller font and indented. References are given in a footnote
- Terms which I supply for the sake of a smooth translation are provided into square brackets
- In the interests of readability, the translation includes emendations made to the text. In the footnotes, I clarify the reasons for such emendations when these need to be justified

- The references[3] to the *Līlāvatī* (henceforth L) are based on the editions by Āpaṭe (1937) and Sarma (1975).[4]
- The references to Śrīdhara's *Pāṭīgaṇita* (henceforth PG) and *Triśatī* (henceforth TŚ) are from Shukla (1959) and Dvivedī (1899), respectively
- When I take a reading suggested by the editor either in the *Addenda et Corrigenda*, in a footnote, or in the running text, it is not indicated unless a different solution is proposed
- The frequently found *āha* has been rendered by "the author enunciates"
- Notwithstanding the punctuation inserted by the editor is problematic, I have avoided to report the anomalies as it is a modern practice not used in Sanskrit manuscripts[5]
- In the SGT, the omission of a *visarga* – or its *sandhi* equivalent – occurs regularly. Since it has been regarded as a characteristic of the manuscript, it has not been corrected
- I have not normalised all the spellings or regularised features which I consider to be oddities of the manuscript or part of the commentator's idiom;[6] this means that in the emendations I follow such spellings
- *Sandhi* irregularities are also a major feature of the SGT;[7] I have provided emendations just in few occasions
- One of my aims has been to bring to light the commentator's own lexicon, phraseology, and literary style. For this reason, lexical entries (substantives, adjectives, and compounds) are given in their corresponding citation forms; the advantage of doing this is that they are more easily recognisable. However, I present inflected forms whenever they are relevant to the commentator's discourse or technical vocabulary
- In the footnotes I provide text-critical analysis as well as emendations
- Sanskrit terms are transliterated according to the International Alphabet of Sanskrit Transliteration

The SGT uses a specific vocabulary which includes mainly technical words not found in standard Sanskrit dictionaries and which are only partly shared

3 In the translation, references to the commentator's explanation of quoted passages from the L are indicated as follows: "SGT ad L", this is followed by the verse number of the L according to the editions mentioned earlier and the verse line of the SGT to the edition by Kāpadīā. Thus, "SGT ad L 50, 9–11" means "Siṃhatilakasūri's commentary upon the L verse 50, lines 9–11, which refer to Kāpadīā's edition.

4 Note that they share the same numbering system.

5 I have not checked against the meters identified by the editor in the footnotes as this has been accomplished by Hayashi (2013).

6 See *dvipakṣa* or *adhi(ka)*.

7 It is worth noting that in Sanskrit manuscripts non-application of sandhi rules often had a specific purpose: to trace word-division and facilitate word-boundaries.

with other treatises on mathematics. A list of technical terms is in Appendix 2; see also the Index.

Kāpadīā's edition is online and available at: www.archive.org.
Unless otherwise stated, all translations are my own.

Editorial conventions

Below I illustrate the conventions and abbreviations used in the translation when editing the text:

- Repairs to the text about which I feel rather little doubt, typically small or obvious corrections, are marked by the abbreviation "em." (emendation). Bolder conjectures are marked by the abbreviation "conj."; the difference is of course subjective. The word or phrase which I have emended is followed by the sign]. For the sake of clarity, emendations are sometimes followed by the reading found in the edition by Kāpadīā (K).
- The sign ° is used to indicate that a word is part of a longer word or compound.
- A term which I consider to be spurious is marked between two stars * *.[8]
- When I have made improvements to numerical layouts, I provide the improved layout in the translation itself. In the footnotes, I explain the reason for the emendation.

8 Some of the terms that I have marked as spurious might represent scribal mistakes or changes.

3 Translation of the *Gaṇitatilaka* and its commentary

The *Gaṇitatilaka* composed by Śrīpati together with the commentary written by Siṃhatilakasūri

3.1 Homage to the Jina[1]

Having bowed to the Jina, the guru whose feet are to be joyfully worshipped by deities, Śrī Siṃhatilakasūri, the disciple of Śrī Vibudhacandra Gaṇabhṛt,[2] who himself rejoices within having experienced the blessing of the *devī* Kuṇḍalinī, composes this commentary[3] on the *Gaṇitatilaka* for the awakening of the Self of others.[4]

3.2 Benedictory verse (*maṅgalācaraṇa*)[5] [GT 1]

Then the author,[6] whose name is Śrīpati, praising the divinity that is accepted by all the systems of philosophy [and] that is one's own Self, declares:[7]

> 1. Having bowed to the supreme true nature of the Self[8]
> whose essence is with and without form
> I will compose [a work on] arithmetic[9] in variegated metrical stanzas
> for the sake of worldly affairs.[10]

1 *namo vitarāgāya* – presumably the invocation by the scribe.
2 *gaṇabhṛt* – a title denoting the head of a group of three or four Jaina wandering monks.
3 *vṛtti*. Another term occurring in the SGT is *gadya* (SGT 27, 27), which denotes a work written in "prose".
4 In these two verses, the commentator pays homage to his spiritual lineage, specifies the name of the text he is going to comment upon, and clarifies the aim of his work.
5 K 1, 11–12.
6 *sūtrakāra*.
7 I take *ātmadevatā* to refer to Śrīpati's religious affiliation rather than to mean "tutelary deity". Further in line 14, the commentator uses again the term *ātma* with the same meaning but compounded in *ātmasvarūpa*.
8 *ātmasvarūpam paramaṃ* – Śrīpati was, in fact, a Śivaite.
9 *gaṇitasya pāṭīm*.
10 In brief, now the treatise can be undertaken because it has a suitable aim and main topic.

For the reason that this verse has many meanings, I am not commenting on it; [it is enough for me to observe that] the treatise [here] begun is concise and causing enjoyment.[11] This [verse] is [in fact] clear for those who know the "real nature of the Self".

3.3 Section on technical terms (*paribhāṣā*)[12] [GT 2–12]

The author, wishing to begin this treatise on mathematics[13] by showing first the rising in measure of the calculation[14] beginning with *eka* via the increase in zeros,[15] enunciates a metrical stanza:[16]

> 2–3. *Eka* ("one"), the [notational] place *daśa* ("ten"), then *śata* ("one hundred"), *sahasra* ("one thousand"), next is *ayuta* ("ten thousand"), *lakṣa* ("one hundred thousand"), *prayuta* ("one million"), *koṭi* ("ten million"), *arbuda* ("one hundred million"), *padma* ("one thousand million"), *kharva* ("ten thousand million"), that called *nikharva*, then *mahāsaroja, śaṅku, samudra, antya, madhya,* [and] *parārdha.*[17] Those learned[18] in the knowledge of the [notational] places[19] explain these numbers[20] which are progressively multiplied by ten.[21]

They explain these **numbers** whose multiplier[22] is ten. To illustrate: [in figures] ***eka*** is 1, ***daśa*** is 10; *eka*[23] multiplied by *daśa* becomes *daśa*. It is always in this way.[24] [The term] ***śata*** is 100; one and three zeros is ***sahasra***, thus: 1000. One and four zeros

11 The commentator establishes the value of the root-text by emphasising that it is properly composed and appropriate for its purpose; as it is concise and pleasant, it meets the criteria expected by a *śāstric* work. "Conciseness" was a highly valued quality for a mathematical text; brevity renders a mathematical verse-treatise compact and therefore easy to be memorised.

12 K 1–3, 18–10.

13 *gaṇitaśāstra* – a compound which also means "the science (*śāstra*) of mathematics (*gaṇita*)", see for instance SGT 14, 3; it denotes the literary corpus of the expert tradition on mathematics. GT 101, 30 uses the expression *saṅkhyāśāstra* to denote the "science of numbers".

14 *gaṇita* – from the root *gaṇ-* "to count, enumerate".

15 *śūnyavṛddhyā*.

16 In the introductory line, the commentator always specifies the number of verses or metrical stanzas composing a rule and a sample problem.

17 Line 21: *parārdham*] em.

18 In both works, the numerous references to the "experts" indicate the authority of the textual corpus of specialised knowledge (i.e., the *śāstric* tradition of *gaṇita*).

19 *sthāna* – literally "place", it has also the technical meaning of "[notational] place".

20 *saṅkhyā*.

21 In *daśaghnīm*, the adjective *ghna* is lit. "killing, striking with". This stanza presents a list of notational places in the decimal number system.

22 *guṇa*.

23 In the neuter form, the noun *eka* is "one" as the "units digit", while in the masculine is the number "one".

24 *evaṃ sarvatra* – a concluding formula often used by the SGT at the end of the explanation of a procedure. It means something like "this holds good everywhere", "bear in mind that this procedure should be always carried out in this way"; here it alludes to the fact that in the same way each notational place should be multiplied by ten (see *daśaguṇa*): *daśa* multiplied by *daśa* becomes *śata*, *śata* multiplied by *daśa* becomes *sahasra*, and so forth.

is ***ayuta,*** that is to say, ten *sahasras*: 10000. One and five zeros is ***lakṣa***: 100000. One and six zeros is ***prayuta*** [or] ten *lakṣas*: 1000000. One and seven zeros is called ***koṭi***: 10000000. One and eight zeros is ***arbuda*** [or] ten *koṭis*: 100000000. One and nine zeros is ***padma*** [or] one hundred *koṭis*: 1000000000. One and ten zeros is ***kharva,*** [or] one thousand *koṭis*: 10000000000. One and eleven zeros is ***nikharva*** [or] one thousand [times] ten *koṭis*: 100000000000. One and twelve zeros is ***mahāsaroja*** [or] one hundred thousand [times] one *koṭi*: 1000000000000. One and thirteen zeros is ***śaṅku*** [or] one hundred thousand [times] ten *koṭis*: 10000000000000. One and fourteen zeros is ***samudra*** [or] one *koṭi* [times] one *koṭi*: 100000000000000. One and fifteen zeros is ***antya*** [or] one *koṭi* [times] ten *koṭis*: 1000000000000000. One and sixteen zeros is ***madhya*** [or] one hundred *koṭis* [times] one *koṭi*: 10000000000000000. One and seventeen zeros is ***parārdha*** [or] one thousand *koṭis* [times] one *koṭi*: 100000000000000000. Therefore, [in this last notational place] together with one, there are eighteen digits.[25]

Even that which is beyond [the last notational place called *parārdha*] is multiplied by ten via the increase in zeros, and its name[26] should be understood as it occurs in [other] treatises.

Then the author enunciates a metrical stanza for technical terms[27] concerning the practice[28] of [measuring units beginning with] cowry-shells:[29]

> 4. There is one *kākiṇī* in five times four cowry-shells, 20. Those who are familiar with this practice say that one *paṇa* is [equal to] four *kākiṇīs*, and one *dramma* is known to be in sixteen of these [*paṇas*].

In other words, **[one *dramma* is known to be in sixteen] of these** "*paṇas*" [which is thus the term to supply].[30]

The author enunciates a metrical stanza with technical terms for the practice of [measuring units of] gold:[31]

> 5. They declare that there is one *niṣpāva* in six barley-corns, one *dharaṇa* is indicated to be in eight of these [*niṣpāvas*]. Those here who are expert in [measuring] gold estimate one *gadyāṇaka* to be in two of these [*dharaṇas*].

There is one ***niṣpāva*** [or] one *suvarṇavalla*[32] in six **barley-corns,** [the expression] **they declare** means "they state"; everything else is clear in meaning.

25 *aṅka* – in the SGT, it is "number, digit, quantity" according to the context. The commentator implies that each digit is a different place value.

26 In brief, the names of the notational places following *parārdha*.

27 Line 12: *sañjñārthaṃ*] em. Stanzas 4–12 list measuring units.

28 *vyavahāra* – it points to the transactions, commercial practices, and usages in business which involve measuring units. In this regard, it is noteworthy that, in Sanskrit, *vyavahāraka* is "merchant". On the use of the term *vyavahāra* in mathematical literature, see further on the discussion concerning *miśravyavahāra* (GT 118).

29 *kaparda*.

30 The commentator singles out the pronoun *taiḥ* to indicate that it qualifies *paṇa*.

31 *suvarṇa*.

32 The commentator provides a measuring unit, familiar to him but not mentioned in the root-text, which is equal to one *niṣpāva*.

The author enunciates a metrical stanza specifying technical terms for the practice of measurable [quantities]:[33]

6. Those who are most proficient in arithmetic[34] call seven pairs of *niṣpāvaka*s one *dhaṭaka*; one *pala* is said to be in ten of these [*dhaṭakas*]. Here is the equivalence for the accomplishment of the practice concerning measurable [quantities].[35]

The author enunciates a metrical stanza specifying technical terms relating to grain measures:[36]

7. They mention the [following] measures: here four *pādikā*s are one *mānaka*, [and] one *setikā* is said to be in four of these [*mānakas*]. There should be one *hārī* in ten *setikā*s.

There is one "*setikā*" in four "*mānaka*s"; the rest is clear. One hundred *mānika*s of four *hārikā*s [each]: that is known to people by its own name.[37] The remaining part [of the stanza] is clear.

The author enunciates two and a half metrical stanzas explaining technical terms for the practice of [measuring units of] field (*kṣetra*):

$\bar{8}$. There is one digit[38] in four [times] six huskless barley-corns. The wise call twenty-four of these digits a cubit.[39]

9. There should be one daṇḍa in four cubits, and one rajju is understood as [being equal to] twenty daṇḍakas. The learned call an equal sided [figure] bounded by the measures of two rajjus, a nivartana.

10. Erudite scholars observe that a pair of one thousand *daṇḍa*s is [equal to] one *krośa*. In this regard, people who are skillful in determining the measure of soil declare that one *yojana* is [equal to] four of these [*krośa*s].

The author enunciates a pair of metrical stanzas specifying technical terms for the practice of [measuring units of] time:[40]

33 *meya*.

34 *pāṭī*.

35 It is unusual that the commentary does not comment at all upon this stanza; presumably in the original SGT a brief passage explaining the technical terms or the expression *spaṣṭaṃ* or "everything is clear" might have occurred.

36 *kaṇa*.

37 I agree with Hayashi who, in a personal communication, has pointed out that the commentator refers to a larger *mānika* and a different equivalence between the two measuring units; Kāpadīā (1937, 104) interprets this passage differently and reads "400 *hārikā*s".

38 *aṅgula*.

39 *hasta*.

40 I have emended the text suggesting that it should read: *atha kālavyavahārārthaṃ sañjñājñāpakaṃ vṛttadvayam āha*] conj. In Kāpadīā's edition, units of time are curiously included under the treatment of measuring units of field. However, a separate treatment to time units, and with it the introductory line by the commentator, must have occurred; in line

11. There is one *vināḍī* in six breaths of a man with vigorous strength, there is one ghaṭikā in sixty [of these *vināḍī*s]. Also, those who are proficient in excellent arithmetic state that there is one day and night[41] in sixty of these [*ghaṭikā*s].

12. The learned call thirty of these [days and nights] a month;[42] there is one year[43] in twelve months. This is the end; [the measures] beginning with cowry-shells should be considered as common standards in this place.

These [measures] are clear [in meaning].

In the *Gaṇitatilaka* written by the illustrious Śrīpati, thus is the section on technical terms (*paribhāṣā*).

3.4 Arithmetical operations with integers [GT $\overline{13}$ – 34]

Addition (saṅkalita)[44] *[GT $\overline{13}$]*

Next is the procedural rule (*karaṇasūtra*), occurring in due order,[45] to arrive at addition;[46] the author enunciates half a metrical stanza:

$\overline{13}$. In [the operation of] addition, as the sum[47] of a number[48] [and others] [each] in its own side[49] should be performed by the regular order,[50] in the same way [it should be performed] in the reverse order.[51]

28, the expression *sārdhaṃ vṛttadvayaṃ* suggests, in fact, that the author Śrīpati devotes two stanzas and a half to measuring units of field only.

41 *ahorātra.*

42 *māsa.*

43 *saṃvatsara.*

44 Here begins the treatment of the eight fundamental arithmetical operations with integers. K 3, 13.

45 The commentator observes that the rule of addition is the first treated because such is the "regular (or conventional, natural) order" of the arithmetical operations. In other words, this treatise rightly begins with the addition because this is the first arithmetical operation.

46 In *saṅkalitānayanāya,* two key terms of mathematical literature are found: *saṅkalita* ("addition") and *ānayana* (from the verbal root *ānī-* "to arrive at, to calculate").

47 *yuti* – lit."union".

48 Here I understand the substantive *aṅka* to refer to the addition at a theoretical level and not at the level of practice; that *aṅka* in *aṅkayuti* is "number" appears clear in the explanation of the SGT.

49 In this context, the possessive adjective *sva* functions as a distributive appositive and its meaning is close to "proper"; *svapakṣa* is lit. "its own side" but it denotes "every side" of a number, each "side" where the digits of the numbers to be added are arranged one below the other following their value in the place-value notation. The commentary expands upon *pakṣa* by explaining that it refers to each vertical row of digits according to their power of ten.

50 *krameṇa.* The term *krama* is lit. "method, way"; the term acquires a specific connotation in the operation of addition, where it indicates the "regular order" of adding digits, in contrast to the "reverse order". In the SGT, *krama* is also often used adverbially in the instrumental case with the meaning of "respectively".

51 *utkramāt.*

Its explanation [is as follows]:[52] [the word] **as** is [used] in the sense of "in that manner";[53] [the expression] **as [each] in its own side** refers to a heap of digits (i.e., a number)[54] which is like[55] a row[56] [composed] of [as many digits as] one, two, and so forth.[57] In respect to this heap [of digits] with a digit occurring first, "its own side" denotes its vertical[58] row of digits; in respect to this heap [of digits] with a digit occurring next,[59] it is its vertical row of digits.[60] By means of the reasoning[61] of the sample problem[62] that will be formulated next, the "numbers" [mentioned] there (i.e., in the rule above) are [the numbers] seven and so forth;[63] their **sum** is the union (*yoga*) [of these numbers]. **In [the operation of] addition,** that is, "in the procedure (*vidhi*)[64] of joining numbers",[65] [the sum] **should be performed,** which means "it should be carried out", **by the regular order**: the regular order is such that the sum[66] of the numbers is from the number which is at the top to [the number which is at] the bottom. **In the same way** means in "that very manner", that is to say, **in the same way in the reverse order:** the reverse order is such that the sum of the numbers is from the number which is at the bottom to [the number

52 *vyākhyā*.

53 See *yena prakāreṇa*...*tena prakāreṇa*; the commentator emphasises the two correlatives *yathā/tathā*.

54 *aṅkarāśi* – lit. "heap, set of digits", this expression is part of Siṃhatilakasūri's own repertoire. It explains a number as being composed of more than one place value, a number is thus made of a group of digits each occupying a specific notational place. It is worth noting that the SGT does not use consistently the compound *aṅkarāśi* to denote a "number" and it often prefers the shorter term *aṅka*. For instance, in the passage on multiplication (SGT 17–18 the number 21586 is indicated by *aṅka* in line 13 and by *aṅkarāśi* in line 15. With respect to the polysemy of the term *aṅka* in the SGT, it is interesting to observe that in line 18 (in the same passage on multiplication), *aṅka* is "digit".

55 The commentator describes a number as a "row" of digits. The sense of "likeness" expressed by °*bhūta* is attested in the Sanskrit works on grammar and semantics *Nirukta* and *Mahābhāṣyadīpika*.

56 *śreṇi*. In the line below, *ūrdhvāṅkaśreṇi* indicates a "vertical (*ūrdhva*) row of digits".

57 In brief, place values. The commentary implies that in the decimal number system, the value of a digit depends on its place.

58 In the SGT, the term *ūrdhva* is also used with the meaning of "above/at the top".

59 In the same line, cf. *pūrvāṅkarāśi* and *agretanāṅkarāśi*. These expressions distinguish the commentator's lexicon and occur sometimes in apposition to verbal numerical expression.

60 This passage (lines 14–15) is an explanatory amplification of *svapakṣa*: *pakṣa* refers to a digit's value, which depends on its position in the "number" (*aṅkarāśi*); in order to carry out their sum, numbers are vertically arranged (see *ūrdhvāṅkaśreṇi*) and their digits are placed one below the other. The term *pakṣa* is thus used in both its general sense ("side") and in its specialised mathematical sense, indicating the "side", the place value shared by digits when numbers are placed in a vertical row.

61 *yukti*.

62 *udāharaṇa* – lit. "illustration, example".

63 The commentator explains the terms given by Śrīpati in the rule by means of the numbers which are found in the next sample problem; this expository style links concepts to their actual scenario and is exemplary of Siṃhatilakasūri's commentarial practice.

64 *vidhi*.

65 *aṅkasammīlana*.

66 *mīlana*; in the SGT, both verbal roots *mil*- ("to add, join") and the less common *parasmaipada* form *mīl*- and their derivatives are found.

which is at] the very top, [such is the meaning of] that [expression] **in the reverse order**. Thus is the connection [of words with their meanings].[67]

In this respect,[68] [the author enunciates] a metrical stanza presenting a pair of examples:

> 14. O learned one, if you know [the operation of addition], tell [me] as quick as possible what number occurs when seven, eight, nine, sixteen, ninety-three, sixty, seventy-six, and fifty are added, and also when twenty-seven and twenty-one are added to thirty-two, fifteen, and five?

In the first half [of the metrical stanza], the first example [is enunciated]; in the second half, the second [example]. Therefore, in the first example [the numbers are] seven, eight, and so forth: |7|8|9|16|93|60|76||50|. On the writing board or on the ground,[69] the seven is [written] first;[70] below it, the eight and so forth are gradually (*kramena*) [written] [each] below [the other]. The row of numbers is written[71] up to [the number] fifty and summed up by both the "regular" and the "reverse order" previously mentioned.

Eight added[72] to seven[73] becomes fifteen, [which in figures is] 15; nine added to it becomes twenty-four: 24, and so forth is thus **by the regular order**. **In the same way**, from the number which is underneath, three is added to six,[74] 6, the six belonging to seventy-six; it becomes nine: 9. When the additive quantity[75] six [of sixteen] is [added] to it, it becomes fifteen: 15. [All the other numbers such as] fifteen and so forth are added in the "reverse order". **O learned one** means "o you skillful in mathematics", **tell [me]** (*vada>brūhi*), **what number occurs as soon as possible,** in the sense of "quickly", **if** is "in case you know": the remaining part [which should be understood in the formulation of the rule] is '[if you know] the science of mathematics'. An answer[76] [to the sample problem] should be obtained, and that [answer] is this: when there is the sum[77] of the numbers beginning with seven, the result is three hundred nineteen, 319.

67 *iti saṃbandha–* in Sanskrit commentarial literature, this is a concluding formula to mark the explanation of features relating to syntax and construction; the commentary has completed the explanation of the meanings of the terms formulated in the rule by the root-text.

68 The *atra* introducing a sample problem is "on this topic, on the topic enunciated in the rule".

69 *paṭṭake bhūmau vā* – this is an interesting reference to the working surface used for computing numbers.

70 The commentator alludes to a vertical arrangement.

71 *likhita*.

72 *kṣipta*. In mathematical works, the verbal root *kṣip-* (lit. "to throw, place in") means "to add".

73 The forms *madhye* (lit. "in the middle") in relation to a sum and *madhyāt* (lit. "from the middle") in relation to a subtraction are part of Siṃhatilakasūri's own idiom.

74 Line 29: *ṣaḍmadhye*] em. This is the first of the several instances of the frequent absence of *sandhi* between *ṣaṭ* and the following word; because it is part of the phonetic irregularities which characterise the manuscript, it will be no longer emended.

75 *kṣepa* – a standard mathematical term denoting the "additive quantity".

76 *uttara*.

77 A common feature of Sanskrit mathematical literature is the alternative use of the feminine *yojanā* ("sum") and the masculine *yojana*; similarly, the masculine *guṇana* ("multiplication") and the feminine *guṇanā* co-occur.

In the same way, in the second example the row of numbers which begins with twenty-seven, |27|21|32|15|5|, [and] of which five is the last number is added. [The questions is] **what number occurs** and so forth [when those numbers are added]? Everything [should be performed] as [explained] before, and the answer should be obtained. The result[78] is one hundred: 100.

In this way, the procedure of addition is concluded (*samāpta*).

Subtraction (vyavakalita)[79] *[GT* $\overline{15}$*]*

Next is the procedural rule for the knowledge of subtraction;[80] the author enunciates half a metrical stanza:

> $\overline{15}$. And in [the operation of] subtraction, for the obtainment of the remainder, the deduction is certainly [carried out] also in that manner.

Its explanation: **and in subtraction**[81] (*vyojana*>*vyavakalita*) means "in that consisting of the deduction,[82] from the [same] numerical place,[83] of a smaller[84] digit from a larger". The word **and** is [used] in the sense of a conjunction; [the word] **also** is [used] with the meaning of "too". **Also in that manner**, that is, by the methods[85] mentioned before, having placed[86] the subtrahend,[87] which is the smaller number, below [the minuend], it is subtracted (*pātyate*) from the larger number (i.e., the minuend);[88] thus is [the meaning of] **in that manner.** In the same way, having placed the number which is the subtrahend above the larger number, it is removed (*niṣkāśyate*); thus is [the meaning of] **in that manner**.[89]

78 *labdha.*
79 K 4, 10.
80 *vyavakalita.*
81 Line 11:**yā**] em.
82 *pātana* – forms of the verbal root *pat* (lit. "to fall") are used in relation to the procedure of deduction.
83 *aṅkasthāna*– "numerical place", a key term of the SGT's mathematical jargon. In the context of a computation, it emphasises a digit in respect to its power of ten. This term also occurs, for instance, in SGT, 18; in SGT 32, 26 and SGT 90, 5 *aṅkapada* shows the same meaning.
84 The adjectives *svalpa* and *bṛhat* are not in the comparative form but this is, in fact, the sense which is epxressed.
85 *rīti.*
86 *nyasya* – the verbal root *nyas* (lit. "to put down") is part of standard Sanskrit mathematical vocabulary.
87 Line 13: *svalpāṅkaṃ pātyaṃ*] em. Being *a* and *b* two numbers, *a* is the minuend and *b* is the subtrahend, the number which is to be subtracted and which is denoted as *pātya.* The commentator specifies that *pātya* is the smaller number.
88 Emendation might be considered; the commentary explains that the subtrahend can be placed either above or below the minuend, one would thus expect *bṛhadaṅkasyādhaḥ* rather than *bṛhadaṅkād adhaḥ.*
89 The two passages ending with *iti krameṇa* (lines 14 and 15) are symmetrical and elaborates on *krameṇa*.

The deduction[90] (*viśodhana>viśuddhi*) is **certainly,** which denotes a certainty, **[carried out also in that manner]**;[91] **for the obtainment,** [also termed] "acquisition", **of the remainder,** that is, "of the number which is left" from [the subtraction of] the number which is the subtrahend.[92] Thus is the connection [of words with their meanings].

[Next is] a sample problem; [the author enunciates] one verse:[93]

16. Ah! If you know [how to carry out] the deduction, tell [me] quickly the remainder having removed (*projjhya*) from one thousand the numbers which have been calculated with respect to the [previous] stanza.

Its explanation: **if you know** [how to carry out] **the deduction** (*viśodhana>pātana*) of a smaller number from a larger number, **tell [me]** (*cakṣva>vada*) **the remainder,** which is the number[94] leftover from one thousand, **having removed**, [also termed] "having subtracted", **from one thousand,**1000, that is, from the number consisting of "one thousand", **the numbers** [obtained] from the calculation[95] [mentioned] in the previous "stanza". In other words, these are namely three hundred nineteen, which has been referred to in the procedure of addition, as well as one hundred. In figures also: one thousand is 1000, from this [the number three hundred nineteen] is detracted by means of the method "above and below".[96] The remainder[97] obtained is 681, six hundred eighty-one. In the same way, when there is the deduction of one hundred from that very number 1000, the remainder is 900, nine hundred.

In this way, [the explanation of] the definition of subtraction is concluded.

90 Line 13: *viśodhanaṃ viśuddhirūpaṃ* |] conj. This conjectural emendation has been preferred to the trasnmitted *viśodhitaṃ viśuddhipadaṃ nītam aṅkasthānam* (lit. "it is deducted, that is, the numerical place is moved into the place of deduction), which does not seem appropriate. It appears neither to clear up an ambiguity not to add any shade of meaning. I have proposed the following emendation for two reasons: firstly, one would expect the substantive *viśodhana* rather than the verbal adjective *viśodhita*; secondly, the passage is semantically awkward. Note that *viśodhana* is the only term of the *karaṇasūtra* not yet glossed by the commentator. This emendation may seem forced by it follows the commentary.

91 *viśuddhi* – lit. "purification" is a technical term for "deduction, subtraction". This passage clarifies that the two numbers are placed with their digits arranged one below the other according to their place value.

92 Line 15: *pātyāṅkā*] em.

93 There is a mistake in the edition: the meter used for this sample problem is, in fact, *anuṣṭubh*.

94 Line 21: *aṅkaṃ*] em.

95 Cf. *uktasaṅkhyāṅkān* (SGT) and *uktasaṅkhyākān* (GT).

96 *ūrdhvādhorītyā* – this expression refers to the way the minuend and the subtrahend are arranged in the layout to carry out the subtraction of their digits. Here the commentator alludes to the fact the subtrahend can be equally placed either above or below.

97 *śeṣa*.

Multiplication (guṇakāra)[98] *[GT 17–$\overline{18}$]*

Next is the procedural rule of multiplication,[99] whose purpose is the increase of small numbers by means of summation[100] [and] whose name is *pratyutpanna* as mentioned in the *Triśatī*.[101]

The author enunciates one and a half metrical stanzas:

> 17–$\overline{18}$: Having placed the multiplicand (*guṇya*) below the quantity called "multiplier",[102] by means of the [method known as the] "junction of two doors"[103] one should multiply (*hanyāt*) having caused [the digits] to shift[104] gradually[105] in the regular order[106] or in the reverse order.[107] Ah! There is also the "exactly so as it stands" [method].[108] Having divided *sthāna* and *rūpa*,[109] one should perform the multiplication[110] called "portion" (*khaṇḍa*).

Its explanation: in this passage, four methods[111] regarding the procedure of multiplication of numbers are mentioned. There[112] **the multiplicand**, [also termed] "that which is to be multiplied" (*guṇanīya*), is [the number] 21586, twenty-one thousand five hundred eighty-six; **the quantity called multiplier** is namely ninety-six, 96. Place[113] [the multiplicand] below it,

98 K 4, 28–K 5, 1.

99 *guṇanā*.

100 *upacaya* – "summation, accumulation"; the commentator uses again this term in SGT 52, 12 with the meaning of "summation". It is not clear whether *upacaya* emphasises the multiplication as a repeated addition of a number or refers to the various sums which the multiplication of two numbers require and which are briefly mentioned by the commentator in his explanation.

101 Siṃhatilakasūri names for the first time the *Triśatī* by Śrīdhara, where in verse 6 one finds the word *pratyutpanna*, which is a technical term for "multiplication". In SGT 132–133, 27, one finds *pratyutpannavidhi* or the "procedure of multiplication".

102 *guṇaka*.

103 *kapāṭadvayasandhi*.

104 In PG 1.23–24, TŚ 6, and L 14 the causative absolutive *utsārya* is also found in the rule of multiplication. The expression *utsāryotsārya* denotes the iteration of the shifting of numbers.

105 *kramaśa*– "consecutively, gradually"; this adverb is also found in Śrīdhara's rule on multiplication and in relation to the same method, which he terms *kavāṭasandhi*.

106 *anuloma*.

107 *viloma*.

108 *tat stham eva*.

109 These are two technical terms whose meaning, as shown by the SGT, vary according to the multiplication method employed.

110 *santāḍana* – lit. "striking".

111 *rīti*. It appears that the four methods meant by the SGT are: the "junction of two doors", the "exactly so as it stands", and the twofold "portion" method (according to *sthāna* and *rūpa*).

112 The SGT begins to explain the rule and its technical terms by taking into account the sample problem formulated next by Śrīpati, where he asks students to multiply 21586 by 96.

113 *nyasya*– here the absolutives *nyasya* and *guṇayitvā* can be interpreted within the scope of the imperatival modality of finite verbs. Such an use of the absolutive is a common construction in the SGT. In line 4, one would expect *taṃ nyasya*.

afterwards[114] multiply first eighty-six by ninety-six **in the regular order,** that is, by the numerical method[115] following the natural course (*anukūlam*), hereafter is [the step] **having caused to shift,** that is, having made ninety-six to move [above the multiplicand]. Or else by the numerical method [denoted as] **in the reverse order,** that is, according to the inverse course (*pratikūlam*), multiply twenty-one by [the multiplier] ninety-six. **Gradually** [also termed] "progressively", according to the regular direction[116] multiply ninety-six[117] above fifteen by means of this procedure, according to the inverse order **one should multiply** (*hanyāt>guṇayet*) fifty-eight by the ninety-six above.[118]

The word **ah!** is [used] for the sake of [denoting] an alternative;[119] the word **also** is [to be understood] with the meaning of "or else". [The expression] **exactly so as it stands** is the qualification of an action.[120] In this very case, according to the regular order **one should multiply** every digit again by [the multiplier] ninety-six, which [this time] is not made to shift,[121] standing indeed above eighty-six, and, according to the reverse order, [one should multiply every digit] by that [multiplier] being static [and] standing above twenty-one.

Alternatively, the "multiplier" such as ninety-six, which is indicated by that word ***sthāna***, stands above the number which is the "multiplicand". **Having divided**[122] it (i.e., the multiplier) into three [equal] parts, it is thirty-two; or else [having divided it] into two [equal] parts, it is forty-eight. Having made it (i.e., *sthāna* or the multiplier 96) [divided] into three [equal parts],[123] **one should perform the multiplication** (*santāḍana>guṇana*) of the "multiplicand", which is the number beginning with twenty-one and which also consists of three [parts][124] (i.e., written three times), by the number which has been made into three [equal parts], [namely by] thirty-two. Or **[having divided** the multiplier] into two [equal parts] [and having thus written the two quotients] **[one should perform the multiplication** of the multiplier] by [the two equal quotients] forty-eight.

114 *pascāt* – throughout the commentary, it is found also as a preposition with the meaning of "behind". This term occurs, in fact, as both space-related and time-related. Interestingly, the SGT applies *paścāt* as "behind" to both directions: from right to left and from left to right.

115 *aṅkarīti.*

116 *gati* – lit. "movement".

117 Line 7: *ṣaṇ ṇavatiṃ*] em.

118 Lines 7–8 explain the movements of the multiplier, which is above the multiplicand, according to both methods, that is, "in the regular order" and "in the reverse order".

119 I have emended the text as the passage *āhośabdo vikalpārthe* (line 5) should occur at the end of line 8. Lines 5–8: *āhośabdo vikalpārthe* | *utaśabdo'thavārthaḥ* | em.]

120 This is a grammatical remark which explains the expression *tat stham eva*. In Sanskrit grammatical literature, the syntagm *kriyāviśeṣaṇa* denotes the determinant or modifier of a verbal activity.

121 Line 9: °*anutsāritayā*] em.

122 The *Addenda et Corrigenda* mistakenly corrects *vibhajya tridhā* to *vibhajyatridhā*.

123 *vāratrayaṃ* – lit. "three times"; *vāra, guṇā,* and *velā* (which occur in this passage too) followed by numerals mean "times".

124 *traya* – lit. "triad".

Alternatively, [the meaning of the word] ***rūpa*** is shown: again that [same] number, which is [also] called *rūpa*, for the purpose of [its] increase[125] by means of the "multiplier" is investigated. **Having divided** it,[126] thus [in this case] having made the number aforementioned beginning with twenty-one into two parts (*kṛtvā dvidhā*), having written two times in two places[127] ten thousand seven hundred ninety-three, **one should perform the multiplication** [of these numbers] by ninety-six.

The fourth method is [as follows]:[128] **having divided** the "multiplier" according to its [notational] places [as much as they are],[129] **one should perform the multiplication** of [the multiplicand, which is] the number beginning with twenty-one once[130] by the digit nine and once by six. Having arranged (*niyoja*) the lower number[131] by being one [notational] place more [with respect to the number above],[132] it should be summed up. The presentation is:[133] 21586 multiplied[134] by 9 becomes 194274; in the same way, [the number starting with 21] multiplied by 6 becomes 129516. Regarding these [two numbers], in this way [having arranged the lower number] by being [one notational place] more [with respect to the number above], when there is the presentation:

194274
 129516

and when there is the sum [of these numbers], the result is 2072256. In this manner, when the number which is the "multiplier" [is split up into as many digits as they are], then have the "multiplicand" multiplied once by one digit [and] once by the second [digit],[135] by means of the method [of positioning numbers] shown [above], from the second digit write (*vilikhya*) the second product [below] by being one [notational] place more, so that, [having added these], it (i.e., the number obtained) is the desired result. This procedure is called "portion". [Now] [the procedure called "portion"] according to the multiplier:[136] have [the number] such as ninety-six added to two. Multiply [then] the number beginning with twenty-one (i.e., the multiplicand) by ninety-eight; afterwards, from the number which is the previous

125 As in the introductory line, here too the commentator underlines that the multiplication implies the increase of numbers.
126 It refers to *rūpa*, here the "multiplicand".
127 *sthāna vilikhya velādvayaṃ*.
128 From line 17, the SGT explains the "fourth method" (this deliberately echoes the first line, where the commentary states *rīticatuṣṭayam uktam* or "four methods are mentioned"), which is subdivided into various procedures.
129 Line 17: *yathāsthānam*] em.
130 *ekavāraṃ…ekavāram*.
131 Line 19: °*aṅkaṃ*] em.
132 *param ekāṅkasthānatayā*.
133 *nyāsa* – lit. "setting down", a key mathematical term denoting the writing down of numbers on the working surface in order to be manipulated.
134 *gu* – this is s an abbreviation standing for the verbal adjective *guṇita* ("multiplied").
135 The use of *dvaya* (lit. a "pair") instead of the ordinal *dvitīya* ("second") is a characteristic of the SGT.
136 Line 24: *yathāsthānaṃ*] em.

product, subtract [the product of] the number beginning with twenty-one[137] which has been multiplied by two.[138] This procedure is called "the portion which makes the multiplier larger".[139] **Having divided** [into two portions] the multiplier (*sthāna*) that consists of ninety-six, having performed **the multiplication** of the [multiplicand, which is the] number beginning with twenty-one once by ninety-two [and] once by four, the sum of the two [obtained] numbers should be made. This procedure is called "the portion which makes the multiplier smaller".[140] Because of the splitting of the number which is the multiplier or of the number which is the multiplicand, the procedure called "portion" is appropriately named.

Next the author enunciates a pair of metrical stanzas together with the presentation:[141]

> 19. O mathematician (*gaṇaka*), if you know [the rule of] multiplication, say quickly what [the result] should be when [the number composed of] two, one, five, eight, [and] six are multiplied by ninety-six: [thus the presentation is] 21586 multiplied by 96. Tell [me] also [what the result should be when] [the number composed of] ninety-three, six, eight, [and] five is multiplied by thirty-two: [thus the presentation is] 93685 multiplied by[142] 32; [and] [the number composed of] nine, eight, five, one, [and] zero [is multiplied] by [the number composed of] two, seven, [and] three: [thus the presentation is] 98510 multiplied by 273.

This [stanza] is clear [in meaning].

> 20. That whose appearance[143] is multiplied by [the number composed of] seven[144] [and] seven[145] is [the number composed of] twelve,[146] nine, eight,[147] seven, zero[148] and thirteen.[149] In this case, the ornament constituted by the clear, radiant, [and] spherical pearls on the neck of Maheśvara will occur.

137 Line 25: °*prabhṛtikaṃ*] em.

138 The commentator refers to the product obtained by multiplying the multiplicand 21586 by 98.

139 Line 26: before *sthānaṃ* a *daṇḍa* should occur.

140 *līnatā* – lit. "the state of being dissolved". This technical term must have been part of the commentator's local mathematical jargon; in the SGT, it is found only in relation to this multiplication method.

141 The commentator underlines that in the sample problem Śrīpati himself provides the setting down of the data.

142 In the edition, cf. *guṇa* (line 3), *gu* followed by a dot (line 5), and *gu* followed by a circle in the explanation of the procedural rule of multiplication (line 19).

143 Note that Śrīpati uses the *bhūtasaṃkhyā* notation, where words of common usage in ancient and medieval Indian culture stand for numbers.

144 *turaṅga* – lit. "horse"; it alludes to the seven horses of the sun's chariot.

145 *śaila* – lit. "mountain"; the geography of the *Pūraṇa*s recognizes seven principal mountains.

146 *arka* – lit. "sun"; it represents the number "twelve" because of its connection to the twelve *āditya*s (solar deities) of Sanskrit literature.

147 *bhujaṅga* – lit. "snakes".

148 *kha* – lit. "sky".

149 *viśva* – it denotes a group of thirteen Vedic deities.

The explanation: by the word *viśva* [the number] "thirteen" is meant, *kha* is "zero", *sapta* ("seven") is clear, by the word *bhujaṅga* eight kinds of snakes are denoted, *nava* ("nine") is clear, [and] *arka* is twelve. By the numerical method [following the natural course],[150] [write] thirteen first, zero behind it, and so forth; these [numbers] following on successively written according to their forms are the "multiplier".[151] [The term] *śaila* is [used to denote] the seven *Kulācalas*;[152] *turaṅga* [stands for the number seven because] the horses of Sūrya are known to be seven; therefore, [the passage should be understood as] that **whose**[153] **appearance** (*deha>svarūpa*) **is multiplied** (*samāhata>guṇita*) by this which consists of the number seventy-seven.[154] The rest is clear [in meaning].

The presentation is: 12987013 multiplied by 77. The presentation of the numbers which are the results is accordingly: twenty hundred thousand seventy-two thousand two hundred fifty-six, [which in figures is] 2072256; twenty-nine hundred thousand ninety-seven thousand nine hundred twenty, also in figures: 2997920; [then] two million sixty-eight hundred thousand ninety-three thousand two hundred and thirty: 26893230, [and lastly] one hundred million plus one: 1000000001.

In this way, the procedure of multiplication is concluded.

Division (bhāgahāra)[155] *[GT 21]*

[Next is] the procedural rule of division;[156] [the author enunciates] a metrical stanza:

> 21. Having reduced,[157] whenever possible, both the divisor and the dividend[158] by a same number, one should gradually divide[159] [the dividend) in the reverse order. Its method (i.e., the method of division) is [so] designated by those who are competent in the knowledge of mathematics.

The explanation: **having reduced** means "having divided" (*khaṇḍitvā*),[160] **by a same number:** briefly, [either] by an even (*sama*) number such as two, four,

150 In order to compose the final number, in the *bhūtasaṃkhyā* notation words are read from right to left.
151 Lines 11–12: *iti krameṇa rūpaiḥ*] conj.
152 The *Kulācalas* are a group of mountains mentioned in the *Pūranas*.
153 *yeṣāṃ te* – this is a grammatical formula to analyse a *bahuvrīhi* compound. The subject is plural since it refers to the word-numerals used to denote the number 12987013. For the sake of a smooth translation, I have rendered the subject singular: "that [number]…".
154 Line 13: **jātaḥ**] em.
155 K 6, 21–24.
156 *bhāgahāra*.
157 *apavartya* – from the verbal root *apavṛt-* lit. "to take away", in mathematical texts it means "to divide, reduce".
158 In *harabhājyau*, *hara* is "divisor", while *bhājya* is the "dividend".
159 *vibhajet*.
160 Lit. "having broken into parts".

and so forth, [or] by an odd (*viṣama*) number such as one, three, and so forth; **both the divisor and the dividend** are [thus] divided. That by which a number that comes by with an increment (i.e., it is larger) is decreased is the quantity [called] "divisor" (*hara>bhāgagrāhāko'ṅka*);[161] in the same way, that number which is divided (*bhājyate>bhāgaṃ pradāpyate*) is "the dividend".[162] The pair **the divisor and the dividend** are placed below (*adhasthita*) and above (*uparitana*) [each other]. **Whenever possible** means "when there is the possibility" that **the divisor and the dividend** are [both equally] suitable (*sahete*) for any aliquot divisor[163] such as two and so forth. [When] the first number is suitable for a [certain] aliquot divisor [but] the second [number] is not, in that case [the reduction of both the divisor and the dividend] should not be made; this [topic] is thus accomplished.[164] **Gradually in the reverse order** means "by the numerical method according to the reverse course", in short, not in the regular order; **one should divide** (*vibhajet>grāhayet bhāgaṃ*). The [meaning of the] remaining part [of the stanza] is clear.

In this respect, [the author enunciates] an exemplifying problem:[165]

> 22. If the definition [of division] has been learnt [by you], tell [me] quickly what the numbers which are the [previous] products will become when divided by their own multiplier?

The explanation: **tell [me]** (*pracakṣa>vada*) **what the** previously mentioned **numbers which are the products** beginning with twenty hundred thousand **will become when divided** (*bhājita>gṛhītabhāga*) "by their own multiplier", that is, by the previous multiplier[166] such as the given ninety-six. In this case, the verification[167] of the procedure is [as follows]: the first "product"[168] is the number twenty hundred thousand[169] seventy-two thousand

161 Lines 26–27: *dvau harabhājyau hṛyete | vṛddhiṃ prāpto'ṅkaḥ prahāṇiṃ nīyate yena sa haro bhāgagrāhāko'ṅkas*] conj. I am grateful to Hayashi for suggesting this emendation. This is the only occurrence of the term *bhāgagrāhāka* in the SGT.

162 In lines 26–28, the commentator provides a definition of "divisor" (*hara*) and "dividend" (*bhājya*).

163 *khaṇḍana* – lit. "the act of breaking".

164 It refers to the explanation of the operands of division and their reduction by a common factor.

165 *uddeśaka*.

166 Line 8: *prāgguṇakena*] conj. I have proposed this emendation on the grounds that a gloss is most likely here to take place and it makes better sense as opposed to *prāgguṇito* occurring in line 9. Another possible reading, but perhaps less appealing, is to consider *guṇakatāṃ* a misspelling for the locative *guṇanāyāṃ* ("in the [previous] multiplication").

167 *ghaṭanā* – lit. "connection, finding", in the SGT a technical term for "verification". This is the first occurrence of the term in the commentary. In the SGT, "verifications" always involve a procedure by reverse: starting from the result obtained, the commentator works back the data of the sample problem to prove that the procedure has been correctly performed. Verifications are mainly found in the treatment of type-problems of fractions.

168 Line 9: *prāg* should be compounded with *guṇito*.

169 Line 9: **eka**; *viṃśatir*] em.

two hundred fifty-six. Next, **having reduced** [the divisor and the dividend], **whenever possible,** that is, when there is the suitability [of both operands] to the division[170] **by a same number,** which in this case is four, when there is the division (*bhāge datte*) of the [number which has been] written down[171] consisting of five hundred thousand eighteen thousand sixty-four, 518064, [and] which is the "dividend", by the divisor (*hāra*)[172] consisting of ninety-six, **having reduced** it [also] by four, [therefore when there is the division] by that which comes to be twenty-four, 24, the result is the primary base (*mūlaprakṛti*)[173] twenty-one thousand [and] five hundred fifty-six. In this manner, if the quantity that is the "dividend" is "reduced" by two, then also the quantity that is the "divisor" [should be reduced] by two; if the "dividend" is [divided] by four, then also the quantity "divisor" should be "reduced" by that very four; that is the principle.[174] Therefore,[175] [according to the procedure of reducing] **whenever possible,** when there is the division of that consisting of twenty-nine hundred thousand ninety-seven thousand nine hundred and twenty, which has been reduced by two, it is 1498960; [***vibhajet*** or "**one should divide**" it] by the divisor thirty-two, **having reduced** it by two, it becomes sixteen. The result is ninety-three thousand six hundred eighty-five.

It is always in this way.[176] As a rule, the highest reduction[177] [should be performed]; [however] even without this very one, the division of the "dividend" is carried out according to its form exactly as it is.[178] Next is the presentation: 26893230 divided[179] by 273, 1000000001 divided by 77. The results are respectively: 98510 [and] 12987013.

In this manner, the procedure of division is concluded.

Square **(varga)**[180] ***[GT 23–$\overline{24}$]***

Next are the procedural rules of square;[181] [the author enunciates] one and a half metrical stanzas:

170 The compound *bhāgayogyatā* elaborates *sati sambhave* and refers to the case in which both operands can be reduced by a same number.
171 The commentator means that the product, here the dividend 2072256, has been reduced by 4 and the obtained product 518064 has been written instead.
172 Line 12: **tathā**] em. In mathematical texts, one finds both forms *hāra* and *hara* for "divisor"; these terms also denote the "denominator" of a fraction which is, in fact, a "divisor".
173 It refers to the multiplicand of the sample problem enunciated in GT 19.
174 *iti tattvam*.
175 The products obtained in the previous sample problem become the dividends, and these are divided by the previous multipliers; the quotients represent the aforementioned multiplicands.
176 Line 18: a *daṇḍa* would fit well after *sarvatra*.
177 *apavartana*.
178 The commentator observes that it is not always possible to carry out the reduction of the two operands by a same factor.
179 The abbreviation *bhā* (line 10) stands for *bhājita* or "divided" (see this term, for instance, in line 8).
180 K 7, 24–25.
181 *varga*.

23. Having carried out the square of the last place (*antyapada*), the last [place] multiplied by two should be multiplied by the remaining places.[182] Having caused it (i.e., the remaining place) to shift[183] from the place, in the same way one should shift[184] [all] the remaining [places] for the execution of [the operation of] square.

The explanation: that digit which is the "last" according to the regular course is the first according to the reverse course; to it, the second is the "last", to the second the third is the "last", thus gradually [one after the other all the digits become the last and are performed as such].[185] Among all digits, that[186] "last place" of one hundred sixty-three, 163, consist of [the digit] one. **Having carried out** its **square** means **[having carried out]** "that which consists of the multiplication of two equal numbers".[187] To illustrate: when there is the "square" of one, it is a single [digit] one; make[188] this [digit] one below the single [digit] one of one hundred sixty-three:

163
1

. Afterwards, **the last** [digit], which consists of the [digit] one standing above, is **multiplied by two** (*dvigna*>*dviguṇa*); one multiplied by two becomes the very two. [This result] **should** then **be multiplied by the remaining places** such as [the digits] six and three; "gradually" is thus the remaining [word to be added in the passage of the *karaṇasūtra*].[189] For instance, two multiplied by six becomes[190] twelve, 12; these two [digits] are written below six and one [respectively]. [The previous digit one is][191] below [the digit] one which is written[192] below the previously written one hundred sixty-three:[193]

163
12
1

. In the same way, that which

182 *śeṣaiḥ padair.* The polysemy of the term *pada* is worth noting: in mathematical texts, it is "place", "square-root", "part", and "one-fourth".

183 *samutsārya* – lit. "to drive away", from the verbal root *samutsṛ-*.

184 The use of *samutsārya* and *utsārayet* emphasises the iteration of the process of the shifting.

185 The idea is that each digit "gradually" becomes the last; once performed, the preceding is, in fact, erased. The commentator mentions a number composed of three digits because he is going to explain the rule by means of the number 163, given by Śrīpati in the next sample problem.

186 Line 27: *yat*] em.; *yaḥ* (K).

187 The commentator provides a definition of "square". Cf. TŚ 11 (*sadṛśadvirāśighāto*) and L 19 (*samadvighātaḥ kṛtir ucyate*); these two works are quoted by the commentator at the end of his explanation.

188 In line 29, *kṛtvā* should precede and not follow the layout.

189 The commentator explains that the last digit is multiplied by each remaining place; he observes that *krameṇa* ("gradually") should be understood as implied in the mathematical step *śeśaiḥ padair dvinighnaṃ*, thus: "it should gradually be multiplied […]".

190 Line 2: *jātā dvādaśa* 12] conj.

191 In brief, the result of the square of the last place.

192 Line 3: *prāg* should be compounded with the following *likhita*°. In the SGT, *prāglikhita* denotes a result which has already been written in the layout.

193 I have emended all the layouts except the layout occurring in line 11; in the edition, these are mostly mistaken (see Kāpadīā 1937, LII).

consists of the "last" [digit] two[194] multiplied by three becomes six; this should be placed in front of (*agre*) two written below one hundred sixty-three: $\left|\begin{array}{l}163\\126\\1\end{array}\right|$. In this manner, because of the object explained (i.e., this computation),[195] that is, because it has been multiplied, the "last place", namely two, is erased (*vinaṣṭa*). Afterwards, **having caused the remaining** [place] such as sixty-three **to shift from the place,** which is the [notational] place where it has been written [in the layout],[196] place[197] it by being one [notational] place[198] more[199] in respect to the lower number obtained, which consists of two[200] hundred twenty-six. It thus is: $\left|\begin{array}{r}63\\126\\1\ \ \ \end{array}\right|$. **In the same way, having carried out the square** means "having performed the procedure of doubling" the previously mentioned "place": in this case, regarding the number sixty-three, six is the "last place"; its **square** is thirty-six. Having executed it, write [its digits] one by one below two hundred twenty-six: $\left|\begin{array}{r}63\\126\\136\end{array}\right|$. Then, "the last place" six **multiplied by two** becomes twelve: 12. [Now] "the remaining place" consists of three; carry out [the step] it **should be multiplied**[201] by it (i.e., by three): twelve multiplied by three is thirty-six. Write it gradually where the previously found[202] [result] is

194 It refers to the last digit which has been doubled: 1×2.

195 *uktārthatvāt* – a linguistic formula often used by the commentator to emphasise that a new number takes place as the result of a step which has been previously explained. The SGT refers to the actual manipulations of the given quantities; it highlights that, at some point of the execution of the algorithm, when a step from the procedural rule has been performed, a new result is obtained, and a previous number is thus erased. Notably, this formula occurs almost exclusively in the first half of the SGT.

196 *likhitasthānakāt* – it elaborates on the term *pada*.

197 In line 7, *vinyasya* should precede *yathā* (in the same line). The layouts (lines 9, 11, and 13) were most likely preceded by *yathā* and followed by a *daṇḍa*.

198 The commentator adds important information: "shifting" means moving the "remaining place" by one place more from left to right in respect to the number below. This remark suggests how the computed quantities should be spatially arranged and aligned to reflect their actual place-value in the computation.

199 *ekādhikasthānakatayā*.

200 Line 7 reads *ṣaḍviṃśatyadhikadviśatalakṣaṇasya* (226), while one would expect *ṣaḍviṃśatya dhikaikaśatalakṣaṇasya* (126).

201 According to the edition (Kāpadīā 1937, 8, f1), one should read *guṇayitvā* (line 10). I suggest, however, the following emendation: *guṇanīyaṃ kṛtvā*, which means that in *guṇanīya* of the given *guṇanīyakṛtvā*, an *anusvāra* is missing. In my understanding, here Siṃhatilakasūri quotes the step *guṇanīyam antyam* from the *karaṇasūtra*. The formula *krtvā* followed by a quoted passage to mean "having performed the mathematical step…" is a common literary device in the SGT (see, for instance, SGT 26, 13–14).

202 *dṛṣṭa* – lit. "seen"; this term is part of Siṃhatilakasūri's own repertoire and denotes a result which has been calculated and written in the layout.

written:

63
1266
136
3

. In this manner, because of the object explained, the "last" [place] six is erased. Afterwards, **for the execution of [the operation of] square, one should shift the remaining [digit]** which consists of three, that is, one should place it by being one [notational] place more [in respect to the number below]:

3
1266
136
3

. Next, in the same way [as before], the expression **having carried out the square** means that one should carry out that previously mentioned [procedure]. For instance, having calculated the "square" of three, which is nine, [arrange it] below the three standing above the previously obtained [result] which has been written:

3
12669
136
3

. Then, in this case because of the absence of a "remaining place", no other procedure [is to be performed]. When there is the sum of the numbers previously written, when there is the "square" of one hundred sixty-three, it becomes twenty-six thousand five hundred sixty-nine: 26569.[203]

The author enunciates the second of the procedures regarding the calculation of square:

> $\overline{24}$. And the multiplication [of the given number] added to and lessened by an assumed [number][204] [and the result] added to the square of the assumed number [yields the square of the given number].

The explanation: for instance, in the calculation of the "square" of five,[205] having written five in two places, one five is "lessened" by an "assumed [number]", which is a number that is convenient (*abhirucita*) [according to the circumstances], such as [here] is two; it becomes three. The other five[206] is added to two, which is [in this case] the selected[207] [number], which is that [mentioned] "assumed" [number]; it becomes seven. Next, there is **the multiplication** (*āhati*>*guṇanā*) of that consisting of seven, which is [the result of the latter five] "added to the

203 Line 16: **iti**] em.
204 *iṣṭa* – lit. "desired"; it denotes a "chosen, assumed" number.
205 Five is one of the numbers mentioned by Śrīpati in the next sample problem.
206 Line 20: *ekaś ca*] em.
207 *ākṛṣṭa* – it explains *iṣṭa*.

assumed" [number two], by that consisting of three, [which is the result of the first five] "lessened by the assumed number".[208] To illustrate: seven multiplied by three becomes twenty-one. Next is the "square" of the "assumed" [number], namely two, that is the [number] which has been selected from one five; it is four. [The product is] **added to** it: twenty-one is thus added to four. [Therefore] when there is the square of five, it becomes twenty-five.

The author enunciates the third procedure [of square];

> 24. Or else, [the square is] the equal multiplication of two alike [numbers].

The explanation: [two] "alike" are two "identical" numbers, for instance twelve [and] twelve.[209] **The equal multiplication** of these [two] is the reciprocal multiplication of twelve by twelve; the "square" of the number twelve consists of one hundred forty-four.[210]

In this respect, having realised that [the expression previously mentioned] **having carried out the square** is a definition which cannot be justified,[211] not having [the author] [first] taught (*anupadiśya*) the procedure of square (i.e., what a square is), [I shall mention that] in the *Līlāvatī,* it is said:

> L 19: "The product of two equal [numbers] is called square.[212] The square of the last digit is to be placed,[213][the remainder then is] multiplied by the last multiplied by two".

In the same way, in the *Līlavātī*:

> L 20: "Or twice the product of two parts [of the given number] added to the sum of the squares of that parts is the square".

A fourth procedure is explained.[214] Therefore, regarding the "square" of five, there are two "parts" of five; on the one hand[215] it is [the number] three, and on the other hand it is two. **The product** of these is [carried out]: two multiplied

208 Line 21: *iṣṭonena*] em; *iṣṭo 'nena* (K).

209 Twelve is also among the numbers mentioned by Śrīpati in the next sample problem.

210 Line 29: **jātaḥ**] em.

211 In this passage, the commentator criticises the expository order of the methods of square chosen by Śrīpati. He observes that Śrīpati mentions the step "having carried out the square" without having first explained what a "square" is; Siṃhatilakasūri hence quotes the L in order to demonstrate a better arrangement of the methods of square; L 19 begins, in fact, with a definition of "square", that is, the product of two equal numbers. This is the first quotation of the L in the SGT.

212 *kṛti*.

213 The edition by Sarma (1975, 22) reads *antyavargo dviguṇo'antyanighnāḥ*; there were slightly different recensions of the L.

214 Śrīpati gives three procedural rules for the operation of square; Siṃhatilakasūri provides a fourth rule by quoting the L 20.

215 In line 4, one would expect the correlatives *ekatas*/*ekatas* instead of *eke*/*ekatas*.

by three becomes six: 6. This is the **product**, which consists of six; [it should be made] **twice,** it becomes twelve. In the same way, that "squares" of the previously mentioned "parts", such as two and three, consist of four and nine. The "sum" of these [should be carried out]: it becomes thirteen; twelve is added to it. The "square" and the "product" of these integers[216] will become the "square"[217] of five. It is always in this way.

In the *Triśatī,*[218] a fourth procedure to calculate the square is found as well; the author enunciates:

> TŚ 11: "Or the sum of terms[219] [in arithmetic progression] of which the first term (*ādi*) is [the number] one (*rūpa*) [and] the common difference[220] is two [gives the square]".

Having made [the number] "one" (*rūpa>eka*) the "first term", afterwards [make the other] "terms" with the "common difference" of "two", that is, they should be arranged[221] being each increased by two. Next, their **sum** (*samāsa>yoga*) should be made. The "square" of that [given] number should be calculated:[222] the numerical places[223] should be counted up according to its form, that is the principle.

As it was hereinbefore [in the TŚ too it is said]: "[The square is] the product of two equal quantities"; this very method is, [by now], perhaps well known.[224]

In this regard, a sample problem [is given]. By the word *uddeśa,* a sample problem is meant.[225]

216 *rūpa.* The commentator observes that the square of the given number has been ascertained by means of the squares of the "parts" and the products of the "parts".

217 Line 7:°*abhihatiḥ*] em.

218 In the SGT, this is the first quotation from the TŚ. The method quoted from verse 11 corresponds, in the edition by Dvivedī (1899, 5), to the third formulated by Śrīdhara (see also PG 24); the commentator, however, emphasises that in the GT this rule becomes the fourth since the methods formulated by the root-text are three.

219 *pada.*

220 *caya.*

221 *maṇḍanīya.* Forms of the root *maṇḍ-* (lit. "decorate, adorn") are often used throughout the SGT with the technical meaning of "to arrange, place".

222 Lines 10–11: *kāryo yathārūpaṃ saṅkhyāny aṅka*° [...]] conj.; I have proposed this conjectural emendation on both semantical and grammatical grounds. Here there is a copula construction; the subject is *aṅkasthānāni,* the adjective *saṅkhyāni* serves as the predicative. The suggested adverbial compound *yathārūpam* fits the context and is a formula often used by the commentator (see, after emendation, *yathāsthānam* in SGT 17–$\overline{18}$, 17 and 24) to express similar concepts. The main idea is that the number of terms varies according to the given number.

223 *aṅkasthāna,* see also *pada* with the same meaning.

224 Siṃhatilakasūri observes that, in TŚ 11, the rule of square which precedes the one just mentioned and which concerns arithmetical series, also defines a square as the product of two equal numbers. In the TŚ, the definition of a square is the second of the four methods formulated.

225 This gloss is redundant and seems spurious, it could also represent a case of scribal transposition; this is not, in fact, the first time that both terms are used by the commentator. One would expect it to occur in the line which introduces GT 22.

On this topic, [a sample problem is given]; [the author enunciates] one verse:[226]

> 25. Tell [me] the square of [the numbers] beginning with one [and] ending with nine, [also the square] of twelve, seventy-two, ninety-three, [and of the number composed of] one hundred and six and three.[227]

Its explanation together with the very presentation: 1, 2, 3, 4, 5, 6, 7, 8, 9, 12, 72, 93. By the word *trisara* mentioned in the fourth quarter[228] [of the verse above], sixty-three is meant,[229] and one hundred. **Tell [me]** denotes a question,[230] **the square** (*kṛti>varga*) of this one hundred sixty-three, [which is in figures] (163).[231] The answer is [the following]: the numbers obtained are one, four, nine, sixteen, twenty-five, thirty-six, forty-nine, sixty-four, eighty-one, one hundred forty-four, five thousand one hundred eighty-four, eight thousand six hundred forty-nine, [and] twenty-six thousand five hundred sixty-nine respectively. Also, in figures: 1, 4, 9, 16, 25, 36, 49, 64, 81, 144, 5184, 8649, 26569.

In this way, the procedure of square is concluded.

Square-root (vargamūla)[232] *[GT 26]*

Regarding that [previously obtained] "squares", there is a procedural rule to calculate their "roots",[233] such as one and so forth;[234] the author enunciates a metrical stanza:

> 26. Having subtracted the square from the [last] odd[235] place, one should divide the remainder by the root[236] moved from [its] place (*sthānacyuta*)[237] [and] multiplied by two. The learned state that, having placed the result in the line[238] [and] having subtracted its square, the square-root (*kṛtipada*) [comes to be] in the half of the doubled [number].

226 The SGT specifies that the sample problem is composed in the meter *anuṣṭubh* and it consists of one (see *eka*) verse (*śloka*).

227 *trirasa* – lit. "three" (*tri*) and "taste" (*rāsa*). In the *bhūtasaṃkhyā* notation, *rasa* (lit. "taste") denotes the number "six" as, in Sanskrit literature, there are six principal tastes.

228 *pāda* – the fourth part of a regular stanza.

229 Line 16: *triṣaṣṭiḥ śataṃ ca*] em.

230 *praśna*.

231 It is not clear whether this number in brackets is an editorial addition.

232 K 9, 24–27.

233 *mūla*.

234 Siṃhatilakasūri underlines the relationship between the two operations of square and the square-root and emphasises that this *karaṇasūtra* concerns the rule to determine the roots of the squares obtained in the previous sample problem. Regarding the "root", see the explanation of the following sample problem one (SGT 27, 29).

235 *viṣama*.

236 See *pada* as "[numerical] place" and *pada* as "root".

237 Partial results were kept in mind or written in a different place of the writing board.

238 *paṅkti*.

The explanation: with respect to the number[239] such as twenty-six thousand five hundred sixty-nine, namely the [previously obtained]square[240] 26569, according to the reverse order[241] starting from [the digit] nine [each digit should be denoted as] odd [and] even,[242] thus is according to this calculation (*gaṇanā*) in which there is a resting place.[243][The first step is] **[having subtracted the square] from** that **[last] odd place:** for instance, in this case two is the last [odd place]; the [highest possible] "square" of this number, such as two, is subtracted. **[One should divide the remainder]** by that **root**: the word "root" signifies the part (*aṃśaka*).[244] For the reason of the division by this **[root],** it (i.e., the quotient) is placed below.[245] To illustrate: **having subtracted,** [also termed] "having deducted", [from the last odd place which is two] **the square** of one, [in this case] **the square** is a single [digit] one, **[one should** thus **divide the remainder]** by the [number] one. Above[246] there is [the number] 26569. **One should divide the remainder** from the number obtained[247] **by** that **root,** namely by one. [The expression] **moved from [its] place [and] multiplied by two** means that what is "moved from its place" is that which is [also] "multiplied by two";[248] [thus, in this case] **one should divide [the remainder]** by that which consists of **two.** Because it should be moved [in the line], the six below is [placed] in front of it[249] (i.e., of the mentioned doubled root). One should perform the division by

239 Line 28: °*prabhṛtyaṅkasya*] em.

240 The number 26569 is the "square" of the "root" 163, which is the number used by the SGT to explain the rule of square enunciated in GT 23, and is the number mentioned by the GT in the sample problem (GT 25).

241 In line 29, one would expect *anuloma* (the "regular order", from right to left) rather than *pratiloma*, since the commentator explains that 9 is the first digit. The same anomaly with the term *pratiloma* is found in SGT 32, 26 and in SGT 94, 7. In all these passages, there is a contrast between the "reverse order" meant by the word *pratiloma* and the commentator's explanations which refer instead to a right from left direction. I do not believe that the SGT uses the word *pratiloma* in a loose way; I suggest to understand here this unusual usage in the light of the way numbers in a layout are commonly expressed by the SGT, which is from left to right with the unit being the last. In brief, the commentary mixes up two methods of denoting numbers: one is the numerical method (which is used by Śrīpati in the rule) and one is the positional method (used by the SGT while explaining the rule). Therefore, although *anuloma* conventionally is from right to left and *pratiloma* is from left to right, in this passage of the SGT *pratiloma* means still "reverse" but "reverse" in respect to the way numbers are written and read according to their power of ten. According to this usage in the SGT, with respect to this procedure, the first digit, which also is the largest power, should be marked as the "last".

242 *sama*.

243 *viśrāma,* it denotes the alternating sequence of odd and even places by which the digits of a number are subdivided in order to obtain its square-root.

244 *aṃśaka,* it appears to emphasise the root as the divisor of the remainder.

245 Line 2: *nyastas*] em.; *nyāsas* (K).

246 The commentator refers to the fact that the given number has not yet undergone changes. In this phase, he describes, without showing the layouts, the first steps of the procedure.

247 Here the commentator uses *aṅkarāśi* (lit. a "heap of digits", number) to denote the remainder 16, which consists, in fact, of more than one digit.

248 Briefly, the root.

249 Line 4: *tatagre*] em. This passage anticipates what is explained in line 7.

means of the suitable[250] number which has been placed below it. For instance, six is [placed] below two; when there is the subtraction of twelve from the number arranged above,[251] in respect to the [previous] number, [now] forty-five [and] sixty-nine are above:

4569
2
6

. Next, **having placed** that **result,** such as the quotient six, **in the line,** that is, **"**in the line" of the two[252] written in front, [hence] next to two, it thus is:

4569
26

. From the number above, which is forty-five, **having subtracted its square** (*kṛti>varga*), which is **[the square]** of "the result", namely six, consisting of thirty-six, it becomes nine hundred sixty-nine: 969. Then, **one should divide** the previously mentioned **remainder** by the number suitable for the division that has been performed, which is that **root** consisting of six; because it has been moved,[253] it is in front [of two]. **[One should divide the remainder]** by the [number] occurring[254] (i.e., six) "moved from its place" [and] "multiplied by two". [When there is the subtraction] of twelve[255] [from sixteen], [four is obtained]. By the joining[256] of the first with the second[257] (i.e., the two *pada*s 1 and 6), [when the result of this union is multiplied by two] it becomes thirty-two, which is [placed] below. For instance in this case, having placed three below thirty-two, the subtraction from ninety-six [should be performed]. Then, perform [the step] **having placed the result,**[258] which is the quotient three, **in the line** in front of thirty-two, [next is the step] **having subtracted (*viśodhya*),** [also termed] "having removed" (*nirgamya*), **the square** (*kṛti>varga*) of this three; it is that consisting of nine. When there is the [previously mentioned] union[259] with the lower (*adhastana*) number, that number [composed of the two *pada*s 1 and 6], as it has been "moved from its place", is [now] "multiplied by two", [when multiplied by two] it is that consisting of thirty-two. When it is halved, it yields sixteen.[260] In this manner,

250 *yogya* – "suitable, proper"; it denotes the quotient six, which is the "chosen" quotient.
251 *niyojyoparyaṅka.*
252 In brief, the doubled root.
253 *upasarpiṇīyatvāt.*
254 *prāpta.*
255 This may have been the intended sense, but it not expressed by the Sanskrit as it stands (line 11). The genitive *dvādaśānāṃ* should be governed by a noun; I suggest that a construction similar to that found in line 6 should be understood here too, hence it should read *dvādaśānāṃ pate.*
256 *yojanā* – here "joining, union" of two numbers one next to the other. One would not expect *madhye*, which in the SGT is mainly used when a sum takes place.
257 Note the use of the adjective *dvika* and the ordinal *eka*.
258 In lines 13–14, see the absolutive *kṛtvā* followed by a quoted step, which is here *niveśya phalaṃ*.
259 The use of *madhye* is awkward; see in line 11 *madhye* […] *yojanayā*.
260 Lines 15–16: *ardhite*] em.; *prāpyate*] em. Note the causative *prāpyate* followed by the abstract form of *ṣodaśa* in the accusative case (lit. "it causes sixteen to arise").

that [quotient] three, by means of the step "moved from its place" [is transferred in front of sixteen], it is not doubled [and hence] should not be halved. [Therefore] it should be kept as it is. Being as such the square,[261] **the root** (*pada>mūla*) of the square, namely of 26569, consists of one hundred sixty-three: 163. Those experts in mathematics **state,** [also termed] "explain", [this procedure]. Thus is the connection [of words with their meanings].

In this respect, a sample problem is given; [the author enunciates] one verse:

> 27. If you know [the procedure of square-root], tell [me] then the roots of the previously obtained squares as well as [the root] of that whose measure[262] is [the number consisting of] four,[263] four,[264] eight,[265] nine,[266] one,[267] eight,[268] [and] eleven.[269]

The explanation: **tell [me] the roots,** namely one, two, and so forth,[270] **of the squares,** which are the **previously** mentioned results, such as one and so forth up to one hundred sixty-three. In that manner, by the word *veda* [and] by the word *abdhi* [the numbers] "four" and "four" are meant; *vasu* is [the number] "eight", by the word *go* "nine" kinds of lands are meant, by the word *candra* [the number] "one" is denoted, [the word] *phaṇi* denotes "eight" types of snakes, [and] *rudra* is "eleven". That of which **the measure** (*miti>pramāṇa*) is [made] by these [numbers]; **tell [me] the root** of that [number] consisting of ten million eighteen one hundred-thousand nineteen thousand eight hundred forty-four, which is a square that has not been mentioned before.[271] The word "say" should be [added to] **as well as** (i.e., "as well as say the square-root of [...]"), such is the connection [of words with their meanings].

[I give below] the presentations of the question and the answer respectively.[272] For the reason that [the procedure of the square-root] has been

261 Line 17: *kṛte*] em.

262 The number 11819844 is expressed by means of the *bhūtasaṃhyā* notation.

263 *veda* – lit. "knowledge"; it denotes the "four" *Veda*s, the corpus of Vedic Sanskrit texts whose canonical division is fourfold.

264 *abdhi* – lit. "ocean". In Sanskrit literature, four oceans are mentioned: eastern, southern, northern, and western.

265 *vasu* – it refers to a class of gods who were originally personifications of natural phenomena.

266 *go* – lit. "earth".

267 *candra* – lit. "moon".

268 *phaṇi* – lit. "snake".

269 *rudra*.

270 Lines 22–23: *ekadvyādilakṣaṇaṃ*] conj.; *ekacaturādilakṣaṇaṃ* (K). I have emended the text, which here is suspect: the commentary specifies as *mūla*s or "roots" the numbers 1, 4, and so forth which are, instead, the "squares" (see SGT 25, 20). The "square-roots" or *mūla*s (the expression *vargamūlāni* is found in line 29 followed by these numbers which) are 1, 2, 3, etc. The emendation follows the commentary.

271 In this passage, the commentary clarifies that Śrīpati asks students to ascertain the roots of the squares mentioned in the preceding sample problem (GT 25), and "also" (see *api*) the root of the square 11819844.

272 The commentator underlines that he provides the *nyāsa* of the squares of which one has to determine the square-root (thus the *nyāsa* of the *praśna* or "question") and the *nyāsa* of the obtained square-root (thus the *nyāsa* of the *uttara* or "answer").

explained before by this prose,[273] both explanations are clear [and are] not [further] investigated: 1, 4, 9, 16, 25, 36, 49, 64, 81, 144, 5184, 8649, 26569, the results are thus accordingly the square-roots: 1, 2, 3, 4, 5, 6, 7, 8, 9, 12, 72, 93, 163, 11819844. The [last] result is the square-root 3438, three thousand four hundred thirty-eight.

In this way, the procedure of square-root is concluded.

Cube (ghana)[274] *[GT 28–30]*

[Next are] a pair of metrical stanzas for the procedural rules of cube:

> 28–29. The cube (*ghana*) of the last [digit], its square multiplied by three and by the preceding[275] [digit], [and] the square of the preceding[276] multiplied by the last and by three should be placed [in the layout]. As there is one [notational] place more,[277] all [these numbers] and the cube of the preceding are added [in this very way]. [By means of this procedure] the cube should occur.

Also, [another procedure for the cube is:] having done [a calculation by mind][278] with a series[279] of integers (*rūpa*) of which [the number] one is the first [term], one should add the [result of the] cube of the preceding quantity plus [the number] one to the last [term] multiplied by three and multiplied by the preceding [term].[280]

Or the [mutual] multiplication of a collection[281] of three equal [numbers] is the cube.

The explanation: regarding the number three hundred seventeen, 317, by means of the numerical method [following the regular order] **the cube of the last [digit],** which consists of [the digit] three, by "the multiplication of a collection of three [equal numbers]"[282] becomes twenty-seven. **As there is one [notational] place more** means that **it should be placed**[283] in a way that it should be so,[284] that is, it should be written below the number previously mentioned:[285]

317
27

.

273 *gadya* – the commentator refers to his own commentary, which is a text written in "prose".
274 K 11, 3–10.
275 *ādi*.
276 *ādima*.
277 *sthānādhikatva* – this abstract (note the use of the suffix *tva*) grammatical construction expresses the state, the condition according to which the obtained numbers should be placed (see the use of *sthāna* or "[notational] place") in the layout.
278 In line 11, the SGT supplies the noun *manasā* to *kṛtvā*.
279 *pracaya* – lit. "multitude, heap".
280 *mūkha* – lit. "face".
281 *rāśi*.
282 Line 12: *samatrirāsi*] em. In order to first clarify what a cube is and then explain how to calculate it, the commentator refers to the last line of the rule. In this regard, see the remarks by Siṃhatilakasūri at the very end of his explanation (lines 22–23).
283 Line 13: *sthāpyaḥ*] em.
284 Here *yathā syāt tathā* marks *sthānādhikatvaṃ* as an expression used adverbially.
285 I have emended the layouts, which in the edition are inaccurate.

[The expression] **its** refers to this [last digit] three, **the square** (*kṛti>varga*) is 9; having placed it elsewhere[286] is **multiplied** (*nighna>guṇita*) **by three and by the preceding** [digit]. To illustrate: nine multiplied by three is twenty-seven, the "preceding" of that root three is [the digit] one; twenty-seven multiplied by it is exactly so. **As there is one [notational] place more,** it should be joined below the previously written three hundred seventeen:[287]

317
277
2

. [The text says] **the square of the preceding:** in relation to three, one is the "preceding" [digit], its **square** (*kṛti>varga*) is one. Having placed it elsewhere, this is **multiplied by the last and by three:** it is multiplied by the "last" [digit] three becoming three and this is multiplied by three becoming nine. **As there is one [notational] place more,** it should be joined below the root three hundred seventeen:

317
2779
2

.

[Regarding the expression] **and the cube of the preceding:** the "cube" of the "preceding", which is one, consists of one; **as there is one [notational] place more,** it should be joined in the *yantraka* of the initial root:[288]

317
27791
2

. In this manner, according to this supplementary rule mentioned in the *Triśatī*:[289] "[in the subsequent rounds] the combined quantity (i.e., the quantity consisting of combining digits) is the last, and it is its cube", the step **the cube of the last**

286 The expression *anyatra sthāpayitvā* (17, 26, and 2; see also *anyatra lekhya*) points out that intermediate results were written "elsewhere" and not in the layout were the algorithm was carried out.

287 In line 16, *prāg* should be compounded with *likhita*°.

288 In this context, *yantraka* is anomalous; in the SGT, this is the first occurrence of the term. It is often found in Siṃhatilakasūri 's work on Jaina rituals entitled *Mantrarājarahasya* where in verse 465 one finds: *yantralikhitam* or "written in a *yantra*". Moreover, in line 20 the expression *prāgmūla* is significant; it appears that the term *yantraka* denotes the "layout" in which the cube of 317, which is the *prāgmūla* or "original, initial root", is to be determined.

289 This is a quotation from TŚ 15. The commentator alludes to the fact that, at this point, the procedure starts from the beginning. Lines 20–25 consist of two paragraphs which are both concerned with the omission, in the subsequent rounds, of the first step of the procedure prescribed in GT 28. For this reason, lines 20–22 cite a supplementary rule from the TŚ; lines 22–24 explain the same omission without relying on the supplementary rule. By this quotation, Siṃhatilakasūri observes that, at this point of the procedure, one does not need to repeat all the steps such as: "the cube of the last should be placed". In the next lines, he observes that, if one adds all the numbers, the cube of the last, which comes to be the number 31 of 317, will arise and this is 29791. The commentator quotes this passage, which ends the first method of cube expounded by the TŚ, to explain that the cube of 31 should not be placed, and that this can be found by adding the numbers occurring in the layout.

[digit] should be placed[290] should not be performed anymore, [while the step] **and its square** and so forth should be applied. Or else, [the cube of the last, which is 31, is obtained] when there is "the cube of the preceding",[291] **all** those [numbers already] calculated, which are the previously obtained numbers,[292] **are added. The cube of the last** [digit] **should be placed**: the "cube" of that consisting of seven, which is the digit that occurs next, is meant;[293] hence this procedure, which takes place by means of the numbers which are "added", should not be applied. And the statement **its square** and so forth "should be performed" is relevant here too.[294] Thus, **its square** (*kṛti>varga*) is [the square] of the **last,** namely thirty-one; it is nine hundred sixty-one: 961. [This is] **multiplied by three and by the preceding:** the mentioned number nine hundred and so forth,[295] having placed it elsewhere, it is multiplied by three; it becomes twenty-eight hundred eighty-three. "The preceding" is seven, [the previous result is] multiplied by it becoming twenty-thousand one hundred eighty-one: 20181. **As there is one [notational] place more,** it is written below twenty-seven seven ninety-one,[296] which is written below the previous three hundred seventeen:

317
27791
2
20181

. [One should then perform] **the square of the preceding:** here "the preceding" is seven, its **square** (*kṛti>varga*) consists of forty-nine: 49. [The text says] **multiplied by the last [digit] and by three:** [forty-nine] multiplied by "the last" thirty-one becomes fifteen hundred nineteen. Having written also this [partial result] elsewhere (*anyatra lekhya*), it is multiplied by three, it becomes four thousand five hundred fifty-seven. **As there is one [notational] place more,** it should be joined:

317
27791
2
20181
4557

. **And the cube of the**

290 Line 21: *ghano'ntyasya*] em. The comma placed by the editor to accommodate the problem with the final *visarga* is misleading as the passage formulated by the commentator includes *antyasya*.

291 It refers to the digit 1.

292 In brief, the partial results already placed in the layout. In line 23, *prāg* should be compounded with *dṛṣṭā*.

293 As in the procedure of square, here too the commentator combines two different ways to denote the digits of the given number. In 317, from a left to right direction the digit 7 is the number that occurs "next" (*agretana*) with respect to the "preceding" (*ādi*) digit, but according to the positional method here applied, which is from right to left, 7 is the "preceding".

294 *iti yukta* – lit. "allowable, appropriate", it is a standard formula in commentarial literature.

295 Line 26: *prāguktanavaśatyādyaṅko*] em.

296 Line 29: **viṃśati**] em.

preceding: the cube "of the preceding", which is seven, consists of three hundred forty three; this should be joined [too]:

317
27791
2
20181
4557
343

. [The expression] **being one [notational] place more** should be always understood [as a way of placing numbers in the layout]. These **all** [numbers] **are added,** [also termed] "joined", [and] **the cube** of three hundred seventeen is thirty million eighteen hundred thousand fifty-five thousand thirteen: 31855013. **The cube should occur,** such is the procedure.[297]

By the [first] three quarters of the second metrical stanza, [the author enunciates] the second procedure to calculate the cube. To illustrate: [the expression beginning with] "one is the first [term]" [is clear], a "series" is a collection (*samuccaya*) of integers, for instance [the numbers] one, two, three, and so forth; **[having done a calculation by mind]** refers to it.[298] In this way, however, the procedure "the sum of terms [in arithmetical progression] of which the first [term] is one and the common difference is two" is not applicable [here].[299] Therefore, **having done a calculation by mind] with a series of integers of which [the number] one is the first** [and] which is placed in a vertical row of numbers, not **with a series** of numbers placed horizontally. [Regarding the expression] **having done:** [here] *kṛtvā* is a verb without an object (*akarmaka*) and with a *kṛt* suffix like [in] *manasā kṛtvā meruṃ gacchanti*.[300] [With respect to the step] "the last [term] is multiplied by three", [of a series] of which "one is the first", that whose **cube** to be arrived at is the desired [term] (*iṣṭa*) is "the last [term]". To illustrate:[301] by means of a series of which "one is the first" [term], **the cube** of three is to be calculated, "the last" [term] which is "multiplied by three" is the very three; it becomes nine. In this case, **the last [term] multiplied by three** consists of nine, [next is the step] **and multiplied by**

297 Line 7: 31855013 *ghanaḥ syād iti kriyā*] conj.

298 The commentator shows the grammatical case of *pracaya*, which is mentioned in the *sūtra*.

299 Siṃhatilakasūri quotes (note that it should read *rūpa*° instead of *eka*°) the method of square from the TŚ already quoted by him in SGT 24, 8. Siṃhatilakasūri observes that, although this method also involves a "series", here the procedure meant is different.

300 The SGT highlights a grammatical feature; it explains that the verb *kṛ-* (lit. "to make") usually governs the accusative case but this is not the case in the *karaṇasūtra*. The use of the verb *kṛ* shown in the GT is the same as in *kṛtvā manasā meruṃ gacchanti* or lit. "having had it in mind, he is going to [the Mount] Meru". This is a grammatical example, thus its relevance is syntactical rather than semantical. It is noteworthy that a similar sentence (*manasā meruṃ gacchati*) is found in the grammatical work *Kātantravṛtti* (ad II, 281) by Durghasiṃha (7th century CE).

301 From line 12, the SGT explains the procedure by means of the number 3, which is among the numbers given by Śrīpati in the next sample problem.

the preceding [term]: "the preceding" [term] of three is two, nine is **multiplied** (*saṅguna>guṇita*) by it; it becomes eighteen. **One should add the cube,** which is the desired [term], **of the preceding quantity,** such as two, since three are the terms occurring: eight **plus one** become nine. When the additive quantity nine is [added] to eighteen, it is "the cube" of three, which is twenty-seven. It is always in this way.

By the fourth quarter [of the second stanza], the author enunciates the third procedure. [The expression] beginning with **three equal [numbers]** denotes a "collection" of three numbers which are "equal"; with respect to it, there is the reciprocal "multiplication" (*prahati>guṇanā*). To illustrate: in order to calculate the "cube" of four, having arranged three times [the number] four, the reciprocal multiplication [should be performed]. Four multiplied by four is sixteen and [four] multiplied by sixteen is sixty-four; this becomes **the cube** of four. It is always in this way. The words *ca* ("and") [and] *uta* ("also") are [used] in a conjunctive sense.[302] The word ***vā* ("or")** denotes a [different] variety (i.e., a different procedure).

In the same way, [as previously in relation to the methods of square], without having mentioned the procedure of cube (i.e., what a "cube" is), [the expression] **the cube of the last** and so forth **should be placed** [is a definition which] cannot be justified. The author should have first considered (*vicintya*) [to explain] "the cube is defined as the multiplication of three equal[303] [numbers]." He should thus have [first] illustrated[304] this [definition] from which that [procedure] **the cube of the last** and so forth **should be placed** can [then] be taught.

Thus [the author enunciates] the fourth procedural rule:

> 30. The [given] number multiplied by [two] parts [and the result] multiplied by three is united to the sum of the cubes of the parts.

The explanation: for instance, to arrive at the "cube" of [the number] five, [one should write in two places] two "parts" of five: [the number] three in one side, [the number] two in the other side. **The** [given] **number** consists of five, **the** [given] **number** [five] is **multiplied** (*āhata>guṇita*) by these [two parts]. Thus [in this case]: five multiplied by two becomes ten;[305] this multiplied by three becomes thirty. Then, [thirty] **multiplied by three** becomes ninety. Next, when there is the "cube" of the "parts", such as two and three, eight and twenty-seven [are obtained] respectively. When there is the "sum" of these [numbers], it becomes thirty-five. That [ninety] is "united" [to it]: it **is united to the sum of**

302 This gloss is mistakenly found here (line 20). This final passage (*ca-utaśabdau*) should occur before, most likely at the end of line 16, since the word *uta* is used in Śrīpati's *sūtra* in what the commentator calls "the second procedure.

303 This passage is found in L 24.

304 Notably, *upadiśya* was used (in the negative form) by the commentator also in SGT 24, 29 where he has also criticised Śrīpati's literary choice in expounding the methods of square in that order.

305 Line 27: *pañca dviguṇā*] em.

the cubes of the parts. Ninety is added to thirty-five, it becomes one hundred twenty-five: 125; this is the cube of five.

It is always in this way.

In this respect, [the author enunciates] a sample problem:

> 31. O friend, think [how to perform the procedure] [and] tell [me] the cubes of [the numbers from] one up to nine, and of eighteen, seventy-three, [and] three hundred seventeen too.

This metrical stanza is clear [in meaning]. The presentation is: 1, 2, 3, 4, 5, 6, 7, 8, 9, 18, 73, 317. Their cubic quantities are obtained; these are accordingly:[306] one, eight, twenty-seven, sixty-four, one hundred twenty-five, two hundred sixteen, three hundred forty-three, five hundred twelve, seven hundred twenty-nine, five thousand eight hundred thirty-two, three hundred thousand eighty-nine thousand seventeen, [and] thirty million eighteen hundred thousand fifty-five thousand plus thirteen. The presentation [in figures] is respectively: 1, 8, 27, 64, 125, 216, 343, 512, 729, 5832, 389017, 31855013.

In this way, the procedure of cube is concluded.

Cube-root (ghanamūla)[307] *[GT 32–33]*

The author enunciates a pair of metrical stanzas for the procedural rule of cube-root:

> 32–33. [The places of the given number should be considered as] one cubic [place],[308] a couple of non-cubic [places] ('*ghanadvandvam*). Having subtracted (*prapātya*) the [highest possible] cube from the [last] cubic [place], one should bring the root below the third place[309] and divide the remainder by its (i.e., of the root) square multiplied by three. Having put the quotient in the line, one should then subtract its square multiplied by the last [and the product] multiplied by three, and [one should also subtract from the result] the cube [of the quotient]. For the obtainment of the cube-root, the mathematician should carry out this [same] procedure again and again.

The explanation: as in the procedure explained before[310] [the digits of the given number were marked as] uneven and even, in the same way here, according to the reverse order,[311] first is the numerical place which is **a cubic** [place]; behind

306 Line 10: **nyāsaḥ**] em.

307 K 13, 18–25.

308 *ghana*.

309 *pada*.

310 The commentator refers to the procedural rule of square-root, where he has explained that the digits of a number are to be considered, following one another, as "even" and "odd".

311 Here the term *pratiloma* is anomalous, one would expect, in fact, *anuloma,* which denotes the "regular order", hence a right to left direction of digits. One finds the same problem in SGT 26, 29; in this regard, see my earlier explanation which holds true here too.

it, two digits are **a couple of non-cubic [places],** behind [these non-cubic places] the numerical place which is one [only digit] is a **cubic** [place]. Behind it, two [places] are **a couple of non-cubic** [places]. Thus is by this method where at the end[312] the resting place is a cubic place.[313]

Therefore, in order to arrive at the "root", which is three hundred seventeen, of the "cube" of the last sample problem which has just been explained,[314] from the [last] "cubic" place, namely one,[315] [*prapātya* or "one should subtract"] **the [highest possible] cube,** that is, **the cube** of the number [one]. The cube of that [mentioned] number [one] is subtracted from the number above. For instance, having placed three below the one of three:[316]

31855013
3

, having subtracted the cube of three, which consists of twenty-seven, from the thirty-one above, the remainder 4 is placed instead. Afterwards, this three is **the root; one should bring** that which [in the *sūtra*] is called "root", below **the third place** of the number above, which is namely five:

4855013
3

. Next [is the step] **its square multiplied by three,** that is, [the square] of the "root" [three]. To illustrate: in this case, three is **the root,**[317] it is **its** "square" (*kṛti*>*varga*) nine to be multiplied **by three**; it is twenty-seven. **One should divide** (*haret*>*bhajet*) **the remainder,** which is the number above, by this [result of the step] **the square multiplied by three,** which is [twenty-seven], 27, [and which] "should be put" by being one [notational] place[318] less with respect to the number above:[319]

4855013
273

. In this case, having made (i.e., brought) one[320] below twenty-seven,

312 This sentence is similar to that found in SGT 26, 29, where the square-root has been explained. There, it is said that in this kind of calculation where digits are denoted as "even" and "odd", there is a "resting place" (*viśrāma*). It seems that here Siṃhatilakasūri uses two languages to refer to digits and, according to a left to right direction, the expression "at the end" (*paryante*) denotes the "last" digit (i.e., the unit place) which is, in fact, that "first numerical place" mentioned two lines above. Another possible understanding is that this remark ("at the end") refers to the number 31855013 (see the use of *tatra*); in the following lines, the commentator shows that the "last cubic place" is 1 but this is a not suitable number and, in order to calculate the cube-root, he takes 31 as the "last".

313 Line 28: *ghanapadaṃ*] em. Otherwise, it is an awkward genitive *tatpuruṣa* compound: "there is the resting place of the cubic place".

314 317 is the *mūla* ("root") of the *ghanamūla* ("cube-root") 31855013.

315 The number 1 is the last cubic place but in order to perform the procedure 31 becomes the last cubic place.

316 This expression underlines that 31 is the last cubic place. In the edition, this layout occurs abruptly in line 29, which does not fit the context. I have hence emended the text and moved the layout in line 2, after *niveśya*.

317 Lines 4–5: *mūlaṃ trikaṃ*] em.

318 This is an important information provided by the commentator to understand how to place the results in the layout.

319 Line 6: *uparyaṅkād*] em.; *uparyaṅko* (K).

320 *ekaṃ kṛtvā*.

[having subtracted it] (i.e., twenty-seven) from the forty-eight above, twenty-seven is erased[321] [and the remainder] twenty-one is placed [instead]. Because of the object explained, due to the division that has been performed, twenty-seven is removed.[322] **Having put,** [also termed] "having placed" (*niveśya*), **the quotient,** which consists of the [previous] result one (*rūpa>eka*), **in the line** in front of three,[323] that [expression] **its square** and so forth means the "square" (*kṛti>varga*) of this quotient one which is placed in the line; it is the very [digit] one. [To be] **multiplied by the last,** it is that [square one], it is multiplied "by the last" three: one multiplied by three becomes the very three. **[And the product] multiplied by three:** three is multiplied by three, it becomes nine. In this way, **one should subtract its square multiplied by the last [and the product] multiplied by three:** above is

2155013
31

; **[one should subtract]** nine from[324] two hundred fifteen, it becomes the remainder two hundred six. **And one should subtract the cube: the cube** of the quotient which consists of one; [hence] **[one should subtract]** [the digit] one from the number above:[325]

2064013
31

.

Then, in order to apply again the procedure **one should bring the root below the third place** [here] too, in this case **one should bring** thirty-one, which is the **root, below the third place** of the number above, which is one:

2064013
31

.

Afterwards, "its square" (*kṛti>varga*), that is, of thirty-one, consists of nine hundred sixty-one; **one should divide the remainder,** which is the number above, by this [square] "multiplied by three" (*trinighna>triguṇa*),[326] namely by twenty-eight hundred eighty-three. **Having put** it by being one [notational] place less below the digits of the upper number, by assigning (*pradattayā*) the three belonging to eighty-three below the three belonging to thirty-one,

321 *gata.*

322 *bhāñjanīya.*

323 Although the way of denoting cubic and not-cubic places follows the regular order (thus from right to left), when the commentator refers to numbers according to their space-relations in the layout, he uses a left to right direction, as it is clear from the observation "in front of three". One finds, in fact, a mixed vocabulary to denote numbers: i) the vocabulary used in the rule by Śrīpati when denoting digits as *ghana* and *aghana* from a right to left direction, and ii) the vocabulary used by the commentator who, in addition to the technical language related to this procedure, also explains the algorithm according to the way numbers are written and read according to their power of ten (from left to right). In fact, even though 1 is placed "in front of three", the three is mentioned as the "last" (*antya*). In brief, there is the positional and the numerical method.

324 At this point, the algorithm goes back to the first step.

325 Line 14: *yathā*] em.; *jātaḥ* (K).

326 Line 16: *tri°*] em.; *vi°* (K).

it is:[327] $\begin{array}{|r|} 2064013 \\ 317 \\ 2883\ \ \ \end{array}$. Therefore, the twenty-eight below [is multiplied] by seven; when there is the subtraction of one hundred ninety-six from the number standing above, which is two hundred six, ten is placed: 10.[328] Afterwards, the eight below [is multiplied] by seven; [the result] fifty-six is subtracted from one hundred four, [and the remainder] forty-eight is placed above. Then, the three below [is multiplied] by seven; twenty-one is subtracted from eighty, which is erased, [and] fifty-nine[329] is placed. Next, because of the object explained, by means of the division performed (*dattabhāgatayā*) twenty-eight hundred eighty-three is removed. Having then placed **the quotient** seven **in the line** in front of thirty-one, as before, **one should subtract the square** of seven[330] "multiplied by three" and by the "last". Thus, the "square" (*kṛti>varga*) of this seven is forty-nine; this multiplied by the "last" thirty-one is fifteen hundred nineteen,[331] which "multiplied by three" is forty-five hundred fifty seven. Without [putting] [in the line] the number-divisor [seven] which has been [already] calculated, **having put** it below by being one [notational] place less [in respect to the number above], **one should subtract** it. The seven belonging to fifty-seven[332] should be below one belonging to thirty-one:[333] $\begin{array}{|r|} 45913 \\ 45317 \\ 57 \end{array}$. Next, because of the likeness (*sadṛśītvāt*) of forty-five, forty-five is erased. [As it is subtracted] from ninety-one, fifty-seven is erased, thirty-four occurs [instead]. Because of the object explained, [the number] forty-five fifty-seven[334] should be removed. In the same way, **one should subtract the cube** of this seven, namely three hundred forty-three. The quotient below is the "cube-root" of the previously mentioned thirty million and so forth, such as three hundred seventeen: 317. **[The mathematician should carry out] this procedure** and so forth, [the meaning of this expression] is [now] clear.

Or else, according to a traditional usage (*āmnāyena*),[335] that "root" of that "cube" thirty million and so forth, which consists of three hundred seventeen,

327 I have emended the layout (line 18) as digits are mistakenly placed below each other. The emendation follows the commentary.

328 It is curious that no *daṇḍa*s occur between lines 15 and 23.

329 Line 22: *ekonaṣaṣṭiḥ*] em.; *ekonapañcaśat* (K). Surprisingly, in the next layout (line 28), in the remainder 45913, one finds 59 (and hence the suggested *ekonaṣaṣṭiḥ* or "fifty-nine", and not the transmitted *ekonapañcaśat* or "forty-nine").

330 Line 24: *sapta*°] em.; *nava*° (K).

331 Line 25: *ekonaviṃśatyadhi(ka)pañcadaśaśatīṃ*; *ekonaviṃśatyā pañcadaśaśasatīṃ* (K).

332 Line 28: °*satkasaptakaḥ*] em.

333 Line 27: °*satkaika*°]em.; °*satka eka*° (K).

334 Line 30: **eko**] em.

335 Here the commentator mentions a method to verify the obtained result having carried out the cube-root of a given number. In a private conversation, Hayashi shared with me

of this 317 the "square" is 100489, namely one hundred thousand four hundred forty-nine.[336] When there is the division of the cube thirty million and so forth by this [square], it is:

31855013
100489
3

. When by a well-known method there is first the multiplication by means of the division[337] of the number above (i.e., 100489) by the three below, the divisor, which is the number standing above, is 1708313. The quotient three, which is the previously mentioned [highest] "cube", is placed somewhere else [on the calculating board]:

1708313
100489

. The number standing above is moved (*sañcārya*) by one [notational] place less [with respect to the number underneath]:

1708313
100489

. When there is the multiplication by means of the division[338] of the number above (i.e., 100489) by one, which is the "quotient",[339]

1708313
100489
1

, the number standing above is 703423, and the "quotient" one, that previous "quotient"[340] in front of the three, should be shown elsewhere [as a partial result written aside]. Similarly, [both] 3 [and] 1 should be placed elsewhere, and the number seven stands below:

703423
100489
7

. When there is the multiplication by means of the division

his analysis of this procedure; the conjectural emendation below is a revised version of Hayashi's, the English translation is entirely my own. Other reconstructions are no doubt possible. Lines 5–8: *prathamaṃ bhāgāṅkasyādhas trikeṇa prasiddharītyoparyaṅkabhāgāpahāre uparisthitāṅkā yathā* 1708313 | *labdhaṃ trikaṃ prāguktaghanam anyatra deyaḥ* | *yathā*

1708313
100489

| *ekasthānonatayā uparyaṅkāḥ sañcaryāḥ* | *yathā*

1708313
100489

| *labdhena*

1708313
100489
1

ekeṇoparyaṅkabhāgāpahāre uparisthitāṅkā 703423 *labdhaṃ caikaḥ sa prāglabdhaṃ trikāgre 'nyatra deyaḥ* | *tathaiva* 31 *anyatra sthāpyau adhaḥsthāṅkāś ca sapta* | *yathā*

703423
100489
7

| *bhāgāṅkasyādhaḥ saptakenoparyaṅkabhāgāpahāre uparyaṅkāḥ sañcāryāḥ* | *yathā*

703423
100489
7

|] conj.

336 Lines 3–4: *ekaṃ lakṣaṃ catuḥśataikonanavati*] em. In line 4, the number in figures should come after the compound ending with *lakṣaṇas*.

337 *apahāra* – lit. "killing"; in the SGT, this is the first occurrence of this term.

338 *bhāgāpahāra*; in the SGT, this is the first occurrence of this term.

339 See *labdha* earlier in lines 9 and 23, in the execution of the cube-root of 31855013.

340 Line 6: *prāglabdham*] em.

of the number above (i.e., 100489) by the seven below the number which is the divisor, [the upper number 703423] is moved:

703423
100489
7

. All the numbers above also are erased and the "quotient", that [number] seven, should be joined in front of the previous quotients three and one; it is 317, and this is the cube-root of three *koṭi*s and so forth.

It is always in this way.[341] When there is the division of the number which is the cube by the square [of its root], the "cube-root" is attained (*āyāti*); [in this way] it has been established.

In this respect, a sample problem occurs; [the author enunciates] one verse:

> 34. O learned one, if your study of the arithmetical operations has been accurate, tell [me] the root of the cubes previously obtained.

Its explanation with the presentation: 1, 8, 27, 64, 125, 216, 343, 712, 729, 5832, 389017, 31855013. Having previously explained the calculation which is the solution of this [latter cube 31855013], [here I am not showing anything further].[342] In accordance with the [obtained] results, by means of the very presentation the roots are: 1, 2, 3, 4, 5, 6, 7, 8, 9, 18, 73, 317.

In this way, the procedure of the cube-root [is concluded]. With the completion of this [subject], the first eight arithmetical operations are accomplished.[343]

3.5 Arithmetical operations with fractions [GT $\overline{35}$–51]

Addition of fractions (bhinnasaṅkalita)[344] *[GT $\overline{35}$]*

Next the author enunciates half a metrical stanza for the procedural rule of addition of fractions:

> $\overline{35}$. [Here] the sum of numerators[345] having equal denominators[346] is specified. [The number] one is to be arranged as the denominator[347] of a number[348] that is devoid (*rahita*) of denominator.

341 A *daṇḍa* after *sarvatra* would fit well.

342 The commentator observes that he has already shown how to derive the cube-root while commenting upon the *karaṇasūtra*.

343 The commentator mentions, for the first time, the expression *aṣṭau parikarmāṇi* to indicate the "eight arithmetical operations"; note that the adjective *pūrvaṇi* qualifies the "operations" (*parikarmāṇi*) as "first" to distinguish from the following topic expounded by the GT: the eight arithmetical operations with fractions.

344 The text expounds the eight fundamental arithmetical operations with fractions. K 15, 20–21.

345 *lava*.

346 *sadṛśahara*.

347 *chedaka* – "denominator, divisor".

348 At the end of the execution of the same problem, the commentator clarifies that here *rāśi* refers to a whole number.

The explanation: [the operation of] addition of whole numbers[349] has been previously explained; now the means for [carrying out] the "addition" of the fractions,[350] which are broken (*khaṇḍita*)[351] integers,[352] which will be mentioned [further] starting with a half of [the whole which is] unity,[353] [are explained]. He states [the expression] **equal: equal denominators** (*sadṛśahara>sadṛśaccheda*) denote the numbers occurring underneath. That [word] **"numerators"** (*ye lavā>aṃśās te*)[354] denote the numbers at the top. In this way, their sum[355] is [their] union;[356] [here] what should be [understood] is the "[operation of] addition" of fractions,[357] thus is the connection [of words with their meanings].

[Next is] a sample problem:[358]

> $\overline{36}$. One-half, one-third, one-ninth, and[359] one-eighteenth:[360] when there is their sum,[361] what [number] should occur?

[I shall first explain] the first half of the metrical stanza. [The fractions are] a **half** of the whole which is unity,[362] "one-third",[363] "one-ninth", and "one-eighteenth". When there is their sum by means of the reasoning previously

349 *pūrṇarūpa* – "whole number, integer".
350 *bhinna*.
351 In this passage (lines 22–23), the SGT elucidates the term *bhinna*, which is lit. "broken, split" and which is a technical term for "fraction". The commentator uses the adjective *khaṇḍita* ("broken") to qualify the noun *rūpa* ("integer"), so that *bhinna* is explained as *khaṇḍitarūpa*: a "fraction" is a "broken integer".
352 Both Śrīpati and Siṃhatilakasūri use the term *rūpa* to mean "integer, a whole unit" as well as "one, unity". Note the contrast between *pūrṇarūpa* and *khaṇḍitarūpa* to indicate a "whole number" and a "fraction" respectively.
353 *rūpa*, here it expresses "one" as "unity, whole". The commentator mentions the fraction "one-half" because this is the first fraction given in the next sample problem by the GT; although the translation may seem clumsy ("a half of unity"), it reflects the commentator's reasoning: fractions are parts of a whole, which is the unity, one.
354 *aṃśa* – a polysemic term meaning in both the GT and the SGT sometimes "numerator", "denominator", "fraction", and "portion".
355 If this is a gloss as it seems, it should be *yojanaṃ* rather than *saṃyojanam*.
356 Line 24: note the orthographic peculiarity *mm* in *sammīlanaṃ*.
357 The SGT clarifies the name of this procedure; the commentary aims to bring students' attention to the fact that now fractions, and not integers, are involved. Notably, although it expounds the new topic of operations with fractions, in the *karaṇasūtra* the root-text has not used the term "fraction", and this is the reason why the SGT supplies and explains it.
358 In the SGT, the abbreviation *udā* is an interesting morphological idiosyncrasy, it clearly stands for *udāharaṇa* and occurs only once. As pointed out by Hayashi (2013, 60), here Siṃhatilakasūri comments upon a sample problem which must have been given by Śrīpati. In line 26, the commentator, in fact, says: *vṛttapūrvārdham* or "[this is] the first half of the metrical stanza", thus pointing out that he is commenting upon a rule given by Śrīpati. Both the editor and Sinha mistakenly consider this passage as prose and part of the commentary.
359 Line 25: *cā°*] em.
360 Line 26: *aṣṭādaśāṃśakaś ca*] em. In this regard, in the line below see the expression *°aṃśaś ca*.
361 In line 26: *eṣāṃ yutau*] em; *yuto* (K). See *yutau* in the line below.
362 Line 26: *rūpasya pūrṇasya*] em. All the mentioned fractions are put into relation to the whole.
363 In line 26, one should supply the term *tri* to *bhāgo*.

explained, the question is: "When there is the union, what [result] should be?" The method for the sum [of these fractions] is explained; by the [aforementioned] word "one" (*rūpa*), [the number] "one" (*eka*) [is meant], [starting] with the number two and so forth the denominators[364] are [written] below it. The presentation is:[365] $\begin{array}{|c||c||c||c|} 1 & 1 & 1 & 1 \\ 2 & 3 & 9 & 18 \end{array}$. Regarding these [fractions],[366] in order to arrive at **numerators having equal denominators,** "one should multiply the numerator and the denominator[367] [of each fraction of a given pair of fractions] reciprocally by the two [exchanged] denominators for [arriving at] the same denominators".[368] [I shall give] the explanation of this half metrical stanza which will be mentioned [later]. To illustrate: in this case [all] the [numbers] one[369] are the "numerators", two[370] and so forth are the "denominators".[371] Then, in the [first] pair [of fractions], [the numerator] one is "reciprocally", [also termed] "mutually", [multiplied by the denominators]. For the sake of [arriving at] equal denominators, by means of the two arranged[372] [denominators], the exchange (*vinimaya*) [should be performed]: for instance, in the one side (*ekatra*) three is [placed] below the "denominator" of the "numerator" [one]; in the other side, two is [placed] below the "denominator" of the "numerator" [one]: $\begin{array}{|c||c|} 1 & 1 \\ 2 & 3 \\ 3 & 2 \end{array}$. Then, "one should multiply the numerator and denominator [of each fraction of a given pair of fractions] by the two [exchanged] denominators" three and two: one multiplied by three becomes three, which is the numerator; in the same way, two multiplied by three becomes six: $\begin{array}{|c|} 3 \\ 6 \\ 3 \end{array}$, which [in the *sūtra*] is called "denominator".[373] In the second [side], one multiplied by

364 In line 28, it is anomalous to find the term *nyāsaḥ* followed by a *daṇḍa*.

365 Line 28: *chedā nyāsa*] em.

366 Line 29: *eṣāṃ sadṛśaharalavānayanāya*] conj.; **kṛte bhāge jāto**] em. The verbal adjective should occur in the line below, thus it should read: *iti vakṣyamāṇasya*] em.

367 *chedana*.

368 The expression *chedasādṛśyahetoḥ* is lit. "for [arriving at] the likeness of the denominators". The SGT has to quote, and hence to anticipate, Śrīpati's rule on the homogenization of the "class of simple fractions" (*bhāgajāti*) which will be explained in GT 53. The reason is that one should apply this rule to carry out the first step mentioned in the rule by Śrīpati: "numerators having equal denominators".

369 All the numerators of the given fractions have the same numerator, i.e., the number 1. These are, in fact, unit fractions.

370 The SGT is going to apply the first step of the procedure on the first two pair of fractions given in the layout (from the left).

371 Line 1: *aṃśāś chedā*] em.

372 *vihita*.

373 In lines 4 and 5, the commentator emphasises that the obtained number 6 corresponds to the *hara* ("denominator") of the *sadṛśahara* mentioned in Śrīpati's *karaṇasūtra*.

two becomes two; three multiplied bytwo becomes six: $\begin{array}{|c|}2\\6\\2\end{array}$, which [in the *sūtra*] is called "denominator". Then, these [two six] are "equal denominators", the two common denominators six are obtained. [Afterwards, one should perform] the "sum"[374] of these [two numbers] consisting of two and three:[375] when the additive quantity two is added to three, it becomes the numerator five whose denominator is six. Because they have been added, the initial[376] denominators and numerators are both removed: $\begin{array}{|c|}5\\6\\2\end{array}$. Below it, because of the object explained, two[377] is removed too: $\begin{array}{|c|}5\\6\end{array}$. Now the quantity which [in the layout] occurs next [is performed]:[378] $\begin{array}{|c|}1\\9\end{array}$. As before, the sum [should be made]: by means of [the procedure] "[one should multiply] the numerator and the denominator"[379] and so forth, when there is the exchange of the denominators, on the one hand nine is below six [and] one the other hand six is below nine: $\begin{array}{|c||c|}5&1\\6&9\\9&6\end{array}$. Next[380] is [the step] "one should multiply [the numerator and the denominator] by the denominators",[381] that is to say, by nine and six: therefore, six multiplied by nine becomes fifty-four [and] five multiplied by nine becomes forty-five: $\begin{array}{|c|}45\\54\\9\end{array}$. In the same way, nine multiplied by six becomes fifty-four, one multiplied by six becomes six:[382] $\begin{array}{|c|}6\\54\\6\end{array}$. **Equal denominators** occur: this

374 As before (line 27), one finds *saṃyojanaṃ*. One would expect *yojanaṃ,* which is found in Śrīpati's rule. In line 8, the feminine *saṃyojanā* occurs.

375 Line 6: *dvitrilakṣaṇayos*] em.

376 Line 7: *prākchedāṃśau*] em.

377 Line 7: *dvikāpy ukta*°] em.

378 Line 8: °*agretanāṅkaḥ*] em. Note that in front of numbers written in figures and in front of the layout the final *visarga* is not always retained.

379 This is a quotation from the rule of the class of simple fractions (GT $\overline{53}$).

380 *tataś*] em.

381 Ibid.

382 I have supplied this layout; I suggest that it should occur in line 12 between *yathā* and the *daṇḍa*.

procedure should be performed [in fact] in order for the "denominators", [and] not the "numerators", to appear **the** very **same.** Regarding these [two] equal denominators, when there is the sum of that consisting of forty-five and six, it becomes fifty-one. Because of the object explained, the denominator is fifty-four, and the previous number six is erased: $\begin{array}{|c|}51\\54\end{array}$. Now the quantity which occurs next [is performed]: [383] $\begin{array}{|c|}1\\18\end{array}$. [One should carry out] the sum: by means of [the procedure] "the numerator and the denominator" and so forth,[384] when there is the exchange of the denominators, eighteen is [placed] below fifty-four; in the same way, fifty-four is [placed] below eighteen: $\begin{array}{|c||c|}51&1\\54&18\\18&54\end{array}$. Next, fifty-four multiplied by eighteen becomes nine hundred seventy-two; in the same way, fifty-one multiplied by eighteen becomes nine hundred and eighteen. In the second [side], eighteen multiplied by fifty-four becomes nine hundred and seventy-two. In the same way, one multiplied by fifty-four becomes fifty-four. Then, as before, because equal denominators occur, when there is the sum of the numerators nine hundred eighteen and fifty and so forth (i.e., fifty-four), it becomes nine hundred seventy-two. Because of the object explained, the previous number fifty-four (i.e., the first denominator) is erased: $\begin{array}{|c|}972\\972\end{array}$. When there is the "addition"[385] of those [fractions] such as one-half, one-third, and so forth, as the numerators have equal denominators, when there is the division of the obtained nine hundred seventy-two by the common denominator [nine hundred seventy-two], the quotient is one (*rūpa>eka*): 1.

[With respect to a given group of quantities] when there are not "denominators" [but] [only] the very "numerators" occur, with respect to fractions (*bhinna>rūpakhaṇḍa*), what one should do? Having raised this doubt,[386] the author thus states: **devoid of denominator** and so forth. That **devoid of denominator,** such as [the numbers] two, three, and so forth,[387] is a very whole number (*rāśiḥ saṃpurṇa*); for instance, by means of the sample problem which will be mentioned [later], [it is explained that] six is a whole number, below it, [the number] "one is to be arranged" as its own "denominator", that is the meaning.

383 I have emended the layout (line 15) as it presents the integer 18 instead of the fraction one-eighteenth. Line 15: °*agretanāṅkaḥ*] em.

384 The *daṇḍa* in line 16 is mistakenly found.

385 Line 23: I have emended the text; *saṅkalite* should occur, in fact, in the line above between °*bhāgādīnāṃ* and *sañjāta*°.

386 *iti āśaṅkya* – this is a typical commentarial formula used to justify an argument put forward by the author of the root-text.

387 Line 25: *dvitricchedādinā*] em; *ekadvibhāgādinā* (K).

The author enunciates the latter half of the stanza which is a sample problem:

> 37. Quickly add [the following quantities]: three plus one-half, six, nine lessened by one-fourth, and seven which is added to one-third.[388]

The explanation: there is a whole number[389] which is the integer "three";[390] [in the layout] **half** of unity[391] should be shown below it. The integer is the whole number "six", below it, **one** is "to be arranged" as **denominator.** In the same way, the integer "nine" is "lessened by one-fourth" (*pādahīna>ekabhāg ahīna*).[392] In the same way, **seven,** which is the integer, is "added" to **one-third** (*tryaṃśa>tribhāga*).

The presentation is:[393] $\begin{array}{|c|c|c|c|}3 & 6 & 9 & 7\\ 1 & & 01 & 1\\ 2 & 1 & 4 & 3\end{array}$. In this case, first the procedure termed *bhāgānubandhajāti*[394] ("the class of fractional increase") should be performed. For instance, [in this regard] in the *Triśatī* [is found]: "The integer multiplied by the denominator [and the result] plus the numerator".[395] Therefore, here regarding the presentation of "three plus one-half", that consisting of three, which is the "integer",[396] multiplied by the "denominator" three,[397] becomes six; this "plus the numerator" below [three],[398] that is, six added to one becomes seven: $\begin{array}{|c|}7\\ 2\end{array}$. For the purpose[399] of the sum [of this result] and six whose

388 Line 30: °*anvitān sapta*] em.

389 *pūrṇa*.

390 The syntagm *rūpatraya* is lit. "a group of three units".

391 Line 1: **caturtha**] em.

392 *ekabhāga* is a "quarter".

393 In Sanskrit mathematical manuscripts, the minus sign is indicated by a zero or a kind of small circle near the number to be subtracted; SGT 61, 18, and SGT 62, 13 refer to it by the term *śūnya* ("zero"). It is unusual that the commentator does not remark on this practice here at its first occurrence.

394 In order to carry out the sum of an integer and a fraction, the commentator has to quote and hence to anticipate the rule of the class of fractional increase (GT 57). He quotes the *karaṇasūtra* of this rule from TŚ 24.

395 In line 4, the editor suggests *rūpaguṇa* instead of *rūpagaṇa*, but this emendation is mistaken as it is based on the corrupt TŚ 24; the correct reading is found in PG 39: *rūpagaṇaś chedasaṅguṇaḥ,* which occurs in relation to the rule of *bhāgānubandhajāti*. This expression is quoted also by the PG's commentator in the context of addition of fractions; see the Sanskrit text in Shukla (1959, 24). The expression *rūpagaṇa* is lit. a "group (*gaṇa*) of units", hence an integer.

396 In line 6, the same *gaṇa* occurs.

397 Line 5: *dvikeṇa*] em.; *trikeṇa* (K). The denominator 2 of $\begin{array}{|c|}1\\ 2\end{array}$ is meant.

398 Line 6: *sāṃśo'*] em.

399 In line 7, it should be *artham* followed by the quotation. The edition shows a peculiar way to separate quotations.

denominator is one, which is the quantity that occurs next, by means of the [procedure] "the numerator and the denominator"[400] and so forth, among the "numerators" above, [two]equal denominators[401] consisting of [the numbers] two and two occur:[402] $\begin{array}{|c|c|} 7 & 12 \\ 2 & 2 \\ 1 & 2 \end{array}$. When there is the sum of twelve and seven,[403] it becomes nineteen whose denominator is two, [and] the first quantity[404] is erased (*nivṛtta*).[405] The [number] two below two also is erased: $\begin{array}{|c|} 19 \\ 2 \end{array}$. Afterwards, when there is the sum of that consisting of nine **lessened by one-fourth**, which is the quantity that occurs next, the class of fractional decrease (*bhāgāpavāhajāti*)[406] should first be performed. The author enunciates: "In the procedure of fractional decrease, once the integer is multiplied by the denominator, one should subtract the numerator from the quantity [obtained]":[407] $\begin{array}{|c|} 9 \\ 01 \\ 4 \end{array}$. When that consisting of nine, which is the "integer", is multiplied by four, which here is the "denominator", it becomes thirty-six; "one should subtract the numerator from the quantity [obtained]", namely from this thirty-six; in this case, when there is the subtraction (*apanayana*) of one [from thirty-six], it becomes thirty-five. One is erased [and] it becomes thirty-five whose denominator is four: $\begin{array}{|c|} 35 \\ 4 \end{array}$. Next is [the step] "the numerator and the denominator [multiplied] by the denominator":[408] in this case, in order to perform the procedure, having made the reduction (*apavarta*) of the denominators, such as two and

400 This is a quotation from the rule of the class of simple fractions (GT $\overline{53}$).

401 Line 7: *sadṛśacchede*] conj. Otherwise it reads: "when the likeness of the denominators occur, when those [denominators] consisting of two and two occur [...]".

402 Line 8: *jāte*] em.; *jñāte* (K). The intermediate passage is: $\begin{array}{|c|c|} 7 & 6 \\ 2 & 1 \\ 1 & 2 \end{array}$.

403 Line 8: *sapta°*] em.

404 Lines 8–9: *prāgrāśir*] em.

405 In line 9, the intermediate layout should look like the following: $\begin{array}{|c|} 19 \\ 2 \\ 2 \end{array}$, the denominator underneath is then erased and the final result is $\begin{array}{|c|} 19 \\ 2 \end{array}$. Line 9: *gāto*] em.; *jāto* (K).

406 In order to carry out this procedure, the commentator has to anticipate the rule of the class of fractional decrease formulated in GT 60.

407 This is a quotation from the rule of the class of fractional decrease (GT 60).

408 This is a quotation from the rule of the class of simple fractions (GT $\overline{53}$).

four, by one and two, one and two are placed below the denominators: $\begin{array}{|c||c|}19 & 35\\ 2 & 4\\ 2 & 1\end{array}$.

Two is multiplied "by the denominator", which is namely two, it becomes four; in the same way, nineteen multiplied by two becomes thirty-eight. In the other place,[409] [that fraction] multiplied by one[410] is thirty-five whose denominator is four: $\begin{array}{|c||c|}38 & 35\\ 4 & 4\end{array}$. Afterwards, because equal denominators occur, when the additive quantity thirty-five is added to the obtained thirty-eight, it becomes seventy-three whose denominator is four. The initial number, namely the denominator one, is removed. In this manner, whenever a number is suitable (*sahate*) for the reduction, in that case one should reduce[411] the number [and] write the reciprocal [results]; [in this regard] [the procedure] "one should multiply [the numerator and the denominator] by the denominators",[412] [which means] by the reduced denominators, has been shown. The presentation is:[413] $\begin{array}{|c|}73\\ 4\end{array}$. Then, for the sake of the sum of seven and one-third, which is the quantity that occurs next, according to [the rule] "the integer multiplied"[414] and so forth, which is the mentioned [rule of] class of fractional increase, seven is multiplied by three; it becomes twenty-one. When the additive quantity, which is the numerator one [is added to it], it becomes twenty-two whose denominator is three: $\begin{array}{|c|}22\\ 3\end{array}$. By means of the [procedure] "the numerator and the denominator"[415] and so forth, when there is the reciprocal multiplication[416] "by the denominators" three and four, when the alike denominator twelve occurs, which is that consisting of the "denominator",[417] when the additive quantity which is the obtained numerator eighty-eight [obtained] from twenty-two multiplied by four is added to the "numerator" consisting of the obtained two-hundred nineteen from seventy-three multiplied by three, it becomes three hundred seven, [and] the denominator below is twelve. The first quantity,[418] which is the denominator three, is erased. Then, because of the

409 *paratra* – it denotes the other side of the layout, where the other fraction to be computed is found.

410 Line 17: *tad eva*] em.; *tādṛśye* (K). Also, in the same line: **ca**] em.

411 *apavartya*. The commentator underlines that, whenever it is possible, one should carry out the reduction of both operands by a same quantity.

412 This is a quotation from the rule of the class of simple fractions (GT $\overline{53}$).

413 In line 21, the use of the term *nyāsa* shows that this does not denote only the initial setting down of the numerical data, but it is a general term to indicate the layout.

414 See lines 5–6; it quotes TŚ 24.

415 This is a quotation from the rule of the class of simple fractions (GT $\overline{53}$).

416 Line 24: *guṇane*] em.; *guṇite* (K).

417 The commentator uses the word *hara* to refer to *sadṛśahara* of Śrīpati's *karaṇasūtra*.

418 Line 26: *prāg* should be compounded with *rāśi*°.

occurrence of the denominator [which is smaller of the numerator],[419] when there is the division of [the numerator] three hundred seven by the twelve below, the quotient is the integer twenty-five, [and] seven whose denominator is twelve: $\begin{array}{|c|}25\\7\\12\end{array}$. It is always in this way.

The [procedure of] addition of fractions is concluded.

Subtraction of fractions **(bhinnavyavakalita)**[420] ***[GT*** $\overline{38}$***]***

[Next is] the procedural rule of subtraction of fractions; the author enunciates half a metrical stanza:

> $\overline{38}$. Those who are familiar with works on arithmetic explain that in the procedure of subtraction [of fractions], [one should perform] the deduction[421] of the numerators of [two] quantities of which equal denominators have been arranged.

The explanation: the author formulates the means for [carrying out] the "subtraction" of the fractions (*bhinna>rūpakhaṇḍa*) beginning with "one-sixth".[422] [Regarding the expression] **equal denominators have been arranged,** there is an expenditure,[423] which involves an [initial] capital (*āya*); therefore, there are two quantities: the quantity-capital [and] the quantity-expenditure.[424] These have "equal denominators arranged", which are the alike determined divisors. There is a pair [of quantities] of which **equal denominators have been arranged:** among these, there is the quantity-capital, from it [one should perform] **the deduction** (*viśleṣa>pāta*) of the "numerators" which have been obtained from

419 The fraction $\begin{array}{|c|}307\\12\end{array}$ is an improper fraction, since the numerator is larger than the numerator. The commentator always transforms this kind of quantity into an integer plus a proper fraction.

420 K 18, 3–4.

421 See the use of *vyavakalita* to denote the operation of "subtraction" and *viśleṣa* to denote the "deduction".

422 See the sample problem which follows.

423 *vyaya* –lit. "loss, expenditure", while *āya* (in the same line) is "profit, income". Here *āya* does not denote an income but instead the initial capital from which different amounts are deducted. This terminology refers to the context of the next sample problem where in line 17 one finds *vyaya* and in line 27 the term *āya.* In this regard, at the very end of his explanation of the sample problem, note the commentator's remark (line 21): "In this commercial practice (*vyavahāre*) [...]". In the SGT, *āya* is often found in relation to the rule of three and its derivatives, and it is usually accompanied by its correlative *vyaya* or "expenditure".

424 *āyarāśi* and *vyayarāśi* expand upon the dual *rāśayor* which occurs in the first line of the *karaṇasūtra*.

the quantity-expenditure,[425] that is, [the deduction] of the quantities resulting from one-sixth and so forth.[426] Afterwards, the number which is the remainder is considered to be the wealth[427] [remaining] from the "subtraction". The rest is clear [in meaning].

In this respect, the author enunciates a couplet of examples by an exemplifying metrical stanza:

> 39. O learned man! Tell [me] the remainder having subtracted (*projjya*) one-sixth, one-half, and one-third[428] from one *dramma*, [and] having removed (*tyaktvā*) one and a half [*drammas*], one minus one-fourth,[429] as well as one plus one-eight from six plus one-third *drammas*.

The explanation: **tell [me] the remainder having removed** from one ***dramma,*** which is a *rūpaka,*[430] a "sixth" of [whole which is] one *dramma;* in the same way a **half** of one *dramma,* in the same way **one-third**, which is (lit.) a "third part" of a *dramma*. This is the first example; its presentation [follows]:[431] having placed that [mentioned "unity"] one,[432] since this is "devoid of denominator",[433] the denominator "one is to be arranged"; the "expenditure" from this[434] will first be executed. The presentation of the parts which are the remainders is in front of it (i.e., of one): $\begin{array}{|c|c|c|c|} 1 & 1 & 1 & 1 \\ 1 & 6 & 2 & 3 \end{array}$. In this case, for the sake of [the step] **equal denominators have been arranged,** in regard to one-sixth, which is the first [fraction] [and] one-half, for the making of equal denominators,[435] by means of [the procedure] "the numerator and the denominator"[436] and so forth, by the previously mentioned exchange of the denominators [and] by means of the multiplication, it becomes [the number] two whose denominator

425 *samutthita.*

426 In this explanatory gloss, the commentator refers to the next sample problem where the quantities-expenditure are first added and the resulting fraction is then deducted from the quantity-capital.

427 *dhana* – here the final "capital" left over when the *vyayarāśi*s are subtracted from the *āyarāśi.*

428 Line 12: *tryaṃśakaṃ*] em.

429 *aṃri* –lit. "foot", in mathematical terminology denotes the number "four".

430 This term could either mean "money, coin" or denote a coin-denomination in use during Siṃhatilakasūri's time and place.

431 The position of *asya nyāsaḥ* between two *daṇḍa*s is awkward, it also seems redundant as *nyāsa* occurs again two lines below and it is followed, in fact, by the presentation of the numerical data.

432 Line 17: **pūrṇāni**] em.

433 As this quotes GT $\overline{35}$, note the loss of the final syllable *ka* in *cheda(ka)*.

434 Here the relative *yasmāt* is anomalous, it is perhaps more natural to interpret it as a misprint (scribal error?) for *tasmāt*; cf. *tasmāt* in line 7.

435 In lines 19, 24, and 27 cf. *sadṛśacchedatākṛtaye, samānacchedatvāt*, and *samaharatvakṛte*.

436 Line 19: *ityādinā*] em. This is a quotation from the rule of the class of simple fractions (GT $\overline{53}$).

is twelve, and above there are two and six.[437] Afterwards, because there is the same denominator, when the additive quantity six is added to two, it becomes eight; twelve, which is the denominator, is underneath. Because of the object explained, the initial[438] quantity having six as denominator is erased,[439] it is: $\begin{array}{|c|}8\\12\end{array}$. Then, regarding the quantity that occurs next which is one-third:[440] $\begin{array}{|c|}1\\3\end{array}$, for the sake of the sum, also again by means of [the procedure] "the numerator and the denominator"[441] and so forth, by means of the exchange of the denominators [and] by the multiplication, equal [denominators] occur. In both places, the denominator is namely thirty-six: $\begin{array}{|c|c|}24&12\\36&36\\3&12\end{array}$. Next, since there are equal denominators, because there are two thirty-six below, when there is the sum of the twelve above and the numerator twenty-four, it becomes thirty-six; underneath, there is thirty-six as well. This is the [total] "quantity-expenditure" which has been summed up,[442] and the initial[443] quantity consisting of that same [number] which is the denominator[444] is erased: $\begin{array}{|c|}36\\36\end{array}$. Afterwards, regarding the quantity-capital, when the equal denominators have been arranged, also again by means of [the procedure] "the numerator and the denominator"[445] and so forth, when there is the exchange of the denominators for [carrying out] the multiplication, it is: $\begin{array}{|c|c|}1&36\\1&36\\36&1\end{array}$. In this case, the two quantities the expenditure and the capital[446] have thirty-six as the equal denominator, above [when the numerator is multiplied by the denominator] two numerators thirty-six occur.[447] Then, when there is the subtraction

437 The mathematical steps partly mentioned in this passage represent the sum of the first two unit fractions, which are the first two *vyayarāśi*s one-sixth and one-half: $\begin{array}{|c|c|}1&1\\6&2\\2&6\end{array}$; $\begin{array}{|c|c|}2&6\\12&12\\2&6\end{array}$; $\begin{array}{|c|}8\\12\end{array}$.

438 Line 21: *prāg* should be compounded with *rāśi*°.

439 *bhājyate*.

440 Line 22: *tryaṃśaḥ*] em.

441 This is a quotation from the rule of the class simple fractions (GT $\overline{53}$).

442 Here Siṃhatilaka specifies that this result is the total quantity-expenditure to be removed from the *āyarāśi* one *dramma*.

443 Line 26: *prāgrāśi*°] em.

444 Line 26: *rāśir sadṛśalakṣaṇacchedaś ca*] em.

445 This is a quotation from the rule of the class of simple fractions (GT $\overline{53}$).

446 Line 28: *dvāv apy āyavyayarāśī*] em.

447 Line 1: *2*] em.

of that consisting of the numerator thirty-six from the [other] numerator,[448] namely thirty-six, the remainder which is obtained is the very zero: |0|.

The author enunciates the second example by the latter half [of the stanza]: **one and a half** and so forth. In this case, having placed the quantity-capital, one should place the quantity-expenditure in front of it, thus is the method. Then, **having removed one and a half [*drammas*],**[449] **one minus one-fourth** (*vyaṃhriṃ rūpaṃ* >*vigataikabhāgaṃ rūpaṃ*), **as well as one** (*eka*>*rūpa*) **plus one-eight from six plus one-third *drammas*, tell [me],** which means "say", its **remainder** which is that "capital" [resulting] from the "subtraction", thus is the connection [of words with their meanings]. The presentation is:

6	1	1	1
1	1	01	1
3	2	4	8

.[450]

In this case, the first quantity is the quantity-capital, the remaining are the quantities-expenditure. Next, having calculated two equal denominators, one should carry out **the deduction** of that consisting of the numerator[451] of the quantity-expenditure from the quantity-capital, this is the purpose of the rule. Having first manipulated the quantity of the expenditure consisting of one and one-half, in order to calculate the "equal denominators", by means of the reasoning of the [procedure] of fractional increase that will be explained [later], **"**one should add the numerator to the integer multiplied by the denominator":[452] one multiplied by two becomes two, which is below, [and] when the additive quantity one below [is added to it], it becomes three whose denominator is two. Regarding the quantity which occurs next consisting of[453] **one minus one-fourth,** by means of the reasoning "in the procedure of fractional decrease, once the integer is multiplied by the denominator, one should subtract the numerator from the quantity [obtained]",[454] one multiplied by four becomes four; when there is the subtraction [of it] from one, it becomes three whose denominator is four. Next, by means of [the procedure] "the numerator and the denominator"[455] and so forth, when there is the exchange of the denominators and when there is the multiplication, the denominators are [two] alike eight, and above, since there is the same denominator, when the

448 *bhāga*.

449 The commentator supplies *rūpa*.

450 I have emended the layout; in the edition, the layout (line 6) is mistaken:

0
1
4

451 The use of *aṃśabhūta* does not seem appropriate here; also, it does not reflect the common phraseology used by the commentator.

452 This is a quotation from the rule of the class of fractional increase (GT 57); in this passage one finds, however, *rūpaṃ* instead of *bhāgaṃ* as in the *karaṇasūtra*.

453 Line 11: *agretanāṅkavyaṃhryekarūpe*] em.

454 This is a quotation from the rule of the class of fractional decrease (GT 60).

455 This is a quotation from the rule of the class of simple fractions (GT 53).

additive quantity six is added to twelve, it becomes eighteen whose denominator is eight. The first quantity whose denominator is two[456] should be erased, it is: $\begin{array}{|c|}18\\8\end{array}$. Then, regarding the quantity that occurs next consisting of one plus one-eight: $\begin{array}{|c|}1\\1\\8\end{array}$, "once the integer is multiplied by denominator",[457] when the additive quantity which is the "integer" [is multiplied by the denominator],[458] one is multiplied by eight, it becomes eight. In this case, when the lower additive quantity one is added to eight, it becomes nine whose denominator is eight. Then, regarding these two quantities, since there is the same denominator, because two [same] eight occur below, when the additive quantity nine is added to the eighteen above, it becomes twenty-seven whose denominator is

456 Line 15: *prāgrāśir dvikacchedaś ca*] em.

457 This is a quotation from the rule of the class of fractional increase (GT 57).

458 The lost 37th folio of the manuscript must have occurred between *chedā* and *labdhaṃ* (line 16); note that one should read *chedāṣṭaguṇa eko* and so forth. Hayashi has kindly shared with me his Sanskrit reconstruction, which I have adopted but also slightly changed; the English translation is my own. Other reconstructions are no doubt conceivable. Some of the passages below are based on the first part of the commentary's explanation of the sample problem. Hayashi has observed (in a private conversation) that according to the editor (Kāpadīā 1937, LXVII), a page of the manuscript contains four lines and one line contains from 52 to 57 letters; the number of letters on the two missing pages are estimated to be about 416–456. The reconstructed passage (on which the translation above is based) is:

> [...] *chedāṣṭaguṇa eko jātāv aṣṭau* | *tatra aṣṭamadhye 'dhastanaikakṣepe jātā navāṣṭacchedāḥ* | *tato dvayo rāśyoḥ sadṛśacchedatvād) 'ṣṭau dvikatvād uparyaṣṭādaśānāṃ madhye navakṣepe jātāḥ saptaviṃśatir aṣṭacchedāḥ prāgrāśiś ca bhajyate* | *yathā* $\begin{array}{|c|}27\\8\end{array}$ *eṣa vyayarāśiḥ* | *āyarāśilakṣaṇe tryaṃśānvitarūpadrammaṣaṭke bhāgānubandhatvāt chedanighneṣu rūpeṣu bhāgaṃ kṣiped iti yuktyā triguṇāḥ ṣaḍ jātā aṣṭādaśa* | *aṣṭādaśānāṃ madhye 'dhastanaikakṣepe jātā ekonaviṃśatis tricchedāḥ* | *yathā* $\begin{array}{|c|}19\\3\end{array}$ *ayaṃ āyarāśiḥ* | *tato dvayor āyavyayarāśyoḥ samaharatvakṛte punar apy aṃśacchedāv ity ādinā chedavinimaye guṇanayā jātāḥ samānā ubhayatra caturviṃśatilakṣaṇāś chedā yathā* $\begin{array}{|c|}152\\24\\8\end{array}\begin{array}{|c|}81\\24\\3\end{array}$ | *atra dvāv apy āyavyayarāśī samānacaturviṃśaticchedau* | *uparyekatrāṃśā dvipañcāśadadhikaṃ śatam ekatraikāśītiś ca jātāḥ* | *tato 'ṃśaikāśītirūpasya dvipañcāśadadhikaśatarūpabhāgāpanay ane śeṣam ekasaptatir labdhāḥ* | *aparārāśir aṣṭalakṣaṇacchedaś coktārthatvād bhañjanīyo yathā* $\begin{array}{|c|}71\\24\end{array}$ | *paścād ekasaptater adhaś caturviṃśatyā chedatvād bhāgaṃ grāhayet labdhaṃ ca* [...] conj.

eight, and the first quantity is erased. Thus $\begin{array}{|c|}27\\8\end{array}$ is the quantity-expenditure.[459] Regarding that consisting of the quantity-capital, which is six plus one-third, because there is a fractional increase, by means of the reasoning "one should add the numerator to the integer multiplied by the denominator",[460] six multiplied by three becomes eighteen. When the lower additive quantity one is added to eighteen, it becomes nineteen whose denominator is three:[461] $\begin{array}{|c|}19\\3\end{array}$, this is the quantity-capital. Then, again in order to arrive at the "equal denominators", among the two which are the capital and the expenditure quantities, by means of [the method] "the numerator and the denominator"[462] and so forth, when there is the exchange of the denominators [and next] by means of the multiplication, in both places equal denominators, namely twenty-four, occur:[463] $\begin{array}{|c|}152\\24\\8\end{array}\begin{array}{|c|}81\\24\\3\end{array}$. In this case, [both] quantities the capital and expenditure have twenty-four as the same denominator; above, on the one hand the numerator is one hundred fifty-two, and on the other hand there is eighty-one. Afterwards, when there is then the subtraction of that consisting of eighty-one, which is the numerator, from the "numerator" consisting of one hundred fifty-two, the remainder seventy-one is obtained. Because of the object explained, the other quantity, which is the denominator consisting of eight, is erased: $\begin{array}{|c|}71\\24\end{array}$. Then, since there is the denominator [which is smaller than the numerator],[464] one should perform the division of seventy-one by the twenty-four underneath and the quotient is first the [number] two.[465] Below it, twenty-three, and below it, twenty-four should be placed; the presentation

459 This fraction denotes the total sum of the various *vyayarāśi*s.

460 This passage quotes the rule of the class of fractional increase (GT 57).

461 $\begin{array}{|c|}19\\3\end{array}$ is the *āyarāśi* once simplified (before applying the *bhāgānubandhajāti*, it was six plus one-third).

462 This is a quotation from the rule of the class of simple fractions (GT $\overline{53}$).

463 Alternatively, another possible emendation could be: *chedavinimaye chedau mitho guṇitau sadṛśau bhavataś caturviṃśatilakṣaṇau yathā* or "When there is the exchange of the denominators, the two denominators reciprocally multiplied become the 'same', which is that consisting of twenty-four, thus [...]". This reconstruction is supported by a sentence occurring in SGT $\overline{53}$, 20.

464 The fraction $\begin{array}{|c|}71\\24\end{array}$ is, in fact, an improper fraction.

465 Line 16: *prāg* should be compounded with *dvau*.

thus is: $\begin{array}{|c|}2\\23\\24\end{array}$. In this manner,[466] the denominator occurs below that consisting of the numerator of the quantity-expenditure. Therefore, the denominator twenty-four[467] is below eighty-one; having then first performed **the deduction** of the numerator such as eighty-one from the quantity-capital one hundred fifty-two, when there is the division of the remainder from the quantity-capital by the denominator such as twenty-four, its[468] quotient is two and so forth; above it, the two remaining parts are namely twenty-three [and] twenty-four, that's the principle.[469] In this commercial practice,[470] when the one plus a half [*drammas*] and so forth are spent (*vyayite*) **from six plus one-third *drammas***, a pair of *drammas* [plus twenty-three whose denominator is twenty-four] are the remainder from the expenditure.[471] It is always in this way.

The procedure of subtraction of fractions is concluded.

Multiplication of fractions (bhinnapratyutpanna)[472] *[GT $\overline{40}$]*

[Next is] the procedural rule of multiplication of fractions; the author enunciates half a metrical stanza:

> $\overline{40}$. The result[473] of the multiplication [of two fractions] always occurs in the multiplication[474] of the numerators[475] and when it (i.e., the result) is divided by [the product of] the multiplication of the denominators.[476]

The explanation: the numbers above are the "numerators" of this passage, [and] the numbers below are termed "denominators". There are two numbers which are the "numerators": **in the multiplication** (*vadha>guṇana*) [means **in the multiplication**] of the [first] "numerator" by the [second] "numerator". [One should multiply] the number below, which is the number arisen as the

466 There is a problem here (lines 17–21): in line 17, the edition shows the result $\begin{array}{|c|}2\\23\\24\end{array}$ but what follows is, unexpectedly, the explanation of how to arrive at this final result.

467 Lines 18 and 20: *caturviṃśati*] conj.; *dvādaśa* (K); the common denominator is, in fact, 24 and not 12. The common denominator is divided by 2 but then it appears back as it was before being reduced.

468 Line 20: *tat*] em.; *yat* (K).

469 Line 21: **tad ardhaṃ**] em.

470 *vyavahāra*.

471 Line 22: *śeṣaṃ*] em. The final expression *tṛtīyaś ca* and so forth is nonsensical.

472 K 19, 26.

473 *phala*.

474 *vadha*.

475 *bhāga*.

476 Śrīpati uses three terms to denote "multiplication": *guṇanā, vadha,* and *tāḍana*.

multiplier,[477] by the "denominator" of the second quantity "denominator". **When it is divided,** that is, when the division is performed, **by** [the product of] **the multiplication,** that is, by the number obtained as the product, the quotient should be that **result of the multiplication.** Thus is the procedure [of this operation].

In this respect, in the exemplifying metrical stanza [which follows] the author enunciates two sample problems:

> 41. O clever one! Tell [me] the product[478] of three plus one-half by nine added to one-third. If you know [how to calculate] the product, [tell me] then what should be [the result of] one-fourth multiplied by one-half?

The explanation: [the expression three] **plus one-half (*sadala*)** is **three** plus one-half (*sārdha*), such as three and one-half,[479] which is [multiplied] **by nine added to one-third** (*trilava*> *tribhāga*). The question is: "What should be **the product?**"

It will be explained in detail.[480] To illustrate, the presentation is:

3	9
1	1
2	3

. Always in a numerical procedure[481] where there is a quantity in association with a fraction,[482] perform (*vidhāya*) the procedure of the class of fractional increase [mentioned] earlier. A different procedure occur when a quantity is involved[483] in a fractional decrease. Carry out (*kṛtvā*) the mentioned procedure [of fractional increase], [and note that] this other procedure [of fractional decrease] should be performed [in a different context]; thus is the essence.[484] In this case, since there is a fractional increase, by means of the reasoning "one should add the numerator to the integer multiplied by the denominator",[485] in the first quantity, three is multiplied by two, it becomes six which added to one becomes seven whose denominator is two. In the same way, in the other place (*paratra*) [of the layout] nine multiplied by three becomes twenty-seven, which added to one becomes twenty-eight whose denominator is three. Afterwards, regarding this [step] **in the multiplication of the numerators** mentioned in the *sūtra*, twenty-eight is multiplied by the "numerator" seven; it becomes one

477 Cf. *guṇaniṣpannāṅka* (line 28) and *guṇitaniṣpannāṅka* (line 1).

478 *guṇita*.

479 *adhyuṣṭa*, it occurs only once in the SGT. The commentary mentions different ways to denote the fraction "three plus one-half".

480 Line 9: *saḥ*] em. This remark is unusual.

481 *aṅkavidhi*.

482 Line 10: *bhāgānubandhāṅke*] em. The transmitted readings *bhāganubandheṣv aṅke* (line 10) and *saṃyateṣv aṅke* (line 11) are an example of *constructio ad sensum*.

483 Line 11: *bhāgāpavāhasaṃyatāṅke*] em. The adjective *saṃyata* is lit. "bounded up".

484 The commentator emphasises the distinction between the procedure of the class of fractional decrease and the procedure of the class of fractional increase. The formula *iti hṛdayam* (*hḍraya* is lit. "heart") occurs only once throughout the SGT.

485 This is a quotation from the rule of the class of fractional increase (GT 57).

hundred ninety-six. It follows **[divided by the product of] the multiplication of the denominators:** there is the "multiplication" of that consisting of three by the "denominator" of the other (*itara*) [fraction] such as two; it becomes six. When the previous[486] [obtained result] one hundred ninety-six **is divided** by this [six], the quotient is thirty-two. When there is the reduction of four [and then] of six by half is respectively (*kramāt*) two and three; these should be written each below the other. The presentation accordingly is:

32
2
3

By means of this commercial practice,[487] thirty-two and two whose denominator is three occur.[488]

Now [I shall explain] the second example. [The expression] **one fourth** [multiplied by] **one- half**: **what should be** [the result when] **one fourth** (*caraṇa*>*caturtha*)[489] **is multiplied** by "one-half" (*dala*>*ardha*)? The presentation is:[490]

1	1
4	2

Whenever above [a fraction] there is an integer, then the [procedure of the] class of fractional increase [should be performed]. However, in this case, regarding the two [quantities], only fractions occur; with this in mind, the class of [simple] fractions[491] [should be performed]. Therefore, here the mentioned procedure of [the class of] fractional increase should not be performed.[492] Next is [the step] **in the multiplication of the numerators,** that is, **in the multiplication** (*vadha*>*guṇana*) of the "numerator" one by the [other] "numerator" one; [the result] is the very [number] one. With regard to that [step] **the multiplication of the denominators:** this is the "multiplication" of [the denominator] two by the "denominator" four, it becomes eight. Since there is the [number] one standing above, it is not possible[493] to divide the numerator by it.[494] Thus, one stands above; below it, among the numerator and the denominator, with respect to that consisting of the denominator, the eight should be placed below that consisting of the numerator above.[495] According

486 In line 16, *prāk* should be compounded with *ṣaṇ*°.

487 The term *vyavahāra* (lines 18 and 26) is to be understood in relation to the observation found in line 26, where the SGT mentions one *dramma* (see also the previous same sample problem). In brief, the commentator understands the quantities involved in this sample problem on the multiplication of fractions to represent monetary quantities.

488 Line 18: *dvau tricchedau dvitriṃśac ca*] em. The final result given in the edition is mistaken.

489 *caraṇa* – lit. "foot". The expression used by the commentator is lit. "which is a fourth part of one", it underlines the fraction as a part of a whole.

490 I have emended the layout (line 20), since the denominator 2 is missing.

491 See GT 53.

492 This wording (lines 20–22) deliberately echoes the commentator's own passage (lines 9–11).

493 *na śakyata iti.*

494 The commentator refers to the second part of the *karaṇasūtra*, where it is said that the result of the two products should be divided; one-eight is, in fact, a unit fraction.

495 The syntax is awkward here (line 25).

to what has been obtained,[496] the presentation is: $\begin{vmatrix}1\\8\end{vmatrix}$. By means of this commercial practice, one eight of a *dramma* [is the result] (…?).[497] It is thus always in this way.

The multiplication of fractions is concluded.

Division of fractions* (bhinnabhāgahāra)[498] *[GT 42]

[Next is] the procedural rule of division of fractions; the author enunciates a metrical stanza:

42. Having performed the interchange (*parīvartana*) of the numerator and the denominator of the divisor[499] as well as the cross-reduction (*kuliśāpavartana*), the procedure of multiplication of the denominators and of the numerators should always be performed by the one who wishes to perform [division of fractions].

The explanation: **the one who wishes to perform [division of fractions]** is the one intending to carry out the procedure of division. [Regarding the expression] **having performed** and so forth: there are again two numbers placed, which are the numerator and the denominator. Moreover, regarding these [two], among the [two] quantities [involved], [there is the interchange] of "the numerator and the denominator"[500] of that [fraction] occurring next, which is designated as "divisor".[501] Then, with respect to **this divisor**, [one should perform the step] **having performed the interchange**[502] **of the numerator** above **and the denominator,** that is, **[the interchange]** of that consisting of the "denominator" below the "numerator" above. According to circumstances,[503] having first performed [some required] procedures, such as the fractional increase, the

496 Line 25: *sthāpyāḥ* ǁ] em. Cf *yathākramaṃ nyāsaḥ* (line 18) and *yathālabdhaṃ nyāsaḥ* (lines 25–26).

497 I cannot make sense of the expression *sārdhaloṣṭikadvayarūpo.*

498 K 21, 3–6.

499 In this stanza, see *hara* as "divisor" and *hāra* as "denominator"; these terms are synonyms. In the line below, *hara* is "denominator", while in the following sample problem, it means "division".

500 This passage is an explanatory amplification of *hara*; note that the commentator clarifies the position of the "divisor" in the layout.

501 The commentary refers to the term *hara* used by Śrīpati in the rule. In line 8, *agretana* indicates that in the layout, with respect to a pair of fractions, the divisor is the quantity that occurs "next" with respect to the fraction-dividend. A similar remark is found in lines 18–19 (in the following page).

502 Line 9: *ca parīvartanam*] em.; *cāpavartanaṃ* (K).

503 *yathāprāptaṃ*. The point of this argument is that before applying the rule of division on the given quantities, students should examine the quantities and understand whether, "according to circumstances" (*yathāprāptam*), other procedures should first be performed, such as the classes of simplification of fractions. The commentary clearly refers to the following sample problem where the class of fractional increase is to be applied before dividing the fractions obtained.

number above should be placed below instead of the second number [and] the number underneath should be placed above.[504] Next is the cross [reduction] (*kuliśatā>vajratā*)[505] of the obtained[506] four numbers by the two sides; if the preceding (*prācya*) or the next number fits [the conditions for] **the reduction,** [one should reduce] the denominator by half and so forth.[507] **As well as** [the step] **having performed the cross-reduction,** there is **the multiplication** too, that is, as [explained][508] before **in the multiplication of the numerators [and] when it is divided by the multiplication of the denominators.**[509] For the obtainment of the result of the division, this procedure should **always** be, which means "at all times", performed. Thus is the connection [of words with their meanings].[510]

In this respect, in the exemplifying metrical stanza [which follows] the author enunciates four examples:[511]

> 43. O mathematician! If you know the division (*hara*) [of fractions], tell [me] quickly what is [the result] when, according to the procedure, ten plus one-fourth is divided by six plus one-third, [the quantity] eighty and one-half [is divided] by five[512] less by one-third, one-half by one-sixth, and one-fourth [is divided] by three.

The explanation: **o mathematician! If you know** the procedure of **division** (*hara>bhāgahāra*) **tell [me] quickly** then **what is [the result] when ten plus one-fourth** (*daśa sacaraṇāḥ>sacaturthabhāgā daśa*) **is divided by six plus one-third.**

This is the first example, the presentation is:

10	6
1	1
4	3

. In this case, as the class of fractional increase occurs, by means of [the procedure] "one should add the numerator[513] to the integer multiplied by the denominator"[514] and so forth,

504 It refers to the interchange of the operands of a fraction.

505 Note the construction with the abstract suffix, where *kuliśatā* is lit. the "crossness". In GSS 3.2, *vajrāpavartana* also denotes the "cross reduction"; Mahāvīrācārya uses this expression in the *karaṇasūtra* of multiplication of fractions.

506 *prāpta*.

507 Cf. line 27, while commenting upon the sample problem the commentator observes that the reduction of the two fractions is not possible.

508 By *pūrvavat* the commentator refers to the preceding rule, which is the multiplication of fractions. The division of fractions is, in fact, the multiplication of fractions after having carried out the "interchange" (*parīvartana*) of the operands and, whenever possible, the cross-reduction (*kuliśāpavartana*).

509 This is a quotation from the rule of multiplication of fractions (GT $\overline{40}$). The commentator links the two rules and underlines that the multiplication mentioned in the rule of division is that procedure explained before (see *pūrvavat*) in the rule of multiplications of fractions.

510 *iti saṇṭaṅka*.

511 The stanza enunciating the sample problem gives four exercises.

512 *śara* – lit. "arrow"; in the *bhūtasaṃkhyā* notation, *śara* denotes the number "five".

513 Line 23: *bhāgam*] em.; *rūpam* (K). Cf. GT 57.

514 This is a quotation from the rule of the class of fractional increase (GT 57).

regarding the first quantity, ten multiplied by four is forty; when the additive quantity one [is added to it], it is forty-one whose denominator is four. In the other place,[515] six multiplied by three is eighteen; when the additive quantity the integer one [is added to it], it becomes nineteen whose denominator is three: $\begin{array}{|c|c|} 41 & 19 \\ 4 & 3 \end{array}$. In this way, the other [quantity which is] termed "divisor" has been simplified (*savarṇita*);[516] afterwards, [the step] **having performed the interchange**, which is the transposition[517] **of the numerator and the denominator of** this **divisor** [should be carried out]:that is to say, having brought[518] nineteen underneath [and] having moved the three above:[519] $\begin{array}{|c|c|} 41 & 3 \\ 4 & 19 \end{array}$. Next, for the reason that the reduction is not possible,[520] [in this case] the very numbers stay as they are (*yathāsthitā*). Now, as before,[521] [the step is] **in the multiplication of the numerator**:[522] forty-one is multiplied by three, it becomes one hundred twenty-three: 123. In this case, the **multiplication of the denominators** [should be carried out]: there is the "multiplication" of the "denominator" four by nineteen, which is namely the "denominator"; it becomes seventy-six.[523] When [the result] **is divided**: when the number one hundred twenty-three [is divided] by it, the quotient is one, which is the integer, [and] forty-seven whose denominator is seventy-six: $\begin{array}{|c|} 1 \\ 47 \\ 76 \end{array}$.

The author enunciates the second example: **according to the procedure, [the quantity] eighty and one-half [is divided] by five less by one-third;** [its meaning] is clear. The presentation is: $\begin{array}{|c|c|} 80 & 5 \\ 1 & 01 \\ 2 & 3 \end{array}$. In this case, regarding the first quantity, since a fractional increase occurs, by means of [the procedure] "multiplied by the denominator"[524] and so forth, eighty multiplied by two is one hundred

515 *paratra*.

516 The SGT emphasises that this composite fraction, composed of an integer plus a fraction, is "simplified" (*savarṇita*) by means of the *bhāgānubandhajāti*.

517 *viparyaya*.

518 *nītvā*.

519 *triṃś copari kṛtvā*.

520 Line 27: *tato'pavartanāsaṃbhavād*] em.

521 It refers to the preceding rule, which is the multiplication of fractions (GT $\overline{40}$).

522 This is a quotation from the rule of multiplication of fractions (GT $\overline{40}$).

523 At this stage, the layout should be the following: $\begin{array}{|c|} 123 \\ 76 \end{array}$.

524 This is a quotation from the rule of the class of fractional increase (GT 57).

sixty, which added to one becomes one hundred sixty-one whose denominator is two: $\begin{array}{|r|}161\\2\end{array}$. In the other place [of the layout], because there is a fractional decrease, by means of [the procedure] "in the procedure of fractional decrease, once the integer is multiplied by the denominator"[525] and so forth, it is: five is multiplied by three, it becomes fifteen. When there is the subtraction of one, it is 14 whose denominator is three:[526] $\begin{array}{|r|}14\\3\end{array}$. When there is the exchange (*parivṛtti*) of the denominator[527] and the numerator of this [fraction] which is termed "divisor", it becomes above three and below fourteen: $\begin{array}{|r|}3\\14\end{array}$. Next is [the step] "in the multiplication of the numerators":[528] "in the multiplication" of one hundred sixty-one by three, it becomes four hundred eighty-three.[529] Afterwards, the **multiplication of the denominators** occurs: two multiplied by fourteen becomes twenty-eight.[530] In like manner, **when** the previous[531] number (i.e., four hundred eighty-three) **is divided** [by the denominator twenty-eight], the quotient is seventeen. Below it, when there is the reduction, when there is the division of seven by seven, it is one; below it, when there is the division of twenty-eight by seven, it is four: $\begin{array}{|r|}17\\1\\4\end{array}$.

The author enunciates the third example: **as well as one-half divided by one-sixth.** This is clear [in meaning], [the presentation is]:[532] $\begin{array}{|r|r|}1&1\\2&6\end{array}$. In this case, regarding the quantity that occurs next, which is one-sixth,[533] when there is the exchange of the denominator and the numerator of that denoted as "divisor", above is six [and] below is one. Next is [the step] "in the multiplication of the numerators":[534] one is multiplied by six, it becomes six. Regarding this [expression] **the multiplication of the denominators**, two is multiplied by one,

525 This is a quotation from the rule of the class of fractional decrease (GT 60).

526 In line 8, it is unusual to find the integer 14 without the preceding verbal numerical expression.

527 In lines 8, 14, and 19 one would expect *aṃśahārayoḥ* instead of *harāṃśayoḥ*; in this regard, cf. the *sūtra* and line 25.

528 This is a quotation from the rule of multiplication of fractions (GT $\overline{40}$).

529 Line 10: *catuḥśatī*] em.

530 At this stage, the layout should be the following: $\begin{array}{|r|}483\\28\end{array}$.

531 *prāktana*.

532 In line 13, before the layout one would expect an introductory term such as *nyāsa* or *yathā*.

533 Lit. "a sixth part of one".

534 This is a quotation from the rule of multiplication of fractions (GT $\overline{40}$).

it is the very two.[535] When there is the division [of the resulting numerator six] by it, the quotient is the integer[536] three: $\left|\begin{array}{c}3\\1\end{array}\right|$.

The author enunciates the fourth example: **one-fourth divided by three.**[537] The explanation:[538] **one-fourth** is (lit). "a fourth[539] part of one",[540] which is "divided" by the integer three, **what is [the result]?** The presentation is: $\left|\begin{array}{c|c}1&3\\4&1\end{array}\right|$.

When there is the exchange of the denominator and the numerator **of the divisor,** which is here the quantity that occurs next, such as three and one, it is one above [and] three underneath. Next is [the step] **in the multiplication of the numerators**: one multiplied by three is the very one. There is then the **multiplication of the denominators:** four is multiplied by three, it becomes twelve. This is the very result, one-twelfth:[541] $\left|\begin{array}{c}1\\12\end{array}\right|$.

In this way, the procedure of division of fractions is concluded.

Square of fractions (bhinnavarga)[542] *[GT $\overline{44}$]*

Next is the procedural rule of square of fractions; the author enunciates half a metrical stanza:

> $\overline{44}$. The square[543] of the numerator is divided by the square[544] of the number of the denominator. It is [so] performed by the experts in squaring fractions.

The explanation: in the very both places,[545] by means of the fractional increase and so forth,[546] **the square,** which consists of the "product of two equal

535 I have emended the text since *dvāv eva* (line 15) should occur before the *daṇḍa* as it is the result of the step *ekaguṇau dvau.*

536 The use of *rūpa* points out that the result is a whole number.

537 Line 17: **bhavati**] em.

538 It is unusual to find *vyākhyā* between *daṇḍas*; moreover, the term *caraṇa* ("foot") has already been elucidated by the commentator while showing the execution of the previous sample problem (see SGT 41, 19).

539 *caturtha.* In the SGT, similar constructions with ordinals occur (see, for instance, SGT 45, 16 and 19).

540 Line 18: *rūpasya*] em.; *rūpaṃ* (K).

541 Lit. "a twelfth part of one".

542 K 22, 24.

543 *varga.*

544 *kṛti.*

545 *Sthānadvaye 'pi* – it occurs in lines 25, 26, 6, and 13. It alludes to the two sides of the layout where the two equal fractions are placed; this is shown by the commentator in the following sample problem.

546 The commentator refers to the numerical data found in the following sample problem where the *bhāgajāti* and the *bhāgānubandhajāti* are to be performed.

quantities",[547] which are [two] same numbers, of the "numerators", which are the numbers above, [is performed in such a manner] **by the experts.** The [expression] **the number of the denominator:** the denominator, in both places again, it is the number underneath; its "square" consists of the "product of two equal quantities". **[The square of the numerator] is divided** (*vihṛta*>*dattabhāga*) by it. **It is [so] performed [by the experts] in squaring fractions**, that is to say, in order to arrive at the square of fractions. Thus is the connection [of words with their meanings].

[In this regard] in the metrical stanza [which follows] the author enunciates four examples:

> 45. O learned one! If you know arithmetic, tell [me] quickly the square of five less by one-fourth, eight plus one-half, as well as [the square of] the quantities one-third [and] one-half.

The explanation: [the meaning of the expressions formulated in the metrical stanza] is clear. [Regarding the expression] **less by one-fourth:** by the first quarter [of the metrical stanza] the first example occurs; the presentation regarding the [examples enunciated in the] metrical stanza is [given] one by one.[548] Henceforward, [since] "the square" by means of a pair of equal numbers should be [arrived at], a double presentation [of the given fraction] concerning the metrical stanza is shown. It thus is:

5	5
01	01
4	4

. In this case, as there is the class of fractional decrease, in the very both places, five is[549] multiplied by four; when there is subtraction[550] of one, in both places it becomes the numerator nineteen whose denominator is four:[551]

19	19
4	4

. Its **square**, because of the multiplication of nineteen by nineteen, it is three hundred sixty-one. Next, **the number of the denominator** consists of four, its **square**,[552] because of the multiplication by the same [number], that it is to say, by a second four, it is such as sixteen. This is the meaning: when there is the division[553]of three

547 This paraphrases TŚ 11, it has been mentioned by the SGT while commenting upon the rule of square of integers. Cf. *kṛtiḥ sadṛśadvirāśighātalakṣaṇā* (line 25), *vargaḥ sadṛśadvirāśighātarūpas* (lines 26–27), and *vargaṃ sadṛśāṅkadvighātalakṣaṇaṃ* (SGT 23, 28).

548 In the introductory line, the commentator has emphasised that this metrical stanza concerns four mathematical exercises.

549 Line 6: *pañca catur°*] em.; *catur°* (k).

550 Line 7: *apanayane*] em.; **sa**] em.

551 I have emended the text and placed the layout found in line 6 after the *yathā* occurring in line 7.

552 Line 9: *vargaḥ*] em.; **tena**] em.

553 Having obtained an improper fraction, the commentator transforms it in a mixed number.

hundred sixty-one by this sixteen, the quotient is twenty-two [and] nine [plus] one-sixteenth: $\begin{array}{|c|}22\\9\\16\end{array}$.

The author enunciates the second example: **eight plus one-half.** The presentation is: $\begin{array}{|c|c|}8&8\\1&1\\2&2\end{array}$. Because there is a fractional increase, by means of [the procedure] "multiplied by the denominator"[554] and so forth, in the very both places, eight is multiplied by two, [when the additive quantity one is added][555] it becomes the numerator seventeen. Its **square**, because of the multiplication of seventeen by seventeen, becomes two hundred eighty-nine. **The number of the denominator** is two; its "square": because of being [two] multiplied by two, it consists of four; [the square of the numerator is] **divided** by it. That is the principle. Two hundred eighty-nine [divided] by four [is seventy-two][556] and there is one-fourth:[557] $\begin{array}{|c|}72\\1\\4\end{array}$.

In the fourth [quarter], by the fourth part of the metrical stanza the author enunciates the third [sample problem]:[558] **[tell me the square] of the quantities one-third [and] one-half.** The presentation of the third example is: $\begin{array}{|c|c|}1&1\\3&3\end{array}$. In this case, the class of simple fraction [should be performed]. [The first step is] **the square of the numerator:** [one] multiplied by one is the very one. **The number of the denominator** consists of three,[559] its "square", since it is multiplied by three, is nine. When there is the division by it, the quotient is a ninth part of one: $\begin{array}{|c|}1\\9\end{array}$.

Now there is the presentation of the fourth example: $\begin{array}{|c|c|}1&1\\2&2\end{array}$. Also in this case, because the class of simple fractions occurs, **the square of the numerator** is the very one. [Next is] **the number of the denominator: the number of the denominator** is such as two; its "square", since it is multiplied by two, is four.

554 This is a quotation from the rule of the class of fractional increase (GT 57).
555 Line 14: one should supply *rūpakṣepe* (see SGT 43, 23).
556 In line 16, after °*śatī*, the result seventy-two is missing; one would thus expect *dvisaptatis*.
557 Line 16: *rūpasya*] em.; *tasya* (K). Lit. "a fourth part of one".
558 Line 17: *tṛtīyaṃ*] em.
559 Line 18: **varga**] em.

Having divided [the numerator] by this [four], regarding the "square" of [this] fraction, the quotient is one-fourth (lit. a fourth part of one): $\begin{vmatrix}1\\4\end{vmatrix}$.

In this way, [the procedure of] the square of fractions is concluded.

Square-root of fractions* (bhinnavargamūla)[560] *[GT $\overline{46}$]

[Next is] the procedural rule of square-root of fractions; the author enunciates half a metrical stanza:

> $\overline{46}$. They explain that when the square-root (*vargamūla*) of the numerator is divided by the square-root of the denominator, the square-root (*kṛtimūla*) of the fraction occurs.

The explanation: [in a given fraction] **the denominator** (*chit>chedana*) is the number [standing] underneath.[561] Its "square" is the multiplication of two equal quantities;[562] of this [square] there is that [termed] **root** (*mūla>bīja*).[563] **[The square-root of the numerator] is divided** (*hṛta>vibhājita*) by it. **When the square-root of the numerator,** that is, the "square-root" of the number above, [is divided by the square-root of the denominator] **the square-root of the fraction** (*vibhinnaṃ kṛtimūlaṃ>bhinnavargamūlam*) occurs;[564] [in the metrical stanza above] supply [the subject] "the learned" [to the verb] **explain.**

In this respect, a sample problem occurs; [the author enunciates] one verse:

> 47. O clever one, if you know the operations with fractions,[565] tell [me] quickly the roots of the squares previously obtained.[566]

The explanation: **previously,**[567] that is, in the *sūtra* of square of fractions, the ascertained (*prāpta*) [results] of the "obtained squares" of [the quantities] five "less by one-fourth" and so forth are those "squares" (*kṛti>varga*) beginning with twenty-two [and] nine whose denominator is sixteen. Their **roots** are [the

560 K 23, 25.

561 The use of *aṅkarāśi* (lit. a "heap of digits") emphasises that the denominator 16 (which is the first denominator occurring in the next sample problem) is a number, thus a mathematical unit composed of digits.

562 Cf. *samadvighāta* from L 19 and *sadṛśadvirāśighāta* from TŚ 11.

563 The term *bīja* is lit. "seed, primary element".

564 The commentator shows that *vibhinna* is used as a predicative adjective ("fractional, relating to the fraction") in Śrīpati's rule; the SGT reformulates this last part of the sentence by means of a *tatpuruṣa* compound where *bhinna* is a substantive. The term *bhinna* occurs again as an adjective in the second line of Śrīpati's next rule and in GT $\overline{48}$ too.

565 *bhinnāni parīkarmaṇi* – lit. "the operations which are fractional", relating to fractions.

566 Line 2: *prāk°*] em. As in the rules of square-root and cube-root of integers, here too Śrīpati asks the student to determine the roots of the squares previously obtained (i.e., in GT 45). See the same pattern in GT 27.

567 Line 4: *prāg°*] em.

quantities] beginning with five less by one-fourth. [Regarding] the expression **tell [me]**: the inflection marker [of this verb] implies the student [as referent].[568] Thus is the connection [of words with their meanings], the rest is clear [in meaning].

On this topic, the first example is: $\begin{vmatrix} 22 \\ 9 \\ 16 \end{vmatrix}$. Since there is a fractional increase, according to [the rule] "multiplied by the denominator"[569] and so forth, twenty-two is multiplied by sixteen, it becomes three hundred fifty-two which added to that [numerator] consisting of nine, it is three hundred sixty-one whose denominator is sixteen. In accordance with what has been taught in the *sūtra*, the "square-root of the numerator" of this simplified square of the numerator is [as follows]:[570] in order to arrive at the "root", according to the procedure[571] previously [explained] "odd and even"[572] and so forth, twenty-nine is the integer thus obtained. Regarding this [number], by the half of that which has been multiplied by two[573] it becomes nineteen, which is the "square-root of the numerator". Next is this [expression] **the square-root of the denominator**: in this case, **the denominator** (*chit>cheda*) is sixteen; its "root" is four, **when [the square-root of the numerator is] divided** by it, [the result is arrived at]. In other words, when there is the division of nineteen by four, the quotient is four, and the remaining number is three whose denominator is four; the quotient four is above, three is below, [and] four is underneath: $\begin{vmatrix} 4 \\ 3 \\ 4 \end{vmatrix}$. This [result] is the "square-root" of the fraction five less by one-fourth.[574] Therefore, [the result] the integer four plus three-fourths is arrived at. Regarding this

568 This grammatical remark appears irrelevant, it neither clears up an ambiguity nor adds any shade of meaning.

569 This is a quotation from the rule of the class of fractional increase (GT 57).

570 *savarṇitāṃśa* or the "simplified fraction" underlines that 361 is the result of the simplification (*savarṇa/savarṇana*), by means of the *bhāgānubandhajāti*, of the compound fraction $22+\frac{9}{16}$. In line 17, see the use of *sarvarṇita* referring to the same fraction. In the previous sample problem, the improper fraction $\frac{361}{16}$, which is the obtained square of the initial quantity $5-\frac{1}{4}$, has been reduced to the compound fraction $22+\frac{9}{16}$. Śrīpati asks the student to ascertain the square-root of this obtained number, now the inversed process is thus involved.

571 *prakriyā*.

572 The commentator quotes his own passage on square-root of integers (SGT 26, 29). The expression *samaviṣamety* is often found throughout the SGT whenever the commentator intends to denote the procedure of square-root of integers.

573 See the rule of square-root of integers.

574 The result $4+\frac{3}{4}$ is equal to the *mūla*-fraction $5-\frac{1}{4}$ and it is also the *vargamūla* of its *varga*-quantity $22+\frac{9}{16}$ which is equal to the improper fraction $\frac{361}{16}$.

commercial practice,[575] [the quantity] five less by one-fourth is said [to be the result]. It is always in this way.

Here a second method, which has not been explained [by the author],[576] is shown as well.

To illustrate: by means of the method of [the class of] fractional increase, from the quantity twenty-two [and] nine whose denominator is six, regarding the [obtained] simplified [fraction], namely three hundred sixty-one whose denominator is sixteen,[577] which precisely is the square of the fraction [nineteen whose denominator is four], [the step] **when** it is **divided by the square-root** [here] denotes that consisting of five less by one-fourth:[578] $\begin{array}{|c|} 5 \\ 01 \\ 4 \end{array}$. By means of the method of fractional decrease, it becomes nineteen whose denominator is four: $\begin{array}{|c|} 19 \\ 4 \end{array}$.

Then, according to [the rule] "having performed the interchange of the numerator"[579] and so forth, **having performed** the exchange [of the two operands], it is: $\begin{array}{|c|} 4 \\ 19 \end{array}$. Afterwards, on the one hand[580] there is three hundred sixty-one whose denominator is sixteen, on the other hand there is [four][581] whose denominator is nineteen:[582] $\begin{array}{|c|c|} 361 & 4 \\ 16 & 19 \end{array}$. [One should carry out] **the reduction**:[583] since it is possible,[584] the reduction by the number four occurs, so that when there is the division of four by four, it becomes one. When there is the division of

575 The use of *vyavahāra* is peculiar since neither in the previous nor in this sample problem monetary units are explicitly mentioned.

576 The commentator provides himself a method regarding the square-root of fractions; it represents a kind of verification.

577 Line 17: **jātam**] em.

578 The idea is that one should divide the square of the initial quantity $5-\frac{1}{4}$ (which is $\frac{361}{16}$) by the root of the square, which is the initial quantity itself. Thus, one should divide $\frac{361}{16}$ by $\frac{19}{4}$.

579 The commentator quotes a passage from the division of fractions (GT 42).

580 See the use of *ekatra* and its correlative *paratra* to denote the two sides of the layouts where the two fractions occur.

581 Line 21: *catvāra ekona°*] em.; *ekona* (K).

582 I have emended the layout; in the edition (line 21), the side where the quantity-dividend should be placed is missing. Moreover, the commentator has just observed that there is the exchange of the two operands but the edition shows the fraction as it was before the exchange.

583 The commentator quotes a passage from the rule of division of fractions (GT 42).

584 In line 21, see the expression *apavṛttisahatvāt*. In a similar context, in SGT 21, 29 one finds the verb *sahate* to denote the same idea, while SGT 22, 10 uses the abstract form of *yogya*.

sixteen by four, it becomes four. When there is the division of three hundred sixty-one [by nineteen],[585] it becomes nineteen; in the same way, when there is the division of nineteen by nineteen, it becomes one:[586] $\begin{array}{|c|c|} 19 & 1 \\ 4 & 1 \end{array}$. Next is **the procedure of multiplication:**[587] by means of this reasoning, nineteen and four are multiplied by one; being multiplied, the multiplier is erased. According to this method,[588] the two [numbers] one are erased. When the division of nineteen by four is performed, the quotient is four and the remainder is three whose denominator is four: $\begin{array}{|c|} 4 \\ 3 \\ 4 \end{array}$.

In this manner, as for the square [of integers], also when the square of a fraction is divided by the root of the square of the [same] fraction, the square-root of the fraction is arrived at.[589] This is well-known: [when][590] that number by which it (i.e., the given number) is multiplied, [the result then] divided by this very one, that [previous number] is then obtained. It has thus been established; it should always be understood in this way.[591]

Now there is the presentation of the second example: $\begin{array}{|c|} 72 \\ 1 \\ 4 \end{array}$. Since a fractional increase occurs, according to [the rule] "multiplied by the denominator"[592] and so forth, seventy-two is multiplied by four becoming [the numerator] two hundred eighty-eight, [this] added to one is two hundred eighty-nine. Regarding this [number], by means of the procedure "odd and even"[593] and so forth, according to the method **in the half of the doubled** [number],[594] the previous result twenty-seven becomes seventeen. **When the square-root of the numerator** [is carried out], it is seventeen; [afterwards] the "square-root of the denominator" [should be performed]. The "square of the denominator" is four, its "root" is two; **when** [the square-root of the numerator is] **divided** by it [the result is arrived at]. In brief, when there is the division of seventeen by

585 Line 23: *ekonaviṃśatyā*] em.; *śati* (?) *tame* (K).
586 Lines 23–24: *ekonaviṃśater ekonaviṃśatyā bhāge*] em.
587 The commentator quotes this passage from the rule of division of fractions (GT 42).
588 *nyāya*.
589 In brief: $6^2 = 36$; $36 \div 6 = 6$.
590 In line 28, there is a syntactical inconsistency as one would expect the correlative *yadā* to formulate the construction: [*yadā*] *yena…tadā sa*.
591 The procedure given by the commentator at the end of the procedure of cube-root of integers is based on a similar principle; see SGT 32–33, 2–10.
592 This is a quotation from the rule of the class of fractional increase (GT 57).
593 The commentator quotes his own passage on square-root of integers (SGT 26).
594 The SGT quotes the final passage from the rule of square-root of integers (GT 26).

two, the quotient is eight, 8; the remainder is one whose denominator is two. The obtained [numbers] are joined, eight is above: $\begin{array}{|c|}8\\1\\2\end{array}$.

The presentation of the third example: $\begin{array}{|c|}1\\9\end{array}$. **When the square-root of the numerator,** which is one, [is calculated], it stands as it is, **the square-root of the denominator** [should be next performed]: the "square of the denominator" is nine, its "root" is three. **When [the square-root of the numerator is] divided** by this [three], the result is one third:[595] $\begin{array}{|c|}1\\3\end{array}$.

The presentation of the fourth example: $\begin{array}{|c|}1\\4\end{array}$. Here too, [first is the step] **when the square-root of the numerator:** as before, the very one is the "square-root" [of the numerator]. Next is **the square-root of the denominator,** that is, the "square of the denominator" is four, its "root" is two. **When [the square-root of the numerator is] divided** by this [two], [in this case] that very one is the quotient: $\begin{array}{|c|}1\\2\end{array}$.

The [procedure of the] square-root of fractions is thus [concluded].

Cube of fractions (bhinnaghana)[596] *[GT $\overline{48}$]*

Next is the procedural rule of cube of fractions; the author enunciates half a metrical stanza:

> $\overline{48}$. The mathematicians[597] explain that when the cube of the numerator is divided by the cube of the denominator, the cube of the fraction [is arrived at].[598]

The explanation: the "denominator" is the number underneath;[599] [the cube of the numerator is divided] **by** its **cube**, which consists of the [mutual] multiplication of a collection of three equal numbers.[600] The "numerator" is the number above, **when** its **cube,** which consists of the [mutual] multiplication of a collection of three equal [numbers], **is divided,** that is to say, when the division is carried out,[601] **the mathematicians,** which are those practising mathematics,

595 Lit. "a third part of one".
596 K 25, 13.
597 *gāṇitikā.*
598 *bhinnaṃ ghanaṃ* – lit. the "fractional cube".
599 *adhastana.*
600 Cf. *samatrirāśi* (GT 28–29).
601 Line 15: *datte bhāge*] em.; *daśabhāge* (K).

explain that **the cube of the fraction** [is arrived at]. Thus is the connection [of words with their meanings].

In this respect, in the exemplifying metrical stanza [below] the author enunciates one [example] depending on the class of fractional increase, a second [example] depending on the class of fractional decrease, as well as two examples depending on the class of simple fractions which is relevant to the previous two [classes]:

> 49. O intelligent one, if you are proficient in arithmetic, tell [me] the cube of nine plus one-fourth, six lessened by one-third, [as well as the cube of] one-sixth and one-third.

This [stanza] is clear [in meaning].[602] The presentation of the first example is: $\begin{array}{|c|}9\\1\\4\end{array}$. In this case, since there is a fractional increase, according to [the procedure] "multiplied by the denominator"[603] and so forth, when it is simplified, it becomes thirty-seven whose denominator is four. Write[604] this number in three places (*sthānatraye*): $\begin{array}{|c|}37\\4\end{array}\begin{array}{|c|}37\\4\end{array}\begin{array}{|c|}37\\4\end{array}$. When there is the reciprocal multiplication, thirty-seven is multiplied by thirty-seven becoming thirteen hundred sixty-nine, 1369; this is again multiplied by thirty-seven becoming fifty thousand six hundred fifty-three: 50653. This number is the "cube of the numerator". [The step] **when the cube of the numerator** [has been performed]; [next is the step] [it is **divided**] **by the cube of the denominator**: the "denominator" is four, its "cube" is meant. For instance, when there is the reciprocal multiplication of three times four,[605] it becomes sixty-four. [The step] **when** it is **divided** by it[606] [should be performed]: in brief, with respect to the number of the "cube" of the previous "denominator",[607] which is sixty-four, when fifty thousand and so forth [is divided by it], when there is the division **by the cube** of the denominator, the quotient is seven hundred ninety-one, [the remainder is] twenty-nine whose denominator is sixty-four: $\begin{array}{|c|}791\\29\\64\end{array}$.

602 In brief, there is no need to clarify technical terms.

603 This is a quotation from the rule of the class of fractional increase (GT 57).

604 *vilikhya* – lit. "having written", as it often occurs in SGT, here too this absolutive can be understood within the scope of the imperatival modality of finite verbs.

605 Lit. a "triple four".

606 Line 29: *anena vibhakte *cchāyā* | ayam arthaḥ pañcāśat sahasrādiṣu*] conj.

607 *aṃśa*.

The author enunciates the second example: **six [lessened by] one-third** and so forth. The presentation is: $\begin{array}{|r|}6\\01\\3\end{array}$. Since there is a fractional decrease, by means of [the procedure] “multiplied by the denominator”[608] and so forth, six is multiplied by three, it becomes eighteen; when there is the subtraction of one, it becomes seventeen whose denominator is three: $\begin{array}{|r|}17\\3\end{array}$. Then, for the sake of [determining] the “cube of the numerator”, the procedure beginning with “the cube of the last [digit] should be placed”[609] [should be applied]. To illustrate, by the numerical method [following the regular order] one is the “last” [digit] of seventeen; its “cube”[610] is the very one, which should be placed [below by being one notational place more in respect to the number above]:[611] $\begin{array}{|r|}17\\1\end{array}$. [Having placed] the “square” of this [number] one elsewhere,[612] this very one “multiplied by three”[613] becomes three. Its “preceding”[614] is seven, three is multiplied by it becoming twenty-one; then [this should be placed by] “being one [notational] place more”[615] in respect to the [notational] place of the root:[616] $\begin{array}{|r|}17\\11\\2\end{array}$. “The square” (*kṛti*> *varga*) “of the preceding [digit]”[617] seven [should be written] in another place. This [resulting] forty-nine multiplied[618] “by the last” (*antyena*) [digit], which is [the number] one, remains as it is[619] and multiplied “by three” becomes one hundred forty-seven. Then, it should be placed “being one [notational] place more” in respect to the [notational]

608 This is a quotation from the rule of the class of fractional decrease (GT 60).

609 This is a quotation from the rule of cube of integers (GT 28). From now onwards, the commentator repeatedly quotes mathematical steps from this same rule.

610 While commenting upon GT 28, Siṃhatilakasūri has explained that, in order to calculate the cube of a number, the digits are to be considered from a right to left direction.

611 I have emended the text (line 6); in the edition the layouts occurring in lines 6, 8, 11, and 12 are mistaken. It is curious that that the SGT quotes the step *sthānādhika* (in GT 28 it occurs, grammatically speaking, in the abstract form) only in the next two lines and not already here as the commentator, while explaining GT 28, observes that the first step is to place the cube of the last digit by one notational place more in respect to the given number.

612 While commenting upon GT 28, the commentator often uses the expression *anyatra sthāpayitvā* to denote that partial results are not placed in the layout but are instead written elsewhere (see SGT 28, 14).

613 See *trighna* in GT 28, which is the rule of cube of integers.

614 *ādi*.

615 This is a quotation from the rule of cube of integers (GT 28).

616 *mūla* – it denotes the given number 17.

617 In GT 28, see the step *kṛtir ādimasya*.

618 *hata*.

619 Line 9: *tadavasthāś*] em.

place of the root:

17
117
24
1

. "The cube of the preceding", which is seven, is three hundred forty-three; this should be placed "being one [notational] place more" in respect to the [notational] place of the root:

17
1173
244
13

. When there is the union of all these [digits], it becomes four thousand nine hundred thirteen. Regarding that [expression] **when the cube of the numerator,** this number is the "cube of the numerator". **[It is divided] by the cube**[620] **of the denominator:** that [mentioned] "cube" **of the denominator** (*chit>cheda*), namely [the number] three, is twenty-seven; when [the cube of the numerator is] divided[621] by this [the result is arrived at]. In brief, when there is the division of the number beginning with four thousand, which is the "cube of the numerator", by twenty-seven, which is the "cube" of the denominator, the quotient is one hundred eighty-one. The number which is the remainder is twenty-six whose denominator is twenty-seven; since the quotient has been placed above,[622] it thus is:

181
26
27

.

The author enunciates the third example: [tell me the cube of] **one-sixth.** The presentation is:

1
6

. In this case, [first is the step] **when the cube of the numerator:** it is the very [number] one. Regarding that **when the cube of the numerator,** namely one, **[is divided] by the cube of the denominator,** it is: the "cube" of the denominator, which is six, because of the [mutual] multiplication of a collection of three equal [numbers], consists of two hundred sixteen. When [the cube of the numerator is] divided by this: because of the absence of that which is to be divided[623] (*bhājyābhāvāt*), this very quotient is the numerator one whose denominator is two hundred sixteen:

1
216

.

620 Line 13: *ghaneneti*] em. This is an example of haplography as, in *ghanene* (from *ghanena iti*), the scribe may have accidentally omitted to repeat the phoneme *ne*.

621 In lines 14 and 19, one would expect *vibhakte* rather than *hṛte*. Note that this latter occurs in the previous *sūtra,* which is dedicated to the square-root of fractions; this may be a case of scribal harmonization to parallels. The same anomaly is found in line 14, while in the previous line 29 and in the next line 23 there is *vibhakte*.

622 Line 16: one would expect *labdhasyopari sthāpyamānatvāt* to precede *yathā*. This type of remark is found again in SGT 51, 26 (differently constructed) and in the following page in line 23.

623 In line 19, the commentator emphasises that this is a proper fraction. Because the dividend (see *bhājya*) is the number one, the division by the denominator is not carried out. The same expression occurs, for instance, in SGT 51, 28 and SGT 61, 4.

The author enunciates the fourth example: **and** [tell me the cube of] **one-third.** The presentation is: $\begin{array}{|c|}1\\3\end{array}$. In this case, [the numerator is] again that consisting of one, thus **when the** very **cube of the numerator [is divided] by the cube of the denominator** [here] means **[divided] by the cube** of three, which is the "denominator". Because the [mutual] multiplication of a collection of three equal [numbers] occurs [underneath], **when** it is **divided** by that consisting of twenty-seven, because of the nature of that which is above (*uparibhāvād*) (i.e., the numerator one), this very quotient is one whose denominator is twenty-seven: $\begin{array}{|r|}1\\27\end{array}$.

In this way, the [procedure of the] cube of fractions [is concluded].

Cube-root of fractions (bhinnaghanamūla)[624] *[GT $\overline{50}$]*

Next is the procedural rule of cube-root of fractions; the author enunciates half a metrical stanza:

> $\overline{50}$. Those who are familiar with mathematical procedures state that the cube-root[625] [of a fraction] occurs when the cube-root[626] of the numerator[627] is divided by the [cube] root of the denominator.

The explanation: among those[628] consisting of the previously[629] obtained cubes of the fractions beginning with seven hundred ninety-one[630] [and] [one hundred eighty-one][631] twenty-six whose denominator is twenty-seven,[632] the numbers above are the "cubes of the numerators". That [word] "root" concerns their ["root"], that is, by means of [the procedure of] the fractional increase and so forth, [the "root"] of the simplified [fractions], for instance seven hundred ninety-one,[633] according to [the rule] "a cubic place, a couple of not-cubic places"[634] and so forth, that [aforementioned] "quotient"[635] is

624 K 26, 27.
625 *ghanapada.*
626 *ghanamūla.*
627 *lava.*
628 *madhye ye*. The commentator mentions the cubes obtained in the previous sample problem (GT 49).
629 In line 28, *prāg* should be compounded with *labdha*°.
630 Line 28: °*śatīprabhṛtīnāṃ*] conj.
631 Line 28: the number one hundred eighty-one, which is the integer obtained, is missing: before *ṣaḍ*°, it should read *ekāśītyadhi(ka)śatam* (see line 15).
632 Line 29: **rūpa**] em.
633 The SGT refers to the fractions obtained while solving the previous sample problem.
634 Line 1: *ghano' ghanadvandvety ādinā*] em. See line 13, where the same quoted passage from GT 32–33 occurs.
635 See *labdha* in GT 32–33 and SGT 32–33, 8 and 23.

the "root" number. There [in the *sūtra* it is said], **those who are familiar with mathematical procedures state that the cube root** (*ghanapada>ghanamūla*) of a fraction occurs **when the cube-root of the numerator**[636] **is divided by the root of the denominator:** in short, the "denominator" is the number underneath, namely sixty-four, its "root" (*pada>mūla*) is four, [therefore] it is divided by this [four].

In this respect, [a sample problem occurs]; [the author enunciates] an exemplifying verse:

> 51. O friend, if you have a good understanding of the execution (*vidhāna*) of the cube-root, tell me the roots of the quantities which are the previous[637] cubes.

[The explanation]: **tell me the roots** (*pada>mūla*), for instance nine plus one-fourth, **of the quantities which are the previous cubes,** that is to say, of the previously mentioned[638] numbers beginning with seven hundred ninety-one. Thus is the connection [of words with their meanings]; the rest [of the verse] is clear [in meaning].

[I shall explain] the first example. The presentation:

791
29
64

. In this case, since there is a fractional increase, by means of [the procedure] "multiplied by the denominator"[639] and so forth, the number above multiplied by sixty-four becomes fifty thousand six hundred twenty-four. When the additive quantity twenty-nine [is added to it], it becomes [that number] ending with fifty-three: 50653. Regarding this [number], in order to arrive at the "root" of the simplified "cube of the numerator", according to [the rule] "one cubic [place] [and] a couple of non-cubic [places]",[640] the "cube-root" of the last [cubic place] is [placed] below the zero belonging to 50. Then, having brought that[641] [cube-root] below, one should subtract the cube [and] the cube of the number above (i.e., 50) is erased. To illustrate, [the given number is] 50653: the "cube" of three[642] is twenty-seven; when it is subtracted from fifty, the remainder twenty-three is placed [instead of 50]. Afterwards, three, which is denoted as "root",[643] "should be placed"[644] below the five belonging to fifty-three:

2	3	6	5	3
			3	

.

636 Line 2: *lavaghanamūle*] em.
637 *prācīna*.
638 Line 8: *prāguktānām*] em.
639 This is a quotation from the rule of the class of fractional increase (GT 57).
640 This is a quotation from the rule of cube-root of integers (GT 32–33).
641 Line 14: *tat*] em; *tam* (K). The suggested *tat* in *tat adhaḥ kṛtvā* refers to the *ghanapadam* mentioned in the same line before the *daṇḍa*.
642 This is the highest possible cube of the last cubic place 50.
643 Cf. *mūlam* in SGT 32–33, 29.
644 From now onwards, the SGT quotes steps of the first procedure of cube of integers (GT 32).

Afterwards, its "square" is nine which "multiplied by three" becomes twenty-seven; from the three, [the obtained twenty-seven should be placed by] by being one notational place less.[645] "One should divide" the number which is the remainder [above] by the previously written [seventeen]: $\begin{vmatrix}23653\\273\end{vmatrix}$. When there is the division by the twenty-seven below, [when the remainder of this division is multiplied] by seven,[646] one hundred eight-nine is arrived at [and] [when there is the subtraction] the remainder forty-seven[647] is placed. Because of the object explained, twenty-seven is removed: $\begin{vmatrix}4753\\3\end{vmatrix}$. "Having put the quotient" seven "in the line" in front of the three, it is: $\begin{vmatrix}4753\\37\end{vmatrix}$. Then, "its square" and so forth[648] means the "square" of this seven; the square is namely forty-nine, having multiplied (*hatvā*) it by the "last" [digit] three [and again by three],[649] it becomes three hundred forty-three: $\begin{vmatrix}343\\37\end{vmatrix}$. "And one should subtract the cube"[650] denotes the "cube" of the preceding seven, which is three hundred forty-three.[651] The quotient is thirty-seven. This is the "cube-root of the numerator" of the simplified [quantity] beginning with seven hundred ninety-one.[652] Therefore, [the step] **when the cube-root of the numerator** [has been accomplished]; [next] the text says **the root of the denominator:** the "denominator" is sixty-four, its root is four, [hence] it is **divided by** this [four]. In brief, when there is the division of thirty-seven by four, the quotient is nine,

645 Line 17: *ekasthānonatayā*] conj.; *tayā* (K). This conjecture is a diagnostic one: the wording of the original cannot be known, but this emendation conveys the sense that we expect. The transmitted *tayā* could represent a casa of scribal abbreviation or lacuna. Note that *ekasthānonatayā* is mentioned in SGT 32–33, 6.

646 This passage is not fully explained by the SGT. Above, the remainder 236 is divided by 27 and the chosen quotient is 7. Although the quotient of the division is 8, a lesser number must be chosen in order to continue the procedure.

647 I have emended the text; in line 19, the *śeṣa* occurring should be *saptacatvāriṃśat* ("forty-seven") and not the *pañcacatvāriṃśat* ("forty-five") found in the text. Interestingly, the layout gives the number 47. The number 47 is the remainder from the division of 236 (from the above 23653) by 27, which is the result of the square of the root 3 multiplied by 3. In fact, $236 - (37 \times 7) = 47$.

648 Line 20: **anya**] em. See *tatkṛtim antyanighnīṃ* in GT 32–33.

649 The quotient is squared, multiplied by the last digit (here the 3 next to the 7) and again by 3, but this last step is not mentioned. Cf. *tatkṛtim antyanighnīṃ trisaṅguṇāṃ* [*cāpanayed ghanaṃ ca*] in GT 32–33.

650 *cāpanayed ghanaṃ ca* is a quotation from GT 32–33.

651 Line 23:**trir gamayet**] em.

652 The passage from *etat* to *ghanamūlam* (lines 23–24) is corrupt; I have proposed the following conjectural emendation (note that the constitution and interpretation of this passage are tentative): *savarṇitaikādhi(ka)navatisaptaśatādīnām etal lavaghanamulam.*

9, the quantity-remainder[653] is one whose denominator is four,[654] and the quotient should be placed above. It should always be understood in this way, it thus is: $\begin{array}{|c|}9\\1\\4\end{array}$.

The second example [is explained]. The presentation: $\begin{array}{|c|}181\\26\\27\end{array}$. Since there is a fractional increase, by means of [the procedure] "multiplied by the denominator"[655] and so forth, one hundred eighty-one is multiplied by twenty-seven, it becomes four thousand eight hundred eighty-seven. Then, when the additive quantity twenty-six [is added to it], it becomes four thousand nine hundred thirteen. Below it, there is the denominator twenty-seven: $\begin{array}{|r|}4913\\27\end{array}$. Regarding the "cube of the numerator" of this [fraction], in order to arrive at the "root" such as six lessened by one-third,[656] [here] another method not explained in the *sūtra* is shown as well.

To illustrate: six subtracted by one-third should be arranged [in the layout]: $\begin{array}{|r|}6\\01\\3\end{array}$. This is the quantity-root of the cube of the fraction [mentioned above]. As there is a fractional decrease, according to [the rule] "multiplied by the denominator"[657] and so forth, six is multiplied by three, it becomes eighteen; when there is the subtraction (*apanayana*) of one, it becomes seventeen whose denominator is three: $\begin{array}{|r|}17\\3\end{array}$. Then, when there is the multiplication of seventeen by seventeen, it becomes the square two hundred eighty-nine, and three multiplied by three is nine: $\begin{array}{|r|}289\\9\end{array}$. Afterwards, by the division,[658]

653 Line 26: *śeṣāṅkaḥ*] em.; *śeṣe* (K).

654 The resulting root $\begin{array}{|r|}37\\4\end{array}$, which is the cube of the improper fraction $\begin{array}{|r|}50653\\64\end{array}$, is transformed into the compound fraction $\begin{array}{|c|}9\\1\\4\end{array}$. This is, in fact, the initial quantity about which, in the previous sample problem, students were asked to find out the cube.

655 This is a quotation from the rule of the class of fractional increase (GT 57).

656 See the previous sample problem.

657 This is a quotation from the rule of the class of fractional decrease (GT 60).

658 Line 5: *bhāgahāreṇa*] conj.

by means of [the procedure] "having performed the interchange"[659] and so forth, "having performed the interchange" of the quantity which is the divisor (*hara*>*bhāgadāyin*),[660] two hundred eighty-nine is underneath and nine is above: $\begin{array}{|c|}9\\289\end{array}$. Next, on the one side[661] [one should write down] four thousand nine hundred [and] thirteen whose denominator is seventeen: $\begin{array}{|r|}4913\\27\end{array}$. On the other side,[662] two hundred eighty-nine [is placed below], [and] the number nine is situated above:[663] $\begin{array}{|r|}9\\289\end{array}$. These four numbers meet the conditions for the reduction (*apavṛtti*), and the reduction of [each pair of] two numbers should be performed by means of a same single number.[664] When there is the reduction[665] of the number beginning with four thousand by two hundred eighty-nine, it becomes seventeen. In the same way, when there is the very reduction [of two hundred eighty-nine] by itself, which means by two hundred eighty-nine, it becomes one. Similarly, when there is the division of [the numerator] nine by nine, it is one. Next, when there is the reduction of [the denominator] twenty-seven by the numerator (*bhāga*) nine, it becomes three: $\begin{array}{|r|r|}17&1\\3&1\end{array}$. When there is the multiplication[666] by [the number] one, [both] the dividend (*bhājya*) and the denominator remain exactly so. Afterwards, when there is the division of seventeen by three, the quotient is five, the remainder two whose denominator is three: $\begin{array}{|c|}5\\2\\3\end{array}$. In this case, the "cross-reduction"[667] has been shown because of the reduction of [both] the numbers on the left (*vāma*) and on the right (*dakṣiṇa*). [The procedure begins again] by means of that not [yet] reduced [quantity], for instance: four thousand nine hundred thirteen whose denominator is twenty-seven.[668] Then, according to [the rule]

659 This is a quotation from the rule of division of fractions (GT 42).
660 The term *bhāgadāyin,* which denotes the "divisor", is part of the commentator's lexicon.
661 *ekatra.*
662 *ekatas,* one would expect the correlative *aparatra.*
663 The fractions $\begin{array}{|r|}4913\\27\end{array}$ and $\begin{array}{|r|}9\\289\end{array}$ are respectively the cube and the square of the root-fraction $\begin{array}{|r|}17\\3\end{array}$.
664 The commentator refers to the fact that all four numbers of the two fractions can be divided, pair by pair, by a common factor without giving a remainder.
665 Line 11: *apavartane*] em.
666 Line 13: *ekaguṇanāyāṃ*] em.
667 This is a quotation form the rule of multiplication of fractions (GT 40).
668 See line 8.

"having performed the interchange"[669] and so forth, nine is above, two hundred eighty-nine is underneath.[670] Next, the number beginning with four thousand is multiplied by nine; it becomes 44217, which is forty-four thousand two hundred seventeen.[671] The multiplier nine is erased. In the same way, two hundred eighty-nine is multiplied by twenty-seven; it becomes seven thousand eight hundred three: 7803. The multiplier twenty-seven is erased. When there is the division of that beginning with forty-four by that beginning with seven thousand, the quotient is five, the number which is the quantity-remainder is five thousand two hundred two: $\begin{array}{|c|}5202\\7803\end{array}$. Then, when there is the reduction of the number above by two thousand six hundred one,[672] it becomes two; when there is the reduction of the number below by this very [number] two thousand and so forth, 2602, it becomes three. Afterwards, regarding the quotient, it is [placed] above, the presentation is:[673] $\begin{array}{|c|}5\\2\\3\end{array}$.

By this [method], as for the cube (*ghanavad*)[674] [of integers], [here] by knowing the root of the cube of a fraction, when there is the division of the number of the cube of the fraction by its (i.e., of the root) square, the root of the cube of the fraction is arrived at. It has thus been established.[675]

Now there is the presentation of the third example: $\begin{array}{|c|}1\\216\end{array}$. In this case, the "cube of the numerator" is one. Regarding this [number], one is also the "root."[676] [The result of the step] **when the cube-root of the numerator** is this one, [next] the text says **the root of the denominator**: the "denominator" is two hundred sixteen, its

669 This is a quotation from the rule of division of fractions (GT 42). In the quotation occurring in line 16, the final syllable *na* of the noun *parivartana* is missing. Cf. lines 5 and 16.

670 This fraction $\begin{array}{|r|}4913\\27\end{array}$ is the divisor mentioned in line 6.

671 Line 18: **yathā**] em.

672 Line 22: *dvisahasraṣaṭśataikenāpavartane*] em.

673 I have emended the layout: in line 23, the edition reads $\begin{array}{|c|}2\\5\\3\end{array}$.

674 In lines 24–25 the commentator states that when the cube of the fraction is divided by the square of the root, the root is arrived at. Similar calculations are presented by the commentary in relation to the cube of integers and the square-root of fractions, see SGT 32–33, 2–10 and SGT 47, 16–28.

675 The wording deliberately echoes the preceding parallel statement: cf. *etena ghanavad* (SGT 51, 24) and *etena vargavad* (SGT 47, 27) and the final formula *iti sthitam* occurring in both places.

676 Line 27: **śunyapadavikāritād**] em.

"root" is six, **[the cube of the numerator is] divided by** it. Because of the absence of that which is to divided, that very one is the quotient: $\begin{vmatrix} 1 \\ 6 \end{vmatrix}$.

The presentation of the fourth example: $\begin{vmatrix} 1 \\ 27 \end{vmatrix}$. As before in this case too, [the result of the step] **when the cube-root of the numerator** is namely one, [next] the text says **the root of the denominator**:[677] the "denominator" is twenty-seven, its "root" is three, **[the cube of the numerator is] divided by** it.[678] As before, that very one is the result.

Having taken into consideration[679] that the operations with fractions are closely connected[680] with the [classes of fractions such as the] class of simple fractions, for the sake of their understanding the author has [chosen to] first expound these [operations with fractions] by explaining the addition of fractions and so forth.[681] Here (i.e., in this work), because of the earlier presentation of [the operations beginning with] addition [of integers], [next] the [operations with fractions beginning with the] addition of fractions have been enunciated [in the same order]. [In this treatise] everything is therefore well situated.[682]

3.6 Arithmetic of zero [GT 52]

The proper form of zero (śunyasvarūpa)[683]

Because it occurs together[684] with numbers, having expected a question[685] regarding the investigation of the proper form[686] of zero, the author

677 Line 29: *harapadeti*] em.

678 Line 30: *vibhakte*] em.; *bhakte* (K).

679 *iti kṛtvā*.

680 *avinābhāvin*.

681 I have emended the text, which is uncertain in this concluding passage. I suggest that, in line 30, there should be a *daṇḍa* between *labdhaṃ* and *bhāga*; it is, in fact, unlikely that the commentator's final remark was originally included in the explanation of the fourth example. Lines 30–2: [...] *labdham* | *bhinnaparikarmāṇi bhāgajātyādy avinābhāvīnīti kṛtvā tadvijñānārthaṃ pūrvaṃ tāny uktvā bhinnasaṅkalitādīny uktavān* | *atra tu saṅkalitādiprastāvāt bhinnasaṅkalitādīni uktvānīti sarvaṃ susthaṃ* |] conj.

682 Here the SGT analyses the structure and the order of presentation of the topics expounded in the root-text. The commentary emphasises that the understanding of the classes of simplification of fractions requires a prior knowledge of fractions, and it is for this reason that in the GT the classes of fractions are rightly explained after the operation with fractions. The commentator also underlines that the root-text explains the operations with fractions starting with addition in the same way as it has previously expounded the operations with integers

683 K 29, 5–8.

684 *sahacārin*.

685 *āśaṅkya*. In other words: "since students will certainly benefit from this knowledge, the author expounds the topic of ...".

686 *svarūpa*.

enunciates [the procedural rule concerning] the relationship[687] of zero [and numbers]:

> 52. When there is the sum of an additive quantity [and zero], zero[688] becomes the same [as the additive quantity]. When there is the deduction or the sum of zero [and a quantity], the quantity is unchanged (*avikārin*), and in the multiplication by zero [the result] is zero. Zero[689] occurs when [a quantity] is divided by zero (*vyoman*), and when zero is divided [*vyomnā* or "by zero"][690] is zero. When there is the square of zero, [the result] should thus be zero (*viyat*), and the cube [of zero] should be zero (*antarikṣa*).

The explanation: **when there is the sum** of [a number], for instance five,[691] which is **the additive quantity, [and zero], zero becomes the same** as the very five. It is in the same way **when there is the deduction** of zero, that is to say, when there is the deduction[692] of zero from [a number], for example ten, as well as **when there is the sum of zero [and a quantity],** that is to say, when the additive quantity zero is added.[693] Similarly [on the topic], in this explanation I will take into consideration (*āmṛśya*) the opinion[694] of the *Līlāvatī*.[695] [In the root-text], the "deduction" [of zero from a quantity] [is mentioned], [also zero as] the "quantity" which is the divisor,[696] [zero as the additive quantity is mentioned too] because [in the *sūtra*] the summation (*upacaya*) [of a quantity and zero] [meant] by the word *milana* ("sum") occurs, [and lastly] [zero as] its multiplier (i.e., of the given quantity).[697] Then, when zero is the multiplier and [subsequently] the divisor should [here] be [also mentioned as it is in the *Līlāvatī*].[698] In brief: the multiplication and division of a certain number [by zero] should[699]

687 *vyāpti*.
688 *śūnya*.
689 *gagana*, together with *kha*, *vyoman*, *antarikṣa* and *vyat*, mean "sky" and are technical terms commonly used in the *bhūtasaṃkhyā* notation to denote "zero".
690 That the missing term of the sentence is the instrumental *vyomnā* or "by zero" is clarified by Siṃhatilakasūri in line 18.
691 The commentator applies the technical terms used by the GT to the sample problem which he fully quotes from the L at the end of his explanation.
692 *ākarṣaṇa* – lit. "pulling, bending", it is a technical term for "deduction".
693 The SGT emphasises that in the first "sum" (*yoga*) mentioned in the rule, the additive quantity is the given number, in the second "sum" (*milana*) mentioned the additive quantity is zero. This statement presents the commutative principle of addition.
694 *abhiprāya*.
695 L 45–46 are dedicated to the arithmetic of zero.
696 Line 11: the term *bhāgadāyin* should be compounded in a *karmadhāraya* compound with *rāśi*.
697 The commentator underlines that the following four operations have been mentioned in both texts: $n-0$, $n\div0$, $n+0$, and $n\times0$. However, note that according to the GT: $n\div0=0$, while the L observes that it gives $\begin{array}{|c|}n\\0\end{array}$.
698 See here the construction of nominal abstract nouns with subjective genitive.
699 Line 13: *na*] em. The suggested emendation follows the commentary when it shows the layouts based on the formulae mentioned in the L which I believe are here implied. In this regard, in SGT ad L 47 see the layout in line 14.

have a zero [placed] above and below.[700] **The quantity** is **unchanged** in three cases.[701] Therefore, by the statement concerning the "deduction" and the "sum of zero", the [arithmetical operations of] subtraction and addition of zero are indicated.[702] Since in the addition [of zero and a quantity] and in the subtraction[703] of zero [from a quantity], the very "quantity" remains unchanged,[704] in the same way **in the multiplication by zero,** that is to say, **in the multiplication** (*ghāta>guṇana*) by "zero",[705] a number, for example five, becomes **zero.** In this manner, if that which occurs as the quantity (*aṅkavat*) multiplied by "zero" is zero, similarly should also be that very "zero" [when zero is the multiplicand].[706] **When [a quantity] is divided by zero** (*vyoman>śūnya*), the given quantity becomes zero [too].[707] In the same way, **when zero** (*vyoman>śūnya*) **is divided** by "zero", the given quantity becomes **zero.**

By the two previous sentences (*vākyadvayena*),[708] the procedures of multiplication and division have been investigated.[709] In the same way, **when there is the square,** which consists of the "product of two same quantities",[710] **of zero** [the result] **should thus be zero.** Similarly, there is the "cube",[711] which consists of the multiplication of a collection of three equal [numbers],[712] of

700 I understand this passage to refer to the rule given by the L according to which: $n \times 0 \div 0 = n$ and to the way the SGT presents the layouts of the multiplication and division by zero while illustrating the sample problem (L 47).

701 The SGT refers to the first three cases mentioned in the GT: n+0, n−0, and 0+n.

702 See the use of *vyavakalita* and *saṅkalita* to denote the operations of subtraction and addition and, in the same sentence, *apagama* and *milana* to denote the "process" of deduction and sum respectively.

703 Line 15: *vyavakalite*] em.; *vyaye* (K). Note that *vyavakalita* is mentioned in the line above; I assume that here the wording deliberately echoes the names of the arithmetical operations. In this regard, see also *guṇana*/*pratyutpanna* (lines 16 and 19), *bhāgahāra* and *varga* (line 19), *ghana* (line 20), *vargamūla* (line 21), and *ghanamūla* (line 22). Although *vyaya* as "loss, disappearance" could perhaps, by stretching its meaning, be "deduction", this use does not occur in the SGT.

704 *avikṛta* – this term is used also in L 46.

705 The commentary analyses the *tatpuruṣa* compound *śūnyaghāta* and shows the instrumental case of its first member.

706 Here (lines 16–17) the commentator expands the *sūtra* and refers to the commutative principle of the multiplication. The SGT explains that if the result is zero when zero is the multiplier, then the result is exactly the same when zero is the multiplicand.

707 Line 17: *aṅkavat śūnyaṃ bhavati*] conj.; *aṅkena…sarve 'pi yāti* (K). I have emended the text as the last part of line 17 is corrupt. The interpretation may seem forced but it follows the commentary (cf. line 18). As I have underlined earlier, differently from the GT, the L observes that the division of a quantity by zero gives $\left|\begin{matrix} n \\ 0 \end{matrix}\right|$.

708 The commentator refers to lines 15–18.

709 The SGT mentions the operations of multiplication (*pratyutpanna*) and division (*bhāgahāra*).

710 Line 19: *sarḍśa*] em. This expression is found in TŚ 11; in the SGT, it is found again in the explanation of square of fractions (see SGT $\overline{44}$, 26 and 27).

711 Line 20: *ghanaḥ*] em.

712 The commentary has already used *sadṛśatrirāśighātalakṣaṇa* in the explanation of the rule of cube of fractions (SGT $\overline{48}$, 15); see *samatrirāśi* in GT 29.

"zero";[713] it should also be **zero** (*antarikṣa>gagana*). Because of the particle *ca* ("and"),[714] [the *sūtra* implies that] "also" when there is the square-root **of zero,**[715] it is the very "zero". The cube-root **of zero** is the very "zero" too. In this manner, the [last] four arithmetical operations [with zero] beginning with the square have been formulated [as well].[716]

In order to show a sample problem[717] on this [topic of the arithmetic of zero], the rule from the *Līlāvatī* is [first] presented:

> L 45–46: "When there is the sum, zero[718] becomes equal to the added quantity, when there is the square and so forth, it is zero. A quantity divided by zero has zero as divisor (*khahara*), [a quantity] multiplied by zero is zero, and the latter is [to be considered] times-zero[719] in the remaining procedures. Moreover,[720] a quantity should be understood to remain unchanged when zero becomes [its] multiplier [and] if zero is subsequently the divisor,[721] similarly if [that quantity] is diminished or increased by zero."

[The meaning of] this pair of *āryā*s has been accomplished by the very explanation of the previous metrical stanza. [I shall present on this topic] one metrical stanza which is a sample problem:[722]

> L 47: "Tell [me] what is [the result of] zero plus five, [and] the square, root, cube, and cube-root of zero, and five multiplied by zero and ten divided by zero?[723] [Also tell me] what [number] multiplied by zero, added to its own half, multiplied by three, and divided by zero gives sixty-three?"

The presentation: |0|; it is added to five becoming five: 5. The presentation of the "square" of "zero" is: |0|. The presentation of the square "root" of zero: |0|.

713 Line 20: *vyomno*] em. The commentator supplies this term to the grammatical construction used by Śrīpati and shows the syntactical relationship with the subject of the sentence.

714 The commentator interprets the conjunction *ca* in *ghanaś ca* ("and the cube") as denoting the remaining operations with zero which have not been explicitly mentioned by the root-text, and these are the square-root and cube-root of zero.

715 The SGT uses the term *vyoman* (see in the GT, *varge vyomno*) to indicate that this same term is implied in *ghanaś ca*.

716 The commentator emphasises that, in this way, Śrīpati rightly has mentioned all the eight arithmetical operations.

717 Since Śrīpati has not provided a sample problem on the rule of zero, the commentary quotes the rule (L 45–46) and the sample problem from the L (L 47).

718 *kha*.

719 *khaguṇa*.

720 The edition by Sarma (1975, 91) reads *punas tadā* instead of *tadā punā*.

721 In brief: $n \div 0 \times 0$.

722 The sample problem supplied by the SGT is the same as L 47.

723 In the edition by Sarma, the adjectives *guṇa* and *uddhṛta* are in the accusative case.

The presentation of the "cube" of zero: $\begin{array}{|c|}0\end{array}$. The presentation of the cube-root of zero is: $\begin{array}{|c|}0\end{array}$. **And five multiplied by zero** is the very "zero", the presentation is: $\begin{array}{|c|c|}0&0\\5&\end{array}$. **And ten divided by zero: ten** divided by "zero" is the very "zero", the presentation is: $\begin{array}{|c|c|}10&0\\0&\end{array}$.

Similarly, there is a number which is an unknown number, this is **multiplied by zero** too, it is added to "its own half" which, by the explanatory prose[724] is [shown to be] namely seven,[725] it is multiplied by three [and][726] **divided by zero**. Since it is the visible quantity,[727] **sixty-three** takes place. To illustrate: the number fourteen is "multiplied by zero", it is:[728] $\begin{array}{|c|}0\\14\end{array}$. It is added to "its own half", which means [added] to the half of fourteen that is seven; it becomes twenty-one. Because it is multiplied by three, it becomes sixty-three. Then, the very sixty-three is "divided by zero" [and] **the quantity remains unchanged**[729] standing with zero, since zero is the multiplier and the divisor[730] above and below, the presentation thus is:[731] $\begin{array}{|c|}0\\63\\0\end{array}$.

The remaining elaboration of this sample problem should be understood [as it is shown] in the explanatory prose[732] of the *Līlāvatī*. In this case, the

724 I assume that *mūlavṛtti* (line 6) alludes to the L; in his work, Bhāskarācārya presents a brief prose commentary after each sample problem.

725 The L shows that final result is 14.

726 Line 7: **san**] em.

727 *dṛśya* – the same term is used in the L.

728 The commentator shows the reverse process: he calculates the numbers backwards by starting from the obtained 14 (as shown in the L).

729 The commentary quotes this expression from Śrīpati's rule. After *rāśīḥ*, one would expect the following layout: $\begin{array}{|c|}63\\0\end{array}$.

730 Here the term *bhāga* is dubious. If, as I have interpreted, the commentary is illustrating the formula of zero as "multiplier" (*guṇaka*) and divisor which is mentioned in L 45–46, one would expect a different term for "divisor", such as *hara* or *bhāgadāyin* which are frequently used in the SGT. In the commentary, *bhāga* is "division", "fraction", and "numerator".

731 The commentary presents this layout to show how to place numbers according to the formula enunciated in the L: $n \div 0 \times 0 = 0$.

732 The SGT presumably refers to a commentary upon the L rather than to the explanatory prose given by the L itself, since the mentioned elaboration of the sample problem does not occur in detail in the latter.

relevant feature[733] concerning the treatise is that the relationship of zero [and numbers] has been realised by means of the supposition operation.[734]

[In this way, the relationship of zero to numbers] is accomplished.

3.7 Classes of simplification of fractions [GT $\overline{53}$–63]

On the simplification of fractions (kalāsavarṇa)[735]

Now those [previously mentioned] parts of whole numbers[736] [which can occur] as fractions of fractions and as additional or subtractive fractions [with respect to an integer or to a fraction] [are elucidated].[737] "How the multiplication and so forth of these [quantities] should be?"[738] Having expected [such a question], the author [thus] expounds [this topic]. For this reason, thereafter the simplification of fractions is undertaken.[739]

733 *gaurava* – lit. "importance".

734 *iṣṭaparikarman* – it denotes the mathematical procedure with an assumed number formulated in L 51 and quoted by Siṃhatilakasūri after its explanation of the visible quantity type-problem with fractions (*dṛśyajāti*) enunciated in GT $\overline{64}$. At the end of his brief prose on the sample problem (L 47), Bhāskarācārya observes: *ato vakṣyamāṇakaraṇena vilomavidhineṣṭakarmaṇā vā labdho rāśiḥ* or "either by means of the inverse procedure, which is a procedure that will be explained later, or by the supposition operation (*iṣṭakarman*), the quantity is obtained". As shown by the commentary *Kriyākramarī* while commenting upon L 47 (see Sarma 1975, 92), by means of the supposition operation, if the *iṣṭarāśi* ("wished quantity") is 1, which "added to its own half" becomes: 1+1/2= 3/2; when it is multiplied by three, it gives 9/2. The *dṛśya* ("visible quantity") is the number 63 and 63 divided by 9/2 gives 14.

735 The same locative *tatpuruṣa* compound *kalāsavarṇe* appears in GT 54. In mathematical terminology, both forms *savarṇa*/*savarṇana* (*savarṇa* is lit. "having the same colour or appearance") are commonly found; in the introductory line, the commentator uses the term *savarṇana* (as well as in SGT 56, line 15). In the PG, Śrīdhara also uses *kalāsavarṇa*, which the anonymous commentator explains to stand for *kalāsavarṇana*. The word *kalā* is a technical term for "fraction". In GSK 6.30 one finds *kalāsavarṇana*, while in GSS 2.54 *kalāsavarṇe ṣaḍ jāti* occurs.

736 In brief, "integers".

737 The text begins the treatment of the classes of simplification, assimilation of fractions. I have mentioned that while commenting upon the previous mathematical topic, which concerns the eight arithmetical operations with fractions, the SGT has to anticipate the explanation of the *jāti*s (the first occurrence is in SGT $\overline{36}$, 29–30). In the sample problems given by the GT in the preceding section, fractions have different denominators and in order to carry out these operations one has first to determine the same denominator by means of the *jāti*s here expounded.

738 The reason why the SGT mentions the multiplication is not fully clear; presumably, it refers to the multiplication of the operands of the quantities involved in the simplification of fractions; for instance, with respect to the *bhāgajāti*, the two operands of the fractions are, in fact, multiplied by the exchanged denominator.

739 Line 15: **spaṣṭam**] em.

The class of simple fractions (bhāgajāti)[740] *[GT $\overline{53}$]*

On this topic, [next] the procedural rule of the class of simple fractions (*bhāgajāti*), which is the first, [is illustrated]; the author enunciates half a metrical stanza:

> $\overline{53}$. One should multiply the numerator and the denominator [of each fraction of a given pair of fractions] reciprocally[741] by the two [exchanged] denominators for [arriving at] the same denominators.[742]

The explanation: [in a given fraction] the "numerator" is the number above, the "denominator" is the number underneath. Because it concerns a pair of quantities,[743] **one should multiply the numerator and the denominator,** which have been written twice,[744] **reciprocally by the two [exchanged] denominators.** [The expression] **by the two denominators** implies that **[one should multiply]** both[745] the first **numerator and denominator** by the "denominator" of the quantity which occurs next (*agretanāṅka*). In the same way, one should multiply **the numerator and the denominator** of the quantity which occurs next by the "denominator" of the first quantity (*prācyāṅka*).[746] In other words, both very "denominators" become the same by being multiplied with one another (*mithas*), while the numerators may or may not [become the same].[747]

In this respect, the author enunciates a sample problem by means of an exemplifying metrical stanza:

> 54. If you have proficiency in [the topic of] the assimilation of fractions,[748] formulate the fractions one-half, one-third, one-fourth, one-sixth, one-fifth, [and] one-seventh as if they had the equal denominators.

The explanation: [the passage] should be construed with "of [the whole] one"[749] [the fractions] **one-half, one-third**, and so forth; the rest is clear.

740 K 30, 16.
741 In brief, one another; here adverb *anyonyasya* acquires a distributive meaning.
742 The syntagm *chedasādṛśyahetoḥ* is lit. "for [arriving at] the likeness of the denominators"; it denotes two equal denominators.
743 Line 17: **bhāga**] em.
744 *sthānadvaya* – lit. a "a pair of places"; it refers to the pair of fractions.
745 The commentator glosses *anyonyasya* ("mutually, reciprocally") with *mithas.*
746 The SGT explains that two fractions are involved in this class and defines the first as the *prācyāṅka*, while the second is the *agretanāṅka*; a similar remark occurs in SGT ad PG 37, 27. The focal point of the terminology employed by the commentator is the actual manipulation of the quantities involved and the notational layout. Note *aṅka* ("number, quantity") referring to a fraction.
747 This passage clarifies the purpose of the procedure of this class of fractions.
748 *kalāsavarṇa.*
749 Here too as before in the section introducing operations with fractions, the term *rūpa* emphasises that fractions are parts of a whole which is "one", unity.

The presentation is: $\begin{array}{|c|c|c|c|c|c|}1&1&1&1&1&1\\2&3&4&6&5&7\end{array}$. In this case, since it is the class of simple fractions, this procedure (*prakriyā*) should be performed with respect to each[750] pair of quantities, 2.[751] To illustrate: regarding the first pair of quantities,[752] **[one should multiply the numerator and the denominator]** reciprocally **by the two denominators:** for instance, three is [placed] below two, two below three; regarding these two [quantities], it is: $\begin{array}{|c|c|}1&1\\2&3\\3&2\end{array}$. Next, one multiplied by three becomes three, two multiplied by three becomes six; in the same way, one multiplied by two becomes two, three multiplied by two becomes six: $\begin{array}{|c|c|}3&2\\6&6\\3&2\end{array}$. Then, as the denominator is an equal pair of six, when the additive quantity two is added to three, it becomes five whose denominator is six. The rest (i.e., the two exchanged denominators) should be erased: $\begin{array}{|c|}5\\6\end{array}$. Afterwards, because of the denominator four of the [numerator] one of the quantity which occurs next, four should be brought (*neya*) below six, and six below four: $\begin{array}{|c|c|}5&1\\6&4\\4&6\end{array}$. In this case, when there is the reduction of six by half, it is three; [when there is the reduction by half] of four, two should be made: $\begin{array}{|c|c|}5&1\\6&4\\2&3\end{array}$. Five multiplied by two becomes ten; six multiplied by two becomes twelve. Regarding the next quantity, one multiplied by three becomes three, four multiplied by three becomes twelve: $\begin{array}{|c|c|}10&3\\12&12\\2&3\end{array}$. Next, when the additive quantity three is added to the above numerator ten, above it becomes thirteen [and] below twelve, the rest is erased: $\begin{array}{|c|}13\\12\end{array}$. Since there is the denominator six of [the numerator] one of the

750 The adverb *mithas* underlines that one should apply the same procedure on a pair of fractions each time.

751 In line 26, the figure 2 to indicate that quantities involved are two each time is anomalous and seems an instance of interpolation.

752 The adjective *prathama* denotes the "first" couple of fractions from the left.

quantity which occurs next, when there is the reduction by means of the division[753] by six, the [resulting quotient] one[754] is [placed] below twelve. In the same way, when there is the reduction of twelve by means of the division by six, the [resulting number] two should be brought below six: $\left|\begin{array}{r|r|}13 & 1\\ 12 & 6\\ 1 & 2\end{array}\right.$. The first quantity is multiplied by one; similarly, in the next quantity (*parāṅka*)[755] one multiplied by two is two, six multiplied by two is twelve: $\left|\begin{array}{r|r|}13 & 2\\ 12 & 12\\ 1 & 2\end{array}\right.$. Because the same denominator occurs, when the additive quantity two is added to thirteen which is the numerator above, it becomes fifteen whose denominator is twelve, the rest is erased: $\left|\begin{array}{r|}15\\ 12\end{array}\right.$.Because of the denominator five of [the numerator] one of the quantity which occurs next, five should be placed below twelve and twelve [should be placed]below five: $\left|\begin{array}{r|r|}15 & 1\\ 12 & 5\\ 5 & 12\end{array}\right.$. Fifteen multiplied by five becomes seventy-five; twelve multiplied by five becomes sixty. In the other quantity, one multiplied by twelve becomes twelve; five multiplied [by twelve] becomes sixty: $\left|\begin{array}{r|r|}75 & 12\\ 60 & 60\\ 5 & 12\end{array}\right.$. Because there is the same denominator, when the additive quantity twelve is added to seventy-five which is the denominator above, it becomes eight-seven whose denominator is sixty; the rest is erased: $\left|\begin{array}{r|}87\\ 60\end{array}\right.$. Because of the denominator seven of the [numerator] one of the quantity which occurs next, now seven [should be placed] below sixty and sixty should be placed below seven: $\left|\begin{array}{r||r|}87 & 1\\ 60 & 7\\ 7 & 60\end{array}\right.$. Eighty-seven multiplied by seven becomes six hundred nine; in the same way, sixty multiplied by seven becomes four hundred twenty. In the other quantity, one multiplied by sixty becomes sixty; seven multiplied by sixty becomes four hundred twenty; $\left|\begin{array}{r||r|}609 & 60\\ 420 & 420\\ 7 & 60\end{array}\right.$. As equal

753 *bhāga*.
754 Line 8: **na naiko**] em.
755 Line 10: *parāṅke*] em.

denominators occur,[756] when the additive quantity sixty is added to the numerator above which is six hundred nine, it becomes six hundred sixty-nine whose denominator is four hundred twenty; the rest is erased: $\left|\begin{array}{c}669\\420\end{array}\right|$. Then,[757] when there is the division of six hundred sixty-nine and of four hundred twenty by three, the quotient becomes two hundred twenty-three whose denominator is one hundred-forty: $\left|\begin{array}{c}223\\140\end{array}\right|$. This is the fraction [which has been obtained].[758]

Although not stated here (i.e., in the GT), an *āryā* stanza is shown [by me] in order to explain an alternative method concerning the class of simple fractions:[759]

> PG 37. "One should multiply the upper numerator (*ūrdhvāṃśa*) by the lower denominator (*adharahara*) and the lower denominator by the upper denominator (*ūrdhvahara*). One should add the [product of the] multiplication[760] of the intermediate numerator and denominator (*madhyāṃśahara*) to the upper numerator (*uparimāṃśa*)".[761]

The explanation:[762] bearing in mind that the "class of simple fractions" concerns a pair[763] of quantities (i.e., of fractions),[764] the pair of numerators

756 Kāpadīā (1937, LXVII) observes that in the manuscript leaves 60b and 61a were illegible. I propose the following reconstruction (see line 20) on which my translation is based: *sadṛśacchedatvāt uparyaṃśanavādhi(ka)ṣaṭśatīmadhye ṣaṣṭikṣepe jātā navādhi(ka)ṣaṣṭiḥ ṣaṭśatī ca*] conj. The proposed emendation reflects the coherence of the text (cf. lines 11 and 15).

757 The text is most uncertain here. The commentator, once completed the assimilation of the six fractions, reduces the numerator and denominator of the obtained fraction by the common factor 3. I suggest the following reconstruction on which my translation is based: *tato ekonasaptatyadhi(ka)ṣaṭśatyāḥ tribhir bhāge viṃśatyadhi(ka)catuḥśatyāś ca jātaṃ labdhaṃ trayoviṃśatyadhi(ka)dviśataṃ catvāriṃśadadhi(ka)śatacchedam*] conj. Also, I have emended the layout (line 23), which is mistaken.

758 *ete rūpabhāgāḥ* is lit. "these are the parts of one, unity". Notably, *rūpa* establishes a specific correspondence between the *rūpa* occurring at the very beginning of the explanation (line 25) and *rūpāṇāṃ ye bhāgā,* which is used by the SGT while introducing the topic of the classes of simplification of fraction.

759 Line 23: *atrānuktāpi bhāgajātau anyayāprakriyāṃ kathayitum āryā pradarśyate*] conj. Other reconstructions are no doubt possible.

760 *ābhyāsa.*

761 This stanza corresponds to PG 37, which is treated in Sinha's translation as a genuine GT's stanza and numbered as 47; I agree with Hayashi (2013) when he observes that this is instead a quoted rule provided by the SGT.

762 The commentator reframes his own vocabulary to elucidate the rule of the class of simple fractions quoted from the PG. Cf. *aṅkadvaya* in SGT $\overline{53}$, 17, SGT ad PG 37, 27, and *aṃśacchedayor yugalakam* in the same line.

763 *yugalaka.*

764 In line 1, *ūrdhvāṅka* denotes the upper quantity (i.e., fraction) in the layout, while *adho'ṅka* is the quantity (i.e., fraction) which is underneath.

and denominators [should be written down each below the other] by a vertical arrangement (*ūrdhvagatyā*).[765] There is the [fraction] denoted as "upper", which is the first pair of numerator and denominator, there is the [fraction] denoted as "lower", which is the second pair of numerator and denominator.[766] **One should multiply the upper denominator by the lower denominator**, which is the denominator underneath.[767] **One should multiply the lower denominator,** which is the denominator of the quantity underneath, **by the upper denominator**, which is the denominator of the upper quantity.[768] In the middle, there is that quantity [composed of] the denominator of the upper quantity and the numerator of the quantity which is underneath.[769] **The multiplication** of these two (i.e., the mentioned intermediate numerator and denominator) occurs, that is, when there is multiplication of the lower numerator by the upper denominator which is above,[770] that arisen number [is the product which] [**should be added to the upper numerator** multiplied by the lower denominator]. **One should** then **add**[771] [this number which is arisen and which is product of] **the multiplication of the intermediate numerator and denominator**[772] **to the upper numerator** multiplied by the lower denominator.[773]

Regarding the metrical stanza of the sample problem [on this topic], [here] is the preceding presentation by a vertical arrangement of the row of

765 The argument focuses on the arrangement of the two fractions. This passage can be better understood in the light of the passage in lines 4–5, where *ūrdhvagatyā* is followed by a layout showing the way the quantities should be placed.

766 This passage expands upon the terms *ūrdhva* and *adhara* occurring in Śrīdhara's rule.

767 See *adharahara* and *adhaścheda*.

768 This statement clarifies further *ūrdhva* or "upper" and *adhara* or "lower" and conveys the idea that the quantities are so defined according to the position they occupy in the layout where they are written.

769 In brief, the quantity which is in the middle; it clearly refers to the quantities arranged in a layout.

770 The commentary gives important information: Śrīdhara defines as "intermediate numerator and denominator" the denominator of the upper fraction and the numerator of the lower fraction respectively; these are the operands of the quantity in the middle. The phraseology used by the SGT turns around Śrīdhara's; *adhastanāṃśa* or the "lower numerator" is the denominator of the quantity in the middle (see the second number 1 in the layout of line 4), while *ūrdhvahara* or "upper denominator" is the numerator of the quantity in the middle (in the layout, this corresponds to the number 2). The stress is on the fact that the operands of the two fractions are placed below one another and that the middle quantity is composed of two quantities belonging to the pair of fractions denoted as *ūrdhvāṅka* and *adho'ṅka* in the preceding line 29.

771 Note the drop of the initial syllable *vi* as, in line 3, one would expect *vinikṣipet* rather than *nikṣipet*. The verb *nikṣipet* occurs in lines 6, 9, 13, 18, and 23.

772 Line 2: °*ābhyāsaṃ*] em.

773 This is an explanatory amplification of the last step of the rule. The commentary rephrases the technical terms used in the passage by the PG and, by supplying *adharaharaguṇiteṣu* expands Śrīdhara's formulation. The SGT specifies that one should add the product of the intermediate numerator and denominator to the upper numerator which has been "multiplied by the lower denominator" (*adharaharaguṇiteṣu*).

[the given] numbers (*ūrdhvagatyā aṅkaśreṇeḥ*):[774] $\begin{array}{|c|}1\\2\\1\\3\end{array}$. In this case, three is the "lower denominator". [The number] one, which is the "upper numerator" is multiplied by it; it becomes three. **One should multiply the denominator** three, which is **the lower**, by two,[775] which is **the upper denominator**; it becomes six. Regarding the "intermediate numerator and denominator", when there is the "multiplication" of two, which is the "upper denominator", by the quantity which is below [such as one], the product[776] is that very two; **one should add** this to three, which is the [product of the multiplication of the] "upper numerator" [by the lower denominator]. It becomes five whose denominator is six: $\begin{array}{|c|}5\\6\end{array}$. Below it, there is one-fourth:[777] $\begin{array}{|c|}5\\6\\1\\4\end{array}$. **One should multiply the upper numerator** [five] **by the lower denominator,** which is four; five [multiplied by four] becomes twenty. **One should multiply the lower denominator** four **by the upper denominator** six, it becomes twenty-four. **One should add the [product of the] multiplication of the intermediate numerator and denominator**: six is multiplied by one, [one should add] that very [product, which is six] to twenty, which is [the product of] the "upper numerator" [multiplied by the lower denominator].[778] It becomes twenty-six whose denominator is twenty-four: $\begin{array}{|c|}26\\24\end{array}$. Below it, there is one-sixth: $\begin{array}{|c|}26\\24\\1\\6\end{array}$. **One should multiply the upper** [denominator], which is twenty-six, **by the lower denominator**, which is six; it becomes one hundred fifty-six. **One should multiply the lower denominator** six **by the upper denominator**

774 The commentator calculates the same fractions of the previous sample problem (GT 54) in order to show the execution of the rule quoted by him from the PG. In this manner, students can compare the way the same fractions area manipulated when applying two different methods of the class of simple fractions.

775 Line 4: *dvikenādharaṃ*] em.

776 Line 6: *guṇitaṃ*] em.; *guṇanaṃ* (K).

777 Line 7: *asyādhaś caraṇo yathā*] em. The transmitted construction without *adhas* does not reflect the explanation of the commentary as "its one-fourth" is not, in fact, the meaning intended. See *adhas* in lines 10 and 19.

778 In lines 2–3, the commentator has specified that one should supply *adharaharaguṇiteṣu,* which qualifies *uparimāṃśeṣu,* to Śrīdhara's passage.

twenty-four; it becomes one hundred forty-four. **One should add [product of] the multiplication of the intermediate numerator and denominator**, that is to say, [the product of] twenty-four[779] multiplied by one [added] **to the [product of the] upper numerator** [multiplied by the lower denominator], which means [that it is added] to one hundred fifty-six. It becomes one hundred eighty whose denominator is one hundred forty-four:[780] $\left|\begin{array}{r}180\\144\end{array}\right|$. Below it, there is one-fifth: $\left|\begin{array}{r}180\\144\\1\\5\end{array}\right|$. **One should multiply the upper numerator,** which is one hundred eighty, **by the lower denominator,** which is five; it becomes nine hundred. **One should multiply the denominator** five, which is **the lower, by the upper denominator,** which is one hundred forty-four; it becomes seven hundred twenty. **One should add [the product of] the multiplication of the intermediate numerator and denominator,** that is to say, [one should add] one hundred forty-four multiplied by one **to the [product of the] upper numerator [multiplied by the lower denominator]**, which means [one should add the previous product] to nine hundred. It becomes one thousand forty-four whose denominator is seven hundred twenty: $\left|\begin{array}{r}1044\\720\end{array}\right|$. Below it, there is one-seventh: $\left|\begin{array}{r}1044\\720\\1\\7\end{array}\right|$. The "upper numerator" one thousand forty-four is multiplied by the "[lower] denominator" seven; it becomes seven thousand three hundred eight.[781] **One should multiply the lower denominator** seven **by the upper denominator** seven hundred twenty; it becomes five thousand forty. **One should add [the product of] the multiplication of the intermediate numerator and denominator**, that is to say [one should add] that consisting[782] of seven hundred twenty multiplied by one **to the [product of the] upper numerator** [multiplied by the lower denominator],

779 Line 13: *caturviṃśatiṃ*] em.

780 Line 14: *yathā* $\left|\begin{array}{r}180\\144\end{array}\right|$ | *asyādhaḥ pañcāṃśako yathā* $\left|\begin{array}{r}180\\144\\1\\5\end{array}\right|$] em. This emendation is supported by the commentary itself; thus far, the commentator has first shown the layout of the final result obtained to only then provide the layout having underneath the next fraction to be performed. Cf. lines 7, 10, and 19.

781 Line 20: *jātā*] em.

782 Line 22: *lakṣaṇam*] em.

which is seven thousand three hundred eight. It becomes[783] eight thousand twenty-eight[784] whose denominator is five thousand forty:[785] $\begin{array}{|c|}8028\\5040\end{array}$. When there is the division of the number above by the number below, the quotient is one (*rūpa>eka*) and above the number-remainder is two thousand nine hundred eighty-eight: 2988. When there is the reduction of this [2988] by thirty-six, it becomes eighty-three. When there is the reduction[786] of the number below, which is five thousand forty, by thirty-six, it becomes one hundred forty:[787] $\begin{array}{|c|}1\\83\\140\end{array}$ *Ete rūpaḥ...*[788]

The multi-part class (prabhāgajāti)[789] *[GT $\overline{55}$]*

Next is the procedural rule of the multi-part class; the author enunciates half a metrical stanza:[790]

> GT $\overline{55}$. In the class called multi-part, the multiplication of the numerators and the product of the denominators [should be carried out]. The numerator and the denominator [of the simplified fraction] will occur.

783 Line 23: *jātam*] em.

784 Line 24: °*aṣṭakaṃ*] em.

785 I have emended the text (line 24) as the transmitted layout $\begin{array}{|c|}8627\\5344\end{array}$ is mistaken.

786 Line 27: *apavarte*] em.

787 The text is most uncertain here. I have emended two layouts occurring at the end of line 27 and at the beginning of line 28 respetively; the emendation is supported by the commentary itself as whenever the final result is a compound fraction, the SGT presents a layout where the integer is at the very top.

788 See a similar final sentence at the end of the explanation of the previous *karaṇasūtra* (SGT 54, 23). The editor (see footnote 1) observes that in the manuscript folio 64 is missing.

789 I assume that folio 65 (K 32, 28) begun with the fraction $\begin{array}{|c|}1\\24\end{array}$. It can be inferred that in folio 64, which is missing, the procedural rule of the *prabhāgajāti* ("the class of multi-part") and its sample problem must have occurred. Note that, before the rule of the *bhāgānubandhajāti*, at the end of the explanation of the sample problem the commentator states: *evaṃ prabhāgajātiḥ samāptā* or "In this way, the procedure of the multi-part class is concluded." I agree with Hayashi (2013, 61–62) who observes that the procedural rule of the *prabhāgajāti* must have been the half *śālinī* stanza coupled with the half *śālinī* stanza which is GT 53 and which formulates the rule of the *bhāgajāti*. Here I follow Hayashi's reconstruction (the translation is my own) of the rule of the *prabhāgajāti,* which is based on the commentator's references and quotations: *aṃśābhyāsacchedasaṃvargam eva aṃśacchedaḥ syāt prabhāgākhyajātau*. It is worth noting that *aṃśābhyāsacchedasaṃvargam eva* is quoted by the commentator himself in SGT 102, 19.

790 My reconstruction of the introductory line to the procedural rule is: *atha prabhāgajātau karaṇasūtraṃ vṛttārdham aha*] conj.

In this respect, in the exemplifying metrical stanza [which follows] the author enunciates a sample problem:[791]

> GT 56. A beggar was given one-quarter of one-third of one-half of one *dramma* by a man, one-sixth of one-half of one-third of one-fifth of one-eighth of one *dramma* by another, and one-tenth of one-fourth of one-eighth of one-seventh of one *dramma* by another. If you are proficient in the multi-part class, having added [these quantities], tell me the result.[792]

[…] one whose denominator is twenty-four:[793] $\begin{array}{|c|}1\\24\end{array}$.

Next is the explanation of the second quarter [of the stanza]: by another [man], one-eight of a *dramma*, one-fifth of that [last result], its one-third, its half, [and] its one-sixth are given [to the beggar].[794] The presentation

791 My reconstruction of the introductory line is: *atroddeśakavṛtte udāharaṇam āha*] conj.

792 Cf. L 33, where a similar sample problem on the *prabhāgajāti* is found. I have slightly changed the English translation by Hayashi (2013, 61–62). This stanza, which must have had four *pada*s (see *dvitīyapada* and *tṛtīyapada* in lines 1 and 9), might have been written in in the meter *śārdūlavikrīḍita,* a meter often used by the GT (see also L 86) for long sample problems.

793 This is the product of the first group of fractions: $\begin{array}{|c|c|c|}1&1&1\\2&3&4\end{array}$. The text is suspect here (see *viṃśatir ekona* in line 28). Below I present a tentative reconstruction of the Sanskrit text of the first part of this sample problem. The conjecture is based on the surviving passages of the commentary:

vyākhyā | atha prathamapadavyākhyā | ekasya drammasya caraṇaś caturbhāgas tasya tryaṃśas tasyārdhaṃ yat kenacit dattaḥ | nyāso yathā $\begin{array}{|c|c|c|}1&1&1\\2&3&4\end{array}$ *| etenaiko'ṅkaḥ sādhyaḥ | tathāhi | aṃśānāṃ ekānām abhyāsaṃ guṇanaṃ kuryāt | jāta eka eva | chedasaṃvargaṃ ca kuryāt | yathā dviguṇās trayo jātāḥ ṣaṭ | tathā ṣaḍguṇāś catvāro jātā caturviṃśatiḥ | tato jāta ekaś caturviṃśatichedaḥ | yathā* $\begin{array}{|c|}1\\24\end{array}$|] conj.

The English translation is: "The explanation: the explanation of the first quarter [of the stanza] follows. One-fourth, which is a fourth part, of one *dramma,* its one-third, [and] its one-half were given by a man. The presentation is: $\begin{array}{|c|c|c|}1&1&1\\2&3&4\end{array}$. In this manner, [by performing this group of fractions], one [only] quantity is to be determined. To illustrate: one should [first] carry out **the multiplication** (*abhyāsa>guṇana*) of the [numbers] one [above], which are the "numerators"; the very one occurs. One should [then] perform **the product** of the "denominators" as well: three multiplied by two becomes six; in the same way, four multiplied by six becomes twenty-four. It becomes one whose denominator is twenty-four: $\begin{array}{|c|}1\\24\end{array}$".

794 In the SGT, the form *bhāgaka* (line 2) is uncommon.

is: $\begin{array}{|c|c|c|c|c|} 1 & 1 & 1 & 1 & 1 \\ 8 & 5 & 3 & 2 & 6 \end{array}$. In this manner, [by calculating this group of fractions], one [only] quantity is to be determined.[795] To illustrate: one should [first] carry out **the multiplication** (*abhyāsa>guṇana*) of the [five numbers] one, which are the "numerators"; the very one occurs. One should [then] perform **the product** of the "denominators" as well: five multiplied by eight becomes forty; in the same way, three multiplied by forty becomes one hundred twenty. Similarly, two multiplied by one hundred twenty becomes two hundred forty, six is multiplied by it giving fourteen hundred forty. Therefore, it becomes one whose denominator is one thousand four hundred forty: $\begin{array}{|r|} 1 \\ 1440 \end{array}$.

Next is the explanation of the third quarter [of the stanza]: one-seventh[796] of one *dramma*, its one-eight, its one-fourth, [and] its one-tenth; he gave this [amount to the beggar]. The other presentation is: $\begin{array}{|c|c|c|c|} 1 & 1 & 1 & 1 \\ 7 & 8 & 4 & 10 \end{array}$. In this manner, [by calculating this group of fractions], one quantity [only] is to be determined. To illustrate: one should [first] carry out **the multiplication** of the [numbers] one, which are the "numerators"; the very one occurs. One should [then] perform **the product** of the "denominators": eight multiplied by seven becomes fifty-six; in the same way, four multiplied by fifty-six becomes two hundred twenty-four, ten multiplied by it becomes two thousand two hundred forty. This is the denominator and one [is the numerator]: $\begin{array}{|r|} 1 \\ 2240 \end{array}$.

Next, in order to [carry out] the assimilation[797] of these three quantities arisen from the [multiplication of the fractions of the] three [*pada*s], the presentation is: $\begin{array}{|r|r|r|} 1 & 1 & 1 \\ 24 & 1440 & 2240 \end{array}$. In this way, there is the class of simple fractions [to be performed] because the procedure of the "multi-part class" [has been carried out].[798] Afterwards, according to [the rule] "the numerator and the denominator"[799] and so forth, when there is the exchange and when there is the reduction of the pair (*yuga*) of denominators of the two fractions, for instance when there is the reduction of twenty-four, it becomes[800] one. This

795 *sādhya* – in the SGT this is the first occurrence of the term.

796 Line 9: *yaḥ saptamas*] em.; *'ṃśaka* (K). We must assume that the commentator is explicitly enumerating all the fractions one by one.

797 *savarṇana*.

798 Line 15: *prabhāgajāty°*] em. Perhaps one should consider emending *jātā* (line 16) to *kāryā* or "it should be performed". The commentator emphasises that first the *prabhāgajāti* has been carried out, the three obtained fractions are added by means of the *bhāgajāti*.

799 The commentator quotes from the rule of the class of simple fractions (GT $\overline{53}$).

800 Line 17: *jāta*] em.; *śata* (K).

is [placed] below one thousand four hundred forty. In the same way, when there is the reduction of fourteen hundred forty by twenty-four, it becomes sixty.[801] This is [placed] below twenty-four:[802] $\begin{array}{|r|r|} 1 & 1 \\ 24 & 1440 \\ 60 & 1 \end{array}$. One multiplied by sixty becomes sixty; twenty-four multiplied by sixty becomes fourteen hundred forty: $\begin{array}{|r|} 60 \\ 1440 \end{array}$. The second [quantity] multiplied by the number [one] remains exactly so.[803] Since the same denominator occurs,[804] when the additive quantity one is added to sixty, it becomes sixty-one whose denominator is fourteen hundred forty: $\begin{array}{|r|} 61 \\ 1440 \end{array}$. The rest is erased (*vinaṣṭa*). For the purpose of the assimilation of the quantity which occurs next, when there is the reduction also of the second denominator fourteen hundred forty by one hundred sixty, it becomes nine. Similarly, [in the reduction] of twenty-two hundred forty [by one hundred sixty], it becomes fourteen. Next, by means of [the procedure] "the numerator and the denominator"[805] and so forth, when there is exchange of the denominators, it is: $\begin{array}{|r|r|} 61 & 1 \\ 1440 & 2240 \\ 14 & 9 \end{array}$. Afterwards, sixty-one multiplied by fourteen becomes eight hundred fifty-four; in the same way, fourteen hundred forty multiplied by fourteen becomes twenty thousand one hundred sixty. Regarding the other number,[806] one multiplied by nine becomes nine; in the same way, twenty-two hundred and so forth multiplied by nine becomes twenty thousand and so forth: $\begin{array}{|r|r|} 854 & 9 \\ 20160 & 20160 \end{array}$. Then, when there is the sum of the upper numerators, it becomes eight hundred sixty-three whose denominator is twenty thousand and so forth. The rest is erased: $\begin{array}{|r|} 863 \\ 20160 \end{array}$.

801 Lines 18: *jātāḥ*] em.
802 Line 19: *caturviṃśater adho yathā*] em. This must have been the intended sense but it is not expressed by the Sanskrit as it stands.
803 This formulation is clumsy; it would be more natural, in accordance with the commentator's phraseology, to find *dvitīyāṅka ekaguṇaḥ sa eva.* See a similar construction in SGT 79, 8; in the SGT, *tathaiva* also occurs.
804 In line 21, *sadṛśacchedatvāt* should follow and not precede the *daṇḍa*.
805 The commentator quotes from the rule of the class of simple fractions (GT 53).
806 In line 27, *parāṅke* should follow and not precede the *daṇḍa*.

Since the number above is smaller [than its denominator],[807] it does not allow division; for, as it is common[808] for [large] whole units (*pūrṇarūpa*),[809] these should be [considered] suitable for division accordingly (i.e., according to their sub-multiplies).[810] For this reason, [in this resulting fraction] that [quantity] consisting of *paṇas*[811] should be calculated first, since there is one *dramma* in sixteen *paṇa*s. Therefore, eighty hundred sixty-three multiplied by sixteen becomes[812] thirteen thousand eight hundred eight: 13808. The division [by the denominator] is again not yet possible, that [quantity] consisting of *kākiṇī*s should be calculated. Four *kākiṇī*s[813] are one *paṇa*, thirteen thousand and so forth multiplied by four becomes fifty thousand two hundred thirty-two.[814] When there is its division by the number beginning with twenty thousand, it is: $\begin{array}{|c|}55232\\20160\end{array}$. The quotient is two *kākiṇī*s: 2, above the number-remainder is fourteen thousand nine hundred twelve. Since this is smaller [than the denominator], it does not allow division; that [quantity] consisting of *kapardaka*s should hence be calculated. For it is said that there is one *kākiṇī* in twenty *kapardaka*s, fourteen thousand and so forth multiplied by twenty becomes two hundred thousand nineteen thousand two hundred forty. When there is its division by twenty thousand and so forth, it is: $\begin{array}{|c|}298240\\20160\end{array}$. The quotient is fourteen *kapardaka*s: 14; above the number remaining is sixteen thousand: 16000. Then, when there is the reduction of the numbers above and underneath by three hundred twenty, it becomes fifty above and sixty-three below: $\begin{array}{|c|}50\\63\end{array}$.

In this way, the procedure of the multi-part class is concluded.

807 *stoka* – this is the first occurrence of the term.

808 *prāyas* – this is the first occurrence of the term. Note that the formula *yatas/tatas* connects the following remark with the previous statement.

809 I have assumed that here *pūrṇarūpa* is not "integer, whole number" as in SGT 35, 22 but it rather denotes a "whole (*pūrṇa*) unit (*rūpa*)". The argument is presumably the division of a large unit of measurement (such as a unit of money, weight, and so forth) into its smaller sub-multiples.

810 The translation reflects the terseness of the passage. Emendation could also be considered, for instance by emending *pūrṇarūpaprāyas* to *purṇarūpaḥ prāyas* the whole passage becomes: "for, as a general rule, whenever there is a [large] whole unit, this should then be […]." Cf. SGT 22, 10 for a similar terminology (*bhāgayogya*) relating to the division.

811 Cf. the construction with the unit of money and *rūpa* in lines 2–3, 5, and 8. For the monetary units mentioned in this passage, see GT 4.

812 Line 4: *jātās*] em.

813 Line 5: *kākiṇīkās*] em.

814 Line 6: *jātāḥ*] em.

The class of fractional increase (bhāgānubandhajāti)[815] *[GT 57]*

Next is the procedural rule of the class of fractional increase; the author enunciates a metrical stanza:[816]

> 57. In the procedure of the class [of fractions] called fractional increase,[817] one should add the numerator (*bhāga*) to the integer multiplied by the denominator. Having multiplied the [upper] denominator by the [lower] denominator, one should multiply the first numerator (*aṃśaka*) by the denominator below added to [its own] numerator.[818]

The explanation: wherever[819] there is [a fraction composed of] a numerator and denominator associated[820] with a fraction, also [where] there is the association of fractions attached[821] to an integer, **one should add**[822] **the numerator, the numerator** which is associated with its integer, **to the integer multiplied by the denominator,** that is to say, **to the integer** standing above[823] multiplied **by the denominator.**[824]

815 K 34, 15–16. The SGT has already quoted and shown the execution of the first method of this rule while solving GT 37.

816 The term *anubandha* is lit. "attachment, junction". Perhaps *bhāgānubandhajāti* may be also translated as the "association of fractions" or "fractions in association", but for the sake of a smooth English reading I have opted for the "class of fractional increase". Here the term *anubandha* implies, in fact, that the methods of this class of fractions involve the sum of a fraction either to an integer or to a fraction of itself. Also, the rendering of *bhāgānubandhajāti* as the "class of fractional increase" facilitates the translation of the definition of the parallel class, which is the *bhāgāpavāhajāti* or the "class of fractional increase". However, whenever in the SGT the term *anubandha* indicates the relationship between a fraction and a quantity, I have translated it as "association".

817 *aṃśānubandhajāti* – this is the name given by Śrīpati to the procedure, which is termed *bhāgānubandhajāti* by the commentator. In *bhāgānubandhajāti*, *bhāga* stands for "fraction" and so does *aṃśa* in *aṃśānubandhajāti*.

818 This stanza gives two procedures of fractional increase: the first is *chedanighneṣu rūpeṣu bhāgaṃ kṣipet*, the second starts with *chedanaṃ chedananiva*. They are both explained in detail by the SGT.

819 In the SGT, *yatra* is not only used with its correlative *tatra* to mark relative clauses but it frequently occurs as a subordinate conjunction or a predicative adverb to denote a transition between two ideas; it serves to contextualise and clarify the circumstances in which a mathematical rule is to be applied.

820 *anubandhin*.

821 *abubaddha*. Perhaps also conceivable would be to emend °*anubaddha* to °*anubaddhānām* as it is, in fact, semantically more appropriate for this adjective to qualify *bhāga* rather than *anubandha*.

822 Cf. *nikṣipet* (line 19) and *kṣipet* in the *karaṇasūtra*; *iti* appears to mark a quotation, hence the initial syllable (*ni*) should not occur.

823 By *uparistheṣu* (line 18), the commentator specifies the place occupied in the layout by the "integer" (*rūpa*) with respect to the fraction.

824 The entire commentarial passage is divided into two parts: lines 17–19 refers to the first method of the class of fractional increase, lines 19–24 introduces the second method. The aim is to distinguish between the two and to establish a correspondence between the methods enunciated by the root-text and the scenario in which they should be applied. The commentary structures differently the two parts: the first begins with a verbal construction where

[Here] it is mentioned[825] that this [procedure] is dependent[826] on the class of simple fractions.[827]

"If the fractions united to an integer are in such a number so that the very fraction of the integer is united to a fraction [of itself] [and] associated to many [other] fractions [as well] [and these] occur as quantities in association with [other] fractions, [in this case] what is then [the procedure]?"[828] [For this reason] the text says: **the denominator and so forth.**[829] **Having multiplied** (*hatvā>guṇayitvā*) **the denominator** of the fraction[830] at the very top by the "denominator" of the fraction underneath, **one should multiply** (*āhanyate>guṇyate*) the "numerator" above **by the denominator below added to [its own] numerator,** which means by the "numerator" below added to the "denominator" underneath. Because it is the multiplier, the quantity (i.e., fraction) which is underneath is erased (*vinaśyati*); the numerator and the denominator above remain.[831] The procedure concerning [each of] the quantities occurring next should be carried out in this way too.[832] [The meaning of the expression] beginning with *nūnam* ("certainly")[833] is clear.

The author enunciates a first sample problem on the first quarter [of the stanza] by the metrical stanza [which follows]:[834]

> 58. O friend, if you have exercised yourself in mathematics, having performed the simplification [of the following fractions], tell me [the result of] ten[835] plus one-fourth, one added to one-half, and two added to one-third.

yatra functions as a predicative adverb, the second is presented by means of a literary device, a typical commentarial formula of question and answer; this serves not only to embellish the sentence and add more information but mostly to justify the treatment of a topic by the root-text, which is thus appropriate for its purpose.

825 In brief, "I the commentator am emphasising that [...]".

826 *āśraya* – it points out that this method of *bhāgānubandhajāti* is based on the *bhāgajāti* or the "class of simple fractions" (GT 53). In the SGT, the adjective *āśraya* is often used to emphasise that a rule is dependant on another rule (see, for instance, the SGT in the introductory line to GT 112).

827 In line 19, the term *śuddha* (lit. "pure") in *śuddhabhāgajāti* characterises as "pure" –thus "simple"– the first method of the "class of [simple] fractions".

828 The SGT presents the second procedure of *bhāgānubandhajāti.*

829 Line 21: *chedanety ādi*] em.

830 *aṃśa* – it may also be interpreted as "numerator" without affecting the intended sense; however, my suggestion is that here for the SGT *aṃśa* is "fraction" as it is for the GT in *aṃśānubandhajāti*. In both the GT and the SGT *aṃśa* is often "fraction"; see for instance, in SGT 64, 11 and 12 and in GT 80.

831 *tiṣṭhata.*

832 The commentary refers to a chain of fractions where the "quantity (i.e., fraction) which occurs next (*agretanāṅka*)" is connected to the preceding fraction (or to the preceding result) as a fraction of itself. This remark can be better understood in the light of the commentator's explanation of the sample problem given in GT 59 (see the chain of fractions occurring in the layout of line 11).

833 In Śrīpati's stanza, the particle *nūnam* serves simply as a device for filling out the meter.

834 By this remark, the commentary underlines that the first sample problem concerns the first method of the class of fractional increase.

835 Line 26: °*daśa rūpam*] em.

[The explanation]:[836] **tell [me]** [the result of] the integer "ten"[837] added to "one-fourth", in the same way **one** (*rūpa>eka*) **added to one-half,** as well as **two added to one-third.** The rest is clear [in meaning].

The presentation is: $\begin{array}{|c|c|c|} 10 & 1 & 2 \\ 1 & 1 & 1 \\ 4 & 2 & 3 \end{array}$. In this case, the procedure is as follows: in the first quantity, the "denominator" is four, the "integer" ten "multiplied" by this becomes forty; when the additive quantity which is the "numerator" one is added to it, it becomes forty-one[838] whose denominator is four: $\begin{array}{|c|} 41 \\ 4 \end{array}$. In the second quantity, one "multiplied by the denominator" two becomes two; when the additive quantity which is the "numerator" one [is added to it], it becomes three whose denominator is two: $\begin{array}{|c|} 3 \\ 2 \end{array}$. In this case, when there is the reduction of the denominators four [and] two by half, when there is exchange of the obtained one and two, it is: $\begin{array}{|c|c|} 41 & 3 \\ 4 & 2 \\ 1 & 2 \end{array}$. By means of [the procedure] "the numerator and the denominator"[839] and so forth, when the preceding quantity is multiplied [by one], it remains as it is; in the next quantity,[840] three multiplied by two becomes six, two multiplied by two becomes four. Since the same denominator occurs, when the additive quantity [six] is added to forty-one, it becomes forty-seven whose denominator is four. The rest is erased: $\begin{array}{|c|} 47 \\ 4 \end{array}$.

[Now] regarding the quantity that occurs next, it is: $\begin{array}{|c|} 2 \\ 1 \\ 3 \end{array}$. In this case, when the "integer" two is "multiplied" by three, which is the denominator, it becomes six; when the additive quantity which is the "numerator" one [is added to it], it becomes seven whose denominator is three: $\begin{array}{|c|} 7 \\ 3 \end{array}$. Then, in this case, according to [the rule] "the numerator and the denominator"[841] and so forth, when there

836 In line 28, the term *vyākhyā* is missing.

837 Here *rūpa* emphasises *daśa* as the integer, in contrast to the fraction which is added to it.

838 Line 2: °*catvāriṃśac*] em.

839 This is a quotation from the class of simple fraction (GT 53).

840 In lines 4, 5, 7, and 10.
see the use of *pūrvāṅka/parāṅka/agretanāṅka* to indicate the quantity which is manipulated. This wording deliberately echoes the previous sentence in line 23 where, while commenting upon the procedural rule, the SGT has observed: "The procedure concerning [each of] the quantities occurring next should be carried out in this way too".

841 Line 8: *aṃśacchedāv ity ādinā*] em.; *chedādāv ity ādinā* (K). The commentary quotes this passage from the class of simple fraction (GT 53).

is the exchange of the denominators four and three, it is: $\begin{array}{|r|r|} 47 & 7 \\ 4 & 3 \\ 3 & 4 \end{array}$. Forty-seven[842] multiplied by three becomes one hundred forty-one, four multiplied by three becomes twelve: $\begin{array}{|r|} 141 \\ 12 \end{array}$. Regarding the next quantity, seven multiplied by four becomes twenty-eight,[843] three multiplied by four becomes twelve: $\begin{array}{|r|} 28 \\ 12 \end{array}$. As the same denominator occurs, when the additive quantity twenty-eight is added to the previous [obtained] one hundred forty-one, it becomes one hundred sixty-nine whose denominator is twelve: $\begin{array}{|r|} 169 \\ 12 \end{array}$. The rest is erased. When there is the division of the number above by the number below, the quotient is fourteen and the remainder is one whose denominator is twelve: $\begin{array}{|r|} 14 \\ 1 \\ 12 \end{array}$.

The author enunciates a metrical stanza to illustrate a sample problem on fractions in association with fractions,[844] the association[845] of fractions devoid of[846] integer, [and] a number joined in a fractional increase.[847]

> 59. Tell [me the result] having simplified[848] [the quantity composed of] one plus one-fourth added to its own half,[849] [added to][850] one-third of itself (i.e.,

842 Line 9: *saptacatvāriṃśaj*] em.

843 Line 11: °*viṃśatiś*] em.

844 *bhāgānubandhabhāga* – in the SGT, the same expression occurs below in line 22 and, while solving the next sample problem, in line 11. See the layout in line 25 where fractions in association with fractions are associated to an initial integer. Note that the commentary provides three definitions to indicate the relationship between the quantities involved in the procedure of *bhāgānubandhajāti*. Now the commentator provides an explanatory amplification of his own elucidation of the *karaṇasūtra*; it reframes the mathematical scenarios mentioned there by employing now a more specific vocabulary.

845 Line 15: °*anubandhaṃ*] em.; °*anubandhi* (K). The transmitted *anubandhi* is not interpretable, one would expect a substantive (*anubandha*) rather than an adjective (*anubandhin*).

846 *rahita* – this verbal adjective often occurs in relation to a subtraction but here is "devoid of" and not "lessened by" as it is, for instance, in SGT 61, 9. The commentator distinguishes between *bhāgānubandhabhāga* and *rūparahitabhāgānubandha,* where the latter refers to the case in which fractions are associated with fractions but without an initial integer, see the layout in line 11.

847 *bhāgānubandhasahitāṅka* – here the term *aṅka* is "whole number, integer" as *rūpa* in *rūparahita*°. With respect to the sample problem, the definition *bhāgānubandhasahitāṅka* represents the case of the quantity "one plus one fourth".

848 *savarṇayitvā.*

849 Line 17: *svadalādhikaṃ*] em.; *ārdhakaṃ* (K). This emendation is supported by the commentary: in line 19, *dalam* is glossed with *ardhaṃ,* and at the very end of the line *tenādhikaṃ* occurs.

850 In line 20, the SGT supplies the verbal adjective *sahitam*.

of the sum) added to one-sixth of itself (i.e., of the last sum), [together with] one-third plus [its] one-sixth added to its (i.e., of the last sum) one-fourth.

The explanation: first **one** plus one-fourth**; its own half** means "its half", that is to say, [half] of "one plus one-fourth", it[851] should be **added** to it. It is added to **one-third of itself,** that is, "one-third" of [the result of] "one plus one-fourth" plus its half; **[one plus one-fourth added to its own half]** is joined to it. In the same way, **one-sixth of itself** is **one-sixth** of "one-third of itself" added to "its own half",[852] which is [half] of "one plus one-fourth"; it (i.e., the last sum) is joined to it. This is the [case] representing fractions attached to an integer.[853]

Henceforth[854] there are fractions in association with fractions. To illustrate: **one-third,**[855] this [is added] to **one-sixth,** that is, **"**one-sixth" of "one-third"**; one third is added to** it. **Its one-fourth** (*pada*>*pāda*) is one-quarter of "one-third" "added to" "one-sixth"; **one-third [plus its one-sixth**] is joined to it. **Having simplified [the following fractions], tell [me the result].** This is the procedure.[856]

Here, by a vertical arrangement (*ūrdhvagatyā*), first is the presentation[857] of the association of fractions made like a chain:[858] $\begin{array}{|c|}1\\1\\4\\1\\2\\1\\3\\1\\6\end{array}$. In this case, the method

851 In brief, the result of one plus one-fourth should be added to its half.

852 Line 21: *sa* in *svārdhayukta*°] em.

853 *rūpapratibaddhabhāga* – this definition appears to summarise the circumstances illustrated by the two definitions given earlier, which are *bhāgānubandhabhāga* and *bhāgānubandhasa hitāṅka.*

854 From line 22 to line 25, the commentator elucidates the second line of the sample problem, where fractions in association with fractions (*bhāgānubandhabhāga*) occur.

855 The SGT explains "one third" as lit. "a third part of one", which emphasises the opposition between the part (*aṃśa*) and the whole (*rūpa*).

856 Note the commentator's expository strategy: he first interprets the given quantities to only then show their manipulation.

857 This layout regards the first half of the sample problem, which has been explained by the SGT in lines 19–22. The commentary demonstrates that the execution (lines 26–10) of this first half of the sample problem involves the two aforementioned definitions *bhāgānubandha sahitāṅka* and *bhāgānubandhabhāga* (see the introductory line to the sample problem). The last half (from line 11 onwards) concerns the *rūparahitabhāgānubandha*, which is executed in the same way as *bhāgānubandhabhāga*; the only difference is that the initial integer is missing.

858 The idea is that each following fraction is associated to the preceding as its one fraction.

[enunciated] in the metrical stanza for the procedure with an integer[859] is shown. Therefore, at first the "denominator" four [is manipulated], one, which is the "integer", is "multiplied" by it giving four. **One should add the numerator** one to it; it becomes five whose denominator is four: $\begin{array}{|c|}5\\4\end{array}$. Half of it is "its own" half: $\begin{array}{|c|}5\\4\\1\\2\end{array}$. Then, **having multiplied the [upper] denominator,** such as the four above, **by the [lower] denominator,** which is the number two below, it becomes eight. That consisting of two, which is the "denominator" underneath, is **added to [its own] numerator,** that is, it is added to the "numerator" one which is its own [numerator]; it becomes three. **One should multiply the first numerator**, which is five, by it; it becomes fifteen whose denominator is eight. Since it was the multiplier,[860] the rest is erased: $\begin{array}{|c|}15\\8\end{array}$. Below it, there is "its" one-third: $\begin{array}{|c|}15\\8\\1\\3\end{array}$. In this case, **having multiplied the [upper] denominator,** such as eight, **by the [lower] denominator**, which is the three underneath, it becomes twenty-four. Similarly, the denominator three is **added to [its own] numerator,** that is, it is **added to** one; it becomes four. **One should multiply the first numerator,** namely fifteen, by this [four]; it becomes sixty whose denominator is twenty-four. The rest is erased: $\begin{array}{|c|}60\\24\end{array}$. Below it, there is "one-sixth of itself", thus: $\begin{array}{|c|}60\\24\\1\\6\end{array}$. In this case, **having multiplied the [upper] denominator** twenty-four[861] **by the [lower] denominator,** which is the number six below, it becomes one hundred forty-four. **One should multiply the first numerator by the denominator below added to [its own] numerator,** that is, **[one should multiply the first numerator]** sixty by the seven obtained from having added one to six; it becomes four hundred twenty whose denominator is one hundred forty-four: $\begin{array}{|c|}420\\144\end{array}$. The rest is erased. This [obtained] quantity should be placed elsewhere (i.e., put aside).[862]

859 Line 26: *sampūrṇa*°] em. The commentator emphasises that hereafter the first procedure of *bhāgānubandhajāti,* which has been explained by Śrīpati in the first quarter of the *karaṇasūtra,* is demonstrated; first (lines 26–27) is the case defined by the SGT as *bhāgānuban dhasahitāṅka,* and then (lines 1–10) is the *bhāgānubandhabhāga.*

860 In brief, since it has been multiplied.

861 Lines 7–8: *caturviṃśatiṃ*] em.

862 In line 10, at that point of the procedure the use of a double *daṇḍa* is anomalous.

Now[863] there is presentation of [the procedure relating to] fractions in association with fractions: $\begin{array}{|c|}1\\3\\1\\6\\1\\4\end{array}$. In this case, because there is no integer (*rūpābhāvāt*), the procedure beginning with **multiplied by the denominator** does not occur.[864] Thus, **having multiplied the [upper] denominator,** which is the three above**, by the [lower] denominator,** which is six, it becomes eighteen. **One should multiply the first numerator by the denominator below added to [its own] numerator**: by being the "numerator" one added to six, [hence one is multiplied] by seven, it becomes seven whose denominator is eighteen. As it was multiplier, the rest is erased: $\begin{array}{|c|}7\\18\end{array}$. Below it, there is its "one-fourth": $\begin{array}{|c|}7\\18\\1\\4\end{array}$. **Having multiplied the denominator**, which is eighteen, **by the denominator**, which is four, it becomes seventy-two. **[One should multiply the first numerator] by the denominator below,** which is five, **added to [its own] numerator**: by being the numerator one added to four, **one should multiply** seven by five;[865] it becomes thirty-five whose denominator is seventy-two: $\begin{array}{|c|}35\\72\end{array}$. When there is the reduction by twelve of the number [which is the result] of the integer in association with fractions which has been previously placed aside,[866] which is four hundred twenty,[867] it becomes thirty-five. When there is the reduction of one hundred forty-four by

863 Cf. the earlier *atra* (line 25) and *idānīm*; these two adverbs provides rhythm to the discourse. The commentator explains how to carry out the second part of the sample problem. According to the commentator's terminology (see the introductory line to the sample problem), the procedure concerns the case termed *rūparahitabhāgānubandha*, as the fractions are associated to each other without integer.

864 This is a quotation from the first passage of the *karaṇasūtra*, which expounds the first method of *bhāgānubandhajāti* concerning an integer in association with a fraction. Since here the given fractions are associated to each other but there is no integer, the commentator points out that the first method should not be applied. Differently, above in lines 26–27 in the computation the first method enunciated in Śrīpati's rule has been applied by the commentator as first, and it is followed by the method concerning association of fractions with fractions.

865 In line 16, one would expect *āhanyate* (see lines 3, 9, and 13) instead of *hanyate*; also, one should supply *ādyaṃ aṃśaṃ*.

866 *sthāpita*.

867 It refers to the layout shown in line 10.

twelve, it becomes twelve: $\begin{array}{|c|}35\\12\end{array}$. The second quantity should be placed in front in the following manner:[868] $\begin{array}{|c|c|}35&35\\12&72\end{array}$. In this case, when there is the reduction of the denominators twelve and seventy-two by twelve, it becomes one and six respectively. Next, by means of [the procedure] "the numerator and the denominator"[869] and so forth, when there is the exchange [of the quotients], it is: $\begin{array}{|c|c|}35&35\\12&72\\6&1\end{array}$. Thirty-five is multiplied by six, it becomes two hundred ten; twelve is multiplied by six becoming seventy-two. The next quantity is multiplied by one, it remains exactly so. Then, when the additive quantity thirty-five is added to two hundred ten, it becomes two hundred forty-five whose denominator is seventy-two; the rest is erased: $\begin{array}{|c|}245\\72\end{array}$. When there is the division of the number above by seventy-two, the quotient is the integer three and twenty-nine whose denominator is seventy-two: $\begin{array}{|c|}3\\29\\72\end{array}$.

In this way, the class of fractional increase is concluded.

The class of fractional decrease (bhāgāpavāhajāti)[870] *[GT 60]*

[Next is] the procedural rule of the class of fractional decrease;[871] the author enunciates a metrical stanza:

> 60. In the procedure of fractional decrease, once the integer is multiplied by the denominator, one should subtract the numerator from the [obtained] quantity.[872] One should multiply the [lower] denominator by the [upper] denominator, [and] one should multiply the foremost numerator by the [lower]

868 I have emended the text as in line 19 the first layout is missing; it is clear, however, that the SGT refers to two fractions placed one next to the other.

869 This is a quotation from the class of simple fraction (GT 53).

870 K 37, 2–5.

871 *apavāha* – lit. "deduction, removal". This procedure involves the deduction of a fraction from an integer or the deduction from a fraction of a fraction of itself. The expression *bhāgāpavāhana* is also used by Śrīpati in the sample problem (see GT 61). The expression *bhāgāpavāha* could also be rendered as "fractions in dissociation".

872 As in the *karaṇasūtra* of the class of fractional increase, here too Śrīpati gives two methods: the first concerns an integer lessened by a fraction, the second the subtraction from a fraction of a fraction of itself.

denominator decreased by the lower[873] numerator. Those who know [the procedure regarding] the reduction [of fractions] explain [this method].

The explanation: wherever a fraction is removed from an integer, or else if from an integer which is diminished by a fraction[874] a fraction [of itself, i.e., of this subtraction] [is removed], or else if from a fraction a fraction [of itself] [is subtracted], in cases as such these then[875][the expression] **in the procedure of fractional decrease** [is relevant].[876] [With respect to the expression] **[once the integer is] multiplied by the denominator,** [regarding the fraction] for instance one-fourth, the "denominator" of the fraction which is above is meant.[877] **Once the integer** is multiplied by it, **one should subtract the numerator** (*lava*>*bhāga*), namely that several[878] [numbers] with a zero standing behind.[879] The products [from the multiplication of the integers] by the "denominators" become the upper numerators instead of the quantity consisting of the numerators above. Or else that quantity which is the "integer" is "multiplied by the denominator"; [one should subtract the numerator] **from the quantity,** which is this [quantity obtained] from the "integer multiplied by the denominator". The meaning is as before.[880]

Afterwards,[881] **one should multiply the [lower] denominator,** which is the "denominator" of the fraction underneath, by the upper (*uparitana*) **denominator**. In other words, **[one should multiply]** the "denominator" of the upper fraction by the "denominator" of the fraction underneath.[882] In the same way, [the expression] **the lower numerator**: **one should multiply the foremost**

873 *adhastana*.

874 Line 6: °*apavāhya*°] conj.; °*saṅkalita*° (K). I cannot interpret the transmitted reading, here *saṅkalita* ("addition") in *bhāgasaṅkalitarūpād* does not fit the context. One would expect a construction with a verbal adjective qualifying *rūpa* ("integer") and indicating its relationship to *bhāga* ("fraction"). I have proposed the following conjectural emendation reproducing the argument put forward by the author.

875 In this nominal sentence, the relative adverb *yatra* and its correlative *tatra* specify the following of reasoning: where the circumstances which have been described occur, then the rule beginning with the expression "in the procedure of fractional decrease" should be applied.

876 The commentator frames the conditions in which to apply the rule.

877 The SGT elucidates the rule by means of the numerical data given by the GT in the next sample problem.

878 *aneka* – in the SGT, this is the first occurrence of the term.

879 In this regard, see the layout of line 19. The SGT underlines (see also SGT 61, 18) that, in order to mark a subtraction, a zero is placed "behind" (from right to left) the quantity to be subtracted.

880 Here ends the explanation of the first method of fractional decrease (lines 7–10). It appears that the commentary refers to the preceding rule; the class of fractional increase is, in fact, the reverse of the class of fractional increase: the algorithms are basically the same, the only difference is the operational sign.

881 The adverb *tatas* introduces the second method.

882 In commenting upon the second method, the commentator paraphrases the root-text and provides synonyms for "numerator", "denominator", and the qualifiers "upper" and "below".

numerator by the [lower] denominator diminished by the numerator which is below. The rest is clear [in meaning].

In this respect, an exemplifying verse:

> 61. O friend, if you know the [procedure of] fractional decrease (*bhāgāpavāhana*), having made their assimilation, tell [me] once more the previously mentioned numbers [but] diminished[883] [this time] by their own fractions.

The explanation: **the previously mentioned [numbers]** are those beginning with "ten plus one-fourth";[884] here one should understand "ten less by one-fourth" and so forth. [Regarding] that previously manipulated[885] "numbers": in this case, the very "numbers" should instead be subtracted[886] by the very "fractions".[887] Since these fractions are subtracted, in order to mark [this operation],[888] a zero should be placed (*deya*) behind these.

The presentation of the first[889] sample problem is:

10	1	2
01	02	01
4	2	3

. In this case, the very procedure of the first quarter of the stanza should be carried out.[890] To illustrate: [with respect to the expression] **the denominator,** in the first quantity, that is, four[891] is **the denominator;** the quantity which is the "integer", such as ten, is multiplied[892] by it; it becomes forty. Then, **one should subtract** the numerator;[893] when there is the subtraction of the numerator one, it becomes thirty-nine whose denominator is four:

39
4

. Regarding the second quantity, [the expression] **[once the integer is] multiplied by the denominator**

883 *vivarjita*.

884 The author refers to the numbers given in the sample problem formulated in GT 58, which concerns the first method of *bhāgānubandhajāti*.

885 *prayukta* – lit. "undertaken, performed". In mathematical literature, this verbal adjective acquires a specific meaning in the context of financial transactions; see, for instance, GT 128–129 and SGT 131, 18 where it means "lent/borrowed".

886 *viyojya*.

887 Line 17: *bhāgais tu*] em. The particle *tu* is significant as it underlines the contrast with the previous circumstances.

888 *upalakṣaṇa* – lit. "mark, designation".

889 The numbers manipulated in this sample problem are the same numbers previously given by Śrīpati (GT 58 and 59). The execution of the second sample problem (GT 59) begins in line 7.

890 In the first quarter of the stanza, the GT has expounded the first method of the class of fractional increase.

891 In line 19, the *daṇḍa* after *prathamāṅke* is mistakenly found. Lines 19–20: *ghnarūparāśir*] em.

892 *ghna*. I suspect that there has been the loss of the initial syllable *ni*, as one would expect the SGT to quote *nighna*.

893 *bhāga*.

[here] implies that one is multiplied by two, it becomes two; when there is the subtraction of one from this [two], it becomes one whose denominator is two: $\begin{array}{|c|}1\\2\end{array}$. Regarding the quantity which is the third, [by applying the step] **[once the integer is] multiplied by the denominator,** two is multiplied by three becoming six; when there is the subtraction of the numerator one [from it], it becomes five whose denominator is three: $\begin{array}{|c|}5\\3\end{array}$. Hence, for the purpose of the sum of these [fractions], by means of [the procedure] "the numerator and the denominator"[894] and so forth, in the first quantity when there is the reduction of the denominator four by half, it becomes two. In the second quantity, when there is the reduction of the denominator two [by half], it becomes one. Next, when there is the exchange [of the resulting reduced denominators], it is: $\begin{array}{|r|r|}39&1\\4&2\\1&2\end{array}$. The first quantity multiplied by one is exactly so; regarding the second quantity, one multiplied by two becomes two; two multiplied by two becomes four. Since there are equal denominators, when the additive quantity two is added to thirty-nine,[895] which is the upper numerator of the first [quantity], it becomes forty-one whose denominator is four. The rest is erased: $\begin{array}{|r|}41\\4\end{array}$. The same [should be carried out] in the third quantity;[896] when there is the exchange of the denominators four and three, it is: $\begin{array}{|r|r|}41&5\\4&3\\3&4\end{array}$. Then, forty-one multiplied by three becomes one hundred twenty-three, four multiplied by three becomes twelve: $\begin{array}{|r|}123\\12\\3\end{array}$. In the next quantity, five multiplied by four becomes twenty, three multiplied by four becomes twelve: $\begin{array}{|r|}20\\12\\4\end{array}$. Afterwards, when the additive quantity twenty is added to the numerator one hundred twenty-three, it becomes one hundred forty-three whose denominator is twelve: $\begin{array}{|r|}143\\12\end{array}$. The rest is erased. When there is then the division of one

894 This is a quotation from the class of simple fraction (GT $\overline{53}$).

895 Line 28: *ekonacatvāriṃśan*] em; *ekacatvāriṃśan* (K). This is an example of dittography (see the same transmitted reading in the same line).

896 Line 29: *tṛtīyāṅke*] em.

hundred forty-three by twelve, the quotient is the integer eleven and eleven-twelfths: $\begin{array}{|c|}11\\11\\12\end{array}$.

Next is the presentation of the sample problem of the second metrical stanza:[897] $\begin{array}{|c|}1\\01\\4\\01\\2\\01\\3\\01\\6\end{array}$. The fractions are:[898] $\begin{array}{|c|}1\\3\\01\\6\\01\\4\end{array}$. In this case, the complete[899] procedure [will be shown].[900] [Regarding the expression] **[once the integer is] multiplied by the denominator**, first the quantity which is the "integer" one is "multiplied" "by the denominator" four becoming four; when there is the subtraction of the "numerator" one [from it], it becomes three. **One should multiply the fore-most numerator** consisting of three **by the denominator** two **decreased by the lower numerator**;[901] since it is lessened by the numerator one, [one should multiply three] hence by one. This is placed in that very manner, thus three whose denominator is eight is: $\begin{array}{|c|}3\\8\end{array}$. Then, below it, the "lower"[902] [fraction] one-third: $\begin{array}{|c|}3\\8\\01\\3\end{array}$. One should multiply[903] **the denominator,** which is the eight above,

897 These are the same numbers given in GT 59. Cf. this layout and the one found in SGT 59, 25.

898 Note that the commentary manipulates those given fractions from line 19 onwards. Cf. this layout and the one given by the SGT in SGT 59, 11.

899 Perhaps also conceivable would be to emend *pūrṇā* to *pūrvā* ("first") so that the passage becomes "here the first procedure [...]". The evidence that *pūrṇā* (lit. "whole, full") could stand for *pūrvā* is based on the commentary itself: the first "procedure" (*prakriyā*) of the class of fractional increase is first executed as the case of an integer decreased by a fraction occurs (lines 8–18); the second procedure follows this result.

900 This first part (in line 8, the use of *prathamam* is indicative) involves the first method of fractional decrease (an integer decreased by a fraction, and so forth), and then the second method (a fraction lessened by a fraction if itself, and so forth) is applied.

901 Line 9: °*lavonahareṇa*] em.

902 Note that here the adjective *adhastana* (from the *karaṇasūtra*) denotes both operands of the fraction as "lower". In the rule, it thus qualifies both *lava* (the "numerator") and *hara* (the "denominator").

903 One would expect *guṇayet* (see in lines 20 and 23) instead of *hanyāt*.

by **the denominator** three;[904] it becomes twenty-four. **One should multiply the foremost numerator by the [lower] denominator** three **decreased by the lower numerator;** since it is subtracted by the numerator one, [one should multiply] that consisting of the three at the top by two. It becomes six whose denominator is twenty-four: $\begin{array}{|c|} 6 \\ 24 \end{array}$. Below it, the "lower" [fraction] one-sixth: $\begin{array}{|c|} 6 \\ 24 \\ 01 \\ 6 \end{array}$. **One should multiply the denominator**, which is the twenty-four above, **by the denominator below**, which is six; it becomes one hundred forty-four. **One should multiply the foremost numerator by the [lower] denominator** six **decreased by the lower numerator;** since it is subtracted by the numerator one, [one should multiply] six by five. It becomes thirty whose denominator is one hundred forty-four: $\begin{array}{|c|} 30 \\ 144 \end{array}$. It should be understood that the multiplier in all cases is erased. When there is the reduction of these [numbers] by means of the division by six, it is five above and twenty-four below: $\begin{array}{|c|} 5 \\ 24 \end{array}$. This should be placed in a [different] place (i.e., put aside).[905]

Next is the procedure with fractions [of fractions].[906] In this case, because there is no integer, the procedure **the integer multiplied by the denominator** does not occur.[907] Thus,[908] the presentation of "one-third" minus [its] "one-sixth": $\begin{array}{|c|} 1 \\ 3 \\ 01 \\ 6 \end{array}$. **One should multiply the denominator**, such as the three above, **by the denominator**, which is the six that is underneath; it becomes eighteen. **One should multiply the foremost numerator by the [lower] denominator** six **decreased by the lower numerator;** as it is subtracted by the numerator one, [one should

904 Although the rule does not specify so, in SGT 60, 10 the commentary clarifies that in *guṇayed hareṇa chedaṃ*, *hara* is to be understood as *uparitana* ("upper") and *cheda* is *adho'ṃśacchedaṃ* or "the denominator of the numerator below". Because of the commutative property of multiplication, it does not make any difference, but it is a curious feature which occurs again in lines 20 and 23.

905 Cf. SGT 59, 10.

906 This remark (see in both cases the use of *bhāga*) refers to *atha bhāgāḥ* and the layout in line 7.

907 The commentary has made a similar remark while solving the previous sample problem (see SGT 59, 11). It clarifies that here only the second method of fractional decrease is involved as no integer is attached to the fraction, while above (lines 7–18) both methods have been carried out.

908 Line 19: **śeṣā tu**] em. Unless *śeṣa* qualifies the preceding *prakriyā* to denote the "remaining" procedure; this usage, however, does not occur in the SGT.

multiply] one by five; it becomes five whose denominator is eighteen: $\left|\begin{array}{c}5\\18\end{array}\right|$. The rest is erased. Below it, there is one-fourth [which] is [to be] subtracted: $\left|\begin{array}{c}5\\18\\01\\4\end{array}\right|$.

One should multiply the above eighteen, which is **the denominator**, by four, which is **the denominator** below; it becomes seventy-two. **One should multiply the foremost numerator by the [lower] denominator** four **decreased by the lower numerator;** since it is subtracted by the numerator one, [one should multiply] five by three; it becomes fifteen whose denominator is seventy-two: $\left|\begin{array}{c}15\\72\end{array}\right|$. When there is the reduction of these [numbers] by three, above is five, below is twenty-four: $\left|\begin{array}{c}5\\24\end{array}\right|$. Afterwards, five added to the quantity derived (*niṣpanna*) from the fraction [attached to] the "integer",[909] which is five whose denominator is twenty-four,[910] since the same denominator occurs, this five is added [to the other numerator five]. It becomes ten whose denominator is twenty-four: $\left|\begin{array}{c}10\\24\end{array}\right|$. When there is the reduction of these [numbers] by half, it becomes five above and twelve below. Because of the absence of a [suitable] dividend,[911] it is: $\left|\begin{array}{c}5\\12\end{array}\right|$.

In this way, the procedure of the class of fractional decrease is concluded.

Chain-simplification class (vallīsavarṇanajāti)[912] *[GT 62]*

Next is the procedural rule of the chain-simplification class; the author enunciates a metrical stanza:[913]

> GT 62. One should multiply the foremost denominator and numerator[914] by the denominator which is underneath (*talasthita*). For a rapid simplification

909 The fraction meant is one minus one-fourth.

910 The commentator refers to the fraction which has been put aside; see line 18.

911 *bhājya* – lit. "to be divided". The SGT observes that the numerator five cannot be divided by its denominator since this is larger; therefore, this is a proper fraction and it t remains as it is. At the end of a procedure, the commentary, in fact, usually transforms an improper fraction into a proper one. See the same expression in a similar context and with the same meaning in SGT 49, 19 and SGT 51, 28.

912 K 39, 7–10.

913 *vallī* – lit. "creeper". This procedure involves the simplification of a creeper-like chain (*vallī*) of fractions into a proper fraction.

914 In line 7, *prāk* should be compounded with *chedabhāgau*.

by the [method called] *vallī*, one should perform (*vidadhyāt*) the numerator standing below either it is positive[915] or negative[916] in relation to the first numerator.[917]

The explanation: this procedure concerns a pair of quantities [each time].[918] There is the quantity-numerator and that consisting of the "denominator". **One should multiply the foremost**[919] **denominator and numerator,** which are the denominator and the numerator at the top, **by the denominator which is underneath**, which is the denominator that is below. Afterwards, **one should perform the numerator standing below,** which is that termed **negative. [One should perform]** that **numerator,** which is defined as "negative" **in relation to the first numerator** since it is subtracted. Behind that "numerator", a zero is written. In other words, one should subtract [the negative numerator] from the uppermost quantity-numerator, which is multiplied **by the denominator** below. Similarly, one should add (*nikṣipet*) the **numerator** standing below, which is that termed **positive.**[920] **[One should perform]** that [termed] "numerator", which is defined as "positive" **in relation to the first numerator**, which is the uppermost quantity-numerator that is multiplied by the denominator below because the [procedure of the] class of fractional increase occurs.[921] Behind the [positive] "numerator", there is no zero.[922]

915 *dhana*.

916 *ṛṇa* – lit. "debt", here this is a technical term used adjectivally to mean "negative". Quantities are defined as "positive" and "negative" according to their operational sign, thus fractions are "positive" when they are "added" and "negative" when they are "subtracted".

917 Cf. PG 41 and TŚ 26.

918 This procedure involves, in fact, the assimilation of fractions two by two.

919 According to the editorial conventions used by Kāpadīā, the em-dash should be placed after the word *prāk* as the commentator glosses *prāk* with *upari* to indicate that the *prākcheda* and *prākbhāga* or the "foremost denominator" and "foremost numerator" are the *uparicheda* and *uparibhāga* or the "denominator at the top" and "numerator at the top". An anomalous em-dash also occurs further in line 1 between *prāk* and *cheda*.

920 Line 15: *dhanam iti*] em.; *dhanaṃ* (K). Cf. *ṛṇam iti* in line 13.

921 The commentary refers to the second procedure of fractional increase, where a fraction is increased by a fraction of itself. In accordance with the commentator's argument, emendation could be considered, in which case *pūrvalave* should be followed by *vivadhyāt* and a *daṇḍa*. Note that the SGT structures the passage which runs from line 12 to line 16 in a symmetric way: each half starts (the second begins with *tathā*, line 14) with the quoted *adhaḥ sthitam aṃśaṃ* and clarifies that this "numerator" (*aṃśa*) can be either "negative" (*ṛṇa*, line 13) or "positive" (*dhana*, line 15), each specifies that a zero may or not occur behind the numerator, and it continues with the same expression *tam aṃśam* followed by *ṛṇasañjñam* the first and by *dhanasañjñam* the second. The final passage ends with the verb which becomes the signifier of the more generic root *vidh-* used in the root-text (see, in this respect, *apanayet* and *nikṣipet*).

922 Lines 12–16 elaborates on *vidadhyāt* or "one should perform". The commentator clarifies that this instruction ("one should perform") acquires a different connotation according

In this case, by the word "negative", the class of fractional increase explained [before] is not [referred to], but it is rather the procedure just seen (i.e., the class of fractional decrease).[923]

In this regard, in the exemplifying metrical stanza [which follows] the author enunciates a sample problem:

> GT 63. O friend, if you wish to learn [mathematics],[924] having performed their simplification, calculate quickly two *dramma*s, also five *paṇa*s and one *kākiṇī* minus one *kapardikā*, and [minus] its[925] one-fourth too.

The explanation: [regarding] **two *drammas*** [and] **also five *paṇas***, [it is said that] there is one *dramma* in such those sixteen *paṇas*.[926] In the *Triśatī,* [a *dramma*] is called *purāṇa*. **Five *paṇa***s are as such.[927] A *kākiṇī* is one-fourth of a *paṇa,* which [therefore] consists of four *kākiṇī*s. This [*kākiṇī*] **minus** one *kapardaka*: a *kaparda*[928] is a twentieth of a *kākiṇī,* [hence one *kākiṇī*] **minus** this [part] as well as **also its one-fourth**. By the word ***tat*** ("its"), according to the rule (*nyāyāt*): "the investigation of that which is implied in a compound by means of a pronoun (*sarvanāmnā*)",[929] "one-fourth" (*aṃhri>caturbhāga*) "of this [mentioned] *kapardaka*" is [meant]; it (i.e., the *kākiṇī*) is subtracted by this [one-fourth *kapardaka*] **too.**[930] This is the reason why **having performed their simplification**[931] is "having joined" [them]; tell [the result] **if you wish to learn** means "if you are learning mathematics".

to the context: it means "one should subtract" if from a fraction a fraction of itself is subtracted, it means "one should add" if the fraction and a fraction of itself are added.

923 When the numerator is "negative", one should not apply the *bhāgānubandhajāti* (as in the case of a "positive" numerator") but the *bhāgāpavāhajāti,* which is, in fact, the previous procedure (GT 60).

924 In line 28, the SGT supplies the object *gaṇita* to the verb *budh-* "to know" and also provides a different verb mood (passive, present indicative).

925 In lines 26–27, the commentator explains that "its" (see *tat*) refers to the *kaparda* unit of money.

926 See the units of money mentioned in GT 4.

927 In brief, according to the equivalence just mentioned. The correlative *yādṛś/tādṛś* emphasise the relationship between the *dramma* and *paṇa* money units.

928 See the forms *kaparda/kapardikā/kapardaka*.

929 The purpose of this quotation is to explain the pronoun *tat* which in Śrīpati's rule occurs in *kapardikonā tadaṃhriṇā*. In brief: it is possible connect a noun which is hidden within a compound (*kapardikonā*) by means of a pronoun (*tat*). This quotation, which is from the *Kāvyālaṅkārasūtra* (5.1.11) by Vāmana (9th century CE), concerns figures of speech and poetry; see the edition of the *Kāvyālaṅkārasūtra* by Acharya (1953).

930 *unā vartite* – lit. "it is treated as subtracted".

931 In line 27, the combination *tat etat* is significant. The SGT points out, after having shown the relationship between quantities expressing different monetary units, that "having simplified" (*savarṇayitvā*) implies to join and assimilate the quantities into to the same unit of measurement.

As the quantities have a chain-like nature,[932] the presentation is by means of a vertical arrangement:

2
1
5
16
1
4
01
20
01
4

. In this case, the procedure concerns a pair of quantities [each time].[933] **One should multiply the foremost** "denominator" one, **by the denominator which is underneath,** which is the "denominator" sixteen; it becomes sixteen. Regarding the "foremost numerator", one **should** [also] **multiply** two [**by the denominator which is underneath**]; it becomes thirty-two:

32
16
5
16

. **One should perform,** which [here] means that "one should add",[934] **in relation to the first numerator** such as thirty-two, **the numerator standing below**, which is five. Since there is not a zero behind it, this is denoted as **"**positive". It becomes thirty-seven whose denominator is sixteen:

37
16

. The multiplier is always erased.[935] Below it, there is one *kākiṇī*, its designation is one-fourth, it (i.e., thirty-seven whose denominator is sixteen) is therefore added to one-fourth:[936]

37
16
1
4

. **One should multiply** the "foremost denominator" sixteen **by the**

932 The commentary emphasises the interdependence between the given quantities, which are placed with the largest money unit being at the very top.

933 Here (line 29) the commentary repeats the same passage formulated in line 11.

934 The verb *vidadhyāt* is glossed by the commentator with either *kṣip-* "to add" or *apanī-* "to remove", depending the sign of the fraction; cf. lines 8, 13, and 19.

935 The passage *guṇako yāti sarvatra* indicates that the fraction underneath, which is the multiplier, once manipulated is erased. The multiplier shows the conversion ratio between the unit of money it represents and the one represented by the number standing above.

936 Line 5: *caturbhāga yuktā*] em. Note the use of the verbal adjective *yukta,* which the commentator has already used in line 27 when he observes that "to simplify" implies "to join" the quantities. Here the commentator refers to the relation between a *paṇa* and a *kākiṇī.*

denominator which is underneath, four; it becomes sixty-four. In the same way, **one should multiply** the "numerator" above, thirty-seven, by four; it becomes one hundred forty-eight. **One should perform,** which means that "one should add", **in relation to** that consisting of **the first numerator, the numerator** underneath, which is one; it becomes one hundred forty-nine whose denominator is sixty-four: $\begin{array}{|c|}149\\64\end{array}$. Below it for the display of the given negative quantity,[937] it is lessened[938] by the quantity one *kapardaka*. Since it is designated as one-twentieth of a *kākiṇī*, a *kapardaka* is [hence] lessened by one-twentieth,[939] a zero should be placed [behind it]:[940] $\begin{array}{|c|}149\\64\\01\\20\end{array}$. **One should multiply the foremost** "denominator" sixty-four **by the denominator which is underneath**, which is twenty; it becomes twelve hundred eighty. In the same way, **one should multiply** one hundred forty-nine, which is **the foremost** "numerator", by twenty; it becomes twenty-nine hundred eighty. In this case, **one should perform,** which [in this case] means that "one should subtract",[941] **in relation to the first numerator,** the "negative" [numerator] one, so that when there is the subtraction of one from twenty-nine and so forth, it becomes twenty-nine hundred seventy-nine whose denominator is twelve hundred eighty: $\begin{array}{|c|}2979\\1280\end{array}$. Below it, which is a [quantity expressed in the unit of money] *kapardaka*,[942] there is one-fourth; it (i.e., the quantity above) is lessened by this: $\begin{array}{|c|}2979\\1280\\01\\4\end{array}$. In this case, **one should multiply the foremost** "denominator" twelve hundred and eighty **by the denominator which is underneath**; it becomes five thousand one hundred twenty.[943] Similarly, **one should multiply the foremost** "numerator"

937 Line 9: °*dattāṅka*°] conj.; °*gatāṅka*° (K). I have proposed this conjectural emendation as *gata* does not seem appropriate in the context. This reading takes into account the terminology normally used by the SGT.

938 The adjective *ūna* refers to the substantive *kākiṇī*. The fraction obtained (see line 8) is, in fact, expressed in *kākiṇīs*.

939 Line 10: *viṃśatibhāgonaḥ*] conj.; *viṃśatim ukta unaḥ* (K). The transmitted reading is not interpretable, this conjecture follows the explanation of the commentary.

940 In line 10, the substantive *śūnya* occurs in the masculine gender while it should be neuter; it should thus read: *śūnyaṃ sthāpyam*. Also, in the same line the pronoun *sa* is spurious.

941 See the drop of the initial syllable *vi* in *[vi]dadhyāt* (line 8).

942 Line 15: *caturbhāgas tenonā*] conj.; *nyūnacaturbhāgā* (K). The passage is corrupt.

943 Lines 16–17: *viṃśatyadhi(ka)śatakena yuktāḥ*] em.

twenty-nine seventy-nine by four; it becomes eleven thousand nine hundred sixteen. [**One should perform** here means that] one should subtract, because it is negative **in relation to the first denominator** consisting of eleven and so forth, **the numerator standing below,** which is one; it becomes [that number] ending with fifteen: $\begin{array}{|r|}11915\\5120\end{array}$. When there is the reduction of these numbers by five, it becomes twenty-three hundred eighty-three above, and ten hundred twenty-four below: $\begin{array}{|r|}2383\\1024\end{array}$. In this case, when the division of the number above by the number below is performed, two *dramma*s are obtained. The remainder is the number above consisting of three hundred thirty-five.

To arrive at the [number of] *paṇa*s, it should be multiplied by sixteen; it becomes[944] fifty-three and sixty: $\begin{array}{|r|}5360\\1024\end{array}$. When there is its division by ten twenty-four, five *paṇa*s are obtained. The number above, which is the remainder, is such as two hundred forty. To arrive at the [number of] *kākiṇī*s **minus one *kapardikā,*** it is multiplied by four; it becomes nine hundred sixty. The division does not occur; with respect to the quotient concerning [the number of] *kākiṇī*s, a zero [is placed instead]: $\begin{array}{|c|}0\end{array}$". Then, to arrive at the [number of] *kapardaka*s, nine hundred sixty is multiplied by twenty; it becomes nineteen thousand two hundred. When there is its division by [its denominator] ten twenty-four, the quotient is eighteen *kaparda*s; the number above is the remainder seven hundred sixty-eight.[945] To arrive at the fraction of *kaparda*s, it is multiplied by four; it becomes thirty seventy-two: 3072.[946] When there is its division by ten twenty four, the result is three-fourths.

In this way, [the procedure of the] chain-reduction class is concluded.

3.8 Type-problems of fractions [GT $\overline{64}$–92]

The visible quantity type-problem (dṛśyajāti)[947] *[GT $\overline{64}$]*

Next is the procedural rule of the visible quantity type-problem; the author enunciates half a metrical stanza:

> $\overline{64}$. In the type-problem [of fractions] called the visible quantity, one should divide the visible quantity (*dṛśya*) by [the number] one lessened by the sum[948] of the fractions.

944 Line 23: *jātā*] em.

945 Line 1: *aṣṭaṣaṣṭiś ca*] em.; *ṣaṣṭiś ca* (K). The remainder above is, in fact, 768 and not the transmitted 760.

946 I have emended the text as in line 2 the edition reads 372 instead of 3072.

947 K 41, 6. Here begins the treatment of the type-problems of fractions.

948 *aikya*.

The explanation: [regarding the expression] **in the type-problem called the visible quantity,** the "visible quantity"[949] is, for instance, "one and a half cubits" of the "pillar" (*stambha*);[950] it is [visible] on the basis of direct perception (*lokapratyakṣa*). The author then specifies, which means "he mentions",[951] an invisible (*adṛśya*) [part of the pillar]. The pillar is[952] precisely six cubits [in measure]; he clarifies that [part of the pillar] which is before the eyes. That which is "called"[953] is [to be found] in a certain position in a *karmadhāraya*;[954] this [occurring here] is a *paranipāta*-compound[955] of the qualifier (*viśeṣaṇa*). Next, **in the type-problem called the visible quantity, [one should divide] the visible,** which will be mentioned later [in the following sample problem] and which is namely "one and a half cubits",[956] **by one; [the sum]** of the "fractions" is the "sum" of the parts[957] [of the pillar] being under water and so forth. Because the sum [of the parts] [should be carried out], [one should calculate it] by means of [the procedure] "the numerator and the denominator"[958] and so forth. **[One it is] lessened** by this [sum]. By the procedure of subtraction[959] [of fractions] beginning with **"**[the deduction of the numerators of two quantities] of which equal denominators have been arranged",[960] having performed the deduction[961] of the [sum of the] parts from one, which consists of the positive quantity,[962] **one should divide [the visible quantity]** by the remainder from [the number] one.[963]

949 Line 7: *dṛśyaḥ*] em.
950 The SGT refers to a "pillar" because it is mentioned in the next sample problem.
951 This remark is curious.
952 Note the narrative use of the imperfect tense (*āsīt* is lit. "was").
953 This observation on grammar alludes to the expression *dṛśyākhyajātau*; the commentator underlines that, differently from the usage shown in the GT, the adjective *ākhya* should normally occur at the end of a compound.
954 This denotes a type of descriptive compound where the relation of the first member to the last is appositional, attributive or adverbial.
955 The term *paranipāta* denotes a compound in which a word occurs irregularly at the end.
956 Line 10: *adiṃ*] em.
957 *aṃśa*. In brief, the fractions representing the parts of the pillar.
958 This passage is a quotation from the rule of simple fractions (GT $\overline{53}$); it is quoted again by the commentator while explaining the sample problem (see lines 17 and 24).
959 *vyavakalita*.
960 This passage is a quotation from the rule of subtraction of fractions (GT $\overline{38}$); it is quoted again by the commentator while explaining the sample problem (see line 28).
961 *viśleṣa*.
962 *āya* – lit. "income, wealth". This term has been already mentioned by the commentary while commenting upon GT $\overline{38}$, which is the procedural rule of subtraction of fractions. While in SGT $\overline{38}$, *āya* is "capital", here it is a technical term denoting a quantity which is "positive" in contrast with the quantity which is subtracted (and thus "negative") from it. This connotation is similar to that acquired by the term *dhana* (and its opposite *ṛṇa*) in the previous *karaṇasūtra* (GT 62). See *āya* used with this same meaning (but not applied to the number one) in the next sample problem on this rule (line 9).
963 By the expression *rūpaśeṣa*, the commentator denotes the result of the deduction of the various fractions from the number one. While commenting the sample problem, he uses the expression *rūparāśi* (line 27) to denote this same quantity.

In this respect, the author enunciates an exemplifying metrical stanza:[964]

> 65. Think [carefully] and tell [me] the measure[965] of the pillar of which one-half is in water, one-twelfth is in mud, one-sixth is immersed in the sand, and one and a half cubits are visible.

[The meaning of this stanza] is clear. The presentation is: $\begin{array}{|c|c|c|c|} 1 & 1 & 1 & 1 \\ 2 & 12 & 6 & 1 \\ & & & 2 \end{array}$. According to [the rule] "the numerator and the denominator"[966] and so forth, when there is the reduction of the denominator twelve by half, when there is the exchange [of the denominators], it is: $\begin{array}{|c|c|} 1 & 1 \\ 2 & 12 \\ 6 & 1 \end{array}$. At first,[967] one multiplied by six becomes six; in the same way, two multiplied by six becomes twelve. Regarding the next number, [when one and twelve are] multiplied by one, it is exactly so. Since there are equal denominators, when the additive quantity one is added to six, it becomes seven whose denominator is twelve: $\begin{array}{|c|} 7 \\ 12 \end{array}$. Next, with respect to [the next quantity] one whose denominator is six,[968] when there is the reduction of the denominator six and of twelve, it becomes one and two respectively. When there is then the exchange [of the denominators], it is: $\begin{array}{|c|c|} 7 & 1 \\ 12 & 6 \\ 1 & 2 \end{array}$. The quantity which occurs first[969] multiplied by one is exactly so. In the next quantity, one is multiplied by two becoming two; six multiplied by two becomes twelve. As there are equal denominators, when the additive quantity [two] is added to seven, it becomes nine whose denominator is twelve:[970] $\begin{array}{|c|} 9 \\ 12 \end{array}$. This is **the sum of the fractions;** it should be subtracted (*pātanīya*) from "one". By means of

964 This stanza is the same as TŚ example 25.

965 *māna*.

966 This passage is a quotation from the rule of simple fractions (GT $\overline{53}$).

967 Emendation could also be considered; because of its position in the sequence (see *parāṅka* and *pūrvāṅka* in lines 19, 22, 25–26), here the adverb *prāk* is suspect, one would expect an adjective qualifying *aṅka*. We could emend *prāk* to *prācyāṅke* and translate it as "in the first quantity"; this expression is commonly used in the commentary. See, for instance, *prācyāṅka* in the explanation of the next sample problem (SGT 66, 23).

968 The sum of the first two fractions is added to the third fraction.

969 *pūrvāṅka*.

970 In line 23, before the layout *yathā* should presumably occur.

[the procedure] "the numerator and the denominator"[971] and so forth, in order to arrive at equal denominators,[972] the [integer] one should be manipulated:[973] $\begin{array}{c}1\\1\end{array}$. Having arranged (*maṇḍayitvā*) the [number] one [next to the other quantity], when there is the exchange of the denominators and when there is the multiplication,[974] it is: $\begin{array}{cc}9&1\\12&1\\1&12\end{array}$. The quantity which occurs first multiplied by one is exactly so. Regarding the next quantity, one multiplied by twelve becomes twelve whose denominator is twelve:[975] $\begin{array}{c}12\\12\end{array}$. From this, which is the quantity [obtained] from [the subtraction from the number] "one" (*rūparāśi*),[976] **the sum of the fractions**, which consists of nine, should be deducted (*viśleṣaṇīya*); the remainder is three, which is the quantity [remaining] from [the subtraction of the sum from] "one", whose denominator is twelve: $\begin{array}{c}3\\12\end{array}$. This becomes the quantity-divisor.[977] According to [the rule][978] "multiplied by the denominator"[979] and so forth, the "visible quantity" becomes three whose denominator

971 This passage is a quotation from the rule of simple fractions (GT 53). Note in the edition the absence of quotation marks.
972 Line 24: *samacchedārtham*] em.
973 Or perhaps, "[the integer] one should be written"; *rūpaṃ…kāryam* is lit. "[the integer] one should be made".
974 The multiplication is the next step carried out.
975 In line 26, it is unusual to find a *daṇḍa* before the layout.
976 *rūparāśi*, see the same expression further on in the same line.
977 In line 28, *hararāśi* underlines that, in the next step, this fraction becomes the "divisor". Notably, this is the same quantity termed *rūparāśi/rūpaśeṣa* in line 13 (while commenting upon the rule) and in line 27.
978 The text is most uncertain here (lines 28–3). After the mentioned *hararāśi* (line 28), it seems that something is missing: the quotation from GT 42 (*kṛtvā parīvartanam* […]) occurs too early, since i) the two quantities have not yet been arranged in the layouts, and ii) the *dṛśya* has not yet been reduced to a simple fraction (by means of the *bhāgānubandhajāti*). I have emended the text as follows (lines 28–3, after *ayaṃ hararāśi jātaḥ*): *chedanighnetyādinā dṛśyaṃ jātās trayo dvicchedā yathā* $\begin{array}{c}3\\2\end{array}$ | *trayāṇāṃ dvicchedānāṃ hararāśitvāt kṛtvā parīvartanam aṃśahārayorityādinā yathā* $\begin{array}{cc}12&3\\3&2\end{array}$ | *kuliśāpavartanaṃ kṛtvā dvādaśadvikayor ardhāpavarte ṣaḍ ekaś ca trayāṇāṃ ca tribhāgāpavarte ekaś ca eko yathā* $\begin{array}{cc}6&1\\1&1\end{array}$ |] conj. This conjecture follows the commentary. Other reconstructions are no doubt possible.
979 This is a quotation from the rule of the class of fractional increase (GT 57).

is two, it is: $\begin{array}{|c|}3\\2\end{array}$. Then, concerning [the quantity] three whose denominator is two, since it is the divisor, by means of [the procedure] "having performed the interchange of the numerator and of the denominator"[980] and so forth, it is: $\begin{array}{|c|c|}12&3\\3&2\end{array}$. Having carried out the "cross-reduction",[981] when there is the reduction of twelve and two by half, it is six[982] and one, and when there is the reduction of the denominator three and of [the numerator] three, it is one and one: $\begin{array}{|c|c|}6&1\\1&1\end{array}$. All [these numbers] multiplied by one remain as they are. Regarding that [number] which has been divided by one,[983] the number which is the quotient remains as it is:[984] six, which is the number that has been divided by [the number] one. The quotient is exactly so: six cubits whose denominator is one: $\begin{array}{|c|}6\\1\end{array}$.

In the *Līlāvatī*, there is a procedural rule regarding the visible quantity type-problem:[985]

> L 51: "The assumed quantity (*iṣṭarāśi*) is treated as the specified quantity[986]; it is multiplied (*kṣuṇṇa*), divided, increased or diminished by the [given] fraction. The visible quantity being multiplied by the assumed

980 This is a quotation from the rule of division of fractions (GT 42). The visible quantity is to be divided by the *hararāśi*/*rūparāśi*/*rūpaśeṣa* $\begin{array}{|c|}3\\12\end{array}$.

981 This is a quotation from the rule of division of fractions (GT 42).

982 Line 2: *ṣaḍ*] em. This is one of the numerous instances of the missing sandhi between *ṣaṭ* and the following word.

983 It refers to the quantity $\begin{array}{|c|}6\\1\end{array}$. It is noteworthy that, in the SGT, similar remarks occur only in the explanation of the type-problems of fractions; see, for instance, SGT 68, 8 and SGT 69, 28.

984 This formulation is clumsy and it is possible that the text requires emendation; one could consider emending *ekabhaktasyāṅko labdhaṃ tathaiva* to *ekabhaktāṅkaḥ sa eva*: "The number which is divided by one is exactly so". This emendation is based on a similar construction found in SGT 91, 10. Also possible would be *ekabhaktāḥ ṣaḍ tathaiva*: "Six is divided by one, it is exactly so"; see SGT 68, 8. Grammatical constructions with *tad* and *eva* instead of *aṅka* as the subject and the adverbial *tathaiva* also occur; see, for instance, *ekabhaktaṃ tad eva* in SGT 82, 13 and SGT 85, 16. Note that the construction with *tad* marks a pronominal clause; *eva* stands in a nominal sentence without copula.

985 The SGT quotes L 51, which is the rule that Bhāskarācārya applies to solve a sample problem on the *dṛśyajāti* (see L 53).

986 *uddiṣṭaka* – lit. "mentioned, described". For a study on the way in Sanskrit literature on musicology this term is related to musicological puzzles and computations, see Petrocchi (2018).

quantity and divided by this [result] yields the number sought. This [procedure] is called *iṣṭakarman*".

[I provide] its (i.e., of the stanza above) explanation by means of the sample problem [below].[987]

The author enunciates the [following] sample problem:

66. [Of a herd of elephants] the first half[988] together with its one-third have fled to the summit of an excellent mountain,[989] one-sixth together with one-seventh are drinking water in a river. One-eighth with its one-ninth are playing near lotuses and a royal elephant followed by three female elephants are among lotuses. What is the number of the herd[990] [of elephants]?

The meaning [of the sample problem] is [clearly] understood (i.e., there is no need of commenting upon it). The presentation is: $\begin{array}{|c|c|c|} 1 & 1 & 1 \\ 2 & 6 & 8 \\ 1 & 1 & 1 \\ 3 & 7 & 9 \end{array}$. |dṛśya 4|.[991] Since it (i.e., the set of data) is devoid of integer, [the procedure of] the class of [fractional increase involving] fractions in association with fractions [should be performed].[992] According to [the rule] "[having multiplied] the [upper] denominator by the [lower] denominator"[993] and so forth, one should thus multiply "the [upper] denominator", which is [the two] above,[994] "by the [lower] denominator", which is the three below; it becomes six. One should multiply "the first numerator by the denominator below added to [its own] numerator", that is to say,[995] [one should multiply it] by three added to one which becomes four, [the first numerator one multiplied by four] becomes four whose denominator is six: $\begin{array}{|c|} 4 \\ 6 \end{array}$. Since it has been multiplied, the rest is always erased (*prayāti*).

987 This sample problem corresponds to TŚ 27. The commentary solves this exercise first by applying the rule given by L 51, and then (from line 27) by applying the rule given by the root-text (GT 64).

988 The edition of the TŚ by Dvivedī (1899) reads *yūthārdhaṃ* instead of *pūrvārdhaṃ*.

989 *girivara* – in Sanskrit literature it is a common expression.

990 *yūtha*.

991 Line 13: *dṛśyaḥ*] em. When the *sandhi* is properly applied, in the SGT the *visarga* is retained before a number in figures and before a layout (see the numerous *nyśaḥ* before the layout in most of the sample problems as well as in SGT 2–3, 26, SGT 11, 11 and SGT 79, 5.

992 See GT 57; the commentator applies the second method of the class of fractional increase, which concerns fractions of fractions.

993 This is a quotation from the rule of the class of fractional increase (GT 57). From now onwards, the commentary quotes steps from this procedure.

994 Perhaps emendation to *uparidvi* could also be considered.

995 Lines 15 and 18: *sāṃśaka° hareṇa*] em. Cf. GT 57.

Regarding the next quantity,[996] one should multiply "the [upper] denominator" six "by the [lower] denominator", which is the seven below; it becomes forty-two. One should multiply one, which is "the first numerator", "by the denominator below added to [its own] numerator", that is to say, [one should multiply it] by seven added to one; it becomes eight whose denominator is forty-five: $\begin{array}{|c|}8\\42\end{array}$. Regarding the third quantity, one should multiply "the [upper] denominator", which is eight, "by the [lower] denominator", which is the nine below; it becomes seventy-two. One should multiply one, which is "the first numerator", "by the denominator below added to [its own] numerator", that is to say, [one should multiply it] by nine added to one; it becomes ten whose denominator is seventy-two: $\begin{array}{|c|}10\\72\end{array}$. Afterwards, according to [the rule] "the numerator and the denominator"[997] and so forth, when there is the reduction of the denominators of the first and second quantities by [means of] the division by six, the division of forty-two by six is seven and [the division of six by six] is one. When there is the exchange [of these results], it is: $\begin{array}{|c|c|}4&8\\6&42\\7&1\end{array}$. In the first quantity, four multiplied by seven becomes twenty-eight; six multiplied by seven becomes forty-two. The second quantity multiplied by one is exactly so. As equal denominators occur, when the additive quantity eight is added to twenty-eight, it becomes thirty-six whose denominator is forty-two: $\begin{array}{|c|}36\\42\end{array}$. Seventy-two is the denominator of the third quantity: $\begin{array}{|c|}10\\72\end{array}$. When there is the reduction [of this denominator] by [means of] the division by six, it gives twelve; when there is the reduction of the denominator forty-two by [means of] the division by six, it is seven. Next, when there is the exchange, it is: $\begin{array}{|c|c|}36&10\\42&72\\12&7\end{array}$. In the first quantity, thirty-six is multiplied by twelve, it becomes four hundred thirty-two; forty-two multiplied by twelve becomes five hundred four. Regarding the next quantity, ten multiplied by seven becomes seventy; in the same way, seventy-two multiplied by seven becomes five hundred four. Since there are equal denominators, when the additive quantity seventy is added to

996 Line 17: *parāṅke*] conj.; *ega* (? *adho'*) (K). This passage is corrupt; the emendation follows the coherence of the text.

997 This is a quotation from the rule of the class of simple fractions (GT $\overline{53}$).

four hundred thirty-two, it becomes five hundred two whose denominator[998] is five hundred four: $\left|\begin{array}{c}502\\504\end{array}\right|$. This quantity is treated as "the specified quantity".

The object of enquiry[999] formulated by the inquirer is:[1000] "the assumed quantity is treated as the specified quantity", that is to say, as it were this.[1001] By assumption (*kalpanā*),[1002] it is four. This [assumed quantity four] is [added] **to the fraction** by means of [the procedure] "the numerator and the denominator"[1003] and so forth, [added] to the obtained (*niṣpanna*) **fraction** which is the "specified quantity" five hundred two,[1004] when the exchange of the denominators[1005] is performed, it is: $\left|\begin{array}{c|c}502 & 4\\504 & 1\\1 & 504\end{array}\right|$. [Four] **multiplied** (*kṣuṇṇa>guṇita*) [by the exchanged denominator] becomes two thousand sixteen whose denominator is five hundred four:[1006] $\left|\begin{array}{c}2016\\504\end{array}\right|$. This is the **assumed quantity**: the obtained positive quantity (*āyarāśi*) having the same denominator [five hundred four].[1007] Then, the quantity of the fraction which is the "specified quantity" consisting of five hundred two [whose denominator is five hundred four] multiplied by the "assumed quantity" four becomes two

998 Line 5: **uttara**] em.

999 *praśna*, here it appears to denote the "point at issue" rather than the "question" as in SGT 25, 17 where it is opposed to *uttara* or "answer".

1000 *pṛccaka*.

1001 *eṣa eva*.

1002 Or perhaps: "by hypothesis, speculation". This is the first occurrence of the term *kalpanā*. Note that here the commentary states that the "visible quantity" 4 composed of the three female elephants which are playing together with one male elephant is the "assumed quantity". In line 13, the number 4 is, in fact, referred to as the "visible quantity".

1003 This is a quotation from the rule of the class of simple fraction (GT $\overline{53}$). However, the procedure carried out is not exactly the same.

1004 Line 7: **tad**] em.

1005 Line 8: *chedayor*] em.

1006 Here there is something anomalous with the execution of the algorithm: this is not the result of the sum, which, in fact, should be $\left|\begin{array}{c}2518\\504\end{array}\right|$. As Hayashi (2013, 58) observes, the SGT applies the rule quoted from the L in a peculiar manner.

1007 The commentator defines the anomalous fraction resulting from what is supposed to be (see earlier) the sum of the "specified quantity" $\left|\begin{array}{c}502\\504\end{array}\right|$ and the "assumed quantity" 4, which is $\left|\begin{array}{c}2016\\504\end{array}\right|$, as "assumed quantity".In the same passage, this same result is termed *āyarāśi* or "positive quantity" (in SGT $\overline{64}$, 12, the SGT has specified the number 1, from which the sum of the parts is subtracted, as *āyarāśi*). In lines 6, 10, and 13, the number 4 is the "assumed quantity".

thousand eight whose denominator is five hundred four: $\begin{vmatrix} 2008 \\ 504 \end{vmatrix}$. The **assumed quantity** two thousand sixteen **lessened by** this **fraction** becomes eight whose denominator is five hundred four: $\begin{vmatrix} 8 \\ 504 \end{vmatrix}$. Afterwards, the "visible quantity" [represented by] "one royal elephant" together with "three female elephants" is **multiplied by the assumed quantity,** that is, **it is** multiplied (*āhata>guṇita*) by the "assumed quantity" four. This [visible quantity] consisting of four, because it is like the assumed quantity, becomes sixteen whose denominator is one: $\begin{vmatrix} 16 \\ 1 \end{vmatrix}$. The [aforementioned quantity], which is the [result of the] "assumed quantity"[1008] "lessened by the fraction", [and] which is the numerator eight whose denominator is five hundred four, is **divided by this** [obtained fraction].[1009] In this case, according to [the rule] "having performed the interchange [of the numerator and the denominator], the cross-reduction [should be carried out]":[1010] when there is the division of eight by eight,[1011] it is one, and when there is the division of sixteen by eight,[1012] it is two : $\begin{vmatrix} 504 & 2 \\ 1 & 1 \end{vmatrix}$. Next, two multiplied by five hundred four becomes one thousand eight, and the denominator multiplied by one is exactly so: $\begin{vmatrix} 1008 \\ 1 \end{vmatrix}$. This is the size of the "herd" [of elephants]: one thousand eight, which has been understood (i.e., obtained) by the number which is the "visible quantity" four.

The verification (*ghaṭanā*):[1013] "half"[1014] of one thousand eight is five hundred four, and its "one-third" is one hundred sixty-eight: $\begin{vmatrix} 504 \\ 168 \end{vmatrix}$. This [one

---

1008 The fraction $\begin{vmatrix} 2016 \\ 504 \end{vmatrix}$, which is the "positive quantity" is also defined as the "assumed quantity" (see line 9).

1009 In this passage, the syntax does not seem to be smooth. The product of the multiplication of the *dṛśya* 4 by the *iṣṭa* 4, which is $\begin{vmatrix} 16 \\ 1 \end{vmatrix}$, becomes the dividend of $\begin{vmatrix} 8 \\ 504 \end{vmatrix}$ (see line 12). This divisor is defined by the commentator as the quantity deriving from the subtraction of $\begin{vmatrix} 2008 \\ 504 \end{vmatrix}$ from the *iṣṭarāśilāyarāśi* $\begin{vmatrix} 2016 \\ 504 \end{vmatrix}$.

1010 This is a quotation from the rule of division of fractions (GT 42).

1011 Line 16: *aṣṭabhāge*] em.

1012 Line 16: *aṣṭabhāge*] em.

1013 In the SGT, this kind of "verifications" occur only in the explanation of the type-problems of fractions. They always involve a backwards procedure.

1014 Line 20: *ardham*] em.

hundred sixty-eight] is the number [of elephants] who went to the mountain. Therefore, [starting] from one thousand eight, [its] one-sixth[1015] is one hundred sixty-eight; then, its "one-seventh" is twenty-four, [which is the number of elephants which] are drinking water in a river: $\begin{array}{|r|}168\\24\end{array}$, In the same way, [starting] from one thousand eight, "one-eight "is one hundred twenty-six. Another [fraction] is "one-ninth" [of this one hundred twenty-six]: [it becomes] fourteen: $\begin{array}{|r|}126\\14\end{array}$. Three female elephants **are playing** together with a [royal] elephant, that is to say, they are four. When there is the sum of these [numbers], it becomes one thousand eight: $\begin{array}{|r|}1008\\1\end{array}$.

[I am going to show] here as well [how to solve this sample problem] by means of the method [explained by the root-text]:[1016] the "sum of the fractions" is exactly the numerator five hundred two whose denominator is five hundred four: $\begin{array}{|r|}502\\504\end{array}$. Then, having placed (*saṃsthāpya*) "one" (*rūpa>eka*) whose denominator is one, by means of [the procedure] "the numerator and the denominator"[1017] and so forth, when there is the exchange of the denominators, it is: $\begin{array}{|r|r|}502&1\\504&1\\1&504\end{array}$. The first number multiplied by one remains as it is. [Regarding] the next number,[1018] one multiplied by five hundred four is exactly so: $\begin{array}{|r|}504\\504\end{array}$. When, from this which is the positive quantity, there is the subtraction (*pāta*) of five hundred two which is the "sum of the fractions", two whose denominator is five hundred four is placed: $\begin{array}{|r|}2\\504\end{array}$. This is the quantity-divisor. Afterwards, the "visible quantity" four whose denominator is one[1019] [is placed]: $\begin{array}{|r|}4\\1\end{array}$. Therefore, according to [the rule] "having performed the interchange of the numerator and the denominator"[1020] and so forth, above it is five

1015 Note that in the sample problem also the GT has used, among others, constructions with ordinals to denote fractions.

1016 Line 27: *atrarītyāpi*] em. Note the repetition of the final syllable *tya*(*ā*) in the transmitted *atratyarītyā* due to scribal oversight; this is an example of dittography within one word. The commentator applies the rule given in GT 64 to solve the same sample problem.

1017 This is a quotation from the rule of the class of simple fractions (GT 53).

1018 Line 30: *parāṅke*] em.

1019 Line 2: *ekacchedāḥ*] em.

1020 This is a quotation from the rule of division of fractions (GT 42).

hundred four, [and] below it is two. Next, when there is again the exchange of the two denominators, it is: $\begin{array}{|r|r|}504&4\\2&1\\1&2\end{array}$. When there is the reciprocal multiplication of the numerator above, which is four, by five hundred four, it becomes two thousand sixteen; the denominator one multiplied by two becomes two. Then, when there is the division of two thousand sixteen by two, it is: $\begin{array}{|r|}2016\\2\end{array}$. The quotient is one thousand eight, which is the [size of the] "herd" of elephants: 1008. Therefore, [in this case] by means of the reasoning[1021] of the class of fractional increase and so on forth, the visible quantity type-problem has been ascertained.

In this way, the visible quantity type-problem is concluded.

The remainder type-problem* (śeṣajāti)[1022] *[GT $\overline{67}$]

Next is the procedural rule of the remainder[1023] type-problem; the author enunciates half a metrical stanza:

> $\overline{67}$. The quantity termed "known" (*prakaṭa*) is to be divided by [the result of] the product of the denominators lessened by the numerators divided by the product of the denominators.

In this case,[1024] such denominators of the numerators above should be placed two times (i.e., written twice).[1025] [In the case presented in the sample problem], the remainder type-problem depends on three quantities:[1026] one-half of the herd [of elephants] are playing, **one-third** of half of the remainder **disappeared**[1027] into the mountain, [and] next is its one-fourth, which is [one-fourth] of that "remainder", [and] which is entertaining itself in removing an "itching sensation" from the upper part of the forehead. **One-fifth from the remainder entered [a river] to drink water.**[1028] Thus in this way the remainder type-problem consists of remainders.

1021 Line 8: °*yuktyā*] em.

1022 K 44, 11.

1023 *śeṣa.*

1024 At the very beginning of the commentator's explanation the term *vyākhyā* is missing.

1025 In this regard, see the layout in line 24.

1026 This remark can be understood in the light of the next sample problem.

1027 Line 13: *praṇaṣṭaḥ*] em. Note that *praṇaṣṭa* is used by Śrīpati in the next sample problem; the transmitted *praviṣṭa*, which is also found in the next sample problem, might be a case of scribal harmonization to parallels.

1028 The text is uncertain here; if, as I have interpreted, the commentator is enumerating the various fraction-remainders of the herd of elephants mentioned in the sample problem, then in line 14 one should find a final sentence denoting the third quantity-remainder. The suggested emendation is: *tataḥ pañcam aṃśaś ca śeṣāt pāthaḥ pātuṃ*] em. Cf. GT 68. It is noteworthy that the commentator has just explained that this rule depends on three quantities, which are the successive remainders of the first fraction one-half.

Regarding this [type-problem], that "product" which is the [result of the] mutual multiplication is the "product of the denominators" which are the denominators written a second time.[1029] [The verbal adjective] **divided** qualifies this [term *ghātena* or "by the product"].[1030] The "numerators" (*lava>aṃśa*) stand above;[1031] because it is subtracted (*apavarjitatvāt*), the set[1032] of the "denominators", which is the set [composed] of the denominators written first,[1033] are "lessened" by these [numerators]. That "product" is their mutual multiplication (i.e., of the denominators written first).[1034] **The quantity termed "known",** namely the sixty elephants that are seen, **should be divided** (*bhājya>bhajanīya*) **by** this **product of the denominators lessened by the numerators [divided by the product of the denominators].**"

In this respect, in the exemplifying metrical stanza [which follows] the author enunciates a [first] sample problem:

> 68. One-half of a herd of intoxicated elephants is playing somewhere; one-third of the remainder disappeared into the caves making noise out of fear of a lion. One-fourth of the remainder (*śeṣāṃhri*) is removing an itching sensation from the cheek, and one-fifth from the remainder entered [a river] to drink water, and sixty elephants were seen. O friend, tell [me the total number of elephants].

[The meaning] is clear.[1035] The presentation is: $\begin{array}{|c|c|c|c|} 1 & 1 & 1 & 1 \\ 2 & 3 & 4 & 5 \\ 2 & 3 & 4 & 5 \end{array}$. In this case, the set of the "denominators" is gradually "lessened" by the "numerators" one by one

1029 The last part of the passage between lines 15 and 16 up to *chid ghāta*s expands upon *ghātena* from the *karaṇasūtra*.

1030 See, from the rule, *bhaktena…ghātena.*

1031 This sentence is awkwardly elliptical as it stands; the main point of the commentator's argument is the contrast between the numerators which stand above (until they are subtracted) and the denominators which are written twice. Also, cf. the layouts shown by the SGT at the very beginning of the sample problem (lines 24 and 25).

1032 In the SGT, the term *rāśi* is, according to the context, "quantity, number, digit" or "heap, set". In my understanding, here *rāśi* emphasises the whole "group, set" of denominators, since each of the four denominators undergoes the same procedure: it is lessened by its own numerator.

1033 The commentator clarifies the product by which the known quantity is to be divided. Cf. *dvitīyavelālikhita* (line 15) and *prathamalikhita* (lines 16–17), both refer to the denominators. These expressions specify that there are two sets of denominators in the same layout.

1034 The passage starting with *yo ghātaḥ* (line 17) is to be read together with the passage in line 15: the SGT observes that, in the *karaṇasūtra,* the term *ghāta,* which occurs twice, refers to two different products: *dvitīyavelālikhitacchidaṃ yo ghātaḥ* and *tair unaḥ* […] *prathamalikhitacchedarāśis tasya yo ghātaḥ*. The SGT provides an explanatory amplification of *chidghāta* and *hāraghāta* by structuring two parallel clauses. The discussion focuses on the term *ghāta* and the presence of two set of denominators (note in the GT the use of two different terms for "denominator" – *chid* and *hāra*– distinction which is fully preserved by the SGT in its explanation).

1035 It can be observed that *vyākhyā* or "explanation" is substituted by *spaṣṭam* or "it is clear"; the SGT presents, in fact, the execution of the sample problem straight away.

(*ekaika*); [therefore] the set of the first denominators becomes one, two, three, and four precisely. The [initial] numerators are removed:[1036] $\begin{array}{|c|c|c|c|}1&2&3&4\\2&3&4&5\end{array}$. The "product" of the set of **the denominators lessened by the numerators** is its (i.e., of this set) multiplication, to illustrate: two multiplied by one is exactly so, three multiplied by two becomes six, four multiplied by six becomes twenty-four: 24. This is the "product of the denominators lessened by the numerators".[1037] In order to carry out its division, the "product" of the lower "denominators" [should be calculated]: for instance, three multiplied by two becomes six, four multiplied by six is twenty-four, twenty-four multiplied by five becomes one hundred twenty.[1038] The "product of the denominators" should be performed [in this very way]. Next, [the product of the denominators lessened by the numerators should be divided] **by the product of the denominators,** that is, by one hundred twenty:[1039] $\begin{array}{|r|}1\\120\end{array}$.[1040] Because it is the dividend, when twenty-four, which is the quantity of the "product of the denominators lessened by the numerators", is divided, when there is the reduction of the two [denominators] by twenty-four, in place of twenty-four it is [the number] one; in the same way, instead of one hundred twenty it is five: 5. Sixty whose denominator is one is the "known" quantity,[1041] this should be divided by this [five]: by means of [the procedure] "the interchange of the numerator and the denominator"[1042] and so forth, above it is five and below it is one:[1043] $\begin{array}{|r|r|}5&60\\1&1\end{array}$. Next, when there is the multiplication of the numerators, sixty multiplied by five becomes three hundred whose denominator is one: $\begin{array}{|r|}300\\1\end{array}$. The denominator multiplied by one is again the very one. In the same way, three hundred divided by one remains exactly as it is:[1044] $|300|$. These are the elephants in the "herd".

1036 *bhagna*. The new numerators are obtained from the difference of the first denominators lessened by the numerators.

1037 Lines 24–27 explain the step of the *karaṇasūtra* denoted by the expression *lavonahāraghātena* and here represented by the number 24.

1038 Line 2: **idam**] em. The number 120 is the product of the "denominators written a second time" (see the SGT in line 15); this becomes the quantity-divisor (see next the layout in line 3).

1039 The layout (line 3) shows already the interchange (*parīvartana*) between the numerator and the denominator of the quantity-divisor 120.

1040 A *daṇḍa* would fit well after the layout (line 3).

1041 Line 5: °*rāśiḥ*] em. The syntagm *prakaṭarāśi* stands in apposition to the verbal numerical expression.

1042 This is a quotation from the rule of division of fractions (GT 42).

1043 I have emended the text; in line 6, the second layout is missing.

1044 The text shows inconsistency as the denominator 1 is not always retained: cf., for instance, SGT 65, 3, where the denominator is retained, and SGT 69, 28 where the denominator 1 does not occur.

Its verification: to illustrate, half of three hundred is one hundred fifty, they are playing: 150. **One-third** of the "remainder" of this hundred fifty is fifty: 50, which [is the number of elephants that] **disappeared**[1045] into the mountain. One-fourth of the "remainder" of one hundred is twenty-five, which [is the number of elephants that] are **removing an itching sensation**. Then, **one-fifth** of seventy-five, which is the remainder, is fifteen: they are drinking water and **sixty are seen:**[1046]

150
50
25
15
60

. When there is their sum, it becomes three hundred which is the measure of the [herd of] elephants: 300.

The author enunciates a second sample problem:

> 69. [With respect to a herd of swans] one-half, two-thirds of the remainder, three-fourths of this [last] remainder, and four-fifths[1047] of this [last] remainder having flown somewhere disappeared. Another (lit.) triad of swans is seen gathering together. O clever one, tell [me] how many swans were in that group.

The explanation together with the very presentation:

1	2	3	4
2	3	4	5
2	3	4	5

. The set of the [first] "denominators" is "lessened by the numerators" one, two, three, and four. The denominators written first are two, three, four, and five; the quantity [obtained from the deduction of these first denominators from the numerators] is everywhere the very [number] one. The [initial] numerators are erased.[1048] Then, the "product", which is the [result of the] reciprocal multiplication, of the set of the [first] "denominators lessened by the numerators", which are [the number] one, becomes the very [number] one. [Afterwards] there is the "product of the denominators" written a second time:[1049] three

1045 Line 11: *praṇaṣṭaḥ*] em.

1046 Line 13: **dṛṣṭi**] em.

1047 *iṣu* – lit. "arrow"; *jalanidhi* – lit. "ocean". Śrīpati uses the *bhūtasaṃkhyā* system of notation; see the commentator's explanation in lines 3–4.

1048 It is noteworthy that in the previous sample problem, a similar passage occurs (cf. *aṃśaś ca bhagnāḥ* in SGT 68, 25 and *aṃśā bhagnās* in SGT 69, 21–22) and this is followed by a layout. Supposing that there was a layout following the expression *aṃśā bhagnās* in this sample problem, it would have been the following:

1	1	1	1
2	3	4	5

. The numbers representing the difference of the numerators and the first denominators are the new numerators, and the second denominators occur underneath.

1049 The text is uncertain here, the emendation follows the commentary. Lines 22–23: *dvitīyavelalikhitacchidāṃ ghātaḥ*] conj.; *asya bhāgam adho° yātaḥ* (K). Also possible would

multiplied by two becomes six. In this way, as before, when there is the completion[1050] [of this calculation], it is one hundred twenty, this is the "product of the denominators". The [quantity of] the set of "denominators" "lessened by the numerators" **should be divided** (*bhājya*> *bhakta*), by this [product of the denominators obtained]. [Next is the expression] **the quantity called "known":** for instance, according to [the rule] "having performed the interchange" and so forth, one hundred twenty should be made (i.e., moved) above, and one is underneath; **the known quantity** is three whose denominator is one: $\begin{array}{|r|r|} 120 & 3 \\ 1 & 1 \end{array}$.

When there is "multiplication of the numerators"[1051], three multiplied by one hundred twenty becomes three hundred sixty, and the denominator [one] multiplied by one is the very one. Three hundred sixty divided by [the denominator] one is exactly so: |360|.

The verification: three hundred sixty, [its] **half** is one hundred eighty, [this is the number of the swans which has] flown up. Then,[1052] regarding the "remainder" one hundred eighty which is divided by three, the numerator is two, it is one hundred twenty, this is [the number of the swans which has] flown up. **Three-fourths** of the "remainder", which is sixty, is forty-five. Next is five (lit. "arrow") of the "remainder": by the word "arrow", since it is the designation of the arrow (*bāṇa*) of the god of love (*manobhava*), [the number] "five" is meant. Five (lit. "arrow") of the "remainder" fifteen,[1053] that is to say, it is divided by five; the numerator is equal to four (lit. "ocean"), it is twelve. The "remainder" from [the total number of swans] flown up is three "swans"; it consists of one-fifth [of the remainder fifteen], these are [the swans that are seen] "gathering together": $\begin{array}{|r|} 180 \\ 120 \\ 45 \\ 12 \\ 3 \end{array}$. When there is the sum of these [numbers], three hundred sixty occurs, which is the measure of the "herd" of "swans": |360|.

In this way, the [procedure of the] remainder type-problem is concluded.

be *bhāgānām adho likhitacchidām ghātaḥ* ("the product of the denominators written below the fractions).

1050 *nirvāha*, this is the first occurrence of the term.

1051 This is a quotation from the rule of multiplication of fractions (GT $\overline{40}$).

1052 The syntax does not seem to be smooth in this last passage (lines 2–4). The remainder is divided by the fraction two-thirds.

1053 The remainder is divided by the fraction four-fifths. Line 4: *pañcadaśānāṃ pañcabhaktasya*] em. The scribe has presumably repeated the final syllable *ṣu* in °*iṣu* (presumably taking it as the case ending of a locative plural) and produced the mistaken *pañcadasu pañcabhakteṣu*.

The difference type-problem (viśleṣajāti)[1054] *[GT 70]*

Next is the procedural rule of the difference type-problem; the author enunciates a metrical stanza:

> 70. In the difference type-problem,[1055] having subtracted the smaller (*vihīna*) from the larger (*adhika*) [numerator], the remaining procedure is the very one [previously] formulated. Having subtracted (*apāsya*) the sum of the fractions from [the number] one, one should divide the fraction (*vibhāga*) of the visible quantity by the [obtained] remainder.

The explanation: **in the difference type-problem,** which consists of the subtraction of the smaller number from the "larger" number, **having subtracted** is having carried out the deduction[1056] of **the smaller,** which is the smaller numerator,[1057] **from the larger** numerator among numerators having equal denominators. [Regarding the expression] **the remaining procedure,** by means of "procedures",[1058] such as [in this case] "the numerator and the denominator"[1059][and] "by the [lower] denominators,[1060] the class of simple fractions[1061] which has been [previously] formulated should be understood [to be relevant][1062] in this case as well. Afterwards, **having subtracted the sum of the fractions from one** (*eka>rūpa*), that is, from the quantity with the equal denominator which has been calculated,[1063] one should then divide **[the fraction] of the visible quantity by the [obtained] remainder** from [the subtraction from] that consisting of [the number] one. This [step] should be understood to be the same as in the visible quantity type-problem.[1064]

1054 K 46, 10–13.
1055 *viśleṣajāti*.
1056 *vivara*.
1057 *aṃśa*, it expands upon *adhika* and *vihīna*. The commentary specifies that these adjectives (which in the rule are used as substantives) qualify *aṃśa* or the "numerator" of fractions having equal denominators.
1058 Note the plural form of *vidhi* ("procedure"), as the procedures quoted are, in fact, two.
1059 This is a quotation from the rule of the class of simple fractions (GT 53).
1060 This is a quotation from the rule of the class of fractional increase (GT 57).
1061 This passage is an explanatory amplification of *śeṣo vidhir ukta eva*; the SGT clarifies that this passage refers to the class of simple fractions. Although the commentator has quoted the steps of two different procedure, he mentions only the class of simple fraction because this is the rule implied by the root-text in *śeṣo vidhir ukta eva*; Śrīpati refers, in fact, to the subtraction of fractions having equal denominators. The SGT mentions, however, the class of fractional increase as this is to be applied in the following sample problem. See SGT 71, 26–27, 12, and 10 (where the class of simple fractions is quoted) and lines 6–7 and 18 (where the class of fractional increase is quoted).
1062 In brief, it should be applied here too.
1063 This remark can be understood in the light of the commentator's execution of the next sample problem (lines 9–13).
1064 The commentator adds important information: the last part of this rule is the same as the visible quantity type-problem. Cf. the last two lines of this rule and GT 64.

In this respect, the author enunciates a sample problem in two exemplifying metrical stanzas: [1065]

> 71–72. Of a swarm[1066] of restless bees, one-fifth went to a mango tree, one-eighth to a lotus, their difference multiplied by two plus its half stayed on a fragrant oleander; six times half of the difference between the fraction of bees staying on the oleander and those on the mango three, together with [its][1067] one-third multiplied by three and lessened by [its] one-third rested on a jasmine flower, and ten bees are seen enjoying the blossom of a *tilak* tree. O clever one, if you know [the difference type-problem] tell me quickly the [total] number of bees in the swarm[1068].

The explanation:[1069]**of a swarm of restless bees, one-fifth [went] to a mango-tree** and one-eighth **to a lotus**: $\begin{array}{|c|c|}1&1\\5&8\end{array}$. By means of [the procedure] "the numerator and the denominator"[1070] and so forth, when there is the exchange of the denominators, it is eight below five [and] five below eight: $\begin{array}{|c|c|}1&1\\5&8\\8&5\end{array}$. One multiplied by eight becomes eight, five multiplied by eight becomes forty: $\begin{array}{|c|}8\\40\\8\end{array}$. Regarding the other quantity, one multiplied by five becomes five, eight multiplied by five becomes forty: $\begin{array}{|c|}5\\40\\5\end{array}$. Then, **having subtracted,**[1071] that is, [*vidhāya* or having carried out][1072] the deduction **from the larger** [numerator] among these [two fractions] having equal denominators; for instance, in this case from that consisting of eight, which is **the larger**, **the smaller** [numerator] should be subtracted, the numerator five is the smaller. It becomes three [and] forty

1065 A similar sample problem occurs in GSS 4.6 and L 55.
1066 *nicaya* – lit. "collection".
1067 The SGT provides the possessive adjective *sva* ("its") while explaining the sample problem in line 18 and in doing so he specifies *tryaṃśena* as *svatryaṃśena*.
1068 *samūha*.
1069 Note that there is no initial *nyāsa*.
1070 This is a quotation from the rule of the class of simple fractions (GT $\overline{53}$).
1071 Line 3: *viśodhya*] em.
1072 See, while commenting upon the *karaṇasūtra*, the SGT glosses *viśodhya* with *vivaraṃ vidhāya* (line 15).

[is its denominator].This [fraction] is written in the third place:[1073] $\begin{array}{|c|c|c|} 1 & 1 & 3 \\ 5 & 8 & 40 \end{array}$.

Afterwards, both (i.e., the first two performed fractions) should be erased. **Their difference,** such as three, **multiplied by two** becomes the [fractions] consisting of six[1074] whose denominator is forty, [this] **plus** [its] **half,** it is:[1075] $\begin{array}{|c|} 6 \\ 40 \\ 1 \\ 2 \end{array}$. In this case, when there is the class of fractional increase,[1076] according to [the rule] "[one should multiply] the [upper] denominator by the [lower] denominator" and so forth, one should multiply the "[upper] denominator" forty "by the [lower] denominator" two; it becomes eighty. In the same way, [the first numerator is multiplied by the denominator below] "added to [its own] numerator", the denominator two becomes three; one should multiply six by it, it becomes eighteen. When there is the reduction of these two by half, nine is above, and forty is below. This unchangeable (*niścala*)[1077] [quantity] should be written in the third place: $\begin{array}{|c|c|c|} 1 & 1 & 9 \\ 5 & 8 & 40 \end{array}$. That [forty] is the size of the "number" [of bees] in the swarm; regarding this [size], which is forty [and] is that which is divided,[1078] when there is a ninth part out of it, that [number of bees] which **stayed on a fragrant oleander** should occur. In order to make **the difference of the fractions of bees staying on the oleander and those on the mango tree,** which consist of the numerators one [and] nine whose denominators are five and forty,[1079] according to [the rule] "the numerator

1073 I have emended the text, which here is uncertain as one would expect the layout (preceded by *yathā*) to occur before the *daṇḍa* after *sthāpyam* (line 5). The proposed emendation is supported by the commentary itself when in line 28 a similar passage occurs (see *sthāpyam*), and this is followed by a layout.

1074 Line 6: *jātaṃ ṣaṭkarūpaṃ*] em.

1075 I have emended the transmitted layout, which is $\begin{array}{|c|} 6 \\ 40 \\ 2 \\ 1 \end{array}$.

1076 Here the second method of the class of fractional increase is meant (GT 57).

1077 The use of the adjective *niścala* is to be understood in the light of what is explained further in the commentary: the number forty represents the total number of bees in the swarm.

1078 Since forty is the total number of bees, it is divided into the groups of bees staying on the mentioned flowers and trees. See a similar remark in line 19: *catvāriṃśad bhāgīkṛtayūthasya.*

1079 Line 11: *ekanavāṃśayor pañcacatvāriṃśacchedayor lakṣaṇayor*] em. The denominator five is not mentioned in the text.

and the denominator"[1080] and so forth, when there is the reduction of the denominators, such as five and forty, by means of the division by five, when there is the exchange of [the obtained denominators], namely one and eight respectively, it is: $\begin{array}{|c|c|}1&9\\5&40\\8&1\end{array}$. In the first quantity, one multiplied by eight becomes eight,[1081] five multiplied by eight becomes forty; the next number multiplied by one is exactly so: $\begin{array}{|c|c|}8&9\\40&40\\8&1\end{array}$. In this case, the "difference" occurs; when there is the deduction of eight from nine, one whose denominator is forty is then placed: $\begin{array}{|c|}1\\40\end{array}$. Regarding this "difference", [its half should be calculated]: **half** of [the numerator] one, [its] **half** is not possible.[1082] When the denominator forty is [hence] multiplied by two, it is eighty: $\begin{array}{|c|}1\\80\end{array}$. **Six times** this [fraction] is: $\begin{array}{|c|}6\\80\end{array}$. [This obtained result] **together with** [its] **one-third** is: $\begin{array}{|c|}6\\80\\1\\3\end{array}$. When there is the class of fractional increase, according to [the rule] "the [upper] denominator by the [lower] denominator"[1083] and so forth, one should multiply the "[upper] denominator", which is eighty, "by the [lower] denominator", which is the three below; it becomes two hundred forty. "[The first numerator is multiplied] by the denominator below added to [its own] numerator]," that is, [it is multiplied] by three added to one means multiplied by four; it becomes six whose denominator is two hundred forty: $\begin{array}{|c|}24\\240\end{array}$. It is **multiplied by three**: twenty-four multiplied by three becomes seventy-two, this is **lessened by [its] one-third**: $\begin{array}{|c|}72\\240\\01\\3\end{array}$. By means of the [previously] mentioned class of fractional

1080 This is a quotation from the rule of the class of simple fractions (GT $\overline{5}3$).

1081 The intermediate passage not explained by the commentator is that the common denominator is 40, thus 40 ÷ 5 becomes 8.

1082 Line 16: *nārdhaṃ sahate*] em. The transmitted *ekonārdhaṃ sahata iti* is nonsensical. The commentary explains that when it is not possible to divide the numerator of a fraction by two, its denominator is multiplied by two. This is explained by the commentary in SGT 76, 5 where one finds [...] *nārdhaṃ sahate* [...].

1083 This is a quotation from the rule of the class of fractional increase (GT 57).

decrease,[1084] according to [the rule] "one should multiply [the lower denominator] by the [upper] denominator", "one should multiply the [upper] denominator", which is two hundred forty,[1085] "by the [lower] denominator", which is the three underneath; it becomes[1086] seven hundred twenty. Next, one should multiply "the foremost numerator" seventy-two "by the [lower] denominator decreased by the [lower] numerator", that is, [one should multiply the foremost numerator] by three lessened by the numerator one, [hence] by the obtained two;[1087] it becomes one hundred forty-four: $\begin{array}{|c|}144\\720\end{array}$. When there is the reduction of these two by means of the division by one hundred forty-four,[1088] one is above is and five is below: $\begin{array}{|c|}1\\5\end{array}$.[1089] With respect to the swarm, the measure (*pramāṇa*)[1090] (i.e., forty) has [previously] been ascertained; when there is its one-fifth, that is [the number[1091] of bees that] **rested on a jasmine flower**. This should be written in the fourth place:[1092] $\begin{array}{|c|c|c|c|}1&1&9&1\\5&8&40&5\end{array}$. Afterwards, [here] also again by means of [the procedure] "the numerator and the denominator"[1093] and so forth, when there is the exchange of the denominators, it is: $\begin{array}{|c|c|}1&1\\5&8\\8&5\end{array}$. One multiplied by the first eight becomes eight, five multiplied by eight becomes forty; in the next quantity,[1094] one multiplied by five becomes five, in the same way [eight] multiplied by five becomes forty. As there are equal denominators, when the additive quantity five is added to eight, it becomes thirteen whose denominator is forty: $\begin{array}{|c|}13\\40\end{array}$. Then, regarding the third quantity which is nine, without [performing] again the [already accomplished] procedure (i.e., of

1084 Here the second method of the class of fractional decrease (GT 60) is meant, and this is quoted below.
1085 Line 23: *catvāriṃśadadhi(ka)dviśatī*] em. The number 240 is, in fact, meant.
1086 Line 24: *jātā*] em.; *jātam* (K).
1087 Line 24: *jātadvikena*] em.
1088 Line 26: *catuścatvāriṃśadadhi(ka)śatena*] em. The number 144 is, in fact, meant.
1089 This is the fraction of bees on the jasmine.
1090 Here the wording deliberately echoes *yat yūthasaṅkhyāpramāṇa* above in line 10.
1091 Line 27: *etaj*] em.; *rūpaṃ* (K). In this context, the transmitted *rūpaṃ* is not interpretable, the emendation is suggested on the grounds of the passages in lines 11 (*tat kunde sthitam*) and 22 (*etat tribhāgarahitam*).
1092 The commentator shows the fractions of bees on the mango tree, lotus, oleander, and jasmine respectively.
1093 This is a quotation from the rule of the class of simple fractions (GT $\overline{53}$).
1094 Line 2: *parāṅke pañca*°] em.

fractional increase)[1095] because equal denominators occur, when the additive quantity [nine] is added to thirteen, it becomes twenty-two whose denominator is forty. Next, when there is the reduction of this denominator and the denominator of the fourth quantity by five,[1096] it becomes eight and one respectively. Then, when there is the exchange of the denominators, it is: $\begin{array}{|r|r|} 22 & 1 \\ 40 & 5 \\ 1 & 8 \end{array}$. The first number multiplied by one is exactly so. In the next quantity, one multiplied by eight becomes eight; five multiplied by eight becomes forty. Since equal denominator occur, when the additive quantity eight is added to twenty-two, it becomes thirty whose denominator is forty: $\begin{array}{|r|} 30 \\ 40 \end{array}$. This is **the sum of the fractions**; **having subtracted** this **from one**, it is one whose denominator is one.[1097] Then, by means of [the procedure] "the numerator and the denominator" and so forth, when there is the exchange of the denominators, it is:[1098] $\begin{array}{|r|r|} 30 & 1 \\ 40 & 1 \\ 1 & 40 \end{array}$.The first quantity multiplied by one is exactly so. The next quantity[1099] multiplied by forty becomes forty: $\begin{array}{|r|} 40 \\ 40 \end{array}$. Afterwards, when there is the subtraction of thirty from the forty obtained (*uttha*) from [the multiplication by] one, ten whose denominator is forty is placed: $\begin{array}{|r|} 10 \\ 40 \end{array}$. "One should divide the visible quantity",[1100] which is ten, "by [one] lessened by the sum of the fractions," that is, **by the remainder** ten [whose denominator is forty]. To illustrate: the "visible quantity" is ten whose denominator is one;[1101] then, according to [the rule] "having performed the interchange"[1102] and so forth of the denominator forty and [the numerator] ten, which is the "remainder" of the "sum of the

1095 In other words, it is not necessary to carry out again the just accomplished fractional increase; because the fractions have equal denominators, they can be simply added.

1096 The fourth quantity is $\begin{array}{|r|} 1 \\ 5 \end{array}$, which is the fraction of bees on the jasmine.

1097 This is the minuend $\begin{array}{|r|} 1 \\ 1 \end{array}$.

1098 I have emended the layout; the numerator should, in fact, be 30 and not 10 as given in the edition (line 11).

1099 Line 11: *eka* in *ekacatvāriṃśad*] em.

1100 The sample problem specifies that ten bees are "seen" enjoying the blossom of a *tilak* three. The commentator quotes the visible quantity type-problem (GT64), rephrases the root-text, and establishes the relationship between the visible quantity type-problem and the difference type-problem; by *śeṣeṇa,* which quotes the *karaṇasūtra,* the SGT combines the terminology used by Śrīpati in both rules.

1101 Line 14: *dṛśyo daśaikacchedā*] em.

1102 This is a quotation from the rule of division of fractions (GT 42).

fractions", forty is above and ten is below:[1103] $\left|\begin{array}{c|c} 40 & 10 \\ 10 & 1 \end{array}\right|$. Next, ten multiplied by forty, which is the first number, becomes four hundred:[1104] $\left|\begin{array}{r} 400 \\ 10 \end{array}\right|$. When there is the division of this four hundred by the denominator ten multiplied by one, it is:[1105] $\left|\begin{array}{r} 40 \\ 1 \end{array}\right|$. The quotient forty is the measure of the swarm of bees.[1106]

Its verification:[1107] when there is one-fifth of forty, it is eight, [this is the number of the bees] "in the mango tree". When there is one-eighth of forty, it is five, [this is the number of the bees staying] on **the lotus**. Regarding the swarm which is that forty [being] divided (*bhāgīkṛta*) [into parts], there are nine[1108] out of forty, thus nine [is the number of the bees] on **the oleander**. When there is one-fifth of forty, eight occurs, which is [the number of bees] "on the jasmine", and ten are [the bees that are] visible:[1109] $|8,5,9,8,10|$. When there is the sum of these [numbers], forty occurs.[1110]

In this way, the [procedure of the] difference type-problem is concluded.

The remainder-root type (śeṣamūlajāti)[1111] *[GT 73]*

Next is the procedural rule of the remainder-root type-problem; the author enunciates a metrical stanza:

> 73. When the visible quantity near the root[1112] is multiplied by four, [the result] added to the root multiplied by itself, the square-root [of the result]

1103 I have emended the text as the layout (line 15) is mistaken; it gives the number 1 instead of 10 as the numerator of the second side.

1104 Lines 16–17 are corrupt; I have emended the text and suggested that: i) in line 16, after *catuḥśatī,* the term *yathā* followed by the layout should occur, and this should be the same $\left|\begin{array}{r} 400 \\ 10 \end{array}\right|$ which in the edition is mistakenly found in line 17 after *bhāge*, and ii) in line 17, the layout, preceded by *yathā,* should be $\left|\begin{array}{r} 40 \\ 1 \end{array}\right|$ rather than $\left|\begin{array}{r} 400 \\ 10 \end{array}\right|$.

1105 See footnote 1112.

1106 *alikula*.

1107 *vāsanā* – in the SGT, this is the first occurrence of the term. It means "verification" and is used as a synonym of *ghaṭanā*. Note that as *ghaṭanā*, in the SGT *vāsanā* too occurs only in the treatment of the type-problems of fractions.

1108 Nine is the numerator.

1109 It is noteworthy that this is layout is horizontal, while in the previous sample problems layouts showing the final quantities obtained are vertical.

1110 It is unusual that the final result is not shown, one would expect: *yathā* $|40|$.

1111 K 48, 24–25.

1112 *pada*. The vocabulary used by the Śrīpati is clearly space-related and it refers to the arrangement of the given quantities in the layout. See the layout in SGT 74, 11 where the "visible quantity" (*dṛśya*) is placed, in fact, next to the "root of the remainder" (*śeṣamūla*).

> extracted, [this result] added to the root, [the obtained quantity] halved, [the result] multiplied by itself, [and the product] divided by one lessened by the fraction,[1113] [the final result is arrived at]. This [same] procedure [should be performed] again.

[The explanation:] the two words *pada* ("root") and *mūla* ("root") have the same meaning. [With respect to the passage] **when [the visible quantity] near the root,** regarding this [root], it is the "visible quantity" two[1114] to be "multiplied by four". There[1115] [it is stated] **when the visible quantity near the root is multiplied by four:** wherein the first number, such as two, does not have a "root", in that case the very [number] one is the "root",[1116] and [therefore] that [expression] **"multiplied by itself"** means "multiplied by one". Then, it is **added to** one, which is [the result of] the "root multiplied by itself", it becomes nine. Next is [the step] **the square-root [of the result is] extracted:** the "extracted square-root" of the [result of the] "visible quantity" "multiplied by four" "added to the root multiplied by itself", which is nine, is namely three. In that case,[1117] one is the "root", it is **added to** this [three]; it is **halved** (*dalita>ardhīkṛta*) becoming two. **[The result is] multiplied by itself: when it is multiplied** by itself, [it is multiplied] namely by two. "One" is decreased by the "fraction" (*lava>aṃśa*);[1118] the quotient, such as twelve, is **divided** by it. That very "visible quantity" should be placed **again** thereafter. In this case, because of the object explained, the "fraction" which [in the layout] stands in the middle[1119] is erased.[1120] Afterwards, therefore[1121] this **procedure** starting with **when [the visible quantity] near the root [should be performed] again,** when the "visible quantity" [so obtained] occurs, **the procedure [should be repeated again]**.[1122] The remaining[1123] [steps of the procedure], because of the "fraction" [which has been manipulated],[1124] should be carried out without the fraction.

1113 *vilava* is "lessened, decreased by (*vi*) the fraction (*lava*)".
1114 The SGT mentions the data of the following sample problem in order to link each phase of the algorithm formulated by the GT.
1115 In brief, in the stanza above.
1116 The commentator clarifies that when the visible quantity is not a perfect square (in the next sample problem the *dṛśya* is 2) the root is 1.
1117 That is, in the case here considered.
1118 See the drop of the initial syllable *vi* in *[vi]lava*.
1119 *antarāla* – this adjective specifies the way the fraction is arranged in the layout.
1120 The fraction, which in the layout is placed in the middle (see the fraction two-thirds in the next sample problem), is erased once calculated.
1121 Note that *tatra* connects this sentence to the previous; what now occurs is, in fact, the result of the preceding steps, more specifically, of the manipulation of the fraction.
1122 This passage expands upon *punar vidhi* and explains that when the procedure starts again, there are no fractions to be carried out; see SGT 74, 20.
1123 *anya*.
1124 The translation reflects the terseness of this passage.

In this respect, in the exemplifying metrical stanza [which follows] the author enunciates a sample problem:

> 74. The [number equal to the] square-root of a collection of blue lotus leaves, used as ear ornaments by a woman with beautiful eye-brows, had broken during the play with [her] lover [and] fell on the bed. Two-thirds[1125]of the remainder and the square-root of the [result of this] remainder fell on the ground. A pair of leaves were seen. Say, then, how many blue lotus leaves were [in total]?

Its explanation together with the very presentation:[1126] $\begin{array}{|r|r|r|r|} \text{mū } 1 & \text{śela } 2 & \text{śemū } 1 & \text{dṛśya } 2 \\ 1 & 3 & 1 & \end{array}$. The "visible quantity" two "near the root", which is the root of the remainder,[1127] "multiplied by four" becomes eight. There (i.e., in the rule) [it is stated that] "the root is multiplied by itself", that is, one is multiplied by one; when [eight] is **added to** it,[1128] it becomes nine. Therefore, the "root" of nine is three. Then, **[this result is] added to the root:** one is the "root"; **when added to** that [three], it becomes four which **halved** (*dalita>ardhīkṛta*) becomes two. [The expression] **multiplied by itself** means [here] multiplied by two, it becomes four.[1129] In this case, regarding [the number] "one" which is lessened by the fraction two-thirds,[1130] it is: $\begin{array}{|c|c|} 2 & 1 \\ 3 & 1 \end{array}$. By means of [the procedure] "the numerator and the denominator"[1131] and so forth, when there is the exchange of the denominators three and one, it is: $\begin{array}{|c|c|} 2 & 1 \\ 3 & 1 \\ 1 & 3 \end{array}$.The first quantity multiplied by one is that very one. In the next number, one multiplied by[1132] three becomes three. When there is the subtraction of the first numerator two from this [three], it

1125 *dvitryaṃśau*] em. See in line 11 the layout given by the SGT and in lines 15 and 27 the given fraction is *tryaṃśadvaya*.

1126 Line 11: *dṛśyaḥ*] em. In the layout, the two abbreviations *śela* and *śemū* denote the "fraction of the remainder" (*śeṣamūla*) and the "root of the remainder" (*śeṣamūla*) respectively. This type-problem of fractions is called, in fact, *śeṣamūlajāti* or the "type-problem of the root of the remainder".

1127 Here the commentator clarifies that, in *padasamīpa,* the term *pada* corresponds to *mūla* in *śeṣamūla*, which in the layout is placed next to (hence s*amīpa*) the *dṛśya* 2.

1128 Line 12: *yute*] em. See the root-text and the line below.

1129 In line 14, *jātacatuṣkake* represents an instance of grammatical attraction; the same can be said with respect to the locative cases in line 23 and in SGT 74, 20. A *daṇḍa* must have occurred before *atra* (line 14), otherwise here there is a problem with the vocalic *sandhi*.

1130 Line 15: *vilava*°] em.

1131 This is a quotation from the rule of the class of simple fractions (GT $\overline{53}$).

1132 Line 16: *triguṇa*] em.

becomes one whose denominator is three. This is [the result of] "one lessened by the fraction". [Next is the step] **when divided** by it; according to [the rule] "having performed the interchange"[1133] and so forth, three is above and one is below: $\begin{array}{|c|}3\\1\end{array}$. Afterwards, the procedure of multiplication:[1134] the previous four which has been[1135] derived from the visible quantity is multiplied by three; it becomes twelve. This multiplied by one [and] divided by the denominator one is again the very twelve. **This procedure [should be performed] again:** [the procedure starts again with the new *dṛśya* which is] this twelve, since the fraction which was in the middle (i.e., the *śeṣalava*) has been erased[1136] [because it has been calculated]. The "visible quantity[1137] "near" the initial "root" [one][1138] "multiplied by four" becomes forty-eight. That [expression] **when added to the root multiplied by itself:**[1139] **when added to** one, when multiplied by one, it becomes forty-nine. The [step] **the square-root extracted:** the "square-root" of that consisting of the square forty-nine is seven. The [step] **when added to the root**: the "root" is one; **when** [seven is] **added to** it, it becomes eight. When the obtained four is **halved,** when it is **multiplied by itself,** that is, multiplied by the very four, it becomes sixteen. Because of the[1140] procedure with the fraction has been performed, there are no fractions [any longer]. Hence, the very quotient sixteen is the [number of the] **blue lotus leaves**: 16.

Its verification: **the root** of the square sixteen is four, [which are the leaves that] **fell on the bed.** Then, "two-thirds of the remainder" twelve is eight, [which are the leaves that] **fell on the ground.** Then, **the root** of the square four, which is the "remainder", is two, [which are the leaves that also] **fell** [on the ground]. **A pair of leaves were seen**: $\begin{array}{|c|}4\\8\\2\\2\end{array}$. When there is their sum, it is sixteen.

Next is a second sample problem; the author enunciates a metrical stanza:

75. Having flown away, three times the square-root of a flock of parrots landed on a field of rice. One-tenth of its obtained remainder were resting on a mango tree full of fruits. Three-times the square-root of the remainder fell to the snare

1133 This is a quotation from the rule of division of fractions (GT 42).

1134 See *saṅguṇanā* in GT 42.

1135 See line 14, the number 4 has been already halved.

1136 Line 20: *bhagna*] em.; *lagna* (K).

1137 The procedure starts again but, as explained by the commentator, without the fraction because it has already been manipulated.

1138 This passage points out that the layout has changed, with the new *dṛśya* 12 now placed near the original *śeṣamūla* 1.

1139 Line 21: *tatrasva°*] em.; *tatratya*] (K).

1140 Line 24: *°hetutvāt*] em.

of a hunter and were in the power of misfortune. O scholar, if you know [the class of the root-remainder], tell [me] the measure of the flock of parrots.

Its explanation with the presentation:[1141]

mū 3	śe 1	mū 3	
1	10	1	dṛ 0

. Here, in respect to the other [sample problem], because [the "visible quantity"][1142] has not been mentioned, the "visible quantity" which is "near the root of the remainder"[1143] is zero; [this] "multiplied by four" is the very zero. That **[when the visible quantity near the root is multiplied by four] added to the root multiplied by itself, three times** the "root" one is three; when it[1144] is **added to** this, there is nine[1145] instead of zero[1146] because of the similarity of the additive quantity, which is zero.[1147] Therefore, the "root" of nine is three, **[this result is] added to the root,** that is, **the root** [one] is **multiplied by three**; it is **added to** it (i.e., the root three), it becomes six. **[When the obtained quantity is] halved** (*dalita*>*ardhita*), when the obtained three is **multiplied by itself,** that is, it is multiplied by three, it becomes nine. When it is **divided by one lessened by the fraction,** in this case, the "fraction" is one whose denominator is ten.[1148] [The next number is] "one" whose denominator is one. Next, by means of [the method] "the numerator and denominator"[1149] and so forth, when there is the exchange of the denominators, it is:

1	1
10	1
1	10

. [One] multiplied by one is exactly so, one multiplied by ten becomes ten; when there is the subtraction of this "fraction" from one, it becomes nine whose denominator is ten. Then, for the purpose of the division of nine whose denominator is one, [which is the quantity] derived from the visible quantity, according to [the rule] "having performed the interchange"[1150] and so forth, the "remainder" [from the subtraction] from one is

1141 When comparing these with the abbreviations found in the initial *nyāsa* of the previous sample problem (SGT 74, 11), here in the second box from the left one would expect the abbreviation *śela*, which stands for *śeṣalava*, and the next abbreviation should be *śemū*, which stands for *śeṣamūla*.

1142 In line 6, the subjective genitive of the abstract construction is missing, it should be °*uktatvāt śeṣasya*.

1143 The grammar is awkward here as the adjective *samīpa* should qualify a noun in the genitive case. Lines 6–7: *śeṣamūlasya*] em.; also possible would be °*mūla*°] em.

1144 In brief, the product of three multiplied by itself.

1145 Line 8: *jātaṃ navakaṃ*] em.

1146 This remark suggests that at this point there is a change in the layout: the obtained number 9 replaces the *dṛśya* 0.

1147 In brief, 0 + (3× 3)= 9.

1148 This is the *śeṣalava*

1
10

(see the layout of line 6).

1149 This is a quotation from the rule of the class of simple fractions (GT 53).

1150 This is quotation from the rule of division of fractions (GT 42).

nine underneath and ten above: $\begin{array}{|r|r|}10 & 9\\ 9 & 1\end{array}$. Because there is the same [number],[1151] which is nine, when there is the reduction by means of the division by nine, is one: $\begin{array}{|r|r|}10 & 1\\ 1 & 1\end{array}$. Afterwards there is the "multiplication":[1152] one multiplied by ten becomes ten, [one] multiplied by one [is one], [ten] divided by the denominator one is again the very ten. **This procedure [should be performed] again:** since the fraction which was in the middle has been erased, [now the procedure is performed without the fraction, which has been manipulated].[1153] This "visible quantity" ten "multiplied by four" becomes forty. In this case, **the root** [one] "multiplied by three" is the very three. The root [three] is "multiplied by itself", that is, [three] "multiplied" by three becomes nine. **When** [the visible quantity multiplied by four][1154] is **added** to that [nine], it becomes forty-nine. In this case, the "root" is seven. **[This result is] added to the root**: the "root" is three and [the denominator] one, when **added to** that [seven], when the obtained ten is **halved,** when the obtained five is **multiplied by itself,** which means multiplied by the very five, it becomes twenty-five. This should be the very **measure** of the "flock of parrots".

Its verification: regarding twenty-five, the root is five, it is **multiplied by three:** fifteen [is the number of parrots that] **landed** on a field of rice. "One-tenth" of the "remainder" ten: [when there is the reduction of the two ten by ten and this is] multiplied by one [the result is one and this is the number of parrots which] went to a mango tree.[1155] Here **the square-root of the remainder** [nine] is three, multiplied by three is nine, **which fell** in misfortune **to the hunter**: $\begin{array}{|r|}15\\ 1\\ 9\end{array}$.

When there is their sum, it is twenty-five: 25.

Because it (i.e., the result) has been derived from the visible quantity near the root of the remainder, the remainder-root type-problem is [so called and is] concluded.[1156]

1151 Line 14: *sadṛśayor*] em.; the transmitted *dṛśyadvayor* does not fit the context. The emendation follows the text.

1152 This is quotation from the division of fractions (GT 42).

1153 Lind 16: **prathamasamīpa*, gatāntaralavatvāt*] conj. I cannot interpret the transmitted reading; this is not a completely convincing guess but it is based on the commentary, see SGT 74, 20: *bhagnāntaralavatvāt*. The sentence remains awkwardly elliptical as it stands, one wonders whether the result was mentioned here too as in SGT 74, 20; one should thus consider emending to *tataḥ punar vidhi iti ete daśa gatāntaralavatvāt*] conj. The translation is "then [the procedure starts again with the new *dṛśya* which is] this ten, since the fraction which was in the middle (i.e., the *śeṣalava*) has been erased [because calculated]."

1154 In other words, forty.

1155 This sentence is particularly terse.

1156 The commentator illustrates that the name of this procedure is associated to the steps involved in determining the final result.

The type-problem of the fraction [executed] with the root and the visible quantity **(mūlāgrabhāgajāti)**[1157]***[GT 76]***

[Next is] the procedural rule of the type-problem of the fraction [executed] with the root and the visible quantity;[1158] [the author enunciates] a metrical stanza:

> 76. When the visible quantity and the root are divided by one lessened by the fraction,[1159] when the square-root of the [result of the] visible quantity is added to the square (*karaṇī*)[1160]of half the root, [the result is] added to half the root [and] raised to the square, the quantity wished for in your heart should occur.

The explanation: [regarding the expression] **when that the visible quantity and the root,** [when compounded in a *dvandva*], the [terms] "visible quantity" (*dṛśyaṃ*) and the "root" (*mūlaṃ*) become "the visible quantity and the root" (*dṛśyamūlaṃ*). There (i.e., in the stanza above) [there is an absolute construction, which] is **when the visible quantity and the root.**[1161] **[When the visible quantity and the root] are divided by one lessened by the fraction:** with respect to this reference,[1162] [the expression] the "visible quantity" and so forth are divided [is a locative absolute construction],[1163] [they are divided] "by one lessened by the fraction" which has been simplified.[1164] The "root" (*pada*>*mūla*), it is its "half" [which has to be calculated] and whenever the "root" is such that [its] "half" is not possible, make then the denominator which is underneath multiplied by two.[1165] **The square** (*karaṇī*>*varga*) is [the square] of that which has

1157 K 50, 27–51, 2.

1158 The term *agra* is lit. "tip, end" and here it indicates the *dṛśya* or "visible quantity", which is the last numerical data mentioned in both sample problems by the GT. This term is used by the commentator to connote the "visible quantity" according to the space it occupies in the layout; the *dṛśya* is found, in fact, at the very end of the *nyāsa*s provided by the commentary in the next sample problems (see the layouts in lines 15 and 20).

1159 From now onwards, in the rule "visible quantity" and "root" denote the quotients of these divisions.

1160 Line 28: *karaṇī*] em.

1161 This is a remark on the grammatical construction used by the root-text; the commentator emphasises that *dṛśyamūle* is a *dvandva* compound, in the nominative case it should be *dṛśyamūlam* (see line 3), in the root-text it occurs in the locative case (see *tatra dṛśyamūle*, line 3).

1162 *prastāva* – this is the first occurrence of the term.

1163 The commentator observes that the root-text uses a construction with a locative absolute where the copula (see *sati*, line 4) is implied.

1164 Line 4: *savarṇita*] em.; *saṃvargita* (K). The reason for this emendation is that the verbal adjective *saṃvargita* lit. "squared, raised to its power" seems inappropriate in this context. Moreover, this term does not occur in the SGT. On the contrary, the suggested *savarṇita* refers to the fraction of the next sample problem, where the second method of the class of fractional increase is applied in order to simplify the fraction.

1165 The commentary elaborates on *padārdha*. Note that in the next sample problem the case which the commentary explains in this same passage occurs (see lines 25–26).

been halved, which is the "root".[1166] By means of the [procedure] "uneven and even"[1167] and so forth, **when the square-root** of the [result of the] number of the "visible quantity"[1168] **added** to this [result of the square of half the root], when [the root is] halved by being multiplied by two, it is added [to it].[1169] [The expression] beginning with **half the root** means that "root" [obtained] in the occasion (*velāyāṃ*) of the division[1170] [of the root] "by one lessened by the fraction", ["root"] which is not the obtained squared [quantity],[1171] it is [instead] the quotient of that which has been divided by two; [hence, ***mūladvibhāgasahite*** or **added to half the root]** means **added** to it. [The expression] **raised to the square** implies "having calculated the square" of the number [obtained] from the square-root derived from the "visible quantity", [which has been added to the square of half the root], "added to half the root", **the quantity** being **wished for should occur;** this [last passage] is clear [in meaning].

In this respect, in the exemplifying metrical stanza [which follows] the author enunciates a sample problem:

77. One-third of a herd of deer together with [its] one-third disappeared out of the fear of a tiger. Its own square-root (i.e., of the herd), longing for *gīta* ("song"),[1172] were gargling with a mouthful of water and remained with closed eyes. Two female deer with tremulous eyes were seen roaming in a forest separated from the herd. Tell [me] soon, if you know the procedure of this calculation, the measure of the group [of deer].

The explanation with the presentation:[1173]

	bhā 1		
rū 1	3	mū 1	dṛ 2
1	1	1	
	3		

. In this case where there is the class of fractional increase, by means of [the method] "the

1166 It denotes the quotient obtained when the *mūla* is divided by one lessened by the fraction.

1167 The commentator quotes his own passage on square-root of integers (see SGT 26, 29).

1168 It denotes the quotient obtained when the given *dṛśya* is divided by one lessened by the fraction.

1169 *samānīta* – this is the first occurrence of the term.

1170 *vihṛti.*

1171 In this passage, the commentator expands upon *mūla* of *mūladvibhāga.*

1172 In Sanskrit literature, this is a common poetic motif; deer are said to enjoy song and music, becoming thus vulnerable to hunters.

1173 I have emended the layout: in the edition, the last fraction in the box denoted by *bhā* is one-half, while it should be one-third. Also, I have added the *mūla* 1 as it is mentioned in the text. In fact, in line 24, the commentator uses the expression *mūlasthāne* ("in place of the root"), which denotes the fact that in the layout the new result replaces the original *mūla*. Note that, in the layouts of the next sample problems, *mūla,* in fact, occurs. This *mūla* should be placed before the *dṛśya*, which is "at the tip" (see the meaning of *agra* in *mūlāgrabhāgajāti,* which is the name of this type-problem).

[upper] denominator by the [lower] denominator"[1174] and so forth, one should multiply the "[upper] denominator" three by the "[lower] denominator" three; it becomes nine. One should multiply the "first numerator", which is one, "by the denominator below added to [its own] numerator",[1175] that is, by three added to one, [thus] by four; it becomes four whose denominator is nine: $\left|\begin{array}{c}4\\9\end{array}\right|$.

When there is the exchange by means of the denominator one of unity,[1176] similarly[1177] according to [the rule] "the numerator and the denominator" and so forth,[1178] it is: $\left|\begin{array}{c}4\\9\\1\end{array}\right|\left|\begin{array}{c}1\\1\\1\end{array}\right|$. The first quantity multiplied by one is exactly so; the second quantity one multiplied by nine becomes nine. When there is the subtraction of the first numerator four from this [nine], it becomes five whose denominator is nine: $\left|\begin{array}{c}5\\9\end{array}\right|$. This is [the result of] one lessened by the fraction.

When the visible quantity and the root are divided by this [result],[1179] in order to "divide" the "visible quantity" two whose denominator is one, with respect to the first [quantity] five which is [the result of] "one lessened by the fraction", by means of [the method] "having performed the interchange"[1180] and so forth, it is: $\left|\begin{array}{c}9\\5\end{array}\right|\begin{array}{c}2\\1\end{array}\right|$. [Next is] the "multiplication": nine multiplied by two is eighteen, five multiplied by one is the very five; instead of the "visible quantity",[1181] there is eighteen whose denominator is five:[1182] $\left|\begin{array}{c}18\\5\end{array}\right|$. In the same way, the "root" one is multiplied by nine; instead of the "root", there is nine whose denominator is five:[1183] $\left|\begin{array}{c}9\\5\end{array}\right|$. [The step] **when the visible quantity and the root are divided by one lessened by the fraction** should be understood in this way.

1174 This is a quotation from the rule of the class of fractional increase (GT 57). The second part of this step is mentioned in the line below.
1175 Line 17: °*adho hareṇa*] em.
1176 *ekarūpa*.
1177 In line 18, the final *m* of *sama[m]* has been strangely assimilated into the quotation.
1178 This is a quotation from the rule of the class of simple fractions (GT 53).
1179 The result of the step *bhāgonarūpa* becomes the divisor of both the *dṛśya* and the *mūla*.
1180 This is a quotation from the rule of division of fractions (GT 42). The second part of this step is mentioned further in the same line.
1181 In line 24, *dṛśyasthāne* and *mūlasthāne* suggest that, at a certain point, the results obtained replace the original quantities in the layout.
1182 This is the quotient resulting from the division of the *dṛśya* 2 by the result of the step indicated by the expression *bhāgonarūpa*.
1183 This is the quotient resulting from the division of the *mūla* 1 by the result of the step indicated by *bhāgonarūpa*. Line 24: *jātā*] em.; *jāta*° (K).

In this case, the "root" is nine,[1184] its "half"[1185] is not possible; when there is the multiplication[1186] of the five underneath by two, it becomes ten. In this manner, the number above becomes "half". Regarding the number above, which is possible to halve,[1187] when there is the multiplication of [the number] standing underneath by two, the number above becomes halved; [this procedure] should be always understood in this way. The "square" (*karaṇī>varga*) of nine, which is "half the root", becomes eighty-one, and when there is the square of ten, it is one hundred: $\begin{array}{|r|}81\\100\end{array}$. In order to add the "visible quantity" to it, by means of [the method] "the numerator and the denominator"[1188] and so forth, when there is the reduction of five, which is the denominator below eighteen which is [the numerator of] the "visible quantity", by five, it is one; when there is the reduction of one hundred by five, it is twenty. Afterwards, when there is the exchange [of the denominators], it is: $\begin{array}{|r|r|}81&18\\100&5\\1&20\end{array}$. The first quantity multiplied by one is exactly so; in the next quantity, eighteen multiplied by twenty becomes three hundred sixty, five multiplied by twenty becomes one hundred. Since equal denominators occur, when the additive quantity eighty-one is added to three hundred sixty, it becomes four hundred forty-one whose denominator is one hundred: $\begin{array}{|r|}441\\100\end{array}$. Regarding these two numbers, according to [the rule] "uneven and even"[1189] and so forth, when the quotient is forty-two, half[1190] of that which has been multiplied by two, it is twenty-one above and ten underneath: $\begin{array}{|r|}21\\10\end{array}$. In this case, "half the root" consists of the previously mentioned nine whose denominator is ten.[1191] Twenty-one "added" to it becomes thirty whose denominator is ten; its "square" is nine hundred above and one hundred below: $\begin{array}{|r|}900\\100\end{array}$. When there is the division of the number above by the number underneath, the quotient is nine, which is the "measure" of the [herd of] deer.

1184 The next step to be carried out is *padārdha*.
1185 Line 25: *tadardhaṃ*] em.
1186 Line 26: °*guṇanāyām*] em.
1187 Line 26: *sahite*] em.
1188 This is a quotation from the rule of the class of simple fractions (GT 53).
1189 The commentator quotes his own passage on square-root of integers (SGT 26, 29).
1190 Line 8: °*ardhaṃ*] em. This passage refers to the procedure of square-root (GT 23), more particularly to see step *dvighne'rdhite*.
1191 See lines 24–26.

Its verification: **one-third** of nine is three. Its one-third:[1192] when there is one-third of three, it is one which added to that [three] becomes four, this [is the number of deer that] disappeared. **Its own square-root** is the "square-root of nine", which is three, [which is the number of deer] **longing for *gīta*.** Two [deer] "were seen": $\begin{array}{|c|}3\\1\\3\\2\end{array}$. When there is their sum of these, [the result] is nine.

The author enunciates a second sample problem:

> 78. Five-ninths of a troop of monkeys together with the square-root [of the total number of the troop] have their minds absorbed with swinging on the branches of a jack-fruit tree. Ten monkeys longing for the fruits [of the tree] are seen in a dispute. O mathematician, tell [me] quickly the measure of the troop.

[The presentation with the very presentation:][1193] $\begin{array}{|c|c|c|c|}\text{rū } 1 & \text{bhā } 5 & \text{mū } 1 & \\ & & & \text{dṛ } 10\\ 1 & 9 & 1 & \end{array}$ In this case, regarding "one" whose denominator is one, in order to [perform] the subtraction of the "fraction" from one,[1194] by means of [the method] "the numerator and the denominator"[1195] and so forth, when there is the exchange of the denominators, it is: $\begin{array}{|c|c|}1 & 5\\1 & 9\\9 & 1\end{array}$. In the first quantity, one multiplied by nine becomes nine whose denominator is nine; the next quantity multiplied by one is exactly so. Then, when there is the subtraction of the numerator five from nine which is the number occurring from [the multiplication by] "one", it becomes four whose denominator is nine: $\begin{array}{|c|}4\\9\end{array}$. This is [the result of] "one

1192 Line 12: *svatrilavas*] em.; *svalavas* (K).

1193 In line 20, *nyāsenaiva vyākhyā* (see the previous sample problem) or *nyāsenaivāsya vyākhyā* (as in the next sample problem) is missing. I have emended the text as the "fraction" five-ninths should be denoted by the abbreviation *bhā* (standing for *bhāga,* "fraction") rather that the *yū* standing for *yūtha* or "troop, herd". Cf. the layout in the previous sample problem.

1194 Line 20: *bhāgaikapatārtham*] em. The transmitted *bhāgaikyārtha* does not fit the context and it is worth noting that it is a step mentioned in GT 70; it seems, in fact, an instance of harmonization to parallels. The mathematical step meant here is *bhāgonarūpam,* mentioned in line 23. The emendation follows the commentary, cf. SGT 79, 25 where one finds *bhāgaikapāte*. In this same line 20, note the anomalous way the *anusvāra* is assimilated into the quotation.

1195 This is a quotation from the rule of the class of simple fractions (GT 53).

lessened by the fraction". **When the visible quantity and the root are divided** by this [result], it is so that the "visible quantity and the root" are ten and one respectively, [in both cases] one is the denominator, [the visible quantity] is divided:[1196] $\left|10\right|$. Regarding the divisor (*bhāgadāyin*) four and nine,[1197] which is the remainder from [subtracting the given fraction from] one, according to [the rule] "having performed the exchange of the numerator and the denominator"[1198] and so forth, nine is above, [and] four is below: $\left|\begin{array}{c}9\\4\end{array}\right|$. Then, regarding the "visible quantity" ten, when multiplied by nine it becomes ninety whose denominator is four: $\left|\begin{array}{c}90\\4\end{array}\right|$. In the same way, regarding the "root", which is one, when multiplied by nine it becomes nine whose denominator is four: $\left|\begin{array}{c}9\\4\end{array}\right|$. [The step] **when the visible quantity and the root are divided by one lessened by the fraction** should be understood in this way.[1199] Here again the "root "[is nine], "half" of nine is not possible, when there is the multiplication of the four below by two, it becomes eight. In this way, "half root" occurs. Then, the "square" (*karaṇī>varga*) is eighty-one above and sixty-four underneath. This is the square of the fraction, **the visible quantity** four whose denominator is ninety[1200] **is added** to it. Therefore, by means of [the method] "the numerator and denominator"[1201] and so forth, when there is the exchange of the denominators, when there is the reduction of the two [denominators] by four, one [and] sixteen occur:[1202] $\left|\begin{array}{c|c}81&90\\64&4\\1&16\end{array}\right|$. The first quantity multiplied by one is exactly so, in the next quantity ninety multiplied by sixteen becomes fourteen hundred forty, four multiplied by sixteen becomes sixty-four. Since there are equal denominators, when the additive quantity eighty-one is added to fourteen hundred forty, it becomes fifteen hundred twenty-one: $\left|\begin{array}{c}1521\\64\end{array}\right|$. Regarding these [two numbers], [in order to arrive at their square-root] according to [the

1196 In line 24, one would expect the term *dṛśyam* to be specified; also in the same line, the number 10 appears abruptly.

1197 In the SGT, this is an unusual way to denote a fraction.

1198 This is a quotation from the rule of division of fractions (GT 42).

1199 Line 28: *jñeyam*] em.] *jāte* (K). Cf. the same sentence in SGT 77, 25.

1200 In line 2, the verbal numerical expression should also occur in the ablative case as *dṛśyāt*, thus *caturbhyo navatichedebhyaḥ*.

1201 This is a quotation from the rule of the class of simple fractions (GT 53).

1202 Line 3: **jāte* ekaḥ ṣoḍaśakaṃ yathā*] em. The transmitted reading seems to show grammatical attraction.

rule] "uneven and even"[1203] and so forth, by means of half of that which has been multiplied by two, the quotient is thirty-nine above and eight below. This is the square-root of the fraction:[1204] $\left|\begin{array}{r}39\\8\end{array}\right|$. In this case, "half the root" is nine whose denominator is four, which is the remainder [of the subtraction of the fraction] from one. It is "added"[1205] to this [result]: when the additive quantity nine is added to thirty-nine, as equal denominator occurs, it becomes forty-eight. Then, when there is the reduction of forty-eight by eight, six is above [and] one is below: $\left|\begin{array}{r}6\\1\end{array}\right|$. Their "square" is thirty-six above and one below: $\left|\begin{array}{r}36\\1\end{array}\right|$. This is the "measure of the troop" of monkeys.

Its verification: thirty-six divided by nine is four, the numerator is five, four [multiplied] by five is twenty, and the square-root of thirty-six is six, twenty [and six] are [intent] on "swinging", and **ten are seen**: $\left|\begin{array}{r}20\\6\\10\end{array}\right|$. When there is their sum, it is thirty-six.

The author enunciates a third sample problem:

> 79. O clever mathematician, one-eighth of a herd of pigs is playing in mud [and] in a small pond. O friend, the square-root together with its half are digging up the *mustā*-grass.[1206] A female with seven young pigs are seen separated from their herd looking around for their own family. Tell [me] quickly, if you know arithmetic, how many hogs form the herd?

Its explanation with the presentation:[1207] $\left|\begin{array}{r}\text{rū } 1\\1\end{array}\right|\begin{array}{r}\text{bhā } 1\\8\end{array}\left|\begin{array}{r}\text{mū } 1\\1\\2\end{array}\right|\begin{array}{r}\text{dṛ } 8\\1\end{array}\right|$. In this case, regarding the quantity of the "square-root", by means of [the method] "multiplied by the denominator"[1208] and so forth, one multiplied by the "denominator"

1203 The commentator quotes his own passage on square-root of integers (SGT 26, 29).

1204 See the step *mūladvibhāgasahite*.

1205 It refers to the fraction $\left|\begin{array}{r}39\\8\end{array}\right|$.

1206 This is the name of a plant commonly known as *cyperus rotundus*.

1207 I have emended the layout (line 21); I have added the *rūpa* 1, which is found in the layout of both previous sample problems, and I have replaced the abbreviation *yū* (standing for *yūtha* or "herd") with *bhā*, denoting the *bhāga* or "fraction" as in the first sample problem on this type-problem (SGT 77, 15).

1208 The quotation, from the class of fractional increase (GT 57), should occur as *chedanighn eṣvityādinā.*

two becomes two, when the additive quantity two[1209] is added to one, it becomes three whose denominator is two. In order to carry out [the step] **[one] lessened by the fraction,** when there is the exchange of the denominators one of "one" and eight, which is the denominator of the "fraction", it is: $\begin{array}{|c|c|} 1 & 1 \\ 1 & 8 \\ 8 & 1 \end{array}$. One multiplied by eight becomes eight, the next quantity multiplied by one is exactly so. Then, when there is the subtraction of one from the fraction, from the eight obtained from "one" it becomes seven whose denominator is eight: $\begin{array}{|c|} 7 \\ 8 \end{array}$. [The next step is] **when the visible quantity and the root** [***vihṛte*** or "are divided"] by it: since it is the quantity-divisor, in order to perform the division, according to [the rule] "having performed the interchange"[1210] and so forth, eight is above [and] seven is underneath: $\begin{array}{|c|c|} 8 & 8 \\ 7 & 1 \end{array}$. Then, there is the "multiplication": eight multiplied by eight becomes sixty-four, the denominator seven is multiplied by one: $\begin{array}{|c|} 64 \\ 7 \end{array}$. [Next is] **the square-root together with its half,**[1211] that is, when[1212] the additive quantity two is added to one, it is three [and the denominator is] two, which is the [new] "root". There is the "multiplication":[1213] three multiplied by the numerator eight becomes twenty-four; similarly, two multiplied by the denominator seven becomes fourteen. In this way, **when the visible quantity and the root are divided**[1214] **by one lessened by the fraction** is accomplished: $\begin{array}{|c|c|} \text{mū } 24 & \text{dṛ } 64 \\ 14 & 7 \end{array}$. The remainder [of the subtraction of the fraction] from "one"[1215] has been erased, thus the [step] beginning with

1209 Line 22: *rūpamadhye dvikṣepe*] em.; *rūpa eko madhye* (K). The transmitted reading is corrupt, the emendation follows the commentary.

1210 This is a quotation from the rule of division of fractions (GT 42); it continues in the line below.

1211 See in the layout the third box on the left.

1212 This may have been the intended sense, but it not expressed by the Sanskrit as it stands. Line 28: *mūlaṃ sārdhaṃ ekamadhye dvikṣepe mūlatridvau*] em. The transmitted *mūlatridvirūpasya madhye* is nonsensical; the emendation follows the commentary.

1213 The SGT is unusually laconic here, the step meant is the division of the mentioned new "root" $\begin{array}{|c|} 3 \\ 2 \end{array}$ by the difference $\begin{array}{|c|} 7 \\ 8 \end{array}$.

1214 Line 1: °*vihṛte*] em.; °*rahite* (K). See in the *karaṇasūtra* the step *bhāgonarūpavihṛte*.

1215 Here *rūpaśeṣa* denotes the fraction $\begin{array}{|c|} 7 \\ 8 \end{array}$, which is the difference previously obtained (see line 25). The syntagm *rūpaśeṣa* also occurs in the previous sample problem (see SGT 78, 24); Notably, it is not found in the first sample problem on this type-problem of fractions.

the "visible quantity" [has been accomplished].[1216] In this case, the "root" is twenty-four, its "half" is twelve, its "square" is one hundred forty-four. The "square" of the fourteen below is one hundred ninety-six. In this case, the "root" is twenty-four, its "half" is possible without the multiplication by two of the denominator fourteen.[1217] Therefore, the square of the numbers above and below occurs, it is: $\begin{array}{|c|}144\\196\end{array}$. Regarding this [fraction], for the sake of the sum,[1218] according to [the rule] the "numerator and the denominator"[1219] and so forth, when there is the reduction of seven, which is the denominator of the "visible quantity" [whose numerator is] sixty-four, by seven, it becomes one. When there is the reduction of one hundred ninety-six by seven, it is twenty-eight. Then, when there is the exchange, it is: $\begin{array}{|r|r|}144&64\\196&7\\1&28\end{array}$. The first quantity multiplied by one is exactly so; in the next quantity, sixty-four multiplied by twenty-eight becomes seventeen hundred ninety-two; seven multiplied by twenty-eight is one hundred ninety-six: $\begin{array}{|c|}1792\\196\end{array}$. Since equal denominators occur, when the additive quantity one hundred forty-four is added to seventeen hundred and so forth, it becomes nineteen hundred thirty-six whose denominator is one hundred ninety-six: $\begin{array}{|c|}1936\\196\end{array}$. Regarding these [numbers], by means of [the method] "uneven and even"[1220] and so forth, when halved that which has been multiplied by two, it becomes forty-four above[1221] [and] fourteen underneath: $\begin{array}{|c|}44\\14\end{array}$. In this case, "half the root" [should be calculated]:[1222] half is twelve, when it is "added" to this [twelve], it becomes fifty-six whose

1216 The commentator observes that the result of the step *bhāgonarūpa* or "one lessened by the fraction", which is the difference $\begin{array}{|c|}7\\8\end{array}$, has been manipulated as the divisor of the *dṛśya* and the *mūla*.

1217 In line 4, the passage from *ekaṃ* to *sañjñā* is corrupt. I have proposed the following reconstruction on which the translation above is based: *atra padaṃ caturviṃśatis tasyārdham sahate vinā caturdaśacchedānāṃ dviguṇanāṃ*] conj. This is a tentative conjecture, which however aims to reflect the intent of the text; other reconstructions are no doubt possible.

1218 This fraction is to be added to the *dṛśya*.

1219 This is a quotation from the rule of the class of simple fractions (GT 53).

1220 The commentator quotes his own passage on square-root of integers (SGT 26, 29).

1221 I have emended the layout (line 13): the square-root of 1936 is 44 and not 144 as found in the edition.

1222 Line 13: *atra*] em.; *agra* (K). Note that the term *agra* occurs below; it appears, in fact, to be an instance of harmonization to parallels.

denominator is fourteen: $\left|\begin{array}{c}56\\14\end{array}\right|$. When there is the reduction of these by fourteen, four is above, one is below. Then, there is their "square": sixteen is above, one is below: $\left|\begin{array}{c}16\\1\end{array}\right|$. This is the measure of the "herd of hogs".

Its verification: when there is one-eighth of sixteen, it is two. **The square-root** of sixteen is four, **together with its half** it is six. Eight [pigs] **are seen**: $\left|\begin{array}{c}2\\4\\2\\8\end{array}\right|$. When there is their sum, it is sixteen.

In this way, because it (i.e., the solution) has been produced from the fraction [manipulated] with the root and the visible quantity, the type-problem of the fraction [executed] with the root and the visible quantity [is so termed and its explanation] is concluded.[1223]

The type-problem of the visible quantity at both tips* (ubhayāgradṛśyajāti)[1224] *[GT 80]

Next is the procedural rule of the type-problem of the visible quantity at both tips;[1225] the author enunciates a metrical stanza:

> GT 80. The [last] visible quantity and the root are divided by [the result of] the multiplication of one devoid of the fractions.[1226] The square-root from the sum of the [first] visible quantity and that which has been obtained from the sum[1227] of the [last visible quantity and the] square of half the root [is extracted], [the square-root is] added to half the root, [and the result is] multiplied by itself. [The result] will be the desired quantity (*iṣṭarāśi*).

The explanation: this [type-problem] concerns a double tip,[1228] that is, the first tip (i.e., occurring at the beginning) [and] the last tip (i.e., occurring at the end).

1223 The SGT elucidates the name of this type-problem.
1224 K 54, 21–24.
1225 In the lines below, the commentator clarifies that this type-problem of fractions concerns two *agras* or "tips": there is the *ādyāgra* (lit. "first tip") and the *antāgra* (lit. "last tip"). As in the explanation of the previous procedural rule, here too the commentator specifies that the procedure is so named according to the position of the "visible quantity" in the layout: there is the "visible quantity at the beginning" and the "visible quantity at the end". With respect to the layout, this terminology clearly refers to a left to right direction; see the layout in SGT 81, 12 where one finds two *dṛśyas*: one *dṛśya* at the *ādyāgra* (lit. "first tip") and the other at the *antāgra* (lit. "last tip").
1226 Hereafter "the [last] visible quantity" and the "root" denote the quotients obtained.
1227 See *yukta*.
1228 *ubhayāgra*. The commentary specifies the main characteristic of this type-problem of fractions.

For this reason,[1229] (i.e., in the sample problem as well as in the layout), a "visible quantity" is at the beginning and a "visible quantity" is at the end.[1230] This type-problem is therefore so made [and thus so named].[1231] [The expression] **one devoid of the fractions** implies that as many as are the "fractions", such are the times that "one" is lessened by the given "fractions".[1232] Their "multiplication" [should be performed]: as in the multi-part class,[1233] regarding [both] the "numerator and denominator", the multiplication of the "numerator" by the "numerator" and the multiplication of the "denominator" by the "denominator" [should be performed]. They[1234] are "divided" by this [result]:[1235] by means of the method of division, the last "visible quantity"[1236] and the "root" are divided. Therefore [this is the meaning of the expression] **the visible quantity and the root are divided by [the result of] the multiplication of one devoid of the fractions. [The expression] [the square-root] from the sum of the [first] visible quantity** means that, since equal denominators occur,[1237] there is the "sum" (*aikya>saṃyoga*) from the two "visible quantities", which are the first and the last;[1238] [the square-root is extracted] from that [sum].[1239] **[Regarding**

1229 In this passage, sentences *tatra* functions as a conjunctive adverb linking the two nominal sentences. The adverb clause it introduces, which is a subordinating-connecting clause, defines the relationship with the statement expressed in the main clause.

1230 The SGT clarifies again that *agra* or "tip" alludes to the *dṛśya* or "visible quantity". Note that first the commentary illustrates that, differently from the previous type-problems, now the visible quantities are two.

1231 Perhaps the *daṇḍa* is wrongly found and *tatra* or "there" (i.e., in the rule) is to be read together with the quotation which follows; this kind of formula occurs often in the commentary.

1232 This remark anticipates the case presented in the third sample problem on this *jāti*, where the fractions are two. See the explanation in SGT 83, 23–1.

1233 It refers to the *prabhāgajāti* (GT 55).

1234 In brief, the last "visible quantity" and the "root"; lines 25–1 are dedicated to explaining the step *niraṃśarūpāhatibhaktadṛśyamūle*.

1235 The pronoun *tayā* refers to *āhati*, which denotes the product of the mentioned multiplication.

1236 *paryantadṛśya*.

1237 This is a reference to the computation carried out in the next sample problems where the abstract form *samacchedatvāt* conveys the same idea; see SGT 81, 27 (*sadṛśacchedatvāt*); 82, 2 and 83, 11.

1238 Line 1: *ādyāntayor*] em.

1239 The commentator expands upon *dṛśyaikato*. That the final ending *yor* in *dṛśyayor* and *ādyāntayor* denotes an ablative dual rather that a genitive dual (as it should be more natural to interpret in relation to *aikya* or "sum") is evident from the execution of the following three sample problems. Also, the argument elaborated by the commentary is well represented by the adjective *uttha* ("obtained") in SGT 81, 27. For instance with respect to the first sample problem, the steps explained allude to the "sum" (*aikya*) of $\begin{vmatrix}15\\1\end{vmatrix}$ (which is the result of the sum of the new "last visible quantity" 14 and the square of half the root, which is 1) and the "first visible quantity" $\begin{vmatrix}1\\1\end{vmatrix}$; the square-root is then extracted. See the next sample problem, lines 26–28; cf. the two constructions with *tasmāt* here and *asmād dṛśyaikyato* there.

the expression] [the square-root] from the sum of the [first] visible quantity:[1240] what does it consist of (*kiṃ rūpāt*)? [The expression] **"half the root"** is **"half"** (*dvyaṃśaka>ardha*)[1241] the "root",[1242] its "square" [is to be calculated]; [the expression] **from that which has been obtained from the sum** [of the last visible quantity and] that [aforementioned result].[1243] This is the meaning: having first made the last "visible quantity" **added to half the root**, afterwards the "sum" of the [first] "visible quantity" and the "root"[1244] is performed. Then, by means of [the method] "uneven and even"[1245] and so forth, **the square-root is added to half the root:** this is clear [in meaning]. **[And the result is] multiplied by itself:** that number [which is the result] of the "square-root added to half the root" is multiplied by itself, that [result] **will be the desired quantity.**

In this respect, in the exemplifying metrical stanza [which follows] the author enunciates a sample problem:

> 81. [Out of a herd of elephants] an elephant is standing near a pillar;[1246] one-sixth of the remainder is playing in a lake. One-fifth of the remainder is grazing on the twigs of the *śallakī* tree on a mountain slope. One-fourth of the remainder along with the square-root of the total number are seen, [being] terrified by the roaring of a lion. Another elephant is following six elephants.[1247] Tell [me] the measure [of the herd], how many are [the elephants]?

The explanation: **near a pillar** is "near the post" to which an elephant is tied. **An elephant:** because of the use of the singular form, one [elephant] **is standing** [hence *ekena stamberameṇa*].[1248] **One-fifth of the remainder** is a fifth part.[1249] The rest is clear [in meaning].

1240 This passage is clumsy, emendation could also be considered; for instance (lines 1–2) *samacchedanād yaḥ tasmāt | dṛśyaikyataḥ* could be emended to *samacchedanād | tasmād dṛśyaikyataḥ mūlam* |] conj. The translation is: "[...] because equal denominators occur. **The square-root** from this **sum of the [first] visible quantity** [should be extracted]". This emendation follows the commentary, see SGT 81, 28 and 83, 12.

1241 The same gloss occurs in the next sample problems; see SGT 81, 24–25; 82, 23, and 83, 5.

1242 The SGT provides a synonym for *dvyaṃśa* and shows that the relationship between *pada* and the compound *dvyaṃśaka* in *padadvyaṃśaka* is that of a *tatpuruṣa* genitive compound.

1243 It refers to the step "the square of half the root" (*padadvyaṃśaka*). The translation reflects the brevity of the statement.

1244 Here "root" denotes the result of the sum of the last visible quantity and the square of half the root.

1245 The commentator quotes his own passage on square-root of integers (SGT 26, 29).

1246 In the layout given by the SGT (line 12), the quantity represented by the elephant standing near the pillar represents the "first visible quantity" on the left.

1247 These 7 elephants denote the "last visible quantity"; in the layout, see the last box on the right.

1248 In this first line (line 11), the commentator gives a remark on syntax and explains the impersonal construction (*stamberameṇa sthitam*) used by Śrīpati.

1249 In SGT 69, 3–4, the commentator has already explained the meaning of the term *iṣu,* (lit. "arrow", here "five") according to the *bhūtasaṃkhyā* notation.

The presentation: $\begin{array}{|c|c|c|c|c|c|}\text{dṛ } 1 & \text{śe } 1 & 1 & 1 & \text{mū } 1 & \text{dṛ } 7 \\ 1 & 6 & 5 & 4 & 1 & 1\end{array}$. [Regarding the expression] **[the multiplication of] one devoid of the fractions** and so forth,[1250] by [means of the [method] "the numerator and the denominator"[1251] and so forth, when there is the exchange of the denominators and when there is the multiplication, in place of "one" it is respectively: six whose denominator is six, five whose denominator is five, [and] four whose denominator is four: $\begin{array}{|c|c|c|}6 & 5 & 4 \\ 6 & 5 & 4\end{array}$. Then, with respect to each[1252] of the numerators [of the fractions] which have just been written, when there is the subtraction, instead of one it is respectively: five whose denominator is six, four whose denominator is five, [and] three whose denominator is four: $\begin{array}{|c|c|c|}5 & 4 & 3 \\ 6 & 5 & 4\end{array}$. With respect to it, which is [the result of] "one devoid of the fractions", there is the reciprocal "multiplication" (*āhati>guṇana*) of the number of the denominator underneath [and] of the number above: four multiplied by five is twenty, three multiplied by twenty is sixty.[1253] In the same way, five multiplied by six is thirty, four multiplied by thirty is one hundred twenty: $\begin{array}{|c|}60 \\ 120\end{array}$. When there is their reduction by sixty, one is above [and] two is below: $\begin{array}{|c|}1 \\ 2\end{array}$. **[The visible quantity and the root]** "are divided by the multiplication of one devoid of the fractions", that is to say, by this [result]. To illustrate: since it is the divisor, according to the rule "having performed the interchange"[1254] and so forth, two is above [and] one is below:[1255] $\begin{array}{|c|}2 \\ 1\end{array}$. Next is the "multiplication" of seven, which is the [numerator of the] "visible quantity": seven multiplied by two becomes fourteen, the denominator one multiplied by one is exactly so. In the same way, [one should divide] the "root",[1256]

1250 Line 13: **rūpam ekacchedagatam**] em. The transmitted reading is uninterpretable.
1251 This is a quotation from the rule of the class of simple fractions (GT 53).
1252 In line 15, the pronominal adjective *ekaika* ("each one, every") has a distributive sense and it underlines the iteration of the process: the numerator 1 of the quantity *rūpa*, which is $\begin{array}{|c|}1 \\ 1\end{array}$, is subtracted from each numerator of the obtained fractions.
1253 The multi-part class is carried out (see GT 55).
1254 This is a quotation from the rule of division of fractions (GT 42); it continues in the line below.
1255 The obtained fraction $\begin{array}{|c|}1 \\ 2\end{array}$ becomes the divisor of the *dṛśya* $\begin{array}{|c|}7 \\ 1\end{array}$.
1256 The number 1 denotes the *mūla* or "root" in *dṛśyamūle*.

which is one. There is the "multiplication": one multiplied by two becomes two; the denominator one multiplied by one is exactly so. Similarly, the "visible quantity" and the "root" divided by [their denominators] one[1257] stay as they are:[1258] $\left|\begin{array}{r|r}14 & \text{mk } 2\\ 1 & 1\end{array}\right|$. Afterwards, "half the root" two is half [of two], which is one; its "square" is again one. The "visible quantity" fourteen is "added" to it: it becomes fifteen whose denominator is one: $\left|\begin{array}{r}15\\ 1\end{array}\right|$. The expression **[the square-root] from the sum of the [first] visible quantity and that which has been obtained from the sum of [the last visible quantity and the square of half the root]:** it[1259] denotes this [result]. The "visible quantity" which is the first (*ādya*) is one whose denominator is one; because equal denominators occur, when the additive quantity one, which is the first "visible quantity", is added to the fifteen obtained from the last visible quantity, it becomes sixteen: $\left|\begin{array}{r}16\\ 1\end{array}\right|$. From this, which is the **sum of the [first] visible quantity [and from that which has been has been obtained from the sum of the last visible quantity and the square of half the root], the square-root [is extracted]:** it is four. The "root" previously mentioned[1260] is two, its "half" is one, four added to it becomes five; it is **multiplied by itself,** that is, five multiplied by five becomes twenty-five: $|25|$. The quotient [of the division of the numerator by the denominator] is the measure of the herd of elephants.

Its verification: among twenty-five, one is **near a pillar**. One-sixth of the "remainder" twenty-four is four, [which is the number of elephants playing] **in a lake**. One-fifth of the "remainder" twenty is four, [which is the number of elephants found] on a mountain. "One-fourth" of the "remainder" sixteen is four, [which is the number of elephants] frightened by a "lion". In the same way, the "square-root" of the aforementioned twenty-five, which is the just obtained[1261] number, is five; **along with** these denotes the "sum", the seven [elephants] that were "seen" are added: $\left|\begin{array}{r}1\\ 4\\ 4\\ 4\\ 5\\ 7\end{array}\right|$. When there is their sum, it is twenty-five.

1257 In line 24, *ekabhaktam* refers to the division of the numerator by the denominator.

1258 I have emended the layout (line 24) as the abbreviation *mū* is curiously in a separate box. Note that the abbreviation *dṛ* denoting the "visible quantity" (*dṛśya*) is missing.

1259 It refers to the number 15.

1260 See line 24.

1261 The use of the two adjectives meaning "previous, preceding" (*pūrva* and *prathama*) emphasises that the number 25 is the last obtained number, which has been earlier mentioned by the commentator.

The author enunciates a second sample problem:

82. [With respect to a swarm of bees] a pair of tawny bees is seen on a lotus. One-half of the remainder together with one-seventh are resting on the temples of a lordly elephant. The square-root of the [whole] swarm are humming on a jasmine flower, and a pair of bees is seen. O brother, tell [me] the [measure of] the swarm of bees.

Its explanation with the presentation:[1262] $\begin{array}{|c|c|c|c|} & \text{śe } 1 & & \\ \text{dṛ } 2 & 2 & \text{mū } 1 & \text{dṛ } 2 \\ 1 & 1 & 1 & 1 \\ & 7 & & \end{array}$. Because here there is a fractional increase, according to [the rule] "the [upper] denominator) by the [lower] denominator"[1263] and so forth, one should multiply the upper "denominator" two by the lower "denominator" seven, it becomes fourteen. One should multiply the "first numerator" one "by the denominator below added to [its own] numerator", that is, by seven "added" to one. It becomes eight whose denominator is fourteen: $\begin{array}{|c|} 8 \\ 14 \end{array}$. Then, regarding the denominator one of "one" (i.e., the unity *rūpa*), by means of [the procedure] "the numerator and the denominator"[1264] and so forth, when there is the exchange of the denominators, it is: $\begin{array}{|c|c|} 8 & 1 \\ 14 & 1 \\ 1 & 14 \end{array}$. [The first quantity] multiplied by one is exactly so; the two [numbers] one multiplied by fourteen become a pair of fourteen, [which is written] instead of one: $\begin{array}{|c|} 14 \\ 14 \end{array}$. When there is the subtraction of the numerator eight from it, it becomes six whose denominator is fourteen:[1265] $\begin{array}{|c|} 6 \\ 14 \end{array}$. When there is their reduction by half, three is above, and seven is below:[1266] $\begin{array}{|c|} 3 \\ 7 \end{array}$. This is [the result of the step] "one devoid of the fractions". In this case, however,[1267] because of the absence of other fractions no "multiplication" occurs. Since it

1262 I have emended the text as in the layout (line 12) the abbreviations are partially missing; cf. the previous sample problem.

1263 This is a quotation from the rule of the class of fractional increase (GT 57).

1264 This is a quotation from the rule of the class of simple fractions (GT 53).

1265 Line 17: *jātāḥ ṣaṭ caturdaśacchedāḥ* |] em. The transmitted reading is aberrant.

1266 The result of the step *niraṃśarūpa* or "one devoid of the fractions" becomes the divisor of the last *dṛśya* and the *mūla*.

1267 Note the significant role played by the conjunctive adverb *tu*.

is the divisor, according to [the rule] "having performed the interchange"[1268] and so forth, seven is above [and] three is underneath: $\begin{array}{|r|}7\\3\end{array}$. Next, the "multiplication" of two, which is the "visible quantity": it becomes fourteen, three multiplied by one is exactly so: $\begin{array}{|r|}14\\3\end{array}$. Similarly, the "root" one[1269] multiplied by seven becomes seven, three multiplied by one is exactly so: $\begin{array}{|r|}\text{mū } 7\\3\end{array}$. In this way, **the [last] visible quantity and the root are divided by [the result of] the multiplication of one devoid of the fractions** occurs.[1270] Then, "half"[1271] the root, which is seven: [its] half is not possible (*ghaṭate*);[1272] its half is [hence the multiplication] of the denominator [by two]: the three below multiplied by two becomes seven whose denominator is six. Afterwards, the "square-root" of these two numbers is forty-nine above, and thirty-six below: $\begin{array}{|r|}49\\36\end{array}$. Then, when there is the reduction of thirty-six by three, it is twelve which is below the denominator of the "visible quantity"; when there is the reduction of three by three, it is one.[1273] Next, by means of [the method] "the numerator and the denominator"[1274] and so forth: when there is the exchange of the denominators, it is: $\begin{array}{|r|r|}49 & \text{dṛ } 14\\36 & 3\\1 & 12\end{array}$. [The first quantity] multiplied by one is exactly so; fourteen multiplied by twelve becomes one hundred sixty-eight, three multiplied by twelve becomes thirty-six: $\begin{array}{|r|}168\\36\end{array}$. As there are equal denominators, when the additive quantity forty-nine is added to sixty-eight, it becomes two hundred seventeen whose denominator is thirty-six: $\begin{array}{|r|}217\\36\end{array}$. In this manner, **[the square-root from**

1268 This is a quotation from the rule of division of fractions (GT 42); it continues in the line below.

1269 Line 21 *ekaḥ*] em.; *ekasya* (K).

1270 Line 23: *siddham*] em. The transmitted *jātam* does not seem appropriate; also, in the SGT a *iti*-type formula followed by the verbal adjective *jāta* is not common. The emendation is based on the commentary, cf. SGT 83, 4.

1271 In this context, the use of the abstract *dvyaṃśakatā* followed by the subjective genitive is unusual; in the other sample problems on this *jāti* one finds *dvyaṃśakam*.

1272 In the SGT, this is the first occurrence of the verbal root *ghaṭ*- "to suit, be possible".

1273 The obtained *dṛśya* $\begin{array}{|r|}14\\3\end{array}$ is added to the result of *padadvyaṃśakavarga* (see the step *padadvyaṃśakavargayuktāt*).

1274 This is a quotation from the rule of the class of simple fractions (GT $\overline{53}$).

the sum of the visible quantity] and that which has been obtained from the sum of the [last visible quantity and the] square of half the root is accomplished. That [*ādya* or "first"] "visible quantity":[1275] in order to perform the "sum"[1276] [of this result] and the first[1277] "visible quantity" two whose denominator is one, by means of [the procedure] "the numerator and the denominator"[1278] and so forth, when there is the exchange of the denominators, it is: $\left|\begin{array}{r|r} 217 & 2 \\ 36 & 1 \\ 1 & 36 \end{array}\right|$. [The first quantity] multiplied by one is exactly so; two multiplied by thirty-six becomes seventy-two, one multiplied by thirty-six becomes thirty-six: $\left|\begin{array}{c} 72 \\ 36 \end{array}\right|$. Since there are equal denominators, when the additive quantity seventy-two is added to two hundred seventeen, it becomes two hundred eighty-nine whose denominator is thirty-six: $\left|\begin{array}{c} 289 \\ 36 \end{array}\right|$. In this manner, the "sum" from the [first] "visible quantity" occurs. Next, regarding these two [numbers], by means of [the method] "uneven and even"[1279] and so forth, **the square-root** of the number above is seventeen, and below it is six: $\left|\begin{array}{c} 17 \\ 6 \end{array}\right|$. Half the root is the previously calculated[1280] seven whose denominator is six. Since there are equal denominators, when the additive quantity seven is added to seventeen, it becomes twenty-four whose the denominator is six. When there is their reduction by six, four is above and one is underneath: $\left|\begin{array}{c} 4 \\ 1 \end{array}\right|$. Their "square" is sixteen above and one below: $\left|\begin{array}{c} 16 \\ 1 \end{array}\right|$. This [numerator sixteen] divided by [the denominator] one remains as it is. The quotient [sixteen] is [hence] the measure of the swarm of "bees".

Its verification: out of [the final result] sixteen, **a pair** of bees **is seen on a lotus. Half** of the "remainder" fourteen is seven. One-seventh of seven is one, [seven] is added to it, it is makes eight, [which is the number of bees resting]

1275 It refers to the number 2; see the layout found at the beginning of the commentator's explanation (line 12).

1276 See the step *dṛśyaikato*.

1277 It is noteworthy that *mūla* in *mūladṛśya* is "first"; the quantity meant is, in fact, the "first visible quantity" (*ādyadṛśya*) $\left|\begin{array}{c} 2 \\ 1 \end{array}\right|$. In the next sample problem, the same syntagm *mūladṛśya* occurs in a similar context (see SGT 83, 9).

1278 This is a quotation from the rule of the class of simple fractions (GT 53).

1279 The commentator quotes his own passage on square-root of integers (SGT 26, 29).

1280 In line 10, one would expect *prākkṛtasapta ṣaḍchedāḥ* |. The number 7 (*sapta*) is mentioned in line 23; with respect to the transmitted *saptadaśa*, it is interesting to note that this scribal error is due to the copyist's repetition of the term, which has just occurred in the line above (line 9, *mūlaṃ saptadaśa*).

on the temples of a [lordly] elephant. The square-root of the preceding, the aforementioned number sixteen, [which is the final result], is four, which [is the number of bees that] are [humming] on **a jasmine flower,** and, at the end, two are "seen":

2
7
1
4
2

.When there is their sum, it is sixteen.

The author enunciates a third sample problem:

83. A wealthy man gave to a twice-born[1281] one-fourth [of his wealth], then one-third of the remainder, one-fourth of the remainder, the square-root of the total wealth, and [its] one-half. In this way, how much was the wealth[1282] of this man who [after giving away such a big amount of money] became miserable?

[The presentation:][1283]

dṛ 1	śe 1	śe 1	mū 1	dṛ 1
4	3	4	1	2

. [In the stanza above] the word "seen" and the number which defines the "visible quantity" occur neither at the beginning nor at the end.[1284] Regarding the denominator one of "one", because of the denominators of the fractions which are the remainders,[1285] by means of [the method] "the numerator and the denominator"[1286] and so forth, when there is the exchange of the denominators, [the first quantity] multiplied by one is exactly so. The [two numbers] one multiplied by three becomes three whose denominator is three.[1287] The [two numbers] one multiplied by four become four. Four whose denominator is four is [placed] instead of one:

3	4
3	4

.

Next, when there is the subtraction of the numerator one from three, it becomes two whose denominator is two. When there is the subtraction of one

1281 In the traditional Hindu social system, a *dvija* is a member of the first three *varṇa*s or "classes"; it is reasonable to assume that here the term refers to a *brāhmaṇa* (the highest ranking of the four classes), possibly a priest or a teacher (*ācārya*).

1282 *dhana.*

1283 In line 23, the word *nyāsa* and the abbreviations in the layout are missing.

1284 The SGT clarifies that the sample problem does not directly mention either the word "seen" or the number for it, meaning that the visible quantities are not specified. The two *dṛśya*s, the *ādyadṛśya* and the *paryantadṛśya,* are one-fourth and one-half, which are the first and last fractions mentioned in the root-text.

1285 Line 24: *śeṣāṃśacchedābhyo aṃśacchedāvityādinā*] em. The transmitted reading is nonsensical.

1286 This is a quotation from the rule of the class of simple fractions (GT $\overline{53}$).

1287 I have emended the text as in line 25, there is an anomaly: *tricchedāś* refers to *jātās trayaḥ,* which occurs before the *daṇḍa*; it should read *jātās trayas tricchedāḥ* |.

from four, it becomes three whose denominator is four: $\begin{array}{|c|c|}2&3\\3&4\end{array}$. This is [the result of] "one devoid of the fractions". [Next is] its "multiplication": three multiplied by two becomes six, four multiplied by three becomes twelve, it is:[1288] $\begin{array}{|c|}1\\2\end{array}$. Since it is the divisor, according to [the rule] "having performed the interchange[1289] and so forth, two is above and one is below: $\begin{array}{|c|}2\\1\end{array}$. Afterwards, there is the "multiplication" by one-half, which is the [*anta* or "last"] "visible quantity" $\begin{array}{|c|}1\\2\end{array}$; two multiplied by one is [two] above, the very two is below:[1290] $\begin{array}{|r|}\text{dṛ } 2\\2\end{array}$. One, which is the "root", multiplied by two becomes two, and underneath the denominator one multiplied by one is exactly so:[1291] $\begin{array}{|r|}\text{mū } 2\\1\end{array}$. In this manner, [the step] **the [last] visible quantity and the root are divided by [the result of] the multiplication of one devoid of the fractions** is accomplished. Next, "half" (*dvyaṃśaka>ardha*) the "root" two is one. Its "square" is also the denominator one, the square of [the numerator] one is again one: $\begin{array}{|c|}1\\1\end{array}$. For the sake of the sum of this [square and the *dṛśya* two],[1292] by means of [the method] "the numerator and the denominator"[1293]and so forth, when there is the exchange of the denominators, it is:[1294] $\begin{array}{|c|}2\\2\\1\end{array}\begin{array}{|r|}\text{mū } 1\\1\\2\end{array}$. The pair of [number] one multiplied by two

1288 The text is elliptical here; it does not mention (perhaps a passage is missing) the *apavarta* or "reduction" of the fraction $\begin{array}{|c|}6\\12\end{array}$ to its lowest terms.

1289 The obtained fraction becomes the divisor of the *dṛśya* and the *mūla*. This passage is a quotation from the rule of division of fractions (GT 42); it continues in the line below.

1290 I have emended the layout, it is odd that one finds the abbreviation s in a separate box on the right. In the edition, here as well as in the previous sample problems on this type-problem, the abbreviations are very inconsistent: they are sometimes inside the layout together with numbers, sometimes outside on the left side in a separate column, sometimes on the right side in a separate column. Cf. the line below (line 4) and SGT 82, 22 and 27.

1291 In the layouts (lines 3 and 4), the two abbreviations indicate that the results represent the new *dṛśya* and the new *mūla*.

1292 Line 6: **dṛśyacchedanasamam**] em. The text is suspect here.

1293 This is a quotation from the rule of the class of simple fractions (GT 53).

1294 I have emended the layout (line 6) as the abbreviation *mū* strangely occurs in a box by itself.

becomes a pair of two, and [the other quantity] multiplied by one is exactly[1295] so:[1296] $\begin{array}{|c|}2\\ \\2\end{array}$. As there are equal denominators, when the additive quantity two is added to two, it becomes four whose denominator is two. In this manner, [the step] **added to half the root** is accomplished. In order to perform the sum of [of this result] and one whose denominator is four which is the first visible quantity,[1297] according to [the rule] "the numerator and the denominator"[1298] and so forth, when there is the reduction by half, the obtained denominators are one and two:[1299] $\begin{array}{|c|c|}1&4\\4&2\\1&2\end{array}$. The first quantity multiplied by one is exactly so; regarding the next quantity, four multiplied by two becomes eight, two multiplied by two becomes four: $\begin{array}{|c|c|}1&8\\ & \\4&4\end{array}$. As there are equal denominators, when the additive quantity one is added to eight, it becomes nine whose denominator is four: $\begin{array}{|c|}9\\ \\4\end{array}$. This is the "sum of the [first] visible quantity"; from it, **the square-root** of nine [is extracted and this] is three, **the square-root** of four is two: $\begin{array}{|c|}3\\ \\2\end{array}$. One whose denominator is one is "half the root"; for the sake of the sum,[1300] by means of [the method] "the numerator and the denominator" and so forth,[1301] when there is the exchange of the denominators, it is: $\begin{array}{|c|c|}3&1\\2&1\\1&2\end{array}$. [The first quantity] multiplied by one is exactly so, the [two numbers] one multiplied by two become a couple of two: $\begin{array}{|c|c|}3&2\\ & \\2&2\end{array}$. Then, when the additive quantity

1295 It is uncommon that the commentator manipulates first the number which he usually defines as *parāṅka* or *agretanāṅka,* and next the number which he denotes as *pūrvāṅka* or *prācyāṅka.* According to the commentator's usage, it should be the reverse case.

1296 I have emended the transmitted layout (line 7), which is $\begin{array}{|c|}2\\2\end{array}$.

1297 Line 9: *mūladṛśyaikena catuśchedenaikyaṃ kartuṃ aṃśacchedavityādinā*] conj.; *samacchedena samam* (K). The transmitted reading is corrupt, the conjecture follows the commentary; see the same passage in SGT 82, 4.

1298 This is a quotation from the rule of the class of simple fractions (GT $\overline{53}$).

1299 Line 10: °*chedau*] em.

1300 This result is added to $\begin{array}{|c|}1\\1\end{array}$ (see line 5).

1301 This is a quotation from the rule of the class of simple fractions (GT $\overline{53}$).

two is added to three, it becomes five whose denominator is two: $\begin{array}{|c|}5\\2\end{array}$. **The** [step] **square-root added to half of the root** is accomplished. It is **multiplied by itself:** [one should write] in two places again [the same fraction], five multiplied by five is twenty-five, two multiplied by two becomes four: $\begin{array}{|c|}25\\4\end{array}$. When there is the division of the number above by the number below, the quotient is six, and the remainder is one whose denominator is four: $\begin{array}{|c|}6\\1\\4\end{array}$. It becomes six plus one-fourth *dramma*s [which was the initial wealth] of that [man who became] **miserable**, that is, a "very poor man".

Its verification: starting from [the final result] six plus one-fourth, "one-third" [of six] is two.[1302] **One-fourth** of the remainder four is one. Regarding the whole [wealth] six plus one-fourth, since the class of fractional increase occurs, by means of [the method] "multiplied by the denominator"[1303] and so forth, it is $\begin{array}{|c|}6\\1\\4\end{array}$; six[1304] multiplied by the denominator four becomes twenty-four. When the additive quantity, such as one, [is added to that], it is twenty-five. Its **square-root** is five, **the square-root** of four is two: $\begin{array}{|c|}5\\2\end{array}$. Because it has been divided by two, it is namely two plus one-half, In the same way, there is **one-half** (*dala*>*ardha*), which is the [last] "visible quantity": $\begin{array}{|c|}1\\4\end{array}\begin{array}{|c|}2\\1\end{array}\begin{array}{|c|}1\\1\end{array}\begin{array}{|c|}2\\1\\2\end{array}\begin{array}{|c|}1\\2\end{array}$. Regarding these, according to [the rule] "the numerator and the denominator"[1305] and so forth, at the end [the result] is twenty-five whose denominator is four. Then, when twenty-five is divided by four, the quotient is six plus one-fourth.

This process has been shown because it does not lead to happiness [as it involves giving away a too large amount of money].[1306]

In this way, the type- problem of the visible quantity at both tips is concluded.

1302 The remainder one-fourth (from six plus one-fourth) is the *ādyadṛśya* or the "first visible quantity"; see the layout in SGT 83, 23.

1303 This is a quotation from the rule of the class of fractional increase (GT 57).

1304 In line 21, there should be no the *daṇḍa* after the layout. Also, the layout occurs abruptly; one would expect it to after the quantity has been mentioned, thus in the same line after °*ṣatkasya*.

1305 This is a quotation from the rule of the class of simple fractions (GT $\overline{53}$).

1306 This non-mathematical remark is intriguing.

The type-problem of the fractions and the visible quantity which is a fraction (bhinnabhāgadṛśyajāti)[1307] *[GT 84]*

[Next is the procedural rule of the type-problem of the fractions and the visible quantity which is a fraction; the author enunciates a metrical stanza]:[1308]

> 84. When one is lessened by the fraction[1309] of the visible quantity [and the result is] divided by the [product of the] multiplication of the parts (i.e., fractions)[1310] of the pillar,[1311] its result (i.e., the length of the pillar) will occur.

The explanation: there are that [parts of the pillar][1312] which are visible,[1313] [the number] "one" is "lessened" by these, therefore [such is the meaning of the expression] **when one is lessened by the fraction of the visible quantity.**[1314] There are [also] those "parts of the pillar" which are not visible,[1315] the "multiplication" (*ghāta> guṇana*) [is the multiplication] of these. **When** the accomplished[1316] [result of the step] **one lessened by the fraction of the visible quantity is divided by the [product of the] multiplication of the parts of the pillar,** that is to say, by that [mentioned multiplication], **its result will occur:** the length of the "pillar", since is that which is required to be determined, is [called lit.] "fruit".[1317]

In this respect, in the exemplifying metrical stanza [which follows] the author enunciates a sample problem:

> 85. [A part equal to] one-tenth of a pillar multiplied by one-seventh [is hidden]; one-half is seen today. Say quickly, o friend, the exact measure of the pillar.

1307 K 58, 28.

1308 The commentator's introductory line is missing and has been added by the editor within round brackets. The SGT terms this procedure *bhinnabhāgadṛśyajāti* at the end of the execution of the second sample problem (SGT 86, 14).

1309 *aṃśa*. Cf. the compounds *dṛśyāṃśaka* (in "the fraction of the visible quantity", *dṛśya* functions as a genitive of specification, note that the meaning is the same as "the fraction which is the visible quantity") and *bhāgadṛśya* ("the visible quantity which is a fraction") in *bhinnabhāgadṛśyajāti* (SGT 86, 14).

1310 *aṃśa* – it corresponds to the term *bhinna* in *bhinnabhāgadṛśyajāti*.

1311 See the next sample problem.

1312 The commentator refers to the next sample problem where a pillar is mentioned.

1313 Line 29: **dṛśye**] em. If we accept the reading as original, *dṛśye* is "with reference to the visible quantity"; I take this word, however, as spurious, a scribal error possibly ascribed to the influence of the nearby pronoun *ye*.

1314 After *rūpe*, a *daṇḍa* would fit well.

1315 The commentator refers to the contrast between the parts (i.e., fractions) of the pillar (given as fractions) which are hidden and which are, in fact, two, and the fraction which is the "visible" part. In this regard, see the next sample problems. Another type-problem of fractions which refers to a "pillar" has been formulated by the GT in stanza 65.

1316 In line 1, *prākkṛta* or "already made, accomplished" denotes the result of the step "one lessened by the fraction of the visible quantity".

1317 *phala* – lit. "fruit, result"; in mathematical literature, *phala* is a polysemic term.

The presentation: $\begin{array}{|c|c|c|} 1 & 1 & \text{dṛ}\ 1 \\ 7 & 10 & 2 \end{array}$. By means of the reasoning that has been just enunciated,[1318] according to [the step] **when one is lessened by the fraction of the visible quantity,** when there is the subtraction of one, which is the [numerator of the] "fraction of the visible quantity",[1319] from "one", from the obtained two whose denominator is two it becomes[1320] one whose denominator is two: $\begin{array}{|c|} 1 \\ 2 \end{array}$. [Therefore] regarding [both] the [visible and invisible] "parts of the pillar", these are: $\begin{array}{|c|c|} 1 & 1 \\ 7 & 10 \end{array}$. Since there is the multi-part class, by means of [the method] "the multiplication of the numerators"[1321] and so forth, when there is the reciprocal multiplication of the "numerators" which are the [two numbers] one, it is the very one. When there is the reciprocal multiplication of the denominators, which are seven and ten, it becomes seventy. It is one whose denominator is seventy, which is the [product of the] "multiplication of the parts of the pillar": $\begin{array}{|c|} 1 \\ 70 \end{array}$; this is the divisor. Then, according to [the rule] "having performed the interchange"[1322] and so forth, seven is above and one is below: $\begin{array}{|c|} 70 \\ 1 \end{array}$. The dividend is "one" (*rūpa>eka*) whose denominator is two: $\begin{array}{|c|} 1 \\ 2 \end{array}$. In this case, because of the "cross-reduction", when there is the reduction by two, it is thirty-five, which is half of seventy: $\begin{array}{|c|} 35 \\ 1 \end{array}$. When there is half of two, it is one: $\begin{array}{|c|} 1 \\ 1 \end{array}$. There is the "the multiplication": [thirty-five] multiplied by one is that very one, which divided by the denominator one is again the very one. The quotient is thirty-five *hasta*s, which is the "measure of the pillar": $\begin{array}{|c|} 35 \\ 1 \end{array}$.

Its verification: when there is "one-seventh" of thirty-five *hasta*s, it is five. When there is one-tenth of thirty-five, it is three and one-half that multiplied [by it] (i.e., by five) becomes seventeen and one-half *hasta*s,[1323] which are [the] visible [part of the pillar], and seventeen and one-half *hasta*s, which are [the]

1318 In brief, the just enunciated procedural rule.

1319 Line 8: *dṛśyāṃśasya*] em.

1320 Line 9: *jāto*] em. The transmitted *jātau dviccheda eka* is an instance of *constructio ad sensum*; also, one would expect *jāta eka dviccheda*.

1321 This is a quotation from the multi-part class (GT 55).

1322 This is a quotation from the rule of division of fractions (GT 42); it continues in the lines below.

1323 Line 18: *sardhasaptadaśahastā*] em.; *sardhadaśahastā* (K). The sample problem specifies that "half" of the pillar is visible, and thus the part which is not visible is the other "half".

"invisible" [part of the pillar]. When there is their sum, it becomes thirty-five *hastas*: |35|.[1324]

The author enunciates a second sample problem:

> 86. That [part] of a bamboo which is five[1325] whose denominator is two [and] five (i.e., twenty-five) multiplied (*hata*) by three-fiftieths [is hidden]. Ah, there is [also] a part seen by me! Taking into account that this is thirteen[1326] twenty-fifths,[1327] calculate quickly, o noble one, the length of the bamboo.

Its explanation together with the very presentation:[1328] $\begin{array}{|r|r|r|} 3 & 5 & 13 \\ 50 & 25 & 25 \end{array}$. Regarding "one" whose denominator is one,[1329] by [means of the method] "the numerator and the denominator"[1330] and so forth, it becomes twenty-five whose denominator is twenty-five in place of "one". Next, regarding this [result], when there is the subtraction of thirteen, which is the [numerator of the] "fraction of the visible quantity", it becomes twelve whose denominator is twenty-five: $\begin{array}{|r|} 12 \\ 25 \end{array}$.

1324 Line 19: **prabhāgajāti iyam**] em. The transmitted reading is nonsensical. If we assume that the commentator is making a remark on the calculation by means of the "multi-part class" (quoted in line 10), then it should read *prabhāgajātyā kriyā samaptā* ("this procedure has been accomplished by means of the multi-part class"); one would expect such a passage to occur at the end of the explanation rather than at the end of the verification.

1325 Line 22: *dvipañcāṃśakas*] em. It is unclear how this stanza was understood by the commentary; I am probably stretching the text but this is, perhaps, the only reasonable reading to explain the fraction five twenty-fifths. It may not be a completely convincing interpretation, but it is based on the commentator's execution of the sample problem. Note that in Sanskrit literature and more generally in Classical Indian culture, the term *guṇa* is connected with different concepts and numbers; in the *bhūtasaṃkhyā* notation, *guṇa* is frequently a synonym of the number three but it is also a term associated with the five elements and five senses. Alternatively, *dvipañcāṃśakas tadguṇo* is "[that part of a bamboo] whose denominator is two [times] five which is (i.e., the five) its multiplier (*guṇa*) [for the numerator too]".

The possibility that the original wording of the GT was slightly different from the one transmitted by the present manuscript cannot be excluded; the commentator may have had before him a verse differently formulated. This, unfortunately, cannot be verified as the SGT does not comment upon key technical terms, which is another odd aspect concerning this sample problem.

1326 According to the *bhūtasaṃkhyā* notation, the term *tattva* also denotes the number twenty-five.

1327 *viśva* – it represents the number thirteen.

1328 The abbreviations in the layout (line 25) are missing.

1329 In the edition, one finds an em-dash to obviate the hiatus before the final *a* and the initial *e*; it should be *rūpasyaikacchedasya*.

1330 This is a quotation from the class of simple fractions (GT 53).

This is [the result of] "one lessened by the fraction of the visible quantity". As there is the multi-part class,[1331] when there is the mutual "multiplication" of five and three, which are [the numerators of] the "parts of the pillar" [that are hidden], it becomes fifteen. When there is the mutual "multiplication" of fifty and twenty-five, which are the denominators, it becomes twelve hundred fifty: $\begin{array}{|r|}15\\1250\end{array}$. This is [the product of] the "multiplication of the parts of the pillar". Afterwards, since this is the divisor, by means [of the method] "having performed the interchange"[1332] and so forth, twelve hundred and so forth is above [and] fifteen is below: $\begin{array}{|r|}1250\\15\end{array}$. There is the "cross reduction": as before, when there is the reduction of twelve, which is the [numerator of the] "visible quantity", by three, it is four; when there is the reduction of fifteen by three, it is five. In the same way, when there is the reduction of twelve hundred fifty by twenty-five, it becomes fifty; when there is the reduction of twenty-five by twenty-five, it is one: $\begin{array}{|r|r|}50&4\\5&1\end{array}$. Next, there is the "multiplication":[1333] fifty multiplied by four becomes two hundred, five multiplied by the denominator one is exactly so. When there is the division of [the numerator] two hundred by five, it is: $\begin{array}{|r|}200\\5\end{array}$. The quotient forty *hasta*s is the size of the "bamboo": 40.

Its verification: one-fiftieth[1334] of forty *hastas*[1335] is nineteen *aṅgula*s and one-fifth of one *aṅgula*. Since there are three parts,[1336] it is multiplied by three; two *hasta*s, nine *aṅgula*s, and three-fifths[1337] *aṅgula*s are obtained. Eight *hasta*s, which are obtained from five twenty-fifth of forty [*hasta*s], are multiplied by this; [the result is] nineteen *hasta*s, four *aṅgula*s, and four-fifths[1338] of one *aṅgula*. Likewise, from thirteen twenty-fifth of forty [*hasta*s] are obtained twenty *hasta*s, nineteen *aṅgula*s, and one-fifth *aṅgula*s.[1339] When there is their sum, it is forty: 40.

1331 Line 28: *prabhāgajātitvāt*] em.

1332 This is a quotation from the rule of division of fractions (GT 42); it continues in the line below.

1333 Ibid.

1334 Line 9: *pañcāśadbhāga*] em.; *bhāga* (K).

1335 In line 9, the editor's suggestion to read *dva* instead of *ha* is mistaken.

1336 The numerator is the number three.

1337 Lines 10–11: *pañcabhāgās trayaḥ*] em.

1338 Line 12: *pañcabhāgāś catvāraḥ*] em.

1339 Line 14: **ekaḥ**] em. The construction *pañcabhāga ekaḥ* is, in fact, rather curious and uncommon in the SGT; to indicate "one-fifth", *pañcabhāga* would be sufficient. There is a similar construction in SGT 109, 1, where one finds *ekas tribhāgas*.

The type-problem of the [coefficient of the] [square] root of the fraction (bhinnamūlajāti)[1340] *[GT 87]*

[Next is] the procedural rule of the type-problem of the [coefficient of the] [square] root of the fraction;[1341] [the author enunciates] a metrical stanza:

> 87. The [square] root [should be extracted] from the [result of the] visible quantity multiplied by four,[1342] [the result is] divided by the fraction,[1343] and added to the square (*saṃvarga*) of its own[1344] [coefficient of the] [square] root. [The result is] joined to the [coefficient of the] [square] root, [the result is] halved,[1345] squared, [and] multiplied by the fraction.[1346] The [total] measure of the whole[1347] should occur.

[The explanation]:[1348] the **[square] root [should be extracted] from the [result of the]** number of **the visible quantity**[1349] **multiplied** (*hata>guṇa*) "by four" (*payorāśi>catur*), [and the result] **divided** (*āpta>bhakta*) by the "fraction" (*lava>aṃśa*). Regarding the "fraction",[1350] that [term] "root" refers to it;[1351] there is its "square", that is, it is "multiplied by that" [number-coefficient] according to the method of square. [The result which is the previous obtained quotient is] **added** to it (i.e., the resulting square). By means of [the method] "odd and even"[1352] and so forth, **the [square] root** occurs, afterwards [the *sūtra*

1340 K 60, 17–20.

1341 See the commentator's remark on this *jāti* at the end of the second sample problem (SGT 89, 19), where he uses the expression *bhāgamūlajāti*.

1342 *payorāśi* – lit. "ocean", which in the *bhūtasaṃkhyā* notation represents the number 4.

1343 *lava*.

1344 In brief, the coefficient of the square-root, that is, that number times the square-root as these are specified in a sample problem (see "eighteen times" in GT 88 and "nine times" in GT 89). The fact that it is the coefficient and not the square-root as operation to be concerned is clarified by the SGT further in line 23, where for this reason the commentator uses the term *aṅka*.

1345 Line 19: °*kṛtaṃ*] em. The emendation is based on the commentary, see line 25.

1346 *bhāga*.

1347 *yūtha* –lit. "herd, flock", it refers to the next sample problems. This is a type-problem of fractions, it therefore involves fractional quantities of a whole.

1348 In line 21, the term *vyākhyā* is missing.

1349 Line 21: °*aṅkāt*] em.; °*aṃśakāt* (K). As *dṛśyāṃśa* is found in the previous procedural rule as well as in the commentator's explanation upon it, the transmitted reading appears an instance of harmonization to parallels. The grounds for the emendation are the following: i) in the second sample problem (SGT 89, 24), one finds *dṛśyāṅka*, and ii) in both next sample problems, the visible quantity is an integer, and not a "fraction" (*aṃśa*).

1350 Line 22: *lavasya*] em.; *dṛśyasya* (K). The emendation is based on both the root-text and the commentary; the square is the square of the coefficient of the square-root of the "fraction" (*lava*), the "visible quantity (*dṛśya*) is not concerned. With this emendation, in the same line *tasya* refers to *lava*, as it is shown, in fact, in the execution of the sample problems by the SGT.

1351 The SGT expands upon the possessive adjective *sva* ("its own") of *svamūla*°, which is related to the coefficient, the number of the square-root. The "square" (*saṃvarga*) is thus the square of the coefficient of the square-root of the given fraction.

1352 The commentator quotes his own passage on square-root of integers (SGT 26, 29).

teaches that] [this result is] "joined to the [given coefficient of the] [square] root",[1353] that is, it is added to "the [given] number[1354] of the [square] root". In this manner, [whenever] the "visible quantity" is composed of two or three [quantities], [then] **the [square] root** also is dual or triple, [and] the "fraction" also is dual or triple; thus should be understood.[1355] Afterwards, [the result of the sum is] "halved", it is then **squared** (*vargita>kṛtavarga*), [and the result is] **multiplied by the fraction,** that is to say, by that [given] "fraction".[1356] By the means of the method of division, it is divided by the very "fraction".[1357] **The measure of the whole** should occur.

In this respect, in the exemplifying metrical stanza [which follows] the author enunciates a sample problem:

> 88. Eighteen times the [square] root of one-eighth of a herd of elephants having the appearance of a cloud full of water and whose cheeks are wet by the flow of rut [from their temples], are wandering on the top of a mountain. Other eighteen [elephants] are seen terrified having heard the roaring of a young lion. O mathematician, if you have worked hard on this [mathematical] subject, calculate the number of the elephants [in the herd].

Its explanation with the very presentation:[1358]

1	mū 18	dṛ[ya 18
8	1	1

. Eighteen is the "visible quantity"; it is "multiplied by four": eighteen multiplied by

1353 Line 23: *samūlaṃ samūla°*] em. Emendation of *sva* to *sa* follows the coherence of the commentary; the meaning of *sva* has been, in fact, already clarified in the commentary. The commentator refers to the next step of the procedural rule (see, in fact, *paścāt* or "afterwards") where the just obtained result is added to the given root (thus *samūla*). An additional evidence to support this reading is in line 25 where the SGT explains the step "halved, squared" (see *ardhīkṛtavargitaṃ*) and so forth from the passage of the *karṇasūtra* [*samūlam*] *ardhīkṛtavargitaṃ*, which demonstrates that the preceding step denoted in the rule by *samūlam* must have been mentionedearlier by the commentator.

1354 The syntagm *aṅkamūla* makes clear that it is the coefficient, the given "number" (*aṅka*) of the "square-root" (*mūla*) to be denoted by *mūla* in the expression *samūlam*, and not the operation of "square-root" (*mūla*). The coefficient is clarified by the root-text itself in the sample problems.

1355 In lines 23–24, one would expect a construction with the correlatives *yadā/tadā* or *yatra/tatra*: *yadā dṛśyaṃ* [...] *tadā mūlam* [...] *tadā lavā* [...]. This passage clearly anticipates the case found in the second sample problem (GT 89), where the second visible quantity is obtained by manipulating the first, which is then erased, and this second is then manipulated with its own coefficient (see SGT 89, 6–7).

1356 The commentary shows the grammatical case (instrumental) of *bhāga* in the *tatpuruṣa* compound *bhāgāhataṃ*.

1357 In both sample problems the final result is an improper fraction and this is transformed into a compound fraction.

1358 It is awkward that the abbreviation *la* to denote *lava* or the "fraction" (or the abbreviation *bhā* for *bhāga*, given the name of this type-problem) is missing here and in the next sample problems as well. Note that although the *dṛśya* in both cases is an integer, here it is specified as having the numerator one; this is not the case in the next sample problem.

four becomes seventy-two: 72.[1359][Next is the step] **divided by the fraction**; to illustrate: the "fraction" has eight as denominator and since this is the divisor, by means of the method "having performed the interchange"[1360] and so forth, eight is above and one is below: $\left|\begin{array}{r}8\\1\end{array}\right|$. Then, there is the "multiplication": seventy-two multiplied by eight becomes five hundred seventy-six, one multiplied by the denominator one is the very [number] one, [the numerator] divided by one is exactly so. In this manner, [the step] such as "divided by the fraction" [is obtained].[1361] Eighteen is **its own [coefficient of the] [square] root**; its square is three hundred twenty-four: 324. Since there are equal denominators, when the additive quantity three hundred twenty-four is added to five hundred seventy-six, 576, it becomes nine hundred whose denominator is one: $\left|\begin{array}{r}900\\1\end{array}\right|$. According to [the rule] "even and odd"[1362] and so forth, its **[square] root** is thirty. **[The result is] joined to the [coefficient of the] [square] root:** as there are equal denominators, [the resulting root] added to the "[coefficient of the] [square] root", which is eighteen, becomes forty-eight. This **halved** becomes twenty-four, which is **squared:** the square of twenty-four is five hundred seventy-six; it is **multiplied by the fraction,** that is, it is **multiplied by** the "fraction" which is one whose denominator is eight. [The number one] **multiplied** by the number above [five hundred seventy-six] is that very one, and underneath one multiplied by eight becomes eight. Then, five hundred seventy-six [is divided] by eight:[1363] $\left|\begin{array}{r}576\\8\end{array}\right|$. When there is the division, the quotient is seventy-two: $|72|$. This is the measure of the "herd"[1364] of "elephants".

Its verification: "one-eighth" of seventy-two "elephants" is nine; its **[square] root** is three. **Eighteen times** that [square-root], it becomes fifty-four, which [is the number of the elephants that] are going to **the top of a mountain;** eighteen **are seen**: $\left|\begin{array}{r}54\\18\end{array}\right|$. When there is their sum, it becomes seventy-two.

1359 Line 4: *yathā* |72|| *lavāptā tathāhi* [...]] conj. The reason for the emendation is that the formula *yathā* followed by a quotation, which represents the next mathematical step to perform, is uncommon in the SGT; moreover, here this formula fits neither the syntactical construction nor the movement of the discourse.

1360 This is a quotation from the rule of division of fractions (GT 42); it continues in the line below.

1361 Line 7: °*ādīti siddham sva*°|] em.; °*ādi*° (K). The transmitted reading is corrupt; the emendation is based on the commentary, see a similar passage in the next sample problem, lines 1 and 12–13.

1362 The commentator quotes his own passage on square-root of integers (SGT 26, 29).

1363 Line 13: °*śatī*] em.; °*śatyā* (K). Also, this sentence suspiciously lacks a verb to denote the division which takes place.

1364 *yūtha* – the wording echoes the *karaṇasūtra* where *yūthamāna* is found.

The author enunciates a metrical stanza for a sample problem regarding a double visible quantity[1365] which comes together with a double root:

> 89. Nine times the [square] root of two-thirds [of a flock of swans] flew away, six times the square-root of three-fifths[1366] of the remainder of the flock of swans perished, [and] three times eight are seen. O friend, how many were they all?

[Its explanation]:[1367] two whose denominator is three, its **[square] root** is multiplied by "nine". **Six times the square-root** of three-fifths of the "remainder"[1368] **[of the flock of swans] perished. Three-times eight are seen**: [thus] twenty-four. The presentation is:[1369] $\begin{array}{|c|c|c|c|c|} 2 & \text{mū } 9 & \text{śeṣa } 3 & \text{mū } 6 & \text{dṛ } 24 \\ 3 & 1 & 5 & & \end{array}$. Twenty-four, which is the number of the "visible quantity", multiplied by four becomes ninety-six. Because three-fifths is the divisor,[1370] by means of the [procedure] "having performed the interchange"[1371] and so forth, five is above [and] three is below: $\begin{array}{|c|} 5 \\ 3 \end{array}$. The "cross-reduction" occurs: when there is the reduction of ninety-six by three, it is thirty-two, when there is the reduction of three by three, it is one: $\begin{array}{|c|c|} 32 & 5 \\ 1 & 1 \end{array}$. There is then the "multiplication": when there is the mutual multiplication of thirty-two [by five], it is one hundred sixty; one multiplied by the denominator one is exactly so. And [the numerator one hundred sixty] divided by [the number] one, it is that very one: $\begin{array}{|c|} 160 \\ 1 \end{array}$. [The step] "divided by

1365 By manipulating the first visible quantity 24 (eight multiplied by three) with the coefficient of the root 6 ("six times") and the fraction three-fifths of the remainder, the second visible quantity 60 is obtained (see line 6). Note the order in which the quantities are placed in the layout, particularly the fact that the visible quantity is manipulated with the first two quantities next to it (from right to the left). The significance of this contiguity is clear when, in lines 1 and 7, the SGT uses the adjective *pratyāsanna* or "adjacent, contiguous". This term, which qualifies the coefficient of the "root" (*mūla*), could also be rendered as "closely related to".

1366 *śara* – lit. "arrow", in the *bhūtasaṃkhyā* notation it denotes the number "five".

1367 In line 23, *vyākhyā* is missing.

1368 Line 23: *śeṣatripañcabhāgasya*] em. The transmitted *śeṣatrilavabhāga* is nonsensical. The emendation follows the commentary, it suggests that the commentator is glossing *śara* with *pañca* or "five" (see *tripañcabhāga* in line 25).

1369 I have emended the layout (line 24) as the *mūla* 6 should stand alone and not in a same box with the *dṛśya* 24.

1370 Line 25: °*bhāgasya haratvāt*] em. This type of abstract construction requires the subject genitive, which is absent in the transmitted reading.

1371 This is a quotation from the rule of division of fractions (GT 42); it continues in the two lines below.

the fraction" and so forth is [thus] accomplished. Since it is contiguous[1372] [to the visible quantity twenty-four], six is "its own [coefficient of the] [square] root". Its square is thirty-six; one hundred sixty whose denominator is one is added to it: $\left|\begin{matrix}196\\1\end{matrix}\right|$. According to [the rule] "odd and even[1373] and so forth, **the square-root** is fourteen: $\left|\begin{matrix}14\\1\end{matrix}\right|$. It is **joined to the [coefficient of the] [square] root:** [when fourteen is] added to six, it becomes twenty, [when it is] **halved,** it is ten, and [when it is] **squared,** it becomes one hundred: $|100|$. This is **multiplied by the fraction**[1374]**,** which is three-fifths. To illustrate: one hundred multiplied by three is three hundred, one multiplied by five is five, it becomes three hundred whose denominator is five: $\left|\begin{matrix}300\\5\end{matrix}\right|$. When there is the reduction of this by five, it is sixty above and one below: $\left|\begin{matrix}60\\1\end{matrix}\right|$.

The intermediate numbers are erased;[1375] since it is contiguous [to this obtained visible quantity], the number which is the first [coefficient of the] [square] "root" remains.[1376] Sixty whose denominator is one, which is the number of the "visible quantity", multiplied by four becomes two hundred forty whose denominator is one. [This is divided] by the "fraction" which is the previously mentioned two-thirds: as it is the divisor, by means of [the

1372 *pratyāsanna* – lit. "contiguous, closely related"; in the SGT, this is the first occurrence of the term. It refers to the layout where the first visible quantity 24 is manipulated with the two quantities (the fraction and its coefficient) which are "contiguous, next to" it (from right to left). The other two quantities (the other fraction and its coefficient), which are not "contiguous" to the first visible quantity and are at the very left of the layout, will be, in fact, manipulated further with the second visible quantity 60 (line 6) obtained when the procedure starts again from the beginning. In this regard, see the use of *pratyāsanna* in line 7 too. This wording reframes the passage in which the commentator has explained that when the visible quantities are two, so are the coefficients; the use of the *pratyāsanna* appears to have an illustrative, explanatory role to help students to understand that each visible quantity has its own coefficient.

1373 The commentator quotes his own passage on square-root of integers (SGT 26, 29).

1374 Line 4: *bhāgais*] em.

1375 It refers to the first set of quantities which have been manipulated: the visible quantity twenty-four, the fraction three-fifths, and the coefficient nine.

1376 In this sentence, see the contrast between *gata* and *sthita* and between *antara* and *pratama*. The commentator observes that, in respect to the initial layout of line 24, the quantities in the middle are erased (the first fraction and its square-root) as they have been manipulated. At this point, the layout should be: $\left|\begin{matrix}2\\3\end{matrix}\right|\left|\begin{matrix}\text{mū}\,9\\1\end{matrix}\right|$ |dṛ 60|.

procedure] "having performed the interchange"[1377] and so forth, three is above and two is below: $\left|\begin{array}{c}3\\2\end{array}\right|$. The "cross-reduction" [should be carried out]: when there is the reduction of two hundred forty by half, it becomes one hundred twenty, when there is the reduction of two by half, it is one: $\left|\begin{array}{c|c}120&3\\1&1\end{array}\right|$. There is then the "multiplication": one hundred twenty multiplied by three becomes three hundred sixty, one multiplied by the denominator one is exactly so. The number [three hundred sixty] divided by that [denominator one] remains as it is. [The step] "divided by the fraction" and so forth is accomplished. "Its own [given coefficient of the] [square] root" is nine, its square is eighty-one, this added to three hundred sixty becomes four hundred forty-one whose denominator is one. According to [the rule] "even and odd"[1378] and so forth, **the root** is twenty-one. It is **joined to the [coefficient of the] [square] root:** [twenty-one] added to nine becomes thirty. **Halved** is fifteen and **squared** is two hundred twenty-five. [It is] **multiplied** by the fraction, which is two-thirds:[1379] two hundred twenty-five multiplied by two is four hundred fifty, one multiplied by three becomes three: $\left|\begin{array}{c}450\\3\end{array}\right|$. When there is the division of the number above by the number below, the quotient is one hundred fifty: $|150|$. This is the measure (*pramāṇa*) of the "flock of swans".

Because the visible quantity is manipulated (*niṣpannatvāt*) with the fraction and [its] root, the type-problem of the [coefficient of the] [square] root of the fraction [is so called] and it is concluded.

Its verification: two-thirds of one hundred fifty is one hundred, its **[square] root** is ten 10, **nine times** this [ten] becomes ninety, [these are the parrots that] **flew away**. When there is the division[1380] of the "remainder" sixty by five, the quotient is twelve. Three[1381] of this [twelve] gives thirty-six. Its **[square] root** is six; **six times** this [root], which is thirty-six, **perished.** Twenty-four **are seen,** thus: $\left|\begin{array}{c}90\\36\\24\end{array}\right|$. When there is their sum, it is one hundred fifty.

1377 This is a quotation from the rule of division of fractions (GT 42); it continues in the two lines below.

1378 The commentator quotes his own passage on square-root of integers (SGT 26, 29).

1379 Line 16: *dvau tryaṃśābhyāṃ bhāgāhatam*] conj.. See *dvau tryaṃśau* at the very beginning of the passage (line 23) and in the first line of the verification (line 20).

1380 Line 21: °*ṣaṣṭyāḥ pañcabhir bhāge*] em.; °*pañcabhaktāyā* (K).

1381 In the SGT, it is unusual to find such a usage of *traya* (lit. "triad, triple").

The type-problem of the subtractive square (hīnavargajāti)[1382] *[GT 90]*

Next is the procedural rule of the type-problem of the subtractive square; the author enunciates a metrical stanza:

> 90. Here[1383] [the fraction having] the denominator brought up (*uddhṛta*) by its own numerator is [written] twice. This [fraction] is multiplied by the subtractive quantity,[1384] [the result is] joined to the square of half of the other [fraction],[1385] [and the result is] lessened[1386] by the visible quantity. The subtractive quantity[1387] is added to half of the other [and the result is added to][1388] the [square] root of that.[1389] [The result is] divided by the fraction, [and the final result is] obtained.

The explanation: [the expression] **[the fraction with] the denominator brought up by its numerator**[1390] means that the "denominator" is "brought up by its own numerator", that is to say, moved up[1391] by means of a downward movement[1392] of its own "numerator".[1393] **Here** is "in the type-problem of the subtractive square", [which is so called] since it (i.e, the square) is associated with the subtrahend (*ūnāṅka*).[1394] Having placed **twice**, that is to say, in two places, the numerator which has been moved underneath, **the denominator** should be moved above. Afterwards, one [place][1395] is **multiplied** (*āhata*> *guṇita*) by the "subtractive quantity" (*ūna*>*hīnāṅka*) specified by the inquirer[1396] [in the sample problem]. That [word] "half" refers to the "other" [fraction] standing in the second place [and] whose numerator is underneath,[1397] it refers to this quantity whose denominator is [multiplied] by two [in order to be halved].[1398]

1382 K 62, 25–28.
1383 In other words, in this type-problem; see the commentary in line 2.
1384 *ūna*.
1385 There are two equal fractions; cf. *dvidhā* or "[written] twice" in the *karaṇasūtra* and *sthānadvaye* ("in two places") in the second line of the commentator's explanation.
1386 *ūnita*.
1387 Here *ūna* is a nominative neuter.
1388 This is the intended sense but it not clearly expressed by the Sanskrit as it stands; this step is clarified by the SGT in lines 8–9.
1389 It refers to the square-root of the result of the step "lessened by the visible quantity"; see the SGT in lines 4–5.
1390 Line 1: a *daṇḍa* is missing before *iha*.
1391 *nīta*.
1392 *gamana* – lit. "moving, setting out"; in the SGT, this is the first occurrence of the term.
1393 Line 29: *svāṃśasya*°] em. The transmitted *svakasya*° is nonsensical.
1394 It is worth noting that the commentary specifies the name of this procedure and the reason why it is so called.
1395 The use of *eka* is to be understood in relation to *dvitīyasthāna*; each "place" is occupied by one of the two equal fractions.
1396 *pṛcchaka* – the one who has formulated the sample problem.
1397 Line 3: *'ṃśakasya*] em.; *'ṃśako* (K).
1398 In the first sample problem, the fraction five-thirds is halved by having the denominator multiplied by two.

Its "square" (*kṛti>varga*) (i.e., of the halved quantity) [should be calculated]. [The result of the first fraction multiplied by the *ūna*) is "joined" to it (i.e., to the square) by means of the made[1399] equal denominators.[1400] Having first calculated the equal denominator,[1401] [the result is] **lessened** then by the "visible quantity"; **the [square] root of that,** that is to say, **the [square] root** (*pada>mūla*) **of that** number [obtained by carrying out this last mentioned step].[1402] **The subtractive quantity,** which is the given subtractive numerical place,[1403] **is** next **added to** "half" of the mentioned "other" [fraction] by having calculated equal denominators, [and the result is] added to **the [square] root of that** by having calculated equal denominators. Afterwards, it is divided by the aforementioned "fraction";[1404] regarding this [fraction], by means of the method of division by it (i.e., the last result), the exchange above and below [of its operands] is made. By means of the "multiplication"[1405] and so forth, it is **divided**. That [word][1406] **obtained:** in brief, that [obtained] quotient is the wished quantity.

It should also be said: "**the [result of the] subtractive quantity added to half of the other** [fraction] "is added to" **the [square] root of that**"; in the *sūtra*, this is not explained.[1407] Because it is also stated that [this result is] **divided by the fraction** [I have elucidated this passage].

In this respect, in the exemplifying metrical stanza [which follows] the author enunciates a sample problem:

> 91. The square of three-fifths of [a muster of] peacocks lessened by six is playing at the end of [a moment of] passion, [and] six are seen inside a forest. Tell [me] quickly the size of the muster.

1399 Emendation to *kṛta* could also be considered, as in the SGT it is uncommon to find *prayukta* to mean "accomplished, obtained". This appears to be a scribal error, it could represent either a case of harmonization to parallels –the same is found in the previous procedural rule (meaning "joined to") – or a case of repetition, since the word *yukta* ("added to") is found in the same line. Notably, in line 6 one finds another instance of error involving *prayukta*.

1400 See the commentator's explanation of this step in the next sample problem (lines 18–23), where the two fractions are added by means of the class of simple fractions.

1401 Line 4: °*chedaṃ pūrvaṃ kṛtvā*] conj. The transmitted reading is not interpretable.

1402 Siṃhatilakasūri clarifies Śrīpati's stanza, which is ambiguous in the expression *tatpadam* or the "root of that". This point is further elucidated by the commentator while explaining the next sample problem where in line 28, he specifies that the "[square] root of that" is the square-root of the "remainder" (*śeṣa*) from the subtraction.

1403 This expression is related to the next sample problem, where the subtractive quantity is the digit (thus the use of *aṅkapada*) 6.

1404 Line 6: *prāguktair*] em. It denotes the fraction given in the sample problem before the interchange of its operands. The emendation is supported by the commentary since, with respect to the same fraction, the very passage *bhāgaiḥ prāguktais* occurs while the commentator explains the sample problem (see SGT 91, 6).

1405 This is a quotation from the rule of division of fractions (GT 42).

1406 Line 7: *yad*] em.; *sat* (K).

1407 In Śrīpati's stanza, the syntax of the last two lines is problematic because, as it is, *ūnam anyadalānvitaṃ* could be qualified by *vibhaktam* or "divided". In order to clarify this point, the commentator provides an explanatory amplification and specifies that the "result of

Its explanation with the very presentation:[1408] $\begin{array}{|c|c|c|} 3 & \bar{u}\,6 & \text{dṛ}\,6 \\ 5 & 1 & 1 \end{array}$. The "denominator", namely five, is "brought up", which means "moved above", by its own "numerator", which is three: five is above [and] three is underneath. This [fraction] is written in two place: $\begin{array}{|c|c|} 5 & 5 \\ 3 & 3 \end{array}$. Among the two, **this**[1409] first quantity consisting of five and three **is multiplied** by the "subtractive quantity", which is six; it becomes thirty whose denominator is three: $\begin{array}{|c|} 30 \\ 3 \end{array}$. Its[1410] "square" is twenty-five, [and] the "square" of the denominator six is thirty-six. Then, by means of the [method][1411] "the numerator and the denominator"[1412] and so forth, when there is the reduction of three, which is the denominator below thirty, by three, it is one; when there is the reduction of thirty-six by three, it is twelve. Next, when there is the exchange of the denominators, it is: $\begin{array}{|c|c|} 30 & 25 \\ 3 & 36 \\ 12 & 1 \end{array}$. Thirty multiplied by twelve becomes three hundred sixty; three multiplied by twelve becomes thirty-six: $\begin{array}{|c|} 360 \\ 36 \end{array}$. The other quantity multiplied by one remains exactly so. As there are equal denominators, when the additive quantity twenty-five is added to three hundred sixty, it becomes three hundred eighty-five whose denominator is thirty-six: $\begin{array}{|c|} 385 \\ 36 \end{array}$. [This result is] "lessened" by the "visible quantity",

the subtractive quantity added to half of the other added to the [square] root of that" is the dividend of the final division by the fraction. In the next sample problem, see the step explained in lines 6–10 where $\begin{array}{|c|} 9 \\ 1 \end{array}$ is divided by $\begin{array}{|c|} 3 \\ 5 \end{array}$. Also, see the SGT while commenting upon the second sample problem in line 10.

1408 In the initial *nyāsa*s of both sample problems, the first quantity is not accompanied by an abbreviation.

1409 The commentary uses the demonstrative *asau* as in Śrīpati's rule.

1410 In line 18, the text is problematic: the passage should mention the "other fraction" because here the "square" referred by the commentator is the square of its (i.e., of the other fraction) half. A conjectural emendation is: *yathā* $\begin{array}{|c|} 30 \\ 3 \end{array}$ | *paścād anyārdhakṛtīty anyārdham* $\begin{array}{|c|} 5 \\ 6 \end{array}$|] conj. The translation is: "$\begin{array}{|c|} 30 \\ 3 \end{array}$. Next is [the step] 'the square of half of the other [fraction]', half of the 'other' is $\begin{array}{|c|} 5 \\ 6 \end{array}$|".

1411 Line 19: **so**] em.

1412 This is a quotation from the rule of the class of simple fractions (GT $\overline{53}$).

which is six whose denominator is one: therefore, according to [the rule] "the numerator and the denominator"[1413] and so forth, when there is the exchange of the denominators, it is: $\begin{array}{|r|r|} 385 & 6 \\ 36 & 1 \\ 1 & 36 \end{array}$. The quantity which occurs first multiplied by one is that very one. In the next quantity, six multiplied by thirty-six becomes two hundred sixteen, one multiplied by thirty-six is thirty-six: $\begin{array}{|r|} 216 \\ 36 \end{array}$. By means of the reasoning beginning with "the deduction of the numerators of [two] quantities of which equal denominators have been arranged [should be performed]"[1414], when there is the "deduction" of two hundred sixteen from three hundred eighty-five, the remainder is one hundred sixty-nine, [and] thirty six [is its denominator]: $\begin{array}{|r|} 169 \\ 36 \end{array}$. **The [square] root of that** refers to these just written numbers; according to what has been taught, by means of [the procedure] "odd and even"[1415] and so forth, the root is thirteen above and six below: $\begin{array}{|r|} 13 \\ 6 \end{array}$. Afterwards, arrange[1416] **the subtractive quantity** six whose denominator is one, and half **of the other** which is the previously mentioned five whose denominator is six: $\begin{array}{|r|r|} 6 & 5 \\ 1 & 6 \end{array}$. According to [the rule] "the numerator and the denominator"[1417] and so forth, when there is the exchange [of the denominators], it is: $\begin{array}{|r|r|} 6 & 5 \\ 1 & 6 \\ 6 & 1 \end{array}$. Six multiplied by six is thirty-six, one multiplied by six is six, the other number multiplied by one is exactly so. As equal denominators occur, when the additive quantity five is added to thirty-six,[1418] it becomes forty-one. Since there is the same denominator [six], this [forty-one] is added to thirteen; it becomes fifty-four whose denominator is six: $\begin{array}{|r|} 54 \\ 6 \end{array}$. When there is reduction of these by six, nine is above, and one is underneath: $\begin{array}{|r|} 9 \\ 1 \end{array}$. As it is the divisor,

1413 This is a quotation from the rule of the class of simple fractions (GT 53).

1414 This is a quotation from the rule of subtraction of fractions (GT 38).

1415 The commentator quotes his own passage on square-root of integers (SGT 26, 29).

1416 Note the absolutive *maṇḍayitvā* used as main verb and the *bahuvrīhi* compounds which agree with the quoted *ūnaṃ* and *anyadalaṃ* instead than the number-numerator.

1417 This is a quotation from the rule of the class of simple fractions (GT 53).

1418 Line 4: *ṣaṭtriṃśanmadhye pañcakṣepe*] em. The transmitted reading is corrupt, the numbers should be 36 and 5 rather than 30 and 6.

[it is divided] by the aforementioned "fraction" consisting of five and three; according to [the rule] "having performed the interchange"[1419] and so forth, five is above [and] three is below: $\begin{vmatrix}5\\3\end{vmatrix}$. By means of the "cross-reduction", when there is the reduction of nine by three, it is three, when there is the reduction of three by three, then it is [one]: $\begin{vmatrix}3&5\\1&1\end{vmatrix}$. Afterwards, the "multiplication": three multiplied by five becomes fifteen, the denominator one multiplied by one remains as it is. The number [fifteen] which is divided by [the denominator] one is exactly so. Fifteen is **the size of the muster** of peacocks.

Its verification: the quotient of fifteen divided by five is three, this multiplied by three is nine, **lessened by six** becomes three, its square is nine, and six **are** [the peacocks that are] **seen**: $\begin{vmatrix}9\\6\end{vmatrix}$. When there is their sum, it is fifteen.

The author enunciates a metrical stanza for a second sample problem:

> 92. Three times the eighth part of a herd [of *viṣka*s] is halved as well as diminished by sixteen *viṣka*s.[1420] [A part of the herd equal to this result] multiplied by itself is playing in the cavity of a mountain, and [the number of *viṣka*s equal to] four times this [sixteen] is wandering in a forest. [Tell me the total number of this group of *viṣka*s].

The explanation: [regarding the expression] **three times the eighth part**[1421] of a "herd", that consisting of three, which is the numerator above, **is halved**; [however] it is not possible to halve the number three. [Hence] when there is the multiplication of the denominator, namely eight, by two, it is sixteen; **as well as [diminished by sixteen] *viṣka*s**: "it is known that a *viṣka* is [an elephant] twenty-years old". **Four times** sixteen is sixty-four. The rest is clear [in meaning].

The presentation:[1422] $\begin{vmatrix}3&&\\16&\bar{u}16&\text{dṛ }64\end{vmatrix}$. The "denominators brought up by their own numerator" are [written] in two places: $\begin{vmatrix}16&16\\3&3\end{vmatrix}$. [Among the two] one quantity[1423] **is multiplied** by the "subtractive quantity", which is sixteen;

1419 This is a quotation from the rule of division of fractions (GT 42); it continues in the two lines below.

1420 In line 20, the SGT explains that the term *viṣka* refers to an elephant twenty-years old; see also at the end of the explanation the term *gaja* or "elephant".

1421 Line 18: *'*ṣṭau**] em.

1422 The author first calculates some of the initial data to then present the *nyāsa*.

1423 Line 22: *eko'ṅkaḥ*] em.

it becomes two hundred fifty-six whose denominator is three: $\left|\begin{array}{r}256\\3\end{array}\right|$. Regarding the "other" [fraction], "half" of the second standing[1424] sixteen is eight whose denominator is three: $\left|\begin{array}{r}8\\3\end{array}\right|$. Its (i.e., of this fraction) "square" is sixty-four above and nine below: $\left|\begin{array}{r}64\\9\end{array}\right|$. By means of the [procedure] "the numerator and the denominator"[1425] and so forth, when there is the reduction of the denominators three and nine by three, [one and three arise]; when there is the exchange of the obtained [numbers], it is:[1426] $\left|\begin{array}{r|r}256&64\\3&9\\3&1\end{array}\right|$. Two hundred fifty-six multiplied by three becomes seven hundred sixty-eight, three multiplied by three becomes nine: $\left|\begin{array}{r}768\\9\end{array}\right|$. The other number 64 multiplied by one is that very one.[1427] When there is the sum of the two [numbers], it becomes eight hundred thirty-two whose denominator is nine: $\left|\begin{array}{r}832\\9\end{array}\right|$. [In this way the step] **[the result is] joined to the square of half of the other** is accomplished. [This result is] "lessened"[1428] by the "visible quantity" sixty-four whose denominator is one; by means of [the procedure] "the numerator and the denominator"[1429] and so forth, when there is the exchange of the denominators, it is: $\left|\begin{array}{r|r}832&64\\9&1\\1&9\end{array}\right|$. [The first quantity] multiplied by one is exactly so; regarding the next quantity, sixty-four multiplied by nine becomes five hundred seventy-six whose denominator is nine: $\left|\begin{array}{r}576\\9\end{array}\right|$. Since there is the same quantity-denominator, when there is the subtraction of five hundred seventy-six from eight hundred thirty-two, it becomes two hundred fifty-six: $\left|\begin{array}{r}256\\9\end{array}\right|$. **The [square root]** of these [two numbers][should be

1424 In line 23, *dvistha* echoes *sthānadvaye* in line 21.

1425 This is a quotation from the rule of the class of simple fractions (GT 53).

1426 Line 25: *jātayor*] em. Emendation to *jātaikatrikayor* ("of the obtained one and three") could also be considered.

1427 In line 27, the formula *parāṅka* followed by a number in figures does not correspond to the SGT's commentarial style.

1428 In line 1, *ūnitā* should be emended either to *ūnitaṃ* (qualifying °*aṣṭakaṃ*) or to *ūnita iti*.

1429 This is a quotation from the rule of the class of simple fractions (GT 53).

calculated];[1430] according to [the rule] "odd and even"[1431] and so forth, the [square] root is sixteen above and three below: $\begin{array}{|r|}16\\3\end{array}$. In the other place, [one should write down] the "subtractive quantity" sixteen whose denominator is one. "Half" of the "other", which is the second standing [quantity] previously mentioned, is namely three and eight. By means of [the method] "the numerator and the denominator"[1432] and so forth, when there is the exchange, it is: $\begin{array}{|r|r|}16&8\\1&3\\3&1\end{array}$. Sixteen multiplied by three becomes forty-eight; one multiplied by three becomes three: $\begin{array}{|r|}48\\3\end{array}$. The other quantity multiplied by one is exactly so. When there is their sum, it becomes fifty-six whose denominator is three: $\begin{array}{|r|}56\\3\end{array}$. This [which is the result of] **the subtractive quantity added to half of the other** is added to the "root of that" which has previously been written,[1433] which is sixteen; as there are equal denominators, it becomes seventy-two whose denominator is three: $\begin{array}{|r|}72\\3\end{array}$. When there is their reduction by three, it is twenty-four above and one below: $\begin{array}{|r|}24\\1\end{array}$. [Because of the division][1434] by the [given] "fraction", as this is the divisor, according to [the rule] "having performed the interchange"[1435] and so forth, [the fraction above is] **divided** by the inverted three and sixteen. When there is the reduction of three by three, it is one; when there is the reduction of twenty-four by three, it is eight. This is the [result of the] "cross reduction": $\begin{array}{|r|r|}8&16\\1&1\end{array}$. There is the "multiplication": eight multiplied by sixteen is one hundred twenty-eight; one multiplied by the denominator one is exactly so. And the number [one hundred twenty-eight] divided by that [number one] is the very quotient: $\begin{array}{|r|}128\end{array}$. This is the measure of the herd of elephants.

1430 Line 5: *tayor*] em.; *talayor* (K).

1431 The commentator quotes his own passage on square-root of integers (SGT 26, 29).

1432 This is a quotation from the rule of the class of simple fractions (GT $\overline{53}$).

1433 In line 10, *prāglikhita* refers to the "previously written" fraction $\begin{array}{|r|}16\\3\end{array}$ (see line 6).

1434 The step meant is the division of the result $\begin{array}{|r|}24\\1\end{array}$ by the given *bhāga* $\begin{array}{|r|}3\\16\end{array}$.

1435 This is a quotation from the rule of division of fractions (GT 42); it continues in the two lines below.

Its verification: one-eight[1436] of one hundred twenty-eight is sixteen, **three-times** that, it is forty-eight; **halved** is twenty-four, which diminished by "sixteen *viṣka*s" is eight. This is **multiplied by itself**: eight multiplied by eight is sixty-four, which [is the number of *viṣka*s] **playing "**in the [cavity of] a mountain": $\left|\begin{matrix}64\\64\end{matrix}\right|$. When there is their sum, it is one hundred twenty-eight: $|128|$.

Thus in this way, the type-problem of the subtractive square is concluded.

3.9 Inverse operation [GT 93–94]

Inverse operation* (viparītoddeśaka)**[1437] ***[GT 93]

[Next is] the procedural rule of inverse operation; the author enunciates a metrical stanza:

> 93. In the case of inverse[1438] [operation], the multiplier is treated as a divisor, the divisor as a multiplier, the root as a square, the square as a root, the subtractive[1439] number as an additive number,[1440] [and] the additive number as a subtractive number.

The explanation: the inverse operation (*viparītoddeśaka*) is such that whatever[1441] the inquirer mentions, for instance the multiplication, the division, and so forth, everything should be made the opposite.[1442] Therefore (*tatra*), the number which is specified as the multiplier should be performed as **a divisor** (*hara>bhāgadāyin*);[1443] that which is defined as **the divisor** (*hara>bhāgadāyin*) becomes a multiplier. In the same way, the number which is **the root** (*pada>mūla*) becomes **a square** (*kṛti>varga*). Similarly, the number which is **the square** becomes **a root** (*pada>mūla*), [and] the number which is **the subtractive number** (*kṣaya>hīna*)[1444] becomes **an additive number** (*dhana>madhyakṣepya*).

1436 Line 17: *aṣṭamo bhāgaḥ*] em.

1437 K 65, 22–23.

1438 *pratīpaka.*

1439 *kṣaya.*

1440 *dhana.*

1441 Line 24: *yad yat*] em.; *tad tat*] em. Here *yad yat* and the correlative *tad tad* acquire a distributive meaning; *yad yat* introduces a clause (*pṛcchaka uddiśati* […]) which explains the contents of the main sentence: *tad tat viparītaṃ kāryam iti.* In line 24–25, one would expect *hārādikaṃ* to precede rather than to follow *tad tat viparītaṃ*.

1442 In this first passage, the commentator establishes the underlying principle of the procedure. It is worth noting that the *tatra* which follows is a conjunctive adverb, it joins the first clause, which is descriptive and conceptual, with the rest of the passage, which has a more demonstrative and practical character.

1443 According to the editorial conventions used by Kāpadīā in lines 25–26, in order to denote the gloss, in line 26 there should be an em-dash between *haro* and *bhāgadāyī*.

1444 The number which should be subtracted, the subtrahend. It could also be rendered as "positive" (*dhana*) and "negative" (*kṣaya*), keeping in mind that these definitions refer to

Also, the number which is **the additive number** (*dhana>madhyakṣepya*)[1445] should be made **a subtractive number** (*kṣaya>hīna*). Thus is the connection [of words with their meanings].

In this respect, in the illustrative metrical stanza [which follows] the author enunciates a sample problem:

> 94. O friend, if you know arithmetic, tell me o mathematician, what is the number that when multiplied by five, increased by nine, reduced to the square-root, decreased by two, then squared, lessened by one, [and] divided by eight becomes three?

Its explanation with the very presentation:[1446]

|gu 5|dha 9|mū 1|ū 2|kṛti 1|hīna 1|bhā 8|dṛśyarūpa 3|. As here there is the [procedure of] inverse operation, according to the reverse order[1447] three multiplied by eight becomes twenty-four. When the additive quantity one[1448] is added to it, it becomes twenty-five, its root is five, which added to two is seven; its square is forty-nine, which lessened by nine becomes forty. When there is its division by five, the quotient is the integer eight: 8.

Its verification: the multiplier is the divisor,[1449] and the divisor and so forth in this very manner are respectively [the opposite].[1450] For instance: eight multiplied by five becomes forty, this **increased by nine** becomes forty-nine, its square-root is seven, this reduced by two becomes five, its square is twenty-five, which lessened by one is twenty-four, this divided by eight is the quotient previously mentioned: the known quantity, the integer three.

the operational sign of the quantity. The terms used by the GT, the SGT, and the L in the following stanzas and passages to mean "positive, additional quantity" are *dhana/adhika/sva*, while the terms used to mean "negative, subtractive quantity" are *kṣaya/hīna/ūna/ṛṇa*.

1445 In lines 27–28, between *dhanaṃ* and *madhyakṣepyas* there should be an em-dash marking the gloss. In the SGT, the technical term *madhyakṣepya* is found here for the first time.

1446 In line 6, the first abbreviation *gu* is followed by a kind of large dot or a zero; this is not found further in the *nyāsa* given in SGT ad L 50. One finds a similar pattern regarding the abbreviation *gu* in GT 17, 19 and 20 but not in GT 19–20, 5 and 6. Also, I have emended the abbreviation *bhāgu*, which is nonsensical, to *bhā*, which could stand either for *bhāgadayīn* (in the commentator's vocabulary this is a common term for "divisor"; see above in SGT 93, 25) or for *bhā* ("division"). The erroneous *bhāgu* repeats the first *gu* on the left.

1447 *pratiloma*. One would expect either *anuloma* or *krama*; however, I understand *pratiloma* to mean "reverse" in respect to the way numbers are read according to their power of ten (from the largest on the left) or according to the way the commentator usually refers to the quantities in the layout when they have to be computed (from the first on the left). Notwithstanding it retains its meaning ("reverse order"), in this context *pratiloma* denotes a right to left direction. In my translation, I have mentioned the anomaly that the SGT occasionally shows in the use of this term in SGT 26.

1448 Line 7: **rūpa**] em.

1449 Line 11: **guṇaḥ** *sa haras*] em. The repetition of *guṇaḥ* represents an instance of dittography.

1450 In brief, "in the manner which has been explained, according to the inverse operation".

In the *Līlāvatī,* a procedure with fractions which are added or subtracted[1451] is shown:

> L 48. "In order to verify[1452] the given number,[1453] one should make the divisor a multiplier, the multiplier a divisor, the square a root, and the root a square, the negative number (*ṛṇa*) into a positive number (*sva*), a positive into a negative".

This is clear [in meaning].[1454]

> L 49. "When a fraction of itself (i.e., of a given number) is added to or subtracted from [the given number], the divisor is increased[1455] or decreased by the numerator [and] it will be the [new] divisor. The numerator is unchanged, and [then] the rest is according to the reverse order as previously explained".

The explanation: **when a fraction of itself** is "added", that is, [the numerator is] increased[1456] by its own [denominator], for instance four,[1457] **the divisor,** which is the denominator, is "increased by the numerator", that is, four is added to three, the upper denominator should be [then] multiplied by it (i.e., the result).[1458] Afterwards, when there is the multiplication of the upper[1459] "numerator", it should be understood that, according to the "inverse" method (*vilomarītyā*), wherever [the fraction] is an additive quantity, it will instead be subtractive.[1460] That quantity which has been obtained from the sum becomes the last[1461] (i.e., at the bottom) denominator, having deducted (*akṛṣya*) [from it] that very number [which is its own numerator], that [number obtained by adding the last denominator and the last numerator] is made the multiplier [of the upper numerator],[1462] such is the essence of this procedure.[1463]

1451 Siṃhatilakasūri quotes the L in order to provide a rule and sample problem on the inverse operation involving fractions.

1452 *prasiddhi* – lit. "accomplishment, obtainment".

1453 *dṛśyarāśi.*

1454 The method of inverse operation has been explained by the commentary while elucidating the procedural rule by Śrīpati.

1455 *āḍhya.*

1456 *anvita.*

1457 The commentator refers to the fraction three-fourths mentioned in L 50 (see further).

1458 See further the explanation of this passage in SGT ad L 50, 16.

1459 *ūrdhva.*

1460 This passage is better understood in the light of the execution of the sample problem; see the way the two fractions one-third and three-fourths (more particularly their numerators) are manipulated in lines 11 and 15–16.

1461 *anta.*

1462 See SGT ad L 50, 9–11.

1463 The whole passage on L 49 elaborates on *svāṃśe'dhikone*. It is divided into two parts: the first (lines 21–24) elucidates the case in which a fraction is added to a give number (see

However, **when a fraction of itself is subtracted**, it is deducted by its own numerator, such as three; in this case, **the divisor**, which is the denominator, is decreased by its own "numerator". The upper denominator[1464] is [then] multiplied by that [result]. Next, when there is the multiplication of the upper "numerator", it should be understood that, according to the reverse order, it (i.e., the fraction) will be an additive quantity in place of a subtractive (*ūnasthāne*).[1465] The denominator below is added to the "numerator", it (i.e., the result) should be made the multiplier, the other "numerator" remains **unchanged**, it should not be removed (*bhañjanīya*). **The rest is as previously explained**: [one should make] "the divisor a multiplier, the multiplier and so forth", it (i.e., the procedure) should be [performed] in this very manner.

The author enunciates a sample problem:[1466]

> L 50. "O girl with tremulous eyes, if you know the correct inverse procedure tell [me] what number multiplied by three, added to its three-quarters, divided by seven, reduced by its own one-third, multiplied by itself, decreased by fifty-two, [having] its square-root [extracted], added to eight, and divided by ten, becomes two"?

The presentation:[1467]

gu 3	svaca 3	bhā 7	svatryaṃśa 1	svagu 1	hīna 52	mū 1	dha 8	bhā 10	dṛśya 2
	4		3						

According to the reverse way,[1468] the given number[1469] two is multiplied by ten, it becomes twenty; it is lessened by eight becoming twelve, which squared is one hundred forty-four. When the additive quantity fifty-two is added to it, it becomes one hundred ninety-six. Its square-root is fourteen, which decreased by "its own one-third" is:

14
1
01
3

. That [obtained quantity] which should be lessened by "its own one-third" is instead added to it;[1470] **the divisor**, which is

svāṃśe adhike, line 21), the second (lines 24–27) the case in which a fraction is subtracted from the given number (see *svāṃśe ūne*, line 24).

1464 Line 25: *ūrdhva°*] em.; *tū°* (K).

1465 Line 26: *adhikam*] em.; *ityukta* (K). The emendation follows the commentary; see line 23 (*tatra hīna iti jñeyam*) and SGT ad L 50, 11 (*hīnasthāne adhikaṃ*). See the construction of the two parallel passages ending with *iti jñeyam* in lines 23 and 26.

1466 The SGT supplies a sample problem which corresponds to L 50.

1467 I have emended *bhāgu* 7 to *bhā* 7 (as I have mentioned the same anomalous *bhāgu* occurs in SGT 94, 6) and *svatryaṃ* to *svatryaṃśa*.

1468 *vaiparītya*.

1469 Cf. *dṛśyarāśi* in L 48.

1470 Line 10: *adhikam*] em.; *ity ukta* (K).

that [denominator] three underneath, should be lessened[1471] by one, which is its own "numerator"; it becomes two. One, which is the upper denominator, is multiplied by it becoming two. When there is the multiplication of the upper "numerator" [fourteen], since [here] the inverse operation takes place, instead of being subtractive that [numerator one] becomes an additive quantity.[1472] Two becomes the denominator below;[1473] the number one is unchanged.[1474] [Two] added to it becomes three; fourteen multiplied by it becomes forty-two whose denominator is two: $\left|\begin{array}{r}42\\2\end{array}\right|$. When there is the reduction of these by half, it becomes twenty-one whose denominator is one: $\left|\begin{array}{r}21\\1\end{array}\right|$. Twenty-one multiplied by seven becomes one hundred forty-seven whose denominator is one: $\left|\begin{array}{r}147\\1\end{array}\right|$; it is added to "its three quarters": $\left|\begin{array}{r}147\\1\\3\\4\end{array}\right|$. [Regarding the expression] **when a fraction of itself** is "added", **the divisor** below is "added to" (*adhika>yuta*) the "numerator" three; it becomes seven. The upper denominator one is multiplied by it becoming seven. Since it is the inverse operation, having made it (i.e., the numerator three) a subtractive (*hīna*) quantity in place of additive, that seven, which should be added to it, is [instead] subtracted by three; it becomes four. One hundred forty-seven multiplied by it becomes five hundred eighty-eight whose denominator is seven: $\left|\begin{array}{r}588\\7\end{array}\right|$. Seven **multiplied by three** (*trighna>triguṇa*) is twenty-one. When there is the division [of both operands] by it, the quotient is the integer twenty-eight.

Its verification: the integer twenty-eight "multiplied by three" becomes eighty-four, which **added to its three-quarters** is: $\left|\begin{array}{r}84\\3\\4\end{array}\right|$. By means of [the procedure] "the denominator by the denominator"[1475] and so forth, when there is the simplification,[1476] it becomes[1477] five hundred eighty-eight whose

1471 *nyūna*.
1472 The text is rather terse here. The numerator 14 is multiplied by (2 + 1).
1473 *tato*] conj.; *gato* (K).
1474 *ākṛṣṭa*.
1475 This is a quotation from the rule of the class of fractional increase (GT 57).
1476 Line 22: *savarṇane*] em.
1477 Line 22: *jātā*] em.

denominator is four: $\begin{array}{|c|}588\\4\end{array}$. Four multiplied by seven becomes twenty-eight. When there is its division, the quotient is twenty-one. It is **reduced by its own one-third:** when there is the deduction[1478] by seven, it becomes fourteen. Fourteen multiplied by fourteen becomes one hundred ninety-six; this **decreased by fifty-two** becomes one hundred forty-four. Its "square-root" is twelve, which "added to eight" is twenty, which divided by ten gives the quotient, the given number two: 2.

In this way, when there is the inverse operation, from the given number the unknown number [is arrived at], [and in the verification] from the unknown number the given number is arrived at.

The inverse operation is concluded. Thus far, thirty-one arithmetical operations have been accomplished.[1479]

3.10 Rules on proportion [GT 95–117]

Rule of three (trairāśika)[1480] *[GT 95]*

Now the rule of three, which is the thirty-second [operation treated in this work], is undertaken.[1481]

[Below] is the procedural rule on this topic; the author enunciates a metrical stanza:

> 95. The measure[1482] and the requisition[1483] are [written] at the beginning[1484] and at the end[1485] [of the layout] and the fruit,[1486] which is of a different kind, is in the middle. Having multiplied the fruit by the requisition,[1487] one should divide [the result] by the measure. If the inverse (*vāma*) [rule of three] occurs, the inverse procedure[1488] [should be performed].

The explanation: it is stated that either the number which is the commodity[1489] or the number which is the cost[1490] is written **at the beginning** [of the layout], [in

1478 *ākarṣaṇa*, this term is also found in SGT 52, 10.

1479 This is a significant remark, which highlights that the commentator considers "operations" (*parikarman*) all the thirty-one procedures so far explained. In GT 118, the treatment of the *vyvahāras* or "practices" begin.

1480 K 68, 4–7.

1481 Here begins the treatment of rules on proportion.

1482 *pramāṇa.*

1483 *abhipsā.*

1484 *ādi.* The commentator demonstrates that this implies a left to right direction.

1485 *virama.*

1486 *phala.*

1487 *samicchā.*

1488 *vyastavidhi.*

1489 *vastu.*

1490 *mūlya.*

this case] that termed "measure" should be [understood]. It is explained that either the number which is the commodity or the number which is the cost is **at the end** (*virama* > *paryanta*), [in this case] that called "requisition" (*abhīpsā* > *icchā*) [should be understood]. [Measures which are of] the very same **kind** should be made **at the** very **beginning** and **at the end** [of the layout]. This is the meaning: whenever **at the beginning** the number which is the commodity occurs, then at the end there is the very number which is the commodity as well. In the same way, whenever **at the beginning** there is the "measure" of the cost, then at the end the very number which is the wealth (*dravya*) should be made too. The [quantity of] a **different kind**, that which is called "fruit", is to be made in **the middle** of these, which are the "measure" and the "requisition". Whenever **at the beginning** and at the end there is the commodity, then **in the middle** [there is the cost], or else whenever **at the beginning** and at the end there is the number which is the cost, then the number which is the commodity should be made **in the middle**. This [quantity which represents either the commodity or the cost] should be written [in the middle],[1491] according to the method[1492] [explained] this [quantity] is [then] divided. Then, having multiplied (*nihatya* > *guṇayitvā*) **the fruit**, which is the number in the middle, **by the requisition**, which is the last number, by means of the method of division, according to [the rule] "having performed the interchange"[1493] and so forth, one should divide (*bhajet* > *bhāgaṃ grāhayet*) [the result] **by the measure**, which is the first number.[1494] That result will be the fruit of the requisition.[1495] Likewise "if the inverse occurs", that is to say, "if the inverse rule of three occurs".[1496]

In the *Līlavatī*, it is said:

> L 77. "Wherever the increase of the requisition [and] the decrease of the cost occur or when there are the decrease [of the requisition] [and] the increase [of the cost], in this case the inverse rule of three [should be applied]".[1497]

For instance:

> L 78. "When the cost of living beings[1498] is [determined] by their age and, in the case of gold,[1499] when the weight of the *varṇa* [determines

1491 This sentence is awkwardly terse as it stands.
1492 Line 14: *rītyā*] conj.
1493 This is a quotation from the rule of division of fractions (GT 42).
1494 The whole passage expands upon the three terms *pramāṇa*, *phala*, and *samicchā*.
1495 *icchāphala*.
1496 *vyastatrairāśika*.
1497 This is a quotation from L 77.
1498 *jīva*.
1499 *heman*.

> the cost][1500] or whenever the [procedure of] subdivision of heaps[1501] takes place, the inverse rule of three should occur".[1502]

In this manner, when there are these particular cases of the "inverse" procedure [of the rule of three], there is the **inverse procedure**. The previously mentioned [procedure][1503] should be turned around,[1504] so that **having multiplied** the number which is in the middle by the "measure", one should divide [the result] by the "requisition", which is the last number.[1505] This is thus the definition to perform.

In this respect, in the exemplifying metrical stanza [which follows] the author enunciates a first sample problem:

> 96. If one plus a half *pala*s[1506] of musk[1507] are obtained by twelve plus one-fourth *dramma*s,[1508] how much seven plus one-third *pala*s will then cost?

Its explanation with the very presentation: $\begin{array}{|r|r|r|} \text{va}\,1 & \text{mū}\,12 & \text{va}\,7 \\ 1 & 1 & 1 \\ 2 & 4 & 3 \end{array}$. Regarding the first quantity, by means of [the procedure] "multiplied by the denominator"[1509] and so forth, one multiplied by two is two, one is added to this [two]; it is three whose denominator is two: $\begin{array}{|r|} 3 \\ 2 \end{array}$. In the second quantity, twelve multiplied by four becomes forty-eight which plus one is forty-nine: $\begin{array}{|r|} 49 \\ 4 \end{array}$. With respect to the third quantity, seven multiplied by three becomes twenty-one which plus one is twenty-two whose denominator is three:[1510] $\begin{array}{|r|} 22 \\ 3 \end{array}$. This [method which has been performed] is the class of fractional increase. Afterwards, one should multiply **the fruit** forty-nine, which is the number in the middle, **by the requisition** twenty-two, which is last quantity; it becomes

1500 *varṇa* – lit "colour, class", here it is a unit of weight of gold denoting the proportion of pure gold in any given piece of gold ("carat"); the highest purity is 16 *varṇa*s.

1501 *rāśi*.

1502 Line 21: *bhāgahāre*] em.; *bhinnahāre* (K). This passage is a quotation from L 78, which in the edition by Sarma (1975) is slightly different.

1503 In brief, the rule of three mentioned in the root-text.

1504 *parītya*.

1505 The SGT elaborates on *vyastavidhiś ca vame*.

1506 See the units of weight mentioned in GT 6.

1507 *kastūrikā*.

1508 Monetary units are listed in GT 4.

1509 This is a quotation from the rule of the class of fractional increase (GT 57).

1510 I have emended the layout (line 4), which mistakenly reads 77.

one thousand seventy-eight: 1078. Regarding these [obtained quantities], three and four are the denominators. Because of the instruction[1511] **having multiplied,** according to the method by which the denominators are multiplied by the denominators [and] the numerators by the numerators,[1512] it is: four multiplied by three becomes twelve, which is joined by being the denominator below the previously mentioned one thousand and so forth: $\left|\begin{array}{r}1078\\12\end{array}\right|$. This quantity is the dividend (*bhājya*). Since the first quantity is the divisor, by means of [the procedure] "having performed the interchange"[1513] and so forth, two is above [and] three is underneath: $\left|\begin{array}{r}2\\3\end{array}\right|$. Then, there is the "cross-reduction": when there is the reduction of two by half, it is one: $\left|\begin{array}{r}1\\3\end{array}\right|$; when there is the reduction of twelve by half, it becomes six: $\left|\begin{array}{r}1078\\6\end{array}\right|$. Next is "the multiplication": the number above multiplied by one is exactly so, six multiplied by the three below becomes eighteen. When there is the division of the number above by it, the quotient is fifty-nine *dramma*s: $|59|$. The remainder above is sixteen: $\left|\begin{array}{r}16\\18\end{array}\right|$. When there is their reduction by half, eight is above [and] nine is underneath: $\left|\begin{array}{r}8\\9\end{array}\right|$. [In this fraction], there are no *dramma*s; to arrive at [the number of] *paṇa*s, eight should be multiplied by sixteen, it becomes one hundred twenty-eight: $\left|\begin{array}{r}128\\9\end{array}\right|$. When there is its division by nine, the quotient is fourteen *paṇa*s: $|14|$, the remainder is two. [In this remainder], there are no *paṇa*s; henceforth to arrive at [the number of] *kākiṇī*s, when there is the multiplication of two by four no quotient is obtained, since nine is the denominator underneath.[1514] Therefore in place of *kākiṇī*s there is a zero: $|0|$. Then, to arrive at the [number of] *kaparda*s, eight[1515] is multiplied by twenty, it becomes one hundred sixty: 160. When there is its division by nine, the quotient is seventeen *varāṭaka*s, the fraction [of *varāṭaka*s] is: $\left|\begin{array}{r}7\\9\end{array}\right|$.

1511 *vacana*.

1512 This remark refers to the multiplication of fractions (GT 40).

1513 This is a quotation from the rule of division of fractions (GT 42); it continues in the two lines below.

1514 Since $\left|\begin{array}{r}8\\9\end{array}\right|$ is a proper fraction, there is no need to divide the numerator by the denominator.

1515 The remainder 2 has been multiplied by 4.

He then enunciates a sample problem concerning a presentation [of data] where there is the [quantity] cost:

> 97. O knowledgeable one, if you know the rule of three, tell [me] how many *pala*s of camphor,[1516] whose colour is magnificent like the ivory tusks of a lordly elephant [and] whose perfume has attracted a swarm of bees, are [obtained] by one hundred *dramma*s when one and a half *pala*s are obtained by sixteen plus one-third *dramma*s?

[The presentation]:[1517] $\begin{array}{|r|r|r|} \text{mū}\,16 & \text{va}\,1 & \text{mū}\,100 \\ 1 & 1 & 1 \\ 3 & 2 & \end{array}$. By means of [the procedure] "multiplied by the denominator"[1518] and so forth, when the first quantity is simplified, it becomes forty-nine whose denominator is three: $\begin{array}{|r|} 49 \\ 3 \end{array}$. In the same way, with respect to the second number, it becomes three whose denominator is two: $\begin{array}{|r|} 3 \\ 2 \end{array}$. Then, three is multiplied **by the requisition,** which is the quantity one hundred whose denominator is one; it becomes three hundred, two multiplied by the denominator one is exactly so: $\begin{array}{|r|} 300 \\ 2 \end{array}$. Regarding the first quantity, because it is the divisor, according to [the rule] "having performed the interchange"[1519] and so forth, three is above [and] forty-nine is below. Next is "the multiplication": three hundred multiplied by three becomes nine hundred; forty-nine multiplied by two becomes ninety-eight. When there is the division of nine hundred by it, it is: $\begin{array}{|r|} 900 \\ 98 \end{array}$. The quotient gives [the number of] *pala*s, nine: 9; the remainder above is eighteen. When there is their reduction by half, nine is above [and] forty-nine is underneath: $\begin{array}{|r|} 9 \\ 49 \end{array}$; this is the fraction of *pala*s. Afterwards, in this case to arrive at the equal [number of]

1516 *karpūra*.

1517 In line 25, the term *nyāsa* is missing. I have emended the layout taking into account the preceding sample problem; the edition reads, in fact, an anomalous *ya* (which is most likely a corruption of *va* for *vastu*), and the final term should be the same as the first, thus it should be *mū* for *mūlya* instead of *dra* for *drammas*. Note the inconsistent way the *nyāsa*s of the following sample problems on this mathematical section (up to SGT 117) present abbreviations.

1518 This is a quotation from the rule of the class of fractional increase (GT 57).

1519 This is a quotation from the rule of division of fractions (GT 42).

dhaṭakas as it has been illustrated in the introductory section,[1520] nine multiplied by ten becomes ninety. This is divided by forty-nine, the quotient is one *dhaṭaka*: 1; the remainder is forty-one. Then, by means of that which has been taught: "[those who are most proficient in arithmetic] call seven pairs of *niṣpāvakas* one *dhaṭaka*",[1521] to arrive at the [number of] *vallas*, forty-one multiplied by fourteen becomes five hundred seventy-four. When there is its division by forty-nine,[1522] the quotient is eleven *vallas* and thirty-five forty-ninths of a twelfth [*valla*]: $\begin{array}{|c|c|} 11 & 35 \\ & 49 \end{array}$.

The author enunciates a special sample problem:[1523]

98. O friend, if six occurs apart out of one hundred, tell [me] how much will it be out of one thousand?

In this case, if there is one hundred, six is added[1524] [to it]. The presentation: $\begin{array}{|c|c|c|} 106 & 6 & 1000 \\ 1 & 1 & 1 \end{array}$. [The middle term six is] multiplied by the last, it is 6000. By means of the interchange of the numerator and denominator [of the first quantity], having multiplied it by the first [quantity], it is: $\begin{array}{|c|} 6000 \\ 106 \end{array}$. When there is the division, the quotient is the integer 56. When above and below the remainder is halved, the fraction[1525] is: $\begin{array}{|c|} 32 \\ 53 \end{array}$.

The author enunciates a sample problem:

99. O mathematician,[1526] if one and a half *dhaṭakas* of saffron[1527] are obtained by five and one-fourth *paṇas*, how much will one plus one-third *palas* cost?

The presentation: $\begin{array}{|c|c|c|} 1 & 5 & 1 \\ 1 & 1 & 1 \\ 2 & 4 & 3 \end{array}$. By means of [the procedure] "multiplied by the denominator"[1528] and so forth, it becomes[1529] three whose denominator is two,

1520 *prastāvana* –the commentator refers to the section on technical terms, which is also termed *paribhāṣā*.
1521 This is a quotation from the stanza on units of weight (GT 6).
1522 Line 8: *ekonapañcaśatā*] em. The number meant is, in fact, 49 and not the transmitted 39.
1523 This sample problem corresponds to TŚ example 37.
1524 *prakṣipya* – lit. "having thrown at, inserted".
1525 *rūpabhāga* – lit. the "fraction, part of the whole"
1526 *gaṇaka*.
1527 *kuṅkuma*.
1528 This is a quotation from the rule of the class of fractional increase (GT 57).
1529 Line 19: *jātā*] em.

twenty-one whose denominator is four, four whose denominator is three; it is respectively: $\begin{array}{|c|c|c|}3&21&4\\2&4&3\end{array}$. [Next is the step] **[having multiplied the fruit] by the requisition**, that is to say, the middle [term] twenty-one multiplied by four becomes eighty-four, four multiplied by three becomes the denominator twelve: $\begin{array}{|c|}84\\12\end{array}$. Regarding the first quantity, since it is the divisor, by means of [the procedure] "having performed the interchange"[1530] and so forth, above is two [and] underneath is three: $\begin{array}{|c|}2\\3\end{array}$. When there is the reduction of two and twelve by half, it gives six and one respectively. In the same way, when there is the reduction of three and eighty-four by three, it gives twenty-eight and one respectively: $\begin{array}{|c|c|}1&28\\1&6\end{array}$. Twenty-eight multiplied by one is exactly so; six multiplied by the denominator one is that very one: $\begin{array}{|c|}28\\6\end{array}$. When there is the division of twenty-eight by six, the quotient is four *dramma*s: 4; the remainder above is four. To arrive at the [number of] *paṇa*s, [four is] multiplied by sixteen; it becomes sixty-four. When there is its division by six, the quotient is ten *paṇa*s, the remainder above is four. To arrive at the *kākiṇī*s [the remainder four is] multiplied by four, it becomes sixteen; when there is its division by six, the quotient is two, the remainder above is four. Afterwards, to arrive at the *varāṭaka*s, [the remainder four] is multiplied by twenty, it becomes eighty; when there is its division by six, the quotient is thirteen *kapardaka*s, the remainder above is two and below is six. When there is the reduction of these by half, above is one and underneath is three: $\begin{array}{|c|}1\\3\end{array}$; this is the fraction of *kapardaka*s.

The author enunciates a verse regarding grain:[1531]

> 100. O friend, if two and a half *mānikā*s[1532] [of grain] are obtained (*prāpyate*) by eight minus one-eight *paṇa*s, tell [me] how much will one hundred and one-third *mānikā*s cost?

1530 This is a quotation from the rule of division of fractions (GT) 42. The abbreviation *parī* of the nominal form *parīvartana* is a peculiarity which occurs, whenever this passage is quoted, up to SGT 106, 20.

1531 *kaṇa*.

1532 See the units of capacity mentioned in GT 7.

The presentation: $\begin{array}{|r|r|r|} 2 & 8 & 100 \\ 1 & 01 & 1 \\ 2 & 8 & 3 \end{array}$. Regarding the first and the last quantities, by means of [the procedure] "multiplied by the denominator"[1533] and so forth, it becomes[1534] five whose denominator is two [and] three hundred one whose denominator is three respectively. With respect to the middle quantity, according to [the rule] "in the procedure of fractional decrease, once the integer is multiplied by the denominator"[1535] and so forth, eight multiplied by eight becomes sixty-four; when there is the subtraction by one, it is sixty-three whose denominator is eight: $\begin{array}{|r|r|r|} 5 & 63 & 301 \\ 2 & 8 & 3 \end{array}$. One should multiply the middle quantity sixty-three **by the requisition**, which is three hundred one; it becomes eighteen thousand nine hundred sixty-three, the denominator eight is multiplied by three becoming[1536] twenty-four: $\begin{array}{|r|} 18963 \\ 24 \end{array}$. Since the first quantity is the divisor, by means of [the procedure] "having performed the interchange"[1537] and so forth, two is above [and] five is below. There is "the cross reduction": when there is the reduction of two and twenty-four by half, it is one [and] twelve respectively. The number above multiplied by one is exactly so, and the twelve below multiplied by five becomes sixty: $\begin{array}{|r|} 18963 \\ 60 \end{array}$. Reading this [quantity], which denotes the number of *paṇa*s, when there is the division [of the number above] by sixty which has been multiplied by sixteen, [thus] by the obtained nine hundred sixty, the quotient is nineteen *dramma*s: $|19|$; the remainder is seven hundred twenty-three. To arrive at the [number of] *paṇa*s, [seven hundred twenty-three] multiplied by sixteen becomes eleven thousand five hundred sixty-eight: $\begin{array}{|r|} 11568 \\ 960 \end{array}$. When there is its division by nine hundred sixty, the quotient is twelve *paṇa*s: 12; the remainder above is forty-eight whose denominator is nine hundred sixty: $\begin{array}{|r|} 48 \\ 960 \end{array}$. To arrive at the [number of] *kākiṇī*s, forty-eight multiplied by four becomes one hundred ninety-two; in this case, the division cannot occur, thus in place of *kākiṇī*s there is a zero: $|0|$.

1533 This is a quotation from the rule of the class of fractional increase (GT 57).

1534 Line 9: *jātā*] em.

1535 This is a quotation from the rule of the class of fractional decrease (GT 60).

1536 Line 13: *jātāś ca°*] em.; *yathā ca°* (K).

1537 This is a quotation from the rule of division of fractions (GT 42); it continues in the line below.

Then, when there is the reduction of one hundred ninety-two[1538] by ninety-six, two is above [and] ten is below: $\left|\begin{array}{r}2\\10\end{array}\right|$. To arrive at the [number of] *kapardaka*s, two is multiplied by twenty, it becomes forty; when there is its division by ten, the quotient is four *kapardaka*s: $|4|$.

The author enunciates a second sample problem concerning grain:

> 101. O scholar, if your intellect has become strong on the science of numbers,[1539] tell me quickly, if two and one-fourth *mānikā*s of corn[1540] are obtained by six and one-third *dramma*s, how much [corn] can be obtained by eighty and one-half *dramma*s?

The presentation: $\left|\begin{array}{r|r|r}6&2&80\\1&1&1\\3&4&2\end{array}\right|$. By means of the [procedure] "multiplied by the denominator"[1541] and so forth, it is nineteen whose denominator is three, nine whose denominator is four, [and] one hundred sixty-one whose denominator is two respectively: $\left|\begin{array}{r|r|r}19&9&161\\3&4&2\end{array}\right|$. Then, one should multiply the middle quantity nine [whose denominator is four] **by the requisition**, which is one hundred sixty-one; it becomes fourteen hundred forty-nine, four multiplied by the denominator two becomes eight: $\left|\begin{array}{r}1449\\8\end{array}\right|$. Regarding the first quantity, because it is the divisor, according to [the rule] "having performed the interchange"[1542] and so forth, three is above and nineteen is underneath: $\left|\begin{array}{r}3\\19\end{array}\right|$. There is "the multiplication": fourteen hundred forty-nine multiplied by three becomes four thousand three hundred forty-seven; in the same way, nineteen multiplied by eight becomes one hundred fifty-two: $\left|\begin{array}{r}4347\\152\end{array}\right|$. When there is the division of the number above by it, the quotient is twenty-eight *mānikā*s: 28; the remainder above is ninety-one: $|91|$. In order to arrive at the [number of] *hārikā*s, [ninety-one] is multiplied by four, it becomes three hundred sixty-four; when there

1538 Line 23: *dvinavatyadhi(ka)śatasya*] em.; *dvānavatyadhi(ka)navaśatasya* (K). The expression contains a scribal doublet.

1539 *saṅkhyāśāstra*.

1540 *dhānya*.

1541 This is a quotation from the rule of the class of fractional increase (GT 57).

1542 This is a quotation from the rule of division of fractions (GT 42); it continues in the line below.

is its division by one hundred fifty-two, the quotient is two *hārikās*: 2, the remainder above is sixty: $\begin{array}{|c|}60\\152\end{array}$.[1543] When there is the reduction of these by four, it is fifteen instead of sixty, [and] it is thirty-eight: $\begin{array}{|c|}15\\38\end{array}$, which is the fraction of *hārikās*, instead of one hundred fifty-two.

The author enunciates a sample problem concerning time[1544] related to distance:[1545]

> 102. O you noble-minded, if a certain lordly elephant having a swarm of bees wandering around its cheeks, remembering the pastime with a female elephant of the Vindhya forests starts to travel and travels one-third of a half *yojana*[1546] in two less by one-half days, how many days it will take [the elephant] to reach to seventy *yojanas*?

Prior to the presentation, the simplification of "one-third of one-half *yojana*" [is carried out] $\begin{array}{|c|c|c|}1&1&1\\1&2&3\end{array}$. Since there is multi-part class, by means of instruction "the multiplication of the numerators and the product of the denominators"[1547] and so forth, when there is the multiplication of the [denominators] one, it is one. This very one is above; two multiplied by the denominator one is the very two, the three below multiplied by two is six. Next is [the presentation]:[1548] $\begin{array}{|c|c|c|}1&\text{di}\,2&\text{yo}\,70\\6&01&1\\&3&\end{array}$. When the second quantity is reduced by means of the [method of the] class of fractional decrease, it becomes three whose denominator is two: $\begin{array}{|c|}3\\2\end{array}$. Next, [this is multiplied] **by the requisition**: three multiplied by seventy becomes two hundred ten, and the two below multiplied by the denominator is the very two: $\begin{array}{|c|}210\\2\end{array}$. Since one-sixth,[1549] which is derived from one-third of a

1543 Line 11: *śeṣam upari ṣaṣṭiḥ | yathā* $\begin{array}{|c|}60\\152\end{array}$ |] conj. In line 11, the text is suspect as it does not mention the remainder. The conjectural emendation follows the commentator's explanation.

1544 *kāla.*

1545 *mārga.*

1546 See units of length in GT 8–10.

1547 This is a quotation from the multi-part class (GT 55).

1548 Line 20: one would expect *tato nyāsa* instead of *tato yathā*.

1549 Line 23: **ekaccheda**] em.

half *yojana,* is the divisor, by means of [the procedure] "having performed the interchange"[1550] and so forth, six is above [and] one is below: $\begin{array}{|c|}6\\1\end{array}$. When there is the reduction of two [and] six by half, it becomes three [and] one respectively. Two hundred ten multiplied by three becomes six hundred thirty; [the denominator one is] multiplied by the denominator one, [and the numerator] divided by the denominator one is exactly so. Then, this quotient [630] is the duration in days[1551] [of the elephant's journey]. When there is its division by three hundred sixty,[1552] the quotient is one year [and] nine months:[1553] *va* 1, *mā* 9.

The author enunciates another sample problem:

> 103. If a snake[1554] whose body is three and a half cubits[1555] enters a hole and covers one and a half fingers[1556] in one-third of a *ghaṭī* [in time], say quickly o mathematician, how long it will take to [fully] enter it?[1557]

The presentation: $\begin{array}{|c|}1\\1\\2\end{array}\begin{array}{|c|}1\\3\\ \end{array}\begin{array}{|c|}84\\1\\ \end{array}$. First are one and a half fingers,[1558] eighty-four [expressed] in *aṅgula*s [obtained] from three and a half cubits are placed [in the third place]. In the first [quantity],[1559] by means of [the method] "multiplied by the denominator"[1560] and so forth, it becomes three whose denominator is two: $\begin{array}{|c|}3\\2\end{array}$. Then, the middle quantity one multiplied **by the requisition** eighty-four becomes eighty-four, and below three multiplied by one is exactly so: $\begin{array}{|c|}84\\3\end{array}$. Since the first number is the divisor, according to [the rule] "having

1550 This is a quotation from the rule of division of fractions (GT 42).
1551 *dina.*
1552 Units of time are given in GT 11–12.
1553 In line 27, the abbreviation *va* stands for *varṣa* ("year"), and *mā* stands for *māsa* ("month").
1554 *bhujaṅga.*
1555 *kara.*
1556 *aṅgula.*
1557 In the edition, footnote 1 gives a quotation, which is from the *Chandomañjarī* by Gaṅgādāsa (ca. 12th century CE), a classical Sanskrit work on prosody. It reads: "It is said that *tāmarasa* has *nagaṇa*, two *jagaṇa*s, and *yagaṇa*"; this is the definition of the meter *tāmarasa* which is used by Śrīpati in the verse, it is not clear whether this note was found in the manuscript or it has been inserted by the editor.
1558 The text is uncertain here, the order of presentation of the term *nyāsa*, the layout, and the explanation of the quantities is unusual; I have emended the text and moved both verbal explanations after the layout.
1559 Line 7: one would expect *prathamāṅke* rather than *prathame*.
1560 This is a quotation from the rule of the class of fractional increase (GT 57).

performed the interchange"[1561] and so forth, two is above and three is underneath:

2
3

. In this case, by means of the cross-reduction of the three below the two, when there is the division by three, it is one. In the same way, when there is the division of eighty-four by three, it is twenty-eight. Afterwards, the twenty-eight above multiplied by two becomes fifty-six. One multiplied by the three below is the very three. Next, when there is the division by three, the quotient is eighteen *ghaṭikā*s: 18; the remainder above is 2, this multiplied by 60 which are the *pala*s[1562] to be arrived at[1563] becomes one hundred twenty. When there is its division by three, the quotient is forty *pala*s: 40.

The author enunciates a sample problem relating to gold (*svarṇa*):

104. If one *gadyanaka*[1564] plus one *dharaṇa* [of gold] are obtained by fourteen and a half *dramma*s, tell [me] o friend, how much gold can be obtained by ninety minus one-third *dramma*s?

[The presentation:][1565]

dra 14	1	90
1	1	01
2	2	3

. By means of the [procedure] "multiplied by the denominator"[1566] and so forth, in the first quantity it is twenty-nine whose denominator is two; in the second quantity it is three whose denominator is two. In the third quantity, by means of the class of fractional decrease, according to [the rule] "multiplied by the denominator"[1567] and so forth, it is two hundred sixty-nine whose denominator is three. It is respectively:

29	3	269
2	2	3

. One should then multiply the number that is in the middle, which is three [whose denominator is two], **by the requisition**, which is two hundred sixty-nine; it becomes eight hundred seven, and below two multiplied by three becomes six:

807
6

. Next, since the first number is the divisor, by means of [the procedure] "having performed the interchange"[1568] and so forth, two is above and twenty-nine is below:

2
29

. Afterwards, by means of the

1561 This is a quotation from the rule of division of fractions (GT 42).
1562 The unit of time *pala* is not in GT 11, where it is said that the submultiple of a *ghaṭikā* is the *vināḍī* (which corresponds to 60 *ghaṭikā*s).
1563 Line 13: *ānīya*°] em.; *pānīya*° (K).
1564 Gold measuring units are given in GT 5.
1565 In line 20, the term *nyāsa* is missing.
1566 This is a quotation from the rule of the class of fractional increase (GT 57).
1567 This is a quotation from the rule of the class of fractional decrease (GT 60).
1568 This is a quotation from the rule of division of fractions (GT 42).

cross-reduction, when there is the reduction of six and two by half, it is one and three: $\begin{array}{|r|r|}1 & 807\\ 29 & 3\end{array}$. [The numerator 807] multiplied by the one above is exactly so; twenty-nine multiplied by the three below becomes eighty-seven. In the same way, when the number above is divided, the quotient is nine *gadyāṇaka*s: 9; the remainder above is twenty-four and below is eighty-seven: $\begin{array}{|r|}24\\ 87\end{array}$. When there is the reduction of these [numbers] by three, eight is above and twenty-nine is underneath: $\begin{array}{|r|}8\\ 29\end{array}$. To calculate the [number of] *dharaṇa*s, eight is multiplied by two, it is sixteen. The division cannot occur, thus in the place of *dharaṇa*s there is a zero: $\begin{array}{|r|}0\end{array}$.Then, in order to arrive at the [number of] *niṣpāva*s, sixteen is multiplied by eight becoming one hundred twenty-eight. When there is its division by twenty-nine, the quotient is 4 *niṣpāva*s: 4; the remainder above is twelve. In order to arrive at the [number of] *yava*s, this is multiplied by six becoming seventy-two. When there is its division by twenty-nine, the quotient is two *yava*s, and the fraction is fourteen twenty-ninths: $\begin{array}{|r|}14\\ 29\end{array}$.

The author enunciates a sample problem on the topic of the inverse rule of three concerning wealth:[1569]

> $\overline{105}$. Sixteen strings of pearls[1570] contain eight *setikā*s [each]. O friend, tell [me] quickly what is the number occurring [out of these] when the strings of pearls measure six *setikā*s [each]?[1571]

[In the first place] the first half [of the metrical stanza] is shown;[1572] the presentation: $\begin{array}{|r|r|r|}8 & 16 & 6\\ 1 & 1 & 1\end{array}$. In this case, sixteen is multiplied by eight, which is the "measure", it becomes one hundred twenty-eight; one multiplied by the denominator one is exactly so. Since it is the inverse rule of three, [*bhajet* or "one should divide")[1573] [the product] by the "requisition", which is the last number: by means of[the procedure] "having performed the interchange"[1574] and so forth, one is above and six is underneath. [That which is] above and below is multiplied by one, it is exactly so: $\begin{array}{|r|}128\\ 6\end{array}$. When there is the division by

1569 *dhānya*. Cf. TŚ example 38, where the number of string of pearls, is however, twenty; the formulation of the sample problem is also slightly different.

1570 *hīra*.

1571 Units of capacity listed in GT 7.

1572 In line 15, *athottarārdhena* ("by means of the latter half") demonstrates that originally the sample problems in GT 105 and GT 106 were coupled together.

1573 See *bhajet* in GT 95.

1574 This is a quotation from the rule of division of fractions (GT 42).

six, the quotient is twenty-one which is [the number of] the string of pearls, the remainder above is two [and] below is six. When there is their reduction by half, one is above and three is underneath: $\begin{array}{|c|}1\\3\end{array}$.

The author enunciates [a sample problem on] the inverse rule of three concerning gold by the latter half [of the metrical stanza]:

$\overline{106}$. How much gold of eleven *varṇika*s [of pure gold in weight] can be obtained when [a piece of] gold[1575] [weighing] ninety *gadyāṇaka*s [and] sixteen *varṇika*s is given (i.e., exchanged with)?[1576]

[The presentation]:[1577] $\begin{array}{|c|c|c|}16&90&11\\1&1&1\end{array}$. As it is the inverse rule of three, one should multiply the number in the middle, which is ninety, by the "measure" sixteen, which is the first number; it becomes fourteen hundred forty; one multiplied by the denominator one is exactly so. Since eleven, which is the quantity "requisition", is the divisor, by means of [the procedure] "having performed the interchange"[1578] and so forth, one is above and eleven is underneath. The number above multiplied by one is exactly so. When there is the division by the eleven which has been multiplied by one, the quotient is one hundred thirty *gadyāṇaka*s; the remainder is ten-elevenths: $\begin{array}{|c|}10\\11\end{array}$. In this case, when there is the "decrease" of the requisition,[1579] the "increase" of the [weight of] gold standing for[1580] the "price" takes places.[1581] [...?][1582]

1575 *kāñcana*.

1576 This sample problem mentions the weight of pure gold (*varṇa*) in a piece of gold.

1577 In line 18, the term *nyāsa* is missing.

1578 This is a quotation from the rule of division of fractions (GT 42).

1579 It refers to the fact that the piece of gold which is taken into exchange is less pure: *varṇika*s 16>11.

1580 Here the syntagm *mūlyasthānīya* is intended to be a reference to L 77 ("when there is the decrease [of the requisition] [and] the increase [of the cost] [...]"); in the SGT the "increase of [the weight of] gold" is what in L 77 is denoted as the "increase of the cost" (*mūlya*). The commentator argues that it has been shown that, with respect to this sample problem, the decrease of the weight of *varṇika*s and the increase of the weight of gold occur at the same time.

1581 The commentary paraphrases L 77–78, which he has fully quoted while commenting upon GT 95. L 78 mentions calculations on gold and states that it is the weight of the *varṇika*s of pure gold to determine the cost of gold, and not thus the weight of the piece itself. Here Siṃhatilakasūri mentions the "decrease of the requisition", which is represented by the weight of the piece of gold expressed in *gadyāṇaka*s. A piece of gold of 16 *varṇika*s weighing 90 *gadyāṇaka*s is exchanged with one of 11 *varṇika*s weighing 130 plus $\begin{array}{|c|}10\\11\end{array}$ *gadyāṇaka*s, which is a piece of gold of less *varṇika*s but weighing more.

1582 The last passage (lines 23–24) is corrupt and I cannot make sense of it.

3.11 Rule of five (*pañcarāśika*)[1583] [GT 107–117]

Next is the procedural rule of five quantities;[1584] the author enunciates a metrical stanza:

> 107. Having moved[1585] the fruit to the other side,[1586] having performed the procedure of the exchange of the denominators and the multiplication of the quantities of each[1587] side, having carried out their mutual multiplication, one should divide [the product of] one side by [the product of] the side of the other quantities.

The explanation: [first is the step] having put (*vinyasya*)[1588] **the fruit,** such as the interest,[1589] in **the other side,** that is to say, from the first into the second "side"[1590] below.[1591] **Having** always **performed** the "exchange" of the "denominators" (*chid>cheda*), **[having performed] the multiplication** of the "quantities of each side", and **their mutual multiplication,** which is the **multiplication** (*ghāta>guṇana*) of the "quantities" of the first "side" and of the "quantities" of the second "side" as well, **having performed the mutual** multiplication (*ghāta>guṇana*), **one should divide one side**, which is the second "side", that is to say, [one should divide] the larger quantity[1592] which is arisen [and] which is the product of the performed "reciprocal" multiplication, **by the side of the other quantities,** which is the first product (i.e., the product of the first side).

In this regard, in the illustrative verse [which follows] the author enunciates a well-known sample problem:[1593]

> 108. What is the interest which will accrue[1594] in a year on seventy-six [*dramma*s] if [the interest-rate] is 5% per month? O knowledgeable one,

1583 K 74, 26–29.
1584 *pañcarāśika*, the "rule of five".
1585 *ānīya.*
1586 *pakṣa.*
1587 *nijapakṣarāśi* – lit. "its own side". Here the possessive *nija* functions as a distributive appositive.
1588 In brief, "having written".
1589 *vyāja.* The SGT refers to the next sample problem, which concerns financial transactions. In the *Lekhapaddhati,* the term *vyāja*/*vyājaka* as "interest-rate" and "accrued interest" is found too; see, for instance, in Strauch (2002, 171). The *Lekhapaddhati* is a collection of documents of state and of everyday life from early medieval Gujarat; as highlighted by Prasad (2007, 22–23), the *Lekhapaddhati* cites principles and norms from the *Dharmaśāstra* literature and other legal texts in matter relating to credit and banking.
1590 The commentator expands upon *pakṣa* and illustrates that the algorithm of the rule of five involves the division of the layout into two sides.
1591 Notably, no *daṇḍa* occurs in the whole passage.
1592 *bahurāśi* – it is found in TŚ 31, which is the *karaṇasūtra* of the rules of five, seven, nine, and eleven. This is quoted by the commentator while solving the next sample problem (see line 24).
1593 The source is not traced.
1594 *kalāntara*, the "accrued interest".

if you know the rule of five, tell [me] the time (*kāla*), the accrued interest (*phala*), and the capital (*mūladhana*) from these two [data].

The presentation: $\begin{array}{|c||r||r|} \text{mā} & 1 & 12 \\ \text{dra} & 100 & 76 \\ \text{vyā} & 5 & \end{array}$. All these [numbers] have one as denominator.[1595] **Having moved the fruit**, which consists of five *dramma*s per month,[1596] from the first "side" **to the other side**, which is the second "side",[1597] it is: $\begin{array}{|r|r|} 1 & 12 \\ 100 & 76 \\ & 5 \end{array}$. **One should divide**[1598] that obtained (*sañjñam*) [product] of "one side" **by [the product of] the side of the other quantities,**[1599] which is [the product of] one hundred multiplied by the first [number] one $\begin{array}{|r|} 4560 \\ 100 \end{array}$. The quotient is 45, which is **the accrued interest**; the remainder above is sixty [and] below is one hundred: $\begin{array}{|r|} 60 \\ 100 \end{array}$. When there is their reduction by twenty, three is above and five is underneath: $\begin{array}{|r|} 3 \\ 5 \end{array}$.

With respect to this [result], when the given "capital"[1600] is not known,[1601] forty-five *dramma*s and three-fifths are brought in[1602] because they represent

1595 It implies that in this case there is no exchange of the denominators; on this feature, see *chedena saha* ("together with the denominators") in line 19 and also the remark on the two exchanged denominators at the end of the commentator's execution of this sample problem (lines 14 and 18).

1596 *māsika* – lit. "monthly".

1597 *dvitīyapakṣa.*

1598 Line 12: **chedām ihaiva**] em.

1599 "One side" (*pakṣam paraṃ*) denotes the product of the second side, which is divided by the product of the "other", the first "side" (*anyarāśipakṣa*).

1600 *prathamadattadhana.* In SGT 131, 18 *prayukta* and *pradatta* mean "borrowed/lent". Briefly: "in order to determine the initial capital which is borrowed/lent". Below 12, the zero occurring instead of 76 indicates that the capital is to be determined.

1601 Or perhaps: "If the given capital was not known"; the capital is, in fact, given in the sample problem but it is the author himself who asks students to determine it and this is the reason why in line 26 the SGT quotes *mūladhana* ("capital") from the *karaṇasūtra*. The locative absolute construction used by the author in lines 27, 5, 13, and 17 formulates a conditional clause (with almost a causal nuance), the main clause is a nominal sentence. The grammatical subject of each absolute phrase is the different quantity which is each time required to be ascertained.

1602 It refers to the fact that the obtained accrued interest is moved to the second side.

the accrued interest; this being the case, the presentation is: $\begin{array}{|r|r|} 1 & 12 \\ 100 & 0 \\ 5 & 45 \\ & 3 \\ & 5 \end{array}$. In this case, by means of [the procedure] "multiplied by the denominator"[1603] and so forth, forty-five multiplied by five becomes two hundred twenty-five; when the additive quantity three is added to it, it becomes two hundred twenty-eight whose denominator is five: $\begin{array}{|r|r|} 1 & 12 \\ 100 & 0 \\ 5 & 228 \\ & 5 \end{array}$. Therefore,[1604] there are two [kinds of] "interest";[1605] in both "sides", these are to be written inverted together with their denominators. First there is the transition[1606] of the [numerators of the] "interest": $\begin{array}{|r|r|} 1 & 12 \\ 100 & 0 \\ 228 & 5 \\ 5 & 1 \end{array}$; there is then the transition of the denominators:[1607] $\begin{array}{|r|r|} 1 & 12 \\ 100 & 0 \\ 228 & 5 \\ 1 & 5 \end{array}$. **Having carried out** this [step], [next is] **having performed the multiplication of the quantities of each side,** that is, one hundred multiplied by one is that very one, two hundred twenty-eight multiplied by one hundred becomes twenty-two thousand eight hundred whose denominator is one: $\begin{array}{|r|} 22800 \\ 1 \end{array}$. Regarding the numbers of "one side", it is: five multiplied by twelve becomes sixty; five multiplied by sixty becomes three hundred. In this case, by means of what is said in the *Triśatī*:[1608] "One should divide the side of the larger number by the other [side]", when there is the division of twenty-two thousand eight

1603 This is a quotation from the rule of the class of fractional increase (GT 57).

1604 Line 19: *tatra*] em.; *yatra* (K).

1605 One is the interest-rate 5% and the other is the accrued interest 45 plus $\begin{array}{|r|} 3 \\ 5 \end{array}$ *drammas*.

1606 *vyatyaya*.

1607 I have emended the text as in the layout, the number 7 is mistakenly found; a zero should occur instead.

1608 The commentator quotes a passage from TŚ 31. In this regard, see also SGT 107, 4.

hundred by three hundred, the quotient is seventy-six. This should be understood as **the capital** which is borrowed:[1609] |76|.

With respect to the very sample problem, when the "time"[1610] is not known, the presentation is:

1	0
100	76
5	45
	3
	5

. In this case [too] as before, by simplifying[1611] [the compound fraction representing the accrued interest] by means of the procedure of fractional increase, it becomes two hundred twenty-eight whose denominator is five. As before, there is the exchange of both [the monthly and the accrued] "interest" and both denominators:

1	0
100	76
228	5
1	5

. As before, there is the multiplication of the numbers of the first "side": it becomes twenty-two thousand eight hundred. Regarding these [numbers], by means of the multiplication of the numbers of "one side", by multiplying seventy-six by the two fives, when there is the division [of this product] by the obtained nineteen hundred, the quotient is the unknown [number of] months, twelve: |12|.

When the "interest" on the measure-value (*pramāṇaphala*) which is the monthly percentage consisting of the interest five is not known,[1612] the presentation is:

1	12
100	76
0	45
	3
	5

. In this case too as before, by means of the simplification, it (i.e., the accrued interest forty-five plus three-fifths) becomes two hundred twenty-eight whose denominator is five. This very one is the "accrued interest" which should be moved to the "side" of the first numbers. By means of the "exchange" of the denominators, as there is no second "fruit", five is moved to the second "side". Then, as before, the first quantities are multiplied; it

1609 *gṛhīta* – lit. "received, obtained"; here it means "borrowed/lent".

1610 Briefly: "in order to determine the time." In this procedure, the commentator demonstrates that the time regarding the obtained interest accruing on the capital 76 is 12 months. In the layout (line 27), above 76, a zero occurs instead of 12; the zero indicates that the time is the quantity to determine.

1611 As in SGT 102, 18 here too one finds the less common feminine form *savarṇanā*; see also line 6.

1612 Briefly: "in order to determine the interest on the capital". In the layout (line 6), below 100 there is a zero since the interest-rate is this time the quantity to determine.

becomes twenty-two thousand eight hundred. Regarding the second "side", seventy-six multiplied by twelve becomes nine hundred twelve; this multiplied by five becomes four thousand five hundred sixty. When there is the division of twenty-two thousand and so forth, which is the product of the first "side", by this [four thousand five hundred sixty], the quotient is the "interest" on the measure-value,[1613] five *drammas* per month (*māsaṃ prati*): 5.[1614]

Regarding the measure-value (*prāmaṇadhana*), when that consisting of the percentage is not known, the presentation is: $\begin{array}{|r||r|} 1 & 12 \\ 0 & 76 \\ 5 & 45 \\ & 3 \\ & 5 \end{array}$. As before, by means of the simplification it becomes two hundred twenty-eight whose denominator is five. Next, when there is the exchange of the denominators five and one of the twofold "interest",[1615] it is: $\begin{array}{|r||r|} 1 & 12 \\ 0 & 76 \\ 228 & 5 \\ 1 & 5 \end{array}$. The first quantity multiplied by one is exactly so. When there is the division of the obtained number twenty-two thousand eight hundred, which is the product of the other "side", by this, the quotient is one hundred: 100.

With respect to the "time" regarding the measure-value (*pramāṇakāla*), when that consisting of [the time expressed in] months is not known, the presentation is: $\begin{array}{|r||r|} 0 & 12 \\ 100 & 76 \\ 5 & 45 \\ & 3 \\ & 5 \end{array}$. In this case too as before, when it is simplified, it becomes two hundred twenty-eight whose denominator is five. When there is the exchange of the denominators five and one of the twofold "interest", it is: $\begin{array}{|r||r|} 0 & 12 \\ 100 & 76 \\ 228 & 5 \\ 1 & 5 \end{array}$. When the first quantities are multiplied by one hundred, it

1613 *pramāṇaphala*, it denotes the "interest" (*phala*) on the "measure-value" (*pramāṇa*).

1614 It can be observed that in the edition, integers which are the result of calculations, sometimes stand alone as in this case the number 5, sometimes they are placed inside boxes as in the case of the previous number 12.

1615 In lines 14 and 18, the same passage occurs but in line 18 the vowel *sandhi* between the final /a/ and the initial /e/ is not applied. In line 18, the passage should be emended to *phaladvayasyaikapañcacchedasya*.

becomes twenty-two thousand eight hundred. In the other "side", by means of the product of the "reciprocal" multiplication which consists of twenty-two thousand eight hundred, when there is its division, the quotient is the time regarding the measure-value, one month: 1.

The author enunciates a sample problem with fractions:

109. If the interest[1616] on hundred plus one half in one-third of a month is two and a half, tell [me] what will [the interest] on twenty less one-fourth in eight and one-fourth months be?

The presentation:[1617]

1	8
3	1
100	4
1	20
2	01
2	4
1	
2	

. One-third[1618] remains exactly as it is. By means of [the procedure] "multiplied by the denominator"[1619] and so forth, below it becomes two hundred one whose denominator is two:

1
3
201
2

. In the same way, by means of [the procedure] "multiplied by the denominator"[1620] and so forth, when there is the simplification of "two and a half", it becomes five whose denominator is two. This [fraction] should be joined below two hundred one whose denominator is two:

1
3
201
2
5
2

. Then, in the second "side"[1621]

1616 *phala*.

1617 I have emended the layout (line 1) as in the edition the layout is horizontal but the numbers are written as if they ought to be vertical. In the introduction, the editor observes that in the manuscript some of the *nyāsa*s were horizontal, but he does not specified where. It may be the case that either the latter scribe have deliberately changed some of the layouts into horizontal for space purposes or they were horizontal in the exemplar(s) for the same reason.

1618 The syntagm *ekas tribhāga* is curious, to express "one-third" *tribhāga* would be sufficient.

1619 This is a quotation from the rule of the class of fractional increase (GT 57).

1620 Ibid.

1621 This remark demonstrates that originally the layout (line 1) must have been subdivided into two sides, each one presenting the quantities "time", "interest", and "capital". In the

by means of [the procedure] "multiplied by the denominator"[1622] and so forth, eight is multiplied by four and [the result] plus one becomes thirty-three whose denominator is four: $\begin{array}{|r|}33\\4\end{array}$. And underneath, because there is the class of fractional decrease, when there is the subtraction of one from twenty multiplied by four, it becomes seventy-nine whose denominator is four: $\begin{array}{|r|}79\\4\end{array}$. When in the second "side" the simplification has also occurred, **having moved the fruit**, which consists of five whose denominator is two, **to the other side,** it is: $\begin{array}{|r|}33\\4\\79\\4\\5\\2\end{array}$. Having therefore carried out the transition of all the denominators in both sides, it is: $\begin{array}{|r|r|}1&33\\4&3\\201&79\\4&2\\&5\\&2\end{array}$. Having made the transition also of the denominator two below the five in the "side" of the first quantities, it is: $\begin{array}{|r|}1\\4\\201\\4\\2\end{array}$. In the first "side", four multiplied by one is exactly so; two hundred one multiplied by four becomes eight hundred four. It is multiplied by four becoming thirty-two hundred sixteen, and this multiplied by two becomes six thousand four hundred thirty-two. When there is its reduction by half, it becomes thirty-two hundred sixteen; this quantity should be understood to be the divisor. Regarding the second "side", thirty-three multiplied by three becomes ninety-nine; seventy-nine is multiplied by it becoming seventy-eight hundred twenty-one. This multiplied by two becomes fifteen thousand six hundred forty-two,

edition, the layout presents a darker line in the middle, which appears to mark a subdivision. In the SGT, the unusual syntagm *dvipakṣa* (note the use of the cardinal) denoting the "second side" instead of the expected form *dvitīyapakṣa* (see the ordinal) is often found up to SGT 117.

1622 This is a quotation from the rule of the class of fractional increase (GT 57).

and this multiplied by five becomes seventy-eight thousand two hundred ten. When there is its reduction by two, it becomes thirty-nine thousand one hundred five. When there is the division of this which is the quantity-dividend by the previously mentioned divisor thirty-two and so forth, the quotient is twelve: 12; the above remainder is five hundred thirteen and below is thirty-two hundred sixteen: $\left|\begin{array}{r}513\\3216\end{array}\right|$. When there is their reduction by three, above is one hundred seventy-one and below is one thousand seventy-two: $\left|\begin{array}{r}171\\1072\end{array}\right|$.

As before, in this case too the calculation [has been executed] because of the unknown "fruit".[1623]

The author enunciates a third sample problem:

110. If three workers earn twenty *paṇa*s in two days,[1624] tell me o learned one,[1625] what in five days will eight men earn?

The presentation: $\left|\begin{array}{lr}\text{di} & 2\\ \text{ka} & 3\\ \text{pa} & 20\end{array}\right|\left|\begin{array}{lr}\text{di} & 5\\ \text{ka} & 8\\ & 0\end{array}\right|$. In this case, **having moved** twenty to the "other side", it is: $\left|\begin{array}{r}5\\8\\20\end{array}\right|$. Because the denominator one occurs all over, there is no the exchange [of the denominators].[1626] "What is the fruit as well?"[1627] [This is the information which the sample problem requires to be ascertained], there is no exchange of the denominators. Then, in the first "side", three multiplied by two becomes six. In the second "side", eight multiplied by five becomes

1623 The commentator underlines that the numerical data given in the sample problem have been carried out in order to arrive at the unknown second "fruit".

1624 *vāsara*.

1625 *paṇḍita*.

1626 Line 2: *vyatyayaḥ*] em. This sentence is repeatedly used (at times *vinimaya* replaces *vyatyaya*, and there is an anomalous locative construction) by the SGT while showing the execution of some of the sample problems on the rules of five, seven, nine, and eleven. See SGT 111, 12–13; SGT ad TŚ 51, 2–3; SGT ad TŚ 52, 12; SGT ad L 87,7 where however the vocalic assimilation between *sarvatra* and *ekasya* takes place. This passage is also found in his explanation of the rule of barter (*bhāṇḍapratibhāṇḍa*, see SGT 114, 11). The commentator emphasises that the step "after having performed the procedure of the exchange of the denominators" formulated by Śrīpati in GT 107 is not performed in this case, since these are but whole numbers. The SGT further emphasises that these integers have 1 as denominator (cf. GT 35); see, for instance, SGT ad TŚ 52, 12.

1627 The text is suspect here, this remark occurs abruptly, and the last passage on the exchange of the denominators is redundant. In the second side, the zero represents the second "fruit" (*phala*), the quantity to be determined. Note that this is not found in the layouts of the following sample problems.

forty; twenty is multiplied by it becoming eighty. When there is its division by six, the quotient is one hundred thirty-three *paṇa*s: 133; the remainder above is two [and] below is six. When there is their reduction by half, it is one and three respectively: $\begin{array}{|c|}1\\3\end{array}$.

The author enunciates a sample problem, which is the fourth, regarding grain:

111. O learned one, if eight mānikas of śalī [rice][1628] are carried over one yojana at the cost of six paṇas, tell [me] the cost to bring sixty-three māṇikas [of rice] over six multiplied by three krośas.

The presentation: $\begin{array}{|lr||lr|}\text{m} & 8 & & 63\\ \text{kro} & 4 & \text{kro} & 18\\ \text{pa} & 6 & & 0\end{array}$. In this case, six is **the fruit**; **having moved it to the other side,** it is: $\begin{array}{|c|}63\\18\\6\end{array}$. Because the denominator one occurs all over, there is no the exchange [of the denominators]. Then, in the first "side", four multiplied by eight becomes thirty-two; in the second "side", eighteen multiplied by sixty-three becomes eleven hundred thirty-four; this multiplied by six becomes six thousand eight hundred four. When there is its division by thirty-two, the quotient is the integer two hundred twelve: $|212|$; the remainder above is twenty and below is thirty-two. When there is their reduction by four, it is five and eight respectively: $\begin{array}{|c|}5\\8\end{array}$.

In this way, the rule of five is concluded.

[Here] by means of the definition[1629] regarding the rule of five such as **having moved the fruit to the other side**[1630] and so forth, sample problems on the rules of seven, nine, and eleven take place.[1631] Thus, the *Līlavātī* [states in this regard]:

L 82: "In the rules of five, seven, nine, eleven, or more terms, transfer the fruit and the divisor to the other side. When the product of the

1628 It refers to a rice variety.

1629 *lakṣaṇa*.

1630 This is a passage from the rule of five (GT 107).

1631 *jāyante* – lit. "are grown, developed". This term highlights the relationship between the rule of five and the rules of seven, nine, and eleven; while solving the sample problems on the rule of seven, nine, and eleven the SGT quotes, in fact, steps form the rule of five formulated in GT 107. The SGT quotes this rule from the L because it explicitly mentions the rules of seven, nine, and eleven.

multiplication of the larger quantity is divided by the product of the smaller quantity,[1632] the fruit occurs".

The author enunciates a sample problem regarding the rule of seven:[1633]

> TŚ 51: "If a piece of cloth two in breadth [and] eighth in length is obtained by ten, tell [me] how much will two others [pieces of cloth] three and nine in length cost"[1634]?

The presentation:[1635] $\begin{array}{|r|r|} 1 & 2 \\ 2 & 3 \\ 8 & 9 \\ 10 & \end{array}$. In one side, there are four quantities; in the other, three [quantities]. Then, **having moved the fruit** ten **to the other side,** it is:[1636] $\begin{array}{|r|} 2 \\ 3 \\ 9 \\ 10 \end{array}$. Since the denominator one occurs all over, there is no exchange [of the denominators]. Then, in the first "side" two multiplied by one is the very two, eight multiplied by two becomes sixteen; in the second "side", three multiplied by two becomes six, nine multiplied by six becomes fifty-four, ten is multiplied by it becoming five hundred forty. When there is its division by sixteen, the quotient is the whole number thirty-three: 33; the remainder above is twelve [and] below is sixteen: $\begin{array}{|r|} 12 \\ 16 \end{array}$. When there is the reduction of these by four, it gives three and four respectively: $\begin{array}{|r|} 3 \\ 4 \end{array}$.

In this way, the rule of seven is concluded.

1632 *svalparāśi*.

1633 This and the following sample problem correspond to TŚ 51 and 52. Sinha (1982, 130) and Kāpadīā (1937, 100) number these stanzas as they were authentic GT's sample problems.

1634 The present dual *āpnutaḥ* (from the verbal root *āp*-) is lit. "the two are obtained".

1635 I have emended the layout since in the edition it presents 11 quantities, while the commentator clearly states: "In one side, there are four quantities; in the other three [quantities]." The number 1 is, inconsistently, the denominator of some of the quantities, a similar pattern characterises the layout in line 2. In this procedure, there is no a clear reason for the number 1 to be the denominator of the quantities in the layout. Also, here I present the layout subdivided into two sides, according to the commentator's explanation; in the edition the layout is horizontal but with the numbers written as they ought to be vertical.

1636 I have emended the layout (line 2), which is horizontal and presenting the numerator 1 only below the first four quantities.

The author enunciates a sample problem regarding the rule of nine:

> TŚ 52: "The length[1637] is 9, the width[1638] is 5, [and] the thickness[1639] is 1. If a stone of nine, five, and one *hasta*s is obtained by eight, [how much] two other stones of ten, seven, and two *hasta*s [in length, width, and thickness respectively] [will cost]?"[1640]

[The presentation]:[1641]

1	2
9	10
5	7
1	2
8	

. In this case, **the fruit** is eight; **having moved** it **to the other side** it is:[1642]

2
10
7
8

. Among all [the numbers], the very [number] one is the denominator, thus there is no transition of the denominators. Then, in the first "side",[1643] when there is the mutual multiplication it becomes forty-five. In the second "side", when there is the mutual multiplication up to eight (*aṣṭaparyantam*), it becomes two thousand two hundred and forty. When there is its division by forty-five, the quotient is forty-nine; the remainder above is thirty-five [and] below is forty-five. When there is the reduction of these by five, the quotient is seven and nine respectively:

45
7
9

.

[In this respect, in the *Līlāvatī* there is the following sample problem]:[1644]

> L 86: "O friend, if thirty pieces of cloth[1645] measuring twelve[1646] *aṅgula*s in thickness, the square of four *aṅgula*s in breadth[1647]

1637 *āyāma*.
1638 *vyāsa*.
1639 *piṇḍa*.
1640 The edition of the TŚ by Dvivedī reads slightly differently.
1641 In line 11, the term *nyāsa* is missing. I have emended the layout to present a layout with two sides. In the edition, the layout gives the numbers in one only horizontal side but the numbers are written as they should be vertical.
1642 I present a vertical layout; in line 11, the layout is horizontal.
1643 Line 12: *prākpakṣe*] em.
1644 The commentator supplies a sample problem which corresponds to L 86; no introductory line is given.
1645 *paṭṭa*.
1646 The L uses the *bhūtasaṃkhyā* notation, according to which the term *arka* (lit. "sun") denotes the number twelve.
1647 *vistṛti*.

and fourteen *hasta*s in length[1648] cost one hundred, tell me what will be the cost for fourteen pieces of cloth whose breadth, length and thickness are each four less [than those mentioned]?"

The presentation:[1649]

12	8
16	12
14	10
30	14
100	

. Here **the fruit** is one hundred; **having moved** it **to the other side,** it is:

8
12
10
14
100

. In the second "side", when there is the mutual multiplication, it becomes thirteen hundred thousand forty-four thousand: 1344000; in the first "side", when there is the mutual multiplication up to thirty, it becomes eight thousand six thousand forty: 80640. Then, according to the instruction "one should divide the [product of] side of the larger quantity by the other",[1650] **one should divide** the "larger number" thirteen hundred thousand and so forth by the smaller number eighty thousand and so forth. The quotient is **the cost**, which is sixteen *dramma*s; the remainder above is thirty-five thousand seven hundred sixty, and eighty thousand and so forth is below:

35760
80640

. When there is their reduction by twenty-six thousand eighty-eight hundred, the quotient above is two and below is three:[1651]

2
3

.

The author enunciates a sample problem regarding the rule of eleven:[1652]

L 87: "If it costs eight *dramma*s to hire carts[1653] to transport pieces of cloth whose [number and] measures are as previously mentioned [in the sample problem] to a distance of one *gavyūti*, tell [me] what should be

1648 *dīrgha.*

1649 It is noteworthy that in the edition, layouts, especially those with large amount of data, are inconsistent in the way numbers are shown; cf. SGT ad TŚ 51, 1 and SGT ad L 86, 20.

1650 This is a quotation from TŚ 31, which is the *karaṇasūtra* on the rules of five, seven, and nine; one finds the same sentence in the commentator's explanation of the next sample problem (see SGT ad L 87, 10). Notably, the commentator quotes the TŚ while explaining a sample problem from the L.

1651 In this passage, differently from the preceding and the following, the exchange of the denominators is not mentioned.

1652 The commentator supplies a sample problem which corresponds to L 87.

1653 *śakaṭa.*

to hire [carts] in case of the other cloths which were mentioned after those[1654] and whose measures were less by four than those [said above] [and to transport them] over a distance of six *gavyūtis*?"

[The presentation:][1655]

12	8
16	12
14	10
30	14
1	6
8	

. **Having moved the fruit**, which here is eight, **to the other side**, which is the second, it is:

8
12
10
14
6
8

. In brief, since the denominator one occurs all over, there is no exchange[1656] [of the denominators]. Therefore, in the first "side", when there is the mutual "multiplication"[1657] from twelve up to one, it becomes eighty thousand six hundred forty.[1658] In the second "side", when there is the mutual "multiplication", it becomes six hundred thousand forty-five thousand one hundred twenty.[1659] Then, according to the instruction "one should divide the [product of the] side of the larger quantity by the other",[1660] when there is the division of the number beginning with six hundred thousand by eight thousand and so forth, it is:

645120
80640

. The quotient is eight *drammas*: 8.

Regarding the rules of five quantities and so forth, [it can be observed that] [in the layout] at the beginning the first "side" has one quantity more [than the second side]; afterwards, by placing the fruit elsewhere,[1661] the second "side" has [instead] a quantity more [than the first side].[1662]

1654 In brief, the pieces of cloth mentioned after the 30 pieces of cloth, it refers to the 14 pieces of cloth mentioned in the previous sample problem.

1655 In line 6, the term *nyāsa* is missing. I have emended the layout since the edition presents a horizontal layout having the numbers written in one only side but as these ought to be, in fact, vertical.

1656 Line 7: *vinimaya*] em.; *iti tātparyam*] em.; *tātparyam* (K).

1657 See *ghāta* in the rule of five (GT 107).

1658 Lines 8: *jātā*] em.

1659 Line 9: *jātāḥ*] em.

1660 This is a quotation from TŚ 31.

1661 *paratra* – here it refers to the second side.

1662 The commentator emphasises that, according to the rule involved, the quantities are either five, seven, nine or eleven in total, with the first side having a quantity more which

In this way, the rule of eleven is concluded.

3.12 Barter (*bhāṇḍapratibhāṇḍa*)[1663] [GT $\overline{112}$–117]

A [rule] depending on the rule of five [is enunciated].[1664] By means of a commodity (*bhāṇḍa>vastu*), the acquisition[1665] of an exchanged commodity[1666] which is a second good is denoted as "barter" (*bhāṇḍapratibhāṇḍa*).[1667] [Next is] the procedural rule explaining such an exchange (i.e., of goods for goods); the author enunciates half a metrical stanza:

> $\overline{112}$. After having performed the exchange[1668] of the prices, the rule of five is to be carried out.

The explanation: there is the quantity of the "price" placed in the first "side" [of the layout]; regarding the second "side", there is the quantity of the "price" placed in the second "side" [too]. According to the "exchange of the prices", it is moved by means of the [exchange with the price of the] first side. With this, it is said that the exchange of the denominators should also be carried out.[1669] As before, according to [the rule] "having moved [the fruit] to the other side"[1670] and so forth, both the procedures of multiplication [first] and division [after] [should be carried out].

In the same way, in the *Līlāvatī* it is said: "In the same way, in [the rule of] barter the same procedure again: having exchanged (*viparyasya*) the divisors and the prices [...]".[1671]

is the "fruit". After the step "having moved the fruit to the other side", the reverse case takes place. It can be observed that Siṃhatilakasūri's pedagogical style always encourages students to put attention to the actual manipulations and to the way numbers are arranged in the notational layout during each phase of the algorithm.

1663 K 80, 16–17.

1664 This remark deliberately echoes Śrīpati in line 16. Terms from the rule of five are, in fact, quoted by the SGT (see terms such as *ānayana*, *ghāta*, and *pakṣa*) while solving the following sample problems on barter.

1665 *grahaṇa*.

1666 *pratibhāṇḍa*.

1667 The commentator gives a definition of "barter"; *bhāṇḍapratibhāṇḍa* is lit. "goods for goods".

1668 *vinimaya*.

1669 When there are fractions, the exchange of the denominators (see the rule of five in GT 107) occurs.

1670 This is a quotation from the rule of five (GT 107).

1671 It appears that Siṃhatilakasūri quotes L 88 to mention the exchange of the denominators, which is not specified in Śrīpati's rule. However, Kāpadīā (1937, 80 f6) observes that there is a different reading of the quoted L. The following is found in the edition by Āpaṭe: *viparyayas tatra sadā hi mūlye*; this passage does not mention the exchange of the denominators.

In this respect, the author enunciates a sample problem:

> 113. O mathematician, if sixteen mangoes[1672] can be obtained by one *paṇa* and one hundred pomegranates[1673] by three *paṇa*s, tell [me] how many pomegranates can be obtained when they are exchanged[1674] with twelve mangoes?

The presentation: $\begin{array}{|c|c|} 1 & 3 \\ 16 & 100 \\ 12 & \end{array}$. In this case, when there is **the exchange of the prices** one and three, and when there is the transition[1675] of the "fruit" twelve to the other "side", it is: $\begin{array}{|c|c|} 3 & 1 \\ 16 & 100 \\ & 12 \end{array}$. In the first "side", when there is the reciprocal "multiplication", it is forty-eight; in the second "side", when there is the reciprocal "multiplication", it becomes twelve hundred. When there its division by forty-eight, the quotient is twenty-five "mangoes": 25.

The author enunciates a second sample problem:

> 114. If two *pala*s of agarwood[1676] can be obtained by six [and] one *pala* of musk[1677] [can be obtained] by nine, tell [me] how much musk can be obtained by [exchanging it with] seven *pala*s of agarwood?

The presentation: $\begin{array}{|c|c|} 6 & 9 \\ 2 & 1 \\ 7 & \end{array}$. In this case, when there is **the exchange of the prices** six and nine [and] when there is the transition of the "fruit" seven to the other "side", it is: $\begin{array}{|c|c|} 9 & 6 \\ 2 & 1 \\ & 7 \end{array}$. Since the denominator one occurs all over, there is no exchange [of the denominators].[1678] In the first "side", when there is the mutual "multiplication", it becomes eighteen; in the second "side", when there is the mutual multiplication up to seven, it is forty-two. When there is its division by eighteen, the quotient is two *pala*s; the remainder above is six

1672 *sahakāra*.
1673 *dāḍima*.
1674 By *vinimayavidhinā*, the author emphasises that this problem is to be solved by applying the enunciated procedure on barter.
1675 *ānayana* – see *ānīya* or "having moved" in the rule of five (GT 107).
1676 *aguru*.
1677 *kuraṅganābhi*.
1678 Line 11: *vinimayaḥ*] em.; **phalam**] em.

and below is eighteen. When there is their reduction by six, one is above and three is below: $\begin{array}{|c|}1\\3\end{array}$.

The [procedure regarding the] exchange of goods[1679] is thus concluded.

3.13 The sale of living beings (*jīvavikraya*)[1680] [GT $\overline{115}$–117]

[Next is] procedural rule of the sale of living beings,[1681] which depends on the very rule of five;[1682] the author enunciates half a metrical stanza:

> $\overline{115}$. In the rule concerning the sale of living beings, when there is the exchange of the ages, once they are arranged[1683] [the rest of the procedure] is here again[1684] as before.[1685]

The explanation: here **the exchange of the ages** of the two living beings should be carried out. The remaining [part of the] procedure is [the following]: "having moved" all [the quantities] "to the other side"[1686] and so forth, this should be performed as before.

In this respect, the author enunciates a sample problem:

116. Tell [me] in case you have trained yourself in mathematics, if a woman[1687] twice [times] eight years old deserves seventy, what another [woman] twenty years old, having the same beauty and appearance, would deserve?

The presentation:[1688] $\begin{array}{|c|}1\\16\\70\end{array}\begin{array}{|c|}1\\20\\ \end{array}$. In this case, when there is **the exchange of the ages**, which are sixteen and twenty, and in the same way when there is the

1679 *bhāṇḍavinimaya*.
1680 K 81, 16–17.
1681 *jīva* is "living being", while *vikraya* is "sale, selling".
1682 The commentator elucidates that, as the rule of barter, this procedure too depends upon the rule of five. Terms from the rule of five are, in fact, quoted by the SGT (see *ānayana*, *ghāta*, and *pakṣa*) while solving the following sample problems.
1683 In lines 16–17, it should be *vayavyatyatye*. The same problematic *vayovyatyatye* is found in the SGT two lines below (line 18), but in the next sample problems the commentator gives a different construction and analyses the compound as a genitive *tatpuruṣa*.
1684 The conjunctive adverb *pūnar* emphasises that, as previously in the mentioned rule of barter, here (note the use of *atra*) too "in the rule concerning the sale of living beings" the rule of five is to be applied. Note that this rule and the rule of barter are each written in half *rathoddatā* meter.
1685 In brief, as explained in the rule of five.
1686 This is a quotation form the rule of five (GT 107). By *sarve'pi*, the commentator underlines that all the quantities are involved in the step *ānīya pakṣam aparaṃ*.
1687 *strī*.
1688 I have emended the layout (see line 25) as there is a zero at the bottom of the second side, and this does not occur in the other layouts where the second side presents a quantity less.

transition of the "fruit" seventy to the other "side", it is: $\begin{array}{|r|r|} 1 & 1 \\ 20 & 16 \\ & 70 \end{array}$. In the first "side",[1689] when there is the reciprocal "multiplication", it is twenty; in the second "side", when there is the reciprocal "multiplication", it becomes eleven hundred and twenty. When there is its division by twenty, the quotient is the price fifty-six: 56. In this case, when the age increases the price conversely diminishes.[1690]

The author enunciates a second sample problem:

> 117. If three camels[1691] ten years old are obtained by one hundred eight, tell [me] quickly what eight camels nine years old, having the same beauty, are obtained for?

The presentation: $\begin{array}{|r|r|} 3 & 8 \\ 10 & 9 \\ 108 & \end{array}$. Here when there is **the exchange** of the "ages", which are ten and nine,[1692] and when there is the transition of one hundred eight to the other "side", it is: $\begin{array}{|r|r|} 3 & 8 \\ 9 & 10 \\ & 108 \end{array}$. In the first "side", when there is the reciprocal "multiplication", it is twenty-seven; in the second "side", when there is the mutual multiplication, it is eight thousand six hundred forty. When there is its division by twenty-seven, the quotient is three hundred twenty: 320.

In the *Līlāvatī*, the sale of living beings is mentioned with respect to the inverse rule of three by [the expression]: "When the cost of living beings is [determined] by their age" and so forth.[1693] [In the *Līlāvatī*] this very one comes as the first topic next to it (i.e., to the inverse rule of three).[1694]

Afterwards, it should be multiplied by the "requisition" eight.[1695] To illustrate, this very sample problem is thus [explained by this very procedure]:

1689 Line 26: *prākpakṣe*] em.; *prāk* (K).

1690 The SGT paraphrases L 77, which has already been quoted while explaining the inverse rule of three. In this regard, see the commentator's remark on the next sample problem where he mentions L 78 (SGT 117, 9–10).

1691 *uṣṭra*.

1692 Line 6: **ākhya**] em.

1693 This is a quotation from L 78; this passage has been also quoted in SGT 95, 20–21.

1694 The SGT underlines that that the rule of sale of living beings is not separately explained in the L but it is given straight after the inverse rule of three. The commentator quotes this rule because he intends to demonstrate (see the other *nyāsa* in line 12) how to solve a similar sample problem by means of this very procedure.

1695 This remark is to be understood in the light of what the SGT explains further in line 15.

When the first [camel] is ten years old and worthy of thirty-six, which is one-third of one hundred and eight,[1696] how much will then a nine years old [camel] be worthy of?

The presentation: |10|36|9|. In this case, thirty-six multiplied by the "measure" ten becomes three hundred sixty. When there is its division by nine, which is the "requisition", the quotient is forty. The price of that [camel] of nine years old is arrived at by means of the inverse rule of three. Then, it is multiplied by its own "requisition" such as eight; in this case, forty multiplied by eight becomes three hundred twenty. In this way, here it should be understood.

Also, in case of division with fractions, having reduced the price of the first by means of the class of fractional increase, having multiplied [the result] by its own "requisition", such as five and eight, when there is the division by the denominator below, the "fruit" is obtained, as in the rule of five.[1697]

In this way, the [procedure of the] sale of living beings is concluded.

3.14 Practices [GT 118–133]

Practice of mixture (miśravyavahāra)[1698] *[GT 118]*

[Next is] the procedural rule of the practice of mixture;[1699] the author enunciates a metrical stanza:

118. One should make the quantity-measure[1700] multiplied by its own time[1701] and the other time multiplied by its interest-rate.[1702] These [products] are multiplied by the mixed quantity[1703] and divided by their sum.[1704] They (i.e.,

1696 The commentator observes that the cost of one camel is 36, Śrīpati gives the price of 3 camels as 108.

1697 The SGT refers to calculation with fractions. The number 8 represents the *svecchā* 8 mentioned in line 15. The expression *pañcarāśikavat* might refer to the sample problem formulated in GT 109, which is a sample problem on the rule of five involving fractions; there the *phala* is $\left|\begin{matrix}5\\2\end{matrix}\right|$ (see lines 7–8). This is probably the reason why the SGT refers to the "requisition" 8 (see "eight camels") and 5 (see the numerator of the aforementioned fraction $\left|\begin{matrix}5\\2\end{matrix}\right|$).

1698 K 82, 20–23.

1699 *miśravyavahāra*. Here begins the treatment of rules on "practices" (*vyavahāra*).

1700 *pramāṇarāśi* – it denotes the value 100 as whole with respect to a given percentage-rate.

1701 *nijakāla*.

1702 *phala*. According to the context, *phala, vṛddhi*, and *vyāja* denote the "interest-rate" or the "accrued interest".

1703 *vimiśra* – here it represents the mixed quantity composed of the "capital plus the accrued interest".

1704 In lines 26–27, the commentary explains the meaning of "their sum".

the two quotients) become the capital (*mūla*) and the accrued interest[1705] respectively.

The explanation: **one should make the quantity-measure,** namely one hundred,[1706] **multiplied** (*hata>guṇita*) by "its own time", such as one month. **One should make the other time,** namely twelve months,[1707] **multiplied by its interest-rate,** that is to say, multiplied by the "interest-rate" five, which is the monthly interest-rate.[1708] Then there is "their sum", that is, [**the sum** of the products] of the "quantity-measure", namely one hundred, "multiplied" by "its own time", and of the "interest-rate" "multiplied" by the "other time",[1709] thus is the connection [of words with their meanings].[1710] [The two products are] **divided** (*hṛta>bhakta*) by that [result of the sum]. In this way, [the result of] "their sum" should be placed apart. The previous product[1711] of the "quantity-measure" [multiplied by its time] as well as the quantity of the "other time" [multiplied by the interest-rate] are placed (i.e., written) as they are, and these [two products] **are multiplied** (*nighna>guṇita*) by the "mixed quantity",[1712] which is the total amount (*samuccaya*) of the original [quantity of] *dramma*s and the accrued interest,[1713] namely ninety-six. [These products are] divided by the previous arisen (*prāgudita*) [result] of "their sum". **They (i.e., the two quotients) become the capital and the accrued interest respectively:**[1714] when the number (i.e., the product of the quantity-measure multiplied by the mixed quantity) is divided by "their sum", the [number of] *dramma*s of the "capital" occur; when the quantity of the "other time" is divided by "their sum", it becomes the "interest which has accrued", thus is the connection.[1715]

1705 *kalāntara.*

1706 Line 24: *śatādīm*] em.

1707 Line 25: °*māsādim*] em. The commentary mentions the mathematical data of the next sample problem.

1708 *ekamāsikavyāja.*

1709 Lines 26–27: *phalasya ca tāḍitasya parakālena iti*] conj. The transmitted reading is not interpretable; one would expect the main clause, which is a noun phrase with no copula, to be followed by two dependent clauses having the same construction and connected by a coordinate conjunction.

1710 The commentary expands upon *nijayoga* or "their sum".

1711 Lines 27–28: *prāgguṇita*] em.

1712 *miśra* – the SGT shows that *miśra* and *vimiśra* are synonyms.

1713 In line 28, *vyājamaulikyadrammasamuccayena* explains the term *miśra* as the sum total of *vyāja/phala* and *dramma/pramāṇa*, this is also clarified by the commentator in the next sample problem. The adjective *maulikya* (lit. "original") qualifies *dramma* and denotes the "original" capital; further in line 2, the SGT uses the expression *mūladramma*.

1714 In line 1, the *daṇḍa* should precede and not follow *mūlakalāntare*.

1715 The SGT explains that the two quotients have been obtained by the division of the two products – the division of the *pramāṇarāśi* by the *miśraka* and of the *parakālarāśi* by the *miśraka*.

In this regard, the author enunciates a sample problem:

> 119. O friend, at the [monthly] rate of 5 per cent [per month] the sum of the capital and the interest (*phala*) which has accrued in a year[1716] is ninety-six. What are the capital [and] the accrued interest (*kalāntara*)?

The presentation: |1||100||5||12||96|. The "quantity-measure" one hundred is "multiplied" by "its own time", which is one; it remains exactly so. The quantity of the "other time", which is twelve, is "multiplied" by the "interest-rate", which is five; it becomes sixty. Then, with respect to these [numbers], when there is the "sum" of one hundred sixty, it is one hundred sixty, which is placed apart. Next, the "quantity-measure" one hundred is multiplied by ninety-six; it becomes ninety-six hundred: 9600. In the same way, the quantity of the "other time" sixty is multiplied by ninety-six; it becomes five thousand seven hundred sixty: 5760. Regarding these [products],[1717] when there is the division[1718] of each[1719] by one hundred sixty, which is "their sum",[1720] the quotients are **the capital** sixty and **the accrued interest** thirty-six respectively. When there is their sum, it is ninety-six [which is the *miśraka*]. In this [same] way it should be understood for fractions too.[1721]

3.15 Commission to the moneylender (*vyājopajīvivṛtti*)[1722] [GT 120]

Next is the procedural rule whose topic is the commission (*vṛtti*)[1723] [paid] to the moneylender;[1724] the author enunciates a metrical stanza:

> 120. The quantity-measure is multiplied by its own time,[1725] [the other quantities such as] the interest-rate and so forth[1726] (*phalādi*) are multiplied by the elapsed time (*vyatītakāla*). These [products] are multiplied by the mixed quantity and divided by their sum. These consecutively become the capital and so forth.[1727]

1716 *abda*.
1717 Line 12: * *pṛthak sthāpitam* | *] conj. This appears a case of scribal harmonization to parallels (see line 9).
1718 Line 12: *bhāge*] em.; *bhāgena* (K).
1719 *krameṇa*, "one by one".
1720 It refers to the sum of the two initial products (see lines 7–9); see *nijayoga* in the *karaṇasūtra*.
1721 This final remark underlines that, whenever there are fractions, the procedure is exactly the same.
1722 K 83, 15–18.
1723 I render *vṛttī* as "fee"; see SaKHYa (2009, 61) and a similar sample problem in GSK 3.2.
1724 *vyājopajīvin* – lit. "the one who lives on (*upajīvin*) moneylending (*vyāja*)". Sinha (1982, 132) translates the whole expression as "usury". In Sanskrit literature, another term used to denote the "usurer" is *vṛddhyupajīvin*.
1725 *nijakāla*.
1726 In line 20, the commentator explains *ādi*; notably, in GSK 3.2 *phalādi* is also found (Apabhraṃśa *phalāīṇi*).
1727 *mūlādayas*; in this way, all the other quantities are arrived at by manipulating frist the *miśraka*.

The explanation: as before,[1728] **the quantity-measure is multiplied by its own time**; [with respect to] the [expression] **[the other quantities such as the] interest-rate and so forth,**[1729] **the interest-rate** (*phala>vyāja*) is the mentioned [monthly interest] five.[1730] Because of the word **"and so forth"** (*ādi*), the "commission of surety" (*bhāvyaka*)[1731] and so forth are meant. These **are multiplied** (*hata>guṇita*) **by the elapsed time,** which is twelve months. **These [products] are multiplied by the mixed quantity,**[1732] that is to say, they are multiplied by the mixed amount which is the full combined amount of wealth.[1733] **[The products obtained are] divided** (*vihṛta>dattabhāga*)[1734] **by their sum,** that is to say, **by the sum** of the [result of the] "interest and so forth" added to the quantity-measure. **These consecutively become the capital and so forth,** thus is the connection [of word with their meanings].

In this respect, the author enunciates a sample problem:

> 121. O learned one, the interest-rate on one hundred is five per month, [there is the fee] for the commission of surety (*bhāvyaka*), half a *dramma* is for the fee[1735] [paid to another professional] and likewise one-fourth is [the charge] for the scribe.[1736] In twelve months, the mixed quantity[1737] is nine hundred and five. If you have proficiency in the practice of mixture,[1738] tell [me] o friend, what are the [other quantities such as the] capital and so forth?

The presentation: $\begin{array}{|c|c|c|c|c|c|c|c|} 1 & 100 & 5 & 1 & 1 & 1 & 12 & 905 \\ 1 & 1 & 1 & 1 & 2 & 4 & 1 & 1 \end{array}$. Here the first four numbers have one as denominator. The "quantity-measure" one hundred is "multiplied" (*nighna>guṇita*) by "its own time" one month; it gives the very one hundred. The "interest-rate" five "multiplied" (*hata>guṇita*) **by the elapsed time** twelve becomes sixty. In the three places, [all the numbers] one multiplied by twelve become a group of three twelves respectively: $\begin{array}{|c|c|c|c|c|} 100 & 60 & 12 & 12 & 12 \\ 1 & 1 & 1 & 2 & 4 \end{array}$. Regarding

1728 Briefly "as in the previous *karaṇasūtra*", which is GT 118.
1729 The commentator expands upon *phalādi*.
1730 This interest refers to the sample problem below.
1731 The editor Kāpadīā unreasonably places a question mark after *bhāvyaka*, probably because, unfamiliar with this technical term, he found it ambiguous; the same anomaly occurs in SGT 121, 24. Note that GSK 3.2 gives the procedural rule of *bhāvyaka*. Also, see the use of *bhāvyaka* with the same meaning and in the same context in PG example 54; it is translated as "commission of the surety" by Shukla (1959, 32). See also SaKHYa (2009, 61).
1732 *miśra*.
1733 In line 21, the commentator specifies that by *miśra* "the full amount of combined wealth (*sarvamilitadravya*)" is meant.
1734 Line 22: *vihṛtā*] em.
1735 *vṛtti*.
1736 *lekhaka*.
1737 *miśraka*.
1738 *miśrakavyavahṛti*.

these [obtained results], there is “their sum”: when there is the sum of twelve whose denominator is one and one hundred sixty, it becomes one hundred seventy-two whose denominator is one: $\begin{array}{|r|}172\\1\end{array}$. For the purpose of its sum together with twelve whose denominator is two, when there is transposition of the denominators, it is: $\begin{array}{|r|r|}172&12\\1&2\\2&1\end{array}$. The first number multiplied by two becomes three hundred forty-four, one multiplied by two is the very two; the next quantity multiplied by one is exactly so. As there are equal denominators, when the additive quantity twelve is added to three hundred forty-four, it becomes three hundred fifty-six whose denominator is two: $\begin{array}{|r|}356\\2\end{array}$. For the sake of the sum of this [fraction] and twelve whose denominator is four, when there is the reduction of the denominators by half, when there is the exchange, it is: $\begin{array}{|r||r|}356&12\\2&4\\2&1\end{array}$. The first number multiplied by two becomes seven hundred twelve; [two] multiplied by two becomes four. The next quantity multiplied by one is exactly so. Since there are equal denominators, when the additive quantity twelve is added to the first number, it becomes seven hundred twenty-four whose denominator is four: $\begin{array}{|r|}724\\4\end{array}$. This is namely “their sum”; it should be known as the quantity-divisor[1739] of all [the quantities] which will be mentioned further [and] which will be divided [by it].

To illustrate: one hundred, which has been in the meantime placed aside, multiplied by nine hundred five, becomes ninety thousand five hundred; this is the quantity-dividend whose denominator is one. Afterwards, when there is the exchange of the denominator and numerator of the previously mentioned seven hundred and so forth, which is the divisor, four is above and seven hundred twenty-four is below: $\begin{array}{|r|r|}4&90500\\724&1\end{array}$. When there is the reduction of four by four, it is one; when there is the reduction of seven hundred and so forth by four, it is one hundred eighty-one. Nine thousand and so forth multiplied by one is exactly so; one hundred eighty-one multiplied by one is exactly so. When there is the division of nine thousand and so forth by it, it is:[1740] $\begin{array}{|r|}90500\\181\end{array}$,

1739 Line 14: *bhāgadāyīrāśir*] em.

1740 In line 20, before the layout, *yathā* is missing.

the quotient is the capital[1741] five hundred: 500. Sixty is multiplied by nine hundred five,[1742] it becomes fifty-four thousand three hundred whose denominator is one. When there is its division by the previously mentioned one hundred eighty-one, the quotient is the accrued interest three hundred: 300.

The first twelve multiplied by nine hundred five becomes ten thousand eight hundred and sixty. When there is its division by the previously mentioned one hundred eighty-one, the quotient is the "commission of surety", which is sixty: 60.

Next is the second twelve whose denominator is two. Twelve is multiplied by nine hundred five, it becomes ten thousand eight hundred sixty. Then, when there is the exchange of the denominator and the numerator of the previously mentioned [divisor] seven hundred and so forth whose denominator is four, four is above [and] seven hundred and so forth is below. Then there is the "cross-reduction":[1743] when there is the reduction of four and two by half, two is instead of four, and one is in place of two: $\begin{array}{|r||r|} 2 & 10860 \\ 724 & 1 \end{array}$. For a second time, when there is the reduction of two by half, it is one, when there is the reduction of seven hundred and so forth by half, it is three hundred sixty-two. All these multiplied by one remain as they are. Afterwards, when there is the division of ten thousand and so forth by three hundred sixty-two, the quotient is the "fee" thirty: 30.

With respect to the third twelve whose denominator is four, this is multiplied by nine hundred five becoming ten thousand eight hundred sixty. When there is [its] division by seven hundred twenty-four, then[1744] [when there is the exchange of the operands and the reduction of] four by four, the two numbers above and underneath which were the denominators are erased (*apayāta*);[1745] the quotient concerning the [fee for the] scribe is fifteen:[1746] $\begin{array}{|r|} 15 \\ 1 \end{array}$.

Then, regarding these [results] beginning with five hundred,[1747] which are respectively $\begin{array}{|r|} 500 \\ 300 \\ 60 \\ 30 \\ 15 \end{array}$, when there is the sum, it becomes nine hundred five: 905.

1741 *mūladhana*.
1742 The other product 60 is multiplied by the *miśraka*.
1743 This is a quotation from the rule of division of fractions (GT 42).
1744 Line 2: *tato*] em.; *yato* (K).
1745 Line 3: **iti kṛtvā**] em.
1746 I have emended the text as, in the layout (line 4), the 1 of 15 is missing.
1747 Here *ādi* ("and so forth") should be understood in the light of *mūladyaṃ* used by Śrīpati in the sample problem and *mūlādayas* found in the *karaṇasūtra*.

3.16 Rule on interest [GT 122]

Next he introduces the [treatment of the] very interest, not the capital.[1748] Therefore, although this has been stated elsewhere, the author enunciates a procedural rule because it is useful:[1749]

> 122. Having made the wealth (*dravya*) multiplied by the [number of] months, having made again [the result] multiplied by the interest-rate (*vṛddhi*),[1750] and when the division by one hundred is performed, scholars [state that] the interest which accrues is arrived at.

This is clear [in meaning].

> 123. If one hundred *dramma*s become five [*dramma*s] [more] per month,[1751] tell [me] then what in twelve months will the interest be?

The presentation: |1|100|5||12|. Here **the wealth one hundred is multiplied by [the number of] months:** as before,[1752] one hundred multiplied by one is exactly so.[1753] Afterwards, it is **multiplied by** twelve **months** becoming twelve hundred. This is **multiplied by the interest-rate,** that is, it is "multiplied" by five becoming six thousand. When there is the division of this **by one hundred,** it is: $\left|\begin{array}{r}6000\\100\end{array}\right|$. With respect to the interest which will accrue [in twelve months], the quotient is sixty: 60. In this case, as many as one hundred [are the given *dramma*s] such are the *dramma*s [occurring as the accrued interest]; that is the principle.

3.17 Rule on time and double capital[1754] [GT 124]

The author enunciates a procedural rule mentioned in the text on arithmetic by Brahmagupta[1755] on the [following] topic: by what time [the capital], for

1748 In brief, in order to determine the interest; the capital is, in fact, given.

1749 The commentator emphasises that this rule has been stated in another work; the source, however, remains unidentified.

1750 *vṛddhi* – lit. "increase"; that *vṛddhi* is the "interest" is clear in SGT 132–133, 27 where the commentator glosses it with *vyāja*. This term shows the same meaning occurs in PG 47.

1751 It denotes the interest rate.

1752 Cf. the step mentioned in GT 120, where the *prāmaṇarāśi* is multiplied by the *nijakāla*.

1753 Line 13: *śatam tad eva*] em.; *śatam eva* (K).

1754 K 85, 8–9.

1755 It refers to the Hindu astronomer-mathematician Brahmagupta; the commentator specifies that this verse is a quotation from the BSS.

instance one hundred, constantly increasing by means of the interest-rate,[1756] such as five,[1757] would double?

> 124. The capital multiplied by the time, [the result] divided by the accrued interest,[1758] [and] multiplied by the multiplier lessened by one: [the result] is the time.[1759]

The explanation: **the capital,** which is two hundred,[1760] is **multiplied** by the "time", which is namely one month, [and the result is] **divided** by the "accrued interest" (*phala>vyāja*),[1761] for instance the mentioned [accrued interest] six.[1762] Then, that [obtained] quotient is the quantity which one wishes [to calculate] **"by what time will it become three times"?** In this case,[1763] the multiplier should be lessened by "one", [the previously obtained quotient is] **multiplied** by it, that is to say, "by the multiplier lessened by one"; **the time** [hence] arises. Thus is the connection.

In this respect, the author enunciates a sample problem:

> 125. If six *dramma*s are the monthly increase (i.e., the accrued interest) on two hundred, by what time will the lent[1764] money become three times?

The presentation: |1|200|6|gu3|. In this case, two hundred "multiplied" by the "time" one month is exactly so. When there is its division by six, which is the "accrued interest", the quotient is thirty-three; the remainder above is two and below is six. When there is the reduction of two and six by two, it is one and three: $\begin{array}{|c|}1\\3\end{array}$. [The result is] thirty-three [plus one-third]: $\begin{array}{|c|}33\\1\\3\end{array}$. The multiplier three is that which is "lessened by one";[1765] it becomes two. [Thirty-three plus one-third] is multiplied by it becoming sixty-six and two whose denominator is three: $\begin{array}{|c|}66\\2\\3\end{array}$. In brief, in this manner, three times two hundred becomes six hundred in five years plus six months and twenty days.

1756 *vyāja*.
1757 The capital 100 and the interest-rate 5 are mentioned in the previous sample problem.
1758 *phala*.
1759 This verse is from BSS 12.14.
1760 Line 20: *dviśatī*] em.; *śatādi* (K). The SGT refers to the sample problem below where the capital is 200.
1761 Lines 20–21: *vyājenokta*°] conj.
1762 See the following sample problem.
1763 Lines 21–22: *ādīty atra*] em.
1764 *prayukta,* it denotes the capital which is lent; the meaning of *prayukta* ("lent") is clear in GT 128.
1765 Line 28: **etat**] em.

[On this topic] the author enunciates a sample problem with fractions:

126. If in two months five *paṇa*s are the increase on twenty *paṇa*s, tell [me] by what time my wealth[1766] will be one and a half times [more than what it is now]?

The presentation:[1767] $|2||20||5|\begin{array}{|r|} 1 \\ gu\ 1 \\ 2 \end{array}$. In this case, twenty is "multiplied" by the "time" two months; it becomes forty. When there is its division by the "accrued interest" five, the quotient is eight. It is **[multiplied by] the multiplier lessened by one,** that is, when there is the subtraction[1768] of one from [the multiplier] one and a half, it is multiplied by [the resulting] half. Therefore, eight is multiplied by one; it is exactly so. When there is [its] division by two, the quotient is four months. In this way, in four months, twenty [*paṇa*s] are one and a half times [more]; in other words, they become thirty.

3.18 Conversion of several bonds into one (*ekapatrakaraṇa*)[1769] [GT 127]

[Next is] the procedural rule regarding the conversion of [several bonds of] one hundred and so forth, with the interest-rate, such as two, [and] with odd or even months, into a single bond.[1770]

The author enunciates a metrical stanza:

127. [With respect to an equivalent single bond] the average time[1771] occurs when the sum of the interest which has accrued during the elapsed time (*gatasamaya*) is divided by the sum of the [various] monthly profits (*māsavṛddhi*).[1772] In the procedure of conversion [of several bonds] into a single bond (*ekapatrīvidhāna*), when that representing the sum of the monthly

1766 *dhana*.

1767 I have emended the text since, with respect to the previous sample problem, here in the layout (line 6) the abbreviation *gu* is missing: the *guṇa* or "multiplier" is the last quantity one plus one-half.

1768 *lopa* – lit. "deprivation, loss"; in the SGT, this is the first occurrence of the term.

1769 K 86, 12–15.

1770 These data refer to the next sample problem, where there are four bonds having different rates of interest and time. Cf. the following expressions occurring in the GT and SGT: *ekapatrīvidhana*, *ekapatrīkaraṇa*, and *ekapatrakaraṇa*. SaKHYa (2009, 61; 134) translates *ekapatrīkaraṇa* as the "conversion of several bonds into one".

1771 The "average time" (*gatakāla*) denotes the single date on which the amount of money borrowed would be paid by one only payment.

1772 There is a monthly profit in relation to each bond.

profits (*māsalābha*)[1773] is multiplied by one hundred [and] divided by the sum of the [lent] capitals,[1774] the percentage rate (*śataphala*) should also occur.

The explanation: the "time" that is "elapsed", which will be mentioned further [in the sample problem], is seven months and so forth. With respect to it, at the interest-rate[1775] of two and forth, the "accrued interest" (*phala*) becomes fourteen and so forth;[1776] in this case, their (i.e., of the accrued interest on each bond) "sum"[1777] is three hundred seventy-four.[1778] In each month (*māse māse*) [and for each bond], there is a [different] percentage rate.[1779]

The procedure [is as follows]: there are that [mentioned multiple] interest-rates[1780] beginning with two, when their "sum" is divided by the [obtained result], such as forty, the quotient, namely nine months, is the "average time". Having equalised (*vilupya*)[1781] all the months, nine months [is obtained]. In **the procedure of conversion [of several bonds] into a single bond, when that representing the sum of the monthly profits,** which consists of the previously mentioned forty, **is multiplied** (*tāḍita*>*guṇita*), **by one hundred [and] divided by the sum of the capitals:** the "sum" (*yuti*>*yoga*) of the capitals[1782](*draviṇa*>*dhana*) one hundred and so forth is namely one thousand. When [the obtained product is] **divided** by it, having equalised one by one (*pṛthak pṛthak*) [the different interest-rates], **the percentage rate** for all [the four bonds] **should also occur**, and this becomes the interest-rate four [per cent]; thus is the connection [of words and their meanings].

In this respect, the author enunciates a sample problem:

GT 128–129. O clever one, a certain amount of money with an increment of one [hundred] and so forth for each [bond] respectively is lent at [the interest-rates of] two, three, four, and five per cent. The months are seven, eight, six, and twelve. Since the bonds[1783] are four, tell me quickly o friend, having

1773 Here *lābha* indicates the "accrued interest". In this context, I use the term "profit" for the accrued interest, taking into consideration the money-lender point of view. Note that both *māsavṛddhyaikya* and *māsalābhaikya* denote the same result; in the sample problem which follows, these two expressions refer to the number 40.

1774 *dravina*.

1775 *vṛddhi*. In line 22, the interest-rate is *vyāja*.

1776 There should be no em-dash between *phalāni* and *dvā*°.

1777 It refers to the "sum of the interest which has accrued during the elapsed time" (*gatasamayaphalaikya*) and not to the "sum of the monthly profits" (*māsavṛddhyaikya*).

1778 See SGT 128–129, 7.

1779 Line 18: *śatavṛddhiḥ* **śatam**] conj. See *śataphala* as "percentage rate" in Śrīpati's *sūtra*. It is worth noting that *vṛddhi* sometimes is "interest-rate" and sometimes is "accrued interest".

1780 Each bond has an individual interest-rate.

1781 In this context, the verb *vilup-* (lit. "finish off, disappear") indicates the process of evening up both the different monthly interest-rates and the elapsed time of the given bonds.

1782 The syntagm *draviṇayuti* denotes the sum of the capitals of the given bonds, hence the total amount of capital which is lent.

1783 *patra*.

made the conversion of all bonds into a single one, what will the equation of payments be?

The presentation:

1	7	1	8	1	6	1	12
100	100	100	200	100	300	100	400
2		3		4		5	

. In this case, with respect to [the first capital] one hundred, at the interest-rate of two over seven months it becomes fourteen. With respect to [the second capital] two hundred, at the interest-rate of three over eight months it becomes forty-eight. With respect to [the third capital] three hundred, at the interest-rate of four over six months it becomes seventy-two. With respect to [the fourth capital] four hundred, at the interest-rate of five over twelve months it becomes two hundred forty. [Each of] these is the "interest which has accrued during the elapsed time" [in regards to the four bonds]:[1784]

14
48
72
240

. Their "sum" is three hundred seventy-four:[1785] 374. Therefore, [next is the step] **when divided by the sum of the monthly profits**: with respect to one hundred, every month (*māsaṃ prati*) [at the interest-rate of two][1786] two [*dramma*s] occur; with respect to two hundred, at the interest of three, six [*dramma*s] occur; with respect to three hundred, at the interest of four, twelve [*dramma*s] occur; with respect to four hundred, at the interest of five, twenty [*dramma*s] occur. The "sum" of these [obtained numbers], which are the "monthly profits"

2
6
12
20

, is forty.[1787] When there is the division of three hundred seventy-four by it, the quotient is nine months and seven whose denominator is twenty:[1788]

9
7
20

. In this case, every month the *dramma*s accumulated (*caṭanti*) (i.e., the accrued interest) are

1784 In line 6, *gatasamayaphalāni* refers to the *gatasamayaphalaikye* mentioned in GT 127. The "sum" (*aikya*) is performed in the next step.

1785 This is the total sum of the interest of the elapsed period of time of each time.

1786 In line 7, before *dvau* one should expect *dvikavṛddhyā*; see *trikavṛddhyā*, *catuṣkavṛddhyā*, and *pañcakavṛddhyā* in the same passage.

1787 This is the "sum of the monthly profits". There is an interest which accrues monthly with respect to each bond having a different capital and interest-rate, and that the lender thus earns.

1788 This fraction denotes the "average time" (*gatakāla*).

forty.[1789] Then, nine multiplied by forty is three hundred sixty; seven whose denominator is twenty multiplied by forty becomes two hundred eighty. When there is its division by twenty, the quotient is fourteen *dramma*s; these added to three hundred sixty become three hundred seventy-four: 374. In this manner,[1790] [in respect to the equivalent single payment] the "average time" [and] the "elapsed time" have been arrived at.[1791]

When that representing the sum of the monthly profits, namely forty, **is multiplied by one hundred,** it becomes four thousand. In this case, the "sum" of one, two, three, four hundred,[1792] which is the "sum of the capitals" [which have been lent], is one thousand. **When** [four thousand] **is divided** by it, the quotient four is the **percentage rate**: 4. Here for every one thousand, which is the capital (*mūladhana*), at the equalised[1793] interest-rate of four [per cent], it (i.e., the profit) is forty [*dramma*s] every month. Then, the equalised [time expressed in] months, nine and seven whose denominator is twenty,[1794] is multiplied by this [forty]; the previous quotient is three hundred seventy-four.[1795] On that certain day [which has been calculated],[1796] having taken (i.e., earned) this [amount of money] at the interest-rate of four [per cent], the creditor[1797] converts the [four] bonds into a single one.

[On this topic] the author enunciates a sample problem with fractions:

130. O friend, make the conversion into one of the very capitals [mentioned above] having their interest increased by one-fourth and their months by one-third.

[The presentation]:[1798]

1	7	1	8	1	6	1	12
100	1	100	1	100	1	100	1
2	3	3	3	4	3	5	3
1	100	1	200	1	300	1	400
4		4		4		4	

. In this case, [the capitals] are one hundred, two hundred, and so forth: it should be understood

1789 In the SGT, the verbal root *caṭ-* (lit. "go away, disappear"), presumably a technical term, is found for the first time here; every month, 40 *dramma*s are earned (see the use of the terms *labdha* and *vṛddhi*) by the creditor (*dhanin*, line 19) by applying the rate of interest of 4% on the capital lent.

1790 The beginning of this passage (line 14) is corrupt.

1791 See a similar sentence in the next sample problem (SGT 130, 22).

1792 Line 16: *catuḥśatānāṃ*] em.; *pañcaśatānāṃ* (K). The fourth capital is 400 and not the transmitted 500.

1793 *samīkṛta*, the obtained equal, same (for all the bonds) interest-rate.

1794 Line 18: *varṣamāsā nava viṃśatibhāgāś ca sapta guṇitāḥ*] em. The transmitted reading is not interpretable, the emendation follows the commentary (line 10).

1795 The *gatasamayaphalaikya*; this calculation represents a kind of verification.

1796 Line 19: *tad dinaṃ ekapatraṃ*] conj.; *etad dinaprabhṛty ekapatraṃ* (K).

1797 *dhanin*. In the *Lekhapaddhati* (Prasad 2007, 26) and in the *Vaiṣṇava Dharmaśāstra* (Olivelle 2009, 65), the "creditor" is *dhanika*.

1798 In line 24, the term *nyāsa* is missing.

that the reduction[1799] is [performed] according to each own integer, namely one, two, [and so forth], [and] not according to the hundreds.[1800] Then, because here there is a fractional increase, by means of [the procedure] "multiplied by the denominator"[1801] and so forth, nine whose denominator is four is below one hundred:

1
100
9
4

. With respect to the second quantity, which is seven months, by means of [the procedure] "multiplied by the denominator"[1802] and so forth, it is twenty-two whose denominator is three:

22
3

. In this way, "the result of the multiplication" arises:[1803] according to [the rule] the "multiplication of the numerators"[1804] and so forth, nine multiplied by twenty-two is one hundred ninety-eight; below it, four multiplied by three becomes twelve:

198
12

. Regarding the original [capital] one hundred, as there is the integer one, [the number one hundred ninety-eight] multiplied by one is exactly so.[1805] When there is the division of one hundred ninety-eight by twelve, the quotient is sixteen; the remainder above is six [and] below is twelve. When there is the reduction of these by sixteen, it is one [and] two respectively:

16
1
2

. By means

1799 Line 24: *apavarta*] em.; *aparivarta* (K).

1800 The term *svarūpa* is found again in lines 10, 16, and 22. In order to arrive at the compound fractions shown in line 24, each numerator of the four obtained improper fractions (lines 2, 9, 15, and 21) is multiplied by its interest-rate. "According to its own integer" means that with respect to 200, the *svarūpa* is 2, hence the fraction obtained while performing the operations relating to the second *patra* is multiplied by 2, and not by one hundred (see *na tu śatam* in line 25). In this regard, see *ekarūpatvād* (line 2) and *svarūpeṇa* in lines 10, 16, and 22.

1801 This is a quotation from the rule of the class of fractional increase (GT 57).

1802 Ibid.

1803 This is a quotation form the rule of multiplication of fractions (GT 40).

1804 Ibid.

1805 Line 2: *maulikyaśatasyaika*°] em. The transmitted reading is nonsensical, the emendation supplies a subjective genitive to the abstract expression *eka*°. In lines 9, 16, 22, with respect to the second, third, and fourth *patra*, this step is differently formulated; for instance, in the calculation of this first *patra* the multiplication of the denominator of the improper fraction by the denominator one is not mentioned, while it is so regarding the other three *patra*s (see *ekacchedaguṇa*, lines 11, 17, and 23). This means the "according to its own integer" refers to the whole numbers one, two, three, and four having the denominator one underneath:

1	2	3	4
1	1	1	1

.

of this group of two numbers,[1806] a single "interest which has accrued during the elapsed time" is obtained, this is the principle.[1807]

Now regarding the second bond: in the first quantity, by means of[the procedure] "multiplied by the denominator" and so forth, below one hundred there is thirteen whose denominator is four: $\left|\begin{array}{r}1\\100\\13\\4\end{array}\right|$. Regarding the second quantity, by means of [the procedure] "multiplied by the denominator and so forth", it is twenty-five whose denominator is three: $\left|\begin{array}{r}25\\3\end{array}\right|$. Thirteen multiplied by twenty-five becomes three hundred twenty-five; below it, four multiplied by three becomes twelve: $\left|\begin{array}{r}325\\12\end{array}\right|$. With respect to the second bond, according to [the rule] "the result of the multiplication"[1808] and so forth, three hundred twenty-five multiplied by two, which is its own integer, becomes six hundred fifty: 650. When there is its division by twelve which has been multiplied by the denominator one,[1809] the quotient is fifty-four; the remainder above is two and below twelve. When there is their reduction by half, it is one [and] six respectively: $\left|\begin{array}{r}54\\1\\6\end{array}\right|$.

Now regarding the third bond: in the first quantity, by means of the [procedure] "multiplied by the denominator"[1810] and so forth, below one hundred [the result] is seventeen whose denominator is four. In the second quantity, by means of [the procedure] "multiplied by the denominator" and so forth, it is nineteen whose denominator is three. Seventeen is multiplied by it becoming three hundred twenty-three; below it, four multiplied by three becomes twelve:[1811] $\left|\begin{array}{r}323\\12\end{array}\right|$. In this case, with respect to this bond, three hundred twenty-three is multiplied by three, which is the integer of three hundred; it becomes nine hundred sixty-nine. When there is its division by twelve which has been multiplied by one, the quotient is eighty; the remainder above is nine

1806 It refers to the quantities manipulated with respect to the first *patra*, which are the interest-rate 2 increased by one-fourth and the time 7 months increased by one-third.
1807 In line 5, the number 2 is anomalous.
1808 This is a quotation form the rule of multiplication of fractions (GT $\overline{40}$).
1809 As I have mentioned earlier, it refers to the denominator 1 of "its own integer".
1810 This is a quotation from the rule of the class of fractional increase (GT 57).
1811 In line 15 before the layout, *yathā* is missing.

and below is twelve. When there is the reduction of these by three, it is three and four respectively: $\begin{array}{|r|}80\\3\\4\end{array}$.

Now regarding the fourth bond: in the first quantity, below one hundred, by means of [the procedure] "multiplied by the denominator"[1812] and so forth, it becomes twenty-one whose denominator is four. In the second quantity, it is thirty-seven whose denominator is three. Twenty-one is multiplied by it, it is seven hundred seventy-seven; below it, four multiplied by three becomes twelve: $\begin{array}{|r|}777\\12\end{array}$. In this case, with respect to the fourth bond, seven hundred and so forth is multiplied by four, which is the integer of four hundred; it becomes thirty-one hundred eight. When there is the division of these by twelve multiplied by the denominator one, the quotient is two hundred fifty-nine: $\begin{array}{|r|}259\end{array}$.

With respect to these [obtained results] which are the "interest which has accrued during the elapsed period of time" concerning the four bonds, it is respectively: $\begin{array}{|r||r||r||r|}16&54&80&251\\1&1&3&1\\2&6&4&\end{array}$. When the "sum" is carried out, in the first quantity by means of [the procedure] "multiplied by the denominator"[1813] and so forth, it becomes thirty-three whose denominator is two; in the second quantity by means [of the procedure] "multiplied by the denominator"[1814] and so forth, it is three hundred twenty-five whose denominator is six; in the third quantity by means of [the procedure] "multiplied by the denominator[1815] and so forth, it is three hundred twenty-three whose denominator is four:[1816] $\begin{array}{|r||r||r||r|}33&325&323&251\\2&6&4&1\end{array}$. According to [the rule] "the numerator and the denominator"[1817] and so forth, when there is the reduction of the denominators by half, when there is the exchange [of the obtained denominators] one and three, it is: $\begin{array}{|r||r|}33&325\\2&6\\3&1\end{array}$. In this case, thirty-three multiplied by three becomes ninety-nine, [and] two multiplied by three becomes the denominator six.[1818] The next quantity multiplied by one is exactly so. Then, when the additive quantity ninety-nine is added

1812 This is a quotation from the rule of the class of fractional increase (GT 57).
1813 Ibid.
1814 Ibid.
1815 Line 26 (at the end): *chedanighnety ādinā*] em.; *chedety ādinā* (K).
1816 The fourth quantity is not mentioned because it is remains as it is.
1817 This is a quotation from the rule of the class of simple fractions (GT 53).
1818 Line 29: **eka**] em.

to three hundred twenty-five, it becomes four hundred twenty-four whose denominator is six: $\begin{array}{|r|}424\\6\end{array}$. When there is the reduction of the denominator four which is below the third quantity,[1819] and the denominator six by half, when there is the exchange, it is: $\begin{array}{|r||r|}424&326\\6&4\\2&3\end{array}$. The first number at the very top multiplied by two becomes eight hundred forty-eight; six multiplied by two becomes twelve: $\begin{array}{|r|}848\\12\end{array}$. The next number above multiplied by three becomes nine hundred sixty-nine and four multiplied by three becomes twelve: $\begin{array}{|r|}969\\12\end{array}$. When the additive quantity eight hundred and so forth is added to it, it becomes eighteen hundred and seventeen whose denominator is twelve: $\begin{array}{|r|}1817\\12\end{array}$. When there is the exchange of the denominator one of the fourth quantity and the denominator twelve, it is: $\begin{array}{|r||r|}1817&259\\12&1\\1&12\end{array}$. The first quantity multiplied by one is exactly so; the next number above multiplied by twelve becomes thirty-one hundred eight, one multiplied by twelve is exactly so as well: $\begin{array}{|r|}3108\\12\end{array}$. When the additive quantity eighteen hundred and so forth is added to it, it becomes forty-nine hundred twenty-five whose denominator is twelve: $\begin{array}{|r|}4925\\12\end{array}$. This is the "sum of the interest which has accrued during the elapsed time".

When the four bonds are [each] divided by [the time increased by one-third"],[1820] the "monthly profits" are respectively:[1821] $\begin{array}{|r||r||r||r|}2&6&12&21\\1&1&3&1\\4&2&4&\end{array}$. When they are reduced according to [the rule] "multiplied by the denominator"[1822] and

1819 It denotes the third "interest which has accrued during the elapsed time" (*gatasamayaphala*) shown in line 24. The results of line 24 are added two by two: the first to the second, the result to the third, and the result to the fourth. The final result is the *etad gatasamayaphalaikyam* $\begin{array}{|r|}4925\\12\end{array}$ (line 11).

1820 The four *patra*s which has been shown (line 27) are each divided by their months increased by one-third as mentioned in the sample problem.

1821 Line 12: **ekadviśatādiguṇā**] em.

1822 This is a quotation from the rule of the class of fractional increase (GT 57).

so forth, it is respectively:[1823] $\begin{array}{|r||r||r||r|} 9 & 13 & 51 & 21 \\ 4 & 2 & 4 & 1 \end{array}$. In the fourth [quantity], there is no fractional increase. By means of [the method] "the numerator and the denominator"[1824] and so forth, when there is the sum of the four [quantities], it becomes one-hundred seventy whose denominator is four.[1825] This is the "sum of the monthly profits". Because it is the divisor, when there is the exchange of the denominator and the numerator, four is above and below is one hundred seventy: $\begin{array}{|r|} 4 \\ 170 \end{array}$. Twelve is the denominator of the previously mentioned dividend; when there is the "cross-reduction" of four [and] twelve, when there is the reduction by four, it is one and three: $\begin{array}{|r||r|} 1 & 4925 \\ 170 & 3 \end{array}$. The "multiplication" occurs:[1826] by means of the multiplication of the numerators and so forth, when [the numerator four thousand and so forth] is multiplied by one, it is exactly so. Below, one hundred seventy multiplied by three is five hundred ten. When there is the division of forty-nine and so forth by it, the quotient is nine months; the remainder above is three hundred thirty-five [and] below is five hundred ten. When there is the reduction of these by five, sixty-seven is above and one hundred two is below: $\begin{array}{|r|} 9 \\ 7 \\ 102 \end{array}$. In this way, the "elapsed time" and the "average time"[1827] are arrived at.

When that representing the sum of the monthly profits, which is one hundred seventy, is multiplied [by one hundred],[1828] it becomes seventeen hundred thousand: 17000, its denominator is four.[1829] The "sum of the capitals" is one thousand;[1830] because there is the denominator one, when there is the exchange of the denominator and numerator, it is one above and one thousand below. Seventeen hundred thousand multiplied by one is exactly so;

1823 Line 13: **caturtha**] em.
1824 This passage is a quotation from the rule of the class of simple fractions (GT 53).
1825 The sum of $\begin{array}{|r||r||r||r|} 9 & 13 & 51 & 21 \\ 4 & 2 & 4 & 1 \end{array}$ is $\begin{array}{|r|} 170 \\ 4 \end{array}$, which is the *māsavṛddhyaikya* or the "sum of the monthly profits".
1826 This is a quotation from the rule of division of fractions (GT 42).
1827 Line 22: °*samaya ity gata*°] em. The quantity $\begin{array}{|r|} 9 \\ 67 \\ 102 \end{array}$ is the "average time" (*gatakāla*).
1828 Line 23: the word *śatena* is missing.
1829 Line 23–24: […] *catuśchedā yathā* 17000 |] em.; […] *yathā* 17000 | *catuścheditvāt* […] (K). It is curious that the layout does not show the denominator 4.
1830 Line 24: **śata**] em.

one thousand multiplied by four becomes four thousand: $\begin{array}{|r|}4\\1\\4\end{array}$. This is **the percentage rate,** which multiplied by ten becomes forty-two plus one-half:[1831] $\begin{array}{|r|}42\\1\\2\end{array}$. By means of [the procedure] "multiplied by the denominator" and so forth, it becomes eighty-five whose denominator is two: $\begin{array}{|r|}85\\2\end{array}$. According to the method "the result of the multiplication",[1832] it is nine months plus sixty-seven [and] one hundred two.[1833] By means of [the method] "multiplied by the denominator"[1834] and so forth, nine hundred eighty-five whose denominator is one hundred two multiplied by it[1835] becomes eighty-three thousand seven hundred twenty-five. When there is the division by two hundred four, which has been obtained by [multiplying] one hundred by the denominator two, the quotient is four hundred ten; the remainder above is eighty-five and below is two hundred four. When there is its reduction by seventeen, five is above and twelve is below: $\begin{array}{|r|}410\\5\\12\end{array}$. This becomes the whole interest[1836] which has accrued in nine months and in order to verify[1837] it, when there is the division of forty-nine hundred twenty-five, which consists of the "sum of the interest which has accrued during the elapsed time", by the previously obtained[1838] denominator twelve, the quotient is four hundred ten and the fraction is five-twelfths: $\begin{array}{|r|}410\\5\\12\end{array}$. On that certain day,[1839] having taken back (i.e., earned) this [amount of money], the creditor converts the bonds into a single one.[1840]

1831 Line 27: **mūle sahasradhanatvāt**] em. This reading does not fit the context.
1832 This is a quotation from the rule of division of fractions (GT 42).
1833 This is the average time, see line 22.
1834 This is a quotation from the rule of the class of fractional increase (GT 57).
1835 In brief, it is multiplied by $\begin{array}{|r|}85\\2\end{array}$.
1836 *samagravyāja*.
1837 Line 5: *saṃvadanāya*] em. The neuter substantive *saṃvadana* is lit. "examination, argument"; in the SGT, this is the first occurrence of the term.
1838 Line 6: *pūrva*°] em. The transmitted *prāya* is not interpretable.
1839 It refers to the average time which has been calculated.
1840 Line 8: *tad dinaṃ ekapatraṃ*] conj.

In this way, the [procedure concerning the] conversion [of several bonds] into a single payment is concluded.

3.19 Equating instalments of capital (*samīkaraṇa*)[1841] [GT 131]

Next is the procedural rule for making equal the instalments[1842] of the lent[1843] unequal capitals and the periods of time which have been [so] divided for a money debtor;[1844] the author enunciates a metrical stanza:

> 131. The [equal] instalments[1845] of the lent quantity[1846] occur when the opposite numerator and denominator [are reciprocally multiplied], [these results] which are the profits[1847] of each unit (i.e., "instalment"), [having been] separately[1848] [arranged],[1849] are multiplied by the mixed quantity (*vimiśrasva*).[1850] These [products] are divided by their own sum (i.e., of the profits).

The explanation: in this case, in each single instalment[1851] there is a two-fold time (*kālasaṅkhyā*),[1852] the quantity-measure,[1853] [and] the interest-

1841 K 90, 11–14.

1842 *khaṇḍa*, in this context it means "instalment". It is found with the same meaning in the *Lekhapaddhati*; see, for instance, in Strauch (2002, 126) 2.5.2: *prathama-skaṃde…dvitīya-skaṃde…tṛtīya-skaṃde*.

1843 In line 18, the commentator glosses *prayukta* with *pradatta*.

1844 *grāhaka* – lit. "the one who receives". One finds the terms *grāhaka* as "borrower, money debtor" and *dhanin* as the "creditor" (see SGT 128–129, 19 and SGT 130, 8) also in the *Vaiṣṇava Dharmaśāstra* (*prakaraṇa* 6) and in the *Lekhapaddhati* (see Strauch 2002, 454). In this latter, forms of the root *gṛh-* meaning "to borrow" are commonly found; see Prasad (2007, 101) and Strauch (2002, 171).

1845 *khaṇḍaka*.

1846 The *prayuktarāśi* or "lent quantity" represents the whole capital which is lent.

1847 *lābha*, the "accrued interest"; see also GT 127.

1848 *pṛthak*. See the use of *pṛthak* in the commentator's passage (line 16); note that in L 92 this same term also is found. This rule is quoted by the SGT further after the sample problem of GT 132–133.

1849 This clarification is given by the commentary in line 18 (see *vyavasthitair*). In the same line, the commentator shows that the adjective *guṇair* is to be supplied to *vyastāṃśahāraiḥ*.

1850 This is not the "mixed quantity" composed of capital plus interest mentioned in GT 118. In lines 17 and 27, the commentator specifies that *vimiśrasva* denotes the *miśradhana* or "mixed wealth" which is the sum of the various instalments. The term *vimiśrasva* indicates the capital which has been lent and not yet divided into equal instalments; the procedure requires to determine equal instalments and a single interest-rate from this mixed amount.

1851 Line 15: *atraikaikakhaṇḍa*] em.

1852 It underlines that, for instance in the sample problem which follows, there are two sets of numbers related to the time: 1 month, which denotes the monthly interest-rate, and the elapsed time (2, 3, and 4 months), during which the interest on the loan is accumulated.

1853 See *pramāṇa* in L 92, quoted further by the commentator.

rate:[1854] this is a fourfold [group of quantities].[1855] [Now] that [quantities] denoted by the words "numerator" and "denominator" are explained: the [expression] **when the numerator and denominator** indicates that [quantities which are] multiplied conversely by means of the method of multiplication;[1856] **[these results] which are the profits of each unit** are **separately** arranged:[1857] they become the interest which accrues on one *dramma*[1858] [and] **are multiplied** by the mixed wealth.[1859] **[The products are] divided** "by their own (i.e., of the profits) sum", that is to say, **divided** by the "sum" of the capitals such as the simplified placed one hundred.[1860] **The instalments of the lent quantity** (*prayuktarāśi>pradattadhanarāśi*) **occur** by means of the separation[1861] of the unequal interest-rates (*viṣamavyāja*), [therefore] an equal interest-rate[1862] is determined; such is the connection [of words and their meanings].

In this respect, the author enunciates a sample problem:

> 132–133: O mathematician,[1863] one hundred ninety, 190, is lent in three amounts at [the [monthly interest-rate of] three, two, and four per cent [respectively]. The interest which has accrued on these [instalments] after two, three, and four months [respectively] has come to be the same [on each]. Tell [me] quickly the separate value[1864] of the [various] instalments.[1865]

Presentation:

1	3	1	2	1	4
100	1	100	1	100	1
2		3		4	

. The mixed wealth[1866] is 190. Now the procedure of multiplication [is performed], next is [the step] **when the opposite numerator and denominator.**[1867] To illustrate: [regarding the first instalment],[1868] one month, which is the number at the top, is multiplied by one

1854 *phaladhana.* Up to now, the commentator has used *phala* to denote the interest; here *phaladhana* does not seem to have, in fact, a different meaning. The expression *vṛddhidhana* as "interest" is found in PG 47 (see Shukla 1959, 31).

1855 A *daṇḍa* would fit well after *catuṣṭayaṃ*.

1856 See the commentator's explanation in the sample problem below, lines 1–4.

1857 *vyavasthita.*

1858 The commentator explains that, in each *khaṇḍa,* the "profit" (*lābha/vyāja*) on one *dramma* is determined by multiplying the elapsed time by its interest-rate. Since the quantity-measure is 100, the further multiplication by 1 is not mentioned. It should be: $2 \times 3 \times 1$; $3 \times 2 \times 1$; and so forth. Cf. SGT 128–129, 3–9.

1859 *miśradhana.*

1860 "Simplified" (*savarṇita*) by means of the class of simple fractions; see the SGT in the following sample problems, lines 6–12.

1861 *parihṛti.*

1862 *samavyāja.*

1863 *gāṇitika.*

1864 Here *pramāṇa* does not have a technical meaning; it means "value, size" and not "principal".

1865 *khaṇḍaka.*

1866 In brief, the lent amount.

1867 Line 28: *vyastāṃśahāraiḥ*] em. The transmitted reading is nonsensical.

1868 In this regard, see *dvikhaṇḍe* and *trikhaṇḍe* (lines 2 and 3 respectively).

hundred, which is underneath and which is the wealth-value;[1869] it becomes one hundred. Then, the interest-rate (*phaladhana*) two, which is below one hundred, is multiplied by the number above, which consists of [the time] three months; it becomes six.[1870] In this manner, the time has been multiplied by the wealth-value and the interest-rate by the [elapsed] time. By means of the method "below by above" this [last] multiplication should [thus] be considered a "reverse" procedure (*vyastavidhi*).[1871] Regarding the second "instalment", one month multiplied by one hundred becomes one hundred; when three is multiplied[1872] by the two above, it becomes six:[1873] $\begin{array}{|r|}100\\6\end{array}$. Regarding the third "instalment", one multiplied by one hundred becomes one hundred; the four below multiplied by the four above becomes sixteen.[1874] In this [same] way, each "unit" one by one [are calculated]:[1875] at the monthly interest-rate beginning with two, sixteen *drammas*[1876] [and] six and so forth occur [as the accrued profits]. [The procedure] has been accomplished up to [the steps] **when the opposite numerator and denominator [are reciprocally multiplied], [these results] which are the profits of a unit.**

1869 In this context, *pramāṇadhana* is a synonym of *pramāṇarāśi* or "quantity-measure".

1870 The layout of the first *khaṇḍa* is missing; it should be $\begin{array}{|r|}100\\6\end{array}$.

1871 Siṃhatilakasūri explains that first the numbers at the top of the layout (hence 1 by 100) and next the numbers below (hence 2 by 3) are multiplied. The expression *vyastavidhi* or "reverse procedure" echoes the *vyastāṃśahāra* mentioned in the *karaṇasūtra*. The commentator shows that the "numerator" (*aṃśa*) and "denominator" (*hāra*) are the numbers above and underneath in the layout and not the operands of a fraction, thus their meanings are to be understood according to the way quantities are placed in the layout. Also, *vyasta* or "opposite" denotes the position of the two numbers involved; for instance, in the first *khaṇḍa* the denominator 2 and the "opposite" numerator 3 are multiplied by a cross-like multiplication.

The layout regarding the third instalment is not given.

1872 Line 3: *trike guṇite*] em.

1873 The layout with the results of the first *khaṇḍa* is missing; it should look exactly like the second: $\begin{array}{|r|}100\\6\end{array}$.

1874 The layout with the result of the third *khaṇḍa* is missing; it should be: $\begin{array}{|r|}300\\16\end{array}$.

1875 Here *ekaikarūpaṃ* (line 5) refers to the *ekarūpa* of the *karaṇasūtra*: see *pṛthag ekarūpalābhair*.

1876 Line 5: *rūpakāḥ ṣoḍaśādayo*] conj. I cannot make sense of the transmitted *rūpakaviṃśopāḥ ṣoḍaśādayo* (K). The conjectural emendation suggests that the SGT refers to the numbers 16, 6, and 6; note that *ṣoḍaśa* or "sixteen" is mentioned in the line above. In the given layout (line 14), these numbers are the denominators and represent the three *lābhas* or "profit, accrued interest". The term *rūpaka* is both a generic term for "coin" as well as a unit of money (although not mentioned in the SGT); see, for instance, in the *Lekhapaddhati* (Prasad 2007, 27). Here I interpret it as "coin" thinking that the commentator refers to the *dramma* mentioned while commentating upon the *karaṇasūtra* (see SGT 131, 17).

With respect to these [quantities], next comes their own sum: in the first two "instalments", since six is the equal denominator, when the additive quantity one hundred is added to one hundred, it becomes two hundred whose denominator is six: $\begin{array}{|r|}200\\6\end{array}$. This is added to the third "instalment": when there is the reduction of six and sixteen by half, when there is the exchange of the obtained denominators three and eight, it is: $\overline{13}$. In the first quantity, two hundred and six multiplied by eight becomes sixteen hundred, six multiplied by eight is forty-eight, which is the denominator.[1877] In the second place, one hundred and sixteen multiplied by three becomes three hundred whose denominator is forty-eight.[1878] As there are equal denominators, when the additive quantity three hundred is added to sixteen hundred, it becomes nineteen hundred whose denominator is forty-eight: $\begin{array}{|r|}1900\\48\end{array}$. This is the quantity-divisor.[1879]

Again in the three "instalments", one hundred is multiplied by the mixed wealth, which is one hundred ninety; it becomes nineteen thousand: $\begin{array}{|r|r|r|}19000&19000&19000\\6&6&16\end{array}$. In this way, [the step] **[the profits] multiplied by the mixed quantity** is accomplished. And these [obtained results] are the dividends. Afterwards, since the previous quantity is the divisor, when there is the exchange of forty-eight by nineteen hundred, it is: $\begin{array}{|r|}48\\1900\end{array}$. In this case, regarding the first "instalment", it is: $\begin{array}{|r|}19000\\6\end{array}$. When there is the "cross-reduction",[1880] when there is the division of forty-eight by six, it is eight, and when there is the division of six by six, it is one. When there is division of nineteen hundred by nineteen hundred, it is one; in the same way, when there is the division of nineteen thousand by nineteen hundred, it is ten:[1881] $\begin{array}{|r|r|}8&10\\1&1\end{array}$. Next, by means of [the procedure] "the result of the multiplication"[1882] and so forth, ten is multiplied by eight becoming eighty, [one] is multiplied by one, and [eighty] divided by the denominator one is the very eighty: 80. Since

1877 Line 10: *aṣṭaguṇāḥ ṣaḍ jātā aṣṭacatvāriṃśac chedā* |] em.; (*ṣaḍ*) *keṣepe jātā* (*aṣṭacatvāriṃśa cchedā* | (K).

1878 Line 11: **tri*(*triḥ*)*] em.

1879 Line 13: **svanir** *bhāgadāyīrāśiḥ*] em.

1880 Line 19: *guṇanā°*] em.; *saṅgu°* (K). This is a quotation from the rule of multiplication of fractions (GT 40).

1881 I have emended the layout (line 19), as it is incorrect.

1882 This is a quotation from the rule of multiplication of fractions (GT $\overline{40}$).

the same procedure occurs, with respect the "second instalment" too it is the very eighty: 80. Regarding the third "instalment", it is: $\begin{array}{|r|}19000\\16\end{array}$. Since it is the divisor, when there is the reduction of forty-eight by sixteen, it is three; when there is the division of sixteen by sixteen, it is one. When there is the division of nineteen hundred by nineteen hundred, it is one. By means of the previous method, [when there is the reduction] of nineteen thousand [by nineteen hundred], it is ten. According to [the procedure] "the result of the multiplication"[1883] and so forth, ten is multiplied by three, it becomes thirty, [one is] multiplied by one, [thirty] divided by the denominator one is that very thirty: 30. When there is their sum $\begin{array}{|r|}80\\80\\30\end{array}$, the mixed wealth (*miśradhana*) one hundred ninety occurs:[1884] 190. Regarding this [amount], the equal profits[1885] consist of the accrued interest: "What do these will become?" He observes: "By means of the rule of five, the profits will become equal".[1886]

To illustrate, the presentation is:[1887] $\begin{array}{|r|}1\\100\\2\end{array}\begin{array}{|r|}3\\80\\\end{array}$;by means of the [procedure] "having moved the [fruit] to the other side"[1888] and so forth, it is: $\begin{array}{|r|}1\\100\\\end{array}\begin{array}{|r|}3\\80\\2\end{array}$. In this case, because [in the first quantity] there is no denominator, no exchange occurs. Then, eighty multiplied by three becomes two hundred forty; this multiplied by two becomes four hundred eighty.[1889] When there is its division by one hundred, which has been multiplied by one in the first side, it is the

1883 Line 24: *guṇanā°*] em. The transmitted reading seems a case of harmonization to parallels (see line 19). This is a quotation from the rule of multiplication of fractions (GT 40).

1884 I have emended the text, as one finds the number 90 (line 26) instead of 190.

1885 *samavṛddhi.*

1886 The commentary refers to L 92, which gives the same rule enunciated in GT 131. See also PG 59. In these texts, the rule is however differently formulated. Note that L 92 is quoted next by the commentator.

1887 I have emended the text; in the layout (lines 27–28), the third box $\begin{array}{|r|}80\\30\end{array}$ should not, in fact, occur. On the left, the first box represents the first *khaṇḍa* (the quantity-measure 100 and the time 1 month), its given interest-rate 3% is next to it and the first amount which has been obtained (80) is underneath.

1888 This is a quotation from the rule of five (GT 107).

1889 Line 30: *asyāḥ*] em.

integer four and four-fifths:[1890] $\begin{array}{|c|}4\\4\\5\end{array}$. With respect to the second instalment, it is also in this way:[1891] $\begin{array}{|c|}4\\4\\5\end{array}$. Regarding the third instalment, by means of the rule of five, it is: $\begin{array}{|r|r|}1&4\\100&30\\4&\end{array}$. Here too, by means of [the procedure] "having moved [the fruit] to the other side"[1892] and so forth, it is: $\begin{array}{|r|r|}1&4\\100&30\\&4\end{array}$. Also in this case the denominators are not [exchanged]. Next, [thirty] multiplied by four [becomes] one hundred twenty; this [four] is also multiplied by four, [it becomes] four hundred and eighty. As before, when there is its division by one hundred, the quotient is forty [and the remainder is] four whose denominator is five: $\begin{array}{|c|}4\\4\\5\end{array}$. In this way, with respect to the three instalments, there is also the same interest.

The formulation in the *Līlavātī* clearly agrees with this exposition (i.e., Śrīpati's); a metrical stanza, which is a procedural rule for calculating [various] instalments [from a given mixed quantity], is illustrated:

> L 92: "The time multiplied by the quantity-measure is divided by the interest-rate multiplied by the elapsed time (*vyatītakāla*). These [results] are multiplied by the mixed quantity (*vimiśra*) and divided by their own sum (i.e., of the previous results of the first step); one by one,[1893] the instalments of the lent [capital] occur".

The presentation of this very sample problem:[1894] $\begin{array}{|r|r|r|r|r|r|}1&3&1&2&1&4\\100&1&100&1&100&1\\2&&3&&4&\end{array}$. The mixed wealth is $\begin{array}{|c|}190\end{array}$. In this case, [for each instalment] **the time,** which consists

1890 Line 1: *pañcabhāgāś*] em.

1891 The first and the second amounts are the same.

1892 This is a quotation from the rule of five (GT 107).

1893 Note the use of *pṛthak* here too as in GT 131.

1894 The commentator solves the same sample problem (GT 132–133) by applying the rule given in the L.

of one month, is **multiplied by the quantity-measure,** which is the number 100; above it becomes one hundred in all places. [Next] the "interest-rate" is multiplied by the "elapsed time", such as two months and so forth; regarding the first "instalment",[1895] when the "interest-rate" three is multiplied by the "elapsed [time]" two months, it becomes six, which is [placed] below one hundred: $\left|\begin{array}{r}100\\6\end{array}\right|$. With respect to the second "instalment", [two] multiplied by three months becomes six, which is [placed] below one hundred: $\left|\begin{array}{r}100\\6\end{array}\right|$. Regarding the third "instalment", when there is the multiplication[1896] of the "interest-rate", namely four [per cent], which is multiplied by the "elapsed [time]" four months, it becomes sixteen, which is [placed] below one hundred: $\left|\begin{array}{r}100\\6\end{array}\right|$. These [results] are [the results of the step] **divided by the interest-rate multiplied by the elapsed time**. The divisions [to be] performed, since they have been carried out below[1897] [are not shown here again]. Having considered the object: "The opposite numerator and denominator", in this case this very [meaning] has been shown as well; the mathematicians teach this principle. The next half procedure should be explained as the very one before, hence [here] it is not shown.[1898]

1895 I have emended the text, the interest which regards the first *khaṇḍa* mistakenly concerns the second *khaṇḍa* instead. I suggest it should read: (line 15) *ādyakhaṇḍe vyatītamāsadvayaguṇaphalatraye jātāḥ śatādhaḥ ṣaḍ yathā* $\left|\begin{array}{r}100\\6\end{array}\right|$ | (line 16) *dvikhaṇḍe māsatrayaguṇitau jātā śatādhaḥ ṣaḍ yathā* $\left|\begin{array}{r}100\\6\end{array}\right|$|] conj.

1896 In lines 17–18, the grammar is awkward; one would expect *vyatītamāsacatuṣkaguṇaphalacatuṣke* ("when the interest rate four is multiplied by elapsed time four months").

1897 In brief, in the previous sample problem.

1898 In lines 19–21, the commentator points out that he supplies this rule form the L in order to clarify the meaning of the expression "the numerator and the denominator" used by Śrīpati in his *karaṇasūtra*. Although differently formulated, the definitions given by the GT and the L have, in fact, the same meaning.

Part III

Text analysis

4 Text analysis

4.1 Preliminaries

The purpose of Part III is to accompany the reading of the translation and understanding of the texts; it explains the rules formulated in Śrīpati's GT and analyses the mathematical procedures presented in Siṃhatilakasūri's commentary. Starting from the arithmetical operations with integers, the scheme followed for each rule is:

- analysis of the rule enunciated by the GT, for instance: "GT $\overline{13}$"
- analysis of the SGT while commenting upon the rule, for instance "The SGT on GT $\overline{13}$"
- analysis of the commentator's explanation and execution of the sample problem, for instance: "The SGT on GT 14 (ex.)"[1]

4.2 Benedictory section (*maṅgalācaraṇa*) [GT 1]

In the *maṅgalācaraṇa,* which reveals that Siṃhatilakasūri belonged to the medieval tantric tradition of Jainism, the commentator offers respect to his spiritual lineage. After having paid homage to Mahāvīra the Jina, Siṃhatilakasūri worships the *devī* Kuṇḍalinī, who is also frequently mentioned in his text on Jaina mantras and rituals entitled *Mantrarājarahasya.* The commentator introduces himself as *śrīvibudhacandragaṇabhṛcchiṣya* or "the disciple of Śrī Vibudhacandra Gaṇabhṛt". He also specifies the title of the root-text upon which he is going to comment; in the SGT, the name *Gaṇitatilaka* occurs a second time at the end of the "section on technical terms" (*paribhāṣā*). Śrīpati does not name his own work; however, Siṃhatilakasūri does so twice. It thus appears that this text has been published by the same name provided by the commentator.

In the introductory line to the *paribhāṣā*, Siṃhatilakasūri defines the GT as a *gaṇitaśāstra* or a "treatise on mathematics"; in the benedictory section the

1 Here in Part III I use the abbreviation "ex." as shorthand for "exercise", which in this case denotes a stanza/verse enunciating a "sample problem".

sūtrakāra Śrīpati refers to his own work by the expression *gaṇitasya pāṭī*. This same expression is also used, for instance, by Śrīdhara in the TŚ (TŚ 1) and by Bhāskarācārya in the L (L 1) to indicate that their mathematical works deal with "arithmetic".[2] Śrīpati points out that his treatise is written in "variegated metrical stanzas" (*vicitravṛtta*), referring to the different Sanskrit meters used; it is, in fact, by the word *vṛtta* that Siṃhatilakasūri introduces the *sūtrakāra*'s stanzas on rules and sample problems, but when the rule is given in *anuṣṭubh* the commentator uses the word *śloka*.

4.3 Section on technical terms (*paribhāṣā*) [GT 2–12]

At the end of the ten verses which introduce technical terms relating to the decimal place-value notation and various measuring units, Siṃhatilakasūri uses the word *paribhāṣā*.[3] This term denotes a part of a Sanskrit mathematical text which is, however, not always specified by authors as such.[4] This section is found at the very beginning of a work, following the benedictory verse; for instance, the term *paribhāṣā* is used by Śrīdhara and by Mahāvīrācārya in TŚ 8 and in GSS 1.24 respectively.[5] In Śrīpati's text, the *paribhāṣā* (GT 2–12) begins with the list of "notational places" (*sthāna*). The commentator highlights that the author[6] properly opens this treatise on mathematics by explaining the calculation based on the power of ten and which starts with *eka* or "one", here denoting the place "unit" in the decimal place-value system. The GT uses the noun *saṅkhyā* ("number") qualified by *daśaghnī* ("multiplied by ten") to concisely describe the computation on the power of ten characterising the decimal place-value system. Śrīpati lists eighteen decimal places;[7] it is noteworthy that some of these terms go back to Vedic literature and reflect different usages throughout time and place. Siṃhatilakasūri provides equivalents to the terms denoting notational places used by Śrīpati; each notational place is expressed by the commentator both in words and in figures, by this expository technique he demonstrates that this numeration works, in fact, "via the increase in zeros" (see *śūnyavṛddhyā*).

In the GT, the passage on the decimal place-value notation is followed by a list of technical terms and their corresponding values. The GT mentions in

2 In TŚ 1, one finds the expression *pāṭyā gaṇitasya*, while in the L *pāṭīṃ sad gaṇitasya vacmi* occurs.

3 In SGT 97, 4–5, the commentator uses the more generic term *prastāvanā* to refer to this introductory section.

4 It is sometimes placed as heading by editors. Authors, in fact, do not explicitly state a topical structure of their mathematical work.

5 Pingree (1981, 61) observes that the GT's *paribhāṣā* is modelled on Mahāvīrācārya's GSS. I do not agree with this observation; in the GSS, verses 1–70 show an entirely different structure, including the order of the topics treated and the terminology used.

6 The commentator specifies only once the name of the author of the text he is commenting upon, namely at the end of the *paribhāṣā*.

7 From *eka* up to *parārdha*.

order: monetary units, units of gold, weight, capacity, length, and units of time. The commentator introduces each set of measuring units by a passage which clarifies what they stand for; for instance, the list of monetary units is introduced by the expression *kapardavyavahāra.* Siṃhatilakasūri uses the expression *meyavyavahāra* to introduce the third set of measuring units; the term *meya* (lit. "measurable, to be measured"),[8] which I have not found in other Sanskrit mathematical texts, does not help the modern reader to understand what these measuring units stand for. Some of the terms given by Śrīpati and classified by Siṃhatilakasūri into the category *meyavyavahāra* are found in Śrīpati's sample problems; for instance, the unit *dhaṭaka* is mentioned in relation to saffron[9] and *pala* in relation to camphor,[10] while *pala* occurs when agarwood and musk are concerned.[11] By *meyavyavahāra,* Siṃhatilakasūri thus categorises units of weight. With respect to the units classified by Siṃhatilakasūri into the category *kaṇamāna,* it is worth noting that the meaning of the word *kaṇa* is "grain, corn"; *kaṇamāna* refers to units of capacity, but some of them may have been used in ordinary life as both units of weight and units of capacity. Although in the stanza of this section on technical terms Śrīpati mentions four measures, in the sample problems given by the author only *mānika* is found, and it is mentioned in sample problems concerning grain.[12]

4.4 The eight arithmetical operations with integers [GT $\overline{13}$–34]

Some remarks

After the *paribhāṣā,* the first topic treated in the GT is the presentation of the eight elementary arithmetical operations with integers. These are in order: addition, subtraction, multiplication, division, square, square-root, cube, and cube-root. Notably, the first four arithmetical operations (addition, subtraction, multiplication, and division) are not explained by the commentator in depth as the other four; this testifies that the students to which this text was meant were able to carry out basic arithmetic. The term which denotes each arithmetical operation is *parikarman*; Śrīpati uses this term in the sample problem of cube-root of integers (GT 34) and in the sample problems of square-root of fractions (GT 49). The commentator mentions the term *parikarman* several times in his work.[13]

8 From the verbal root *mā-* or "to measure".
9 GT 99, 17.
10 GT 97, 22. In Sanskrit literature, *pala* is found in relation to different kinds of measuring units.
11 GT 114, 7.
12 GT 100, 6 and GT 101, 27.
13 See SGT 34, 17; SGT ad L 50, 2 (concerning the rule of "inverse operation"); SGT 52, 22, and SGT ad L 45–47, 11 (concerning the rule of zero).

Before looking at each *parikarman*, it is noteworthy that, from the very beginning, both texts adhere to a clearly defined structure: Śrīpati's rules are preceded by a brief introduction in which Siṃhatilakasūri specifies the number of *vṛttas* or "metrical stanzas" and the subject treated. In this regard, a significant expression used by the commentator is *karaṇasūtra*, which I translate as "procedural rule".[14] The expression *karaṇasūtra* denotes the set of mathematical steps enunciated in each rule which are to be performed in order to carry out the algorithm formulated in the rule.[15]

Addition (saṅkalita) *[GT $\overline{13}$]*

GT $\overline{13}$: The first *parikarman* treated by the GT is "addition" (*saṅkalita*). This rule does not occur in the earlier GSS, TŚ, and PG,[16] but it is found in the later L. This feature may denote that works on *pāṭīgaṇita* were generally addressed to pupils with already an elementary education in mathematics. The knowledge of this basic operation was probably tacit or perhaps taught in a different context; in this regard, it is interesting to note that the multiplication of numbers smaller than ten seems to have had the status of assumed knowledge too.

In Śrīpati's formulation of the *karaṇasūtra* of *saṅkalita*, the commutative principle of addition is expressed by the correlatives *yathā...tathā,* which introduce two methods of adding digits: the "regular order" (*krama*) and the "reverse order" (*utkrama*). In the half verse, one finds: *yathā... krameṇa* and *tathā...utkramāt.* Also, two other terms are found in relation to addition, namely *saṅkalita* and *yuti*. However, it is important to understand that while *saṅkalita* is "addition" as the process of finding the sum of two or more numbers, *yuti* – as well as its synonyms, such as *yoga* and *melana* – mean "sum", the joining itself, the result of the process of combining two or more quantities.[17] Another interesting feature regards the two technical terms *krama* and *utkrama,* both used in relation to the operations of addition and subtraction and denoting the direction in which digits are to be considered in order to carry out the procedure. The same terms are found in L 12, where Bhāskarācārya enunciates the rules of the operations of addition and subtraction.[18] Interestingly, in relation to the "procedure of multiplication"

14 Regarding passages introducing a *karaṇasūtra*, texts and commentaries show different styles. In both the GSS and the L, the authors themselves introduce, briefly, the procedural rules.

15 SaKHYa (2009, XXXII) observes that "each individual algorithm, that is, a series of mathematical operations designed for a specific type of problems is called *karaṇa,* and a rule for it *karaṇasūtra*".

16 All these works begin with addition and subtraction of arithmetical series.

17 In this context, the term "result" does not denote the amount obtained, which is usually denoted instead by the term *phala*.

18 In the L, in the operation of addition the "regular order" (*krama*) is from right to left, while the "reverse order" (*utkrama*) is from left to right.

(*guṇakāravidhi*) and to the "procedure of division" (*bhāgahāravidhi*) Śrīpati, Mahāvīra, Śrīdhara, and Bhāskarācārya use the terms *anuloma* and *viloma* to indicate the "regular order" and the "opposite order" respectively, and not *krama* and *utkrama*. Therefore, it seems that these two sets of terms – *krama/utkrama* on the one hand, and *anuloma/viloma* on the other hand – have their standard connotation according to the context of different operations.

In Śrīpati's *karaṇasūtra*, an interesting feature is the occurrence of the term *pakṣa* (lit. "wing, side"); one would rather expect *sthāna* or [notational "place", which is found in L 12 where addition is described as the sum of digits according to their positions in the place-value notation (i.e., units with units, tens with tens, and so forth). In the L, the term *sthāna* makes clear that the sum is an operation executed by means of notational places.

The SGT on GT 13: While explaining *svapakṣa*, Siṃhatilakasūri displays a rich terminology which embodies key-concepts essential not only to understanding the rule of addition but to entirely grasp the commentator's lexicon. In lines 14–15, the commentator explains that *svapakṣa* (lit. "its own side") refers to the row occupied by each digit of an *aṅkarāśi* ("number"),[19] which is lit. a "heap of digits"; a number consists, in fact, of a set of digits each representing one specific notational place, and thus place value, in the decimal system.[20] Among the terms used by Siṃhatilakasūri to elaborate on *svapakṣa* is *śreṇi* ("row, line"); in *agretanāṅkarāśes tadūrdhvāṅkaśreṇiḥ* and *pūrvāṅkarāśes tadūrdhvāṅkaśreṇiḥ* (line 15), the term *śreṇi* is both times preceded by the adjective *ūrdhva* (here "upright, vertical"). The compound *ūrdhvāṅkaśreṇi* denotes a "vertical" (*ūrdhva*) "row" (*śreṇi*) of digits; it refers to the setting down of digits one below the other in order to carry out their sum. Siṃhatilakasūri employs the terms *śreṇi, aṅkarāśi, ūrdhvāṅkaśreṇi, agretanāṅkarāśi,* and *purvāṅkarāśi* to elucidate the meaning of *svapkṣa*. The pluralistic use of the term *aṅka* in Siṃhatilakasūri's vocabulary is noteworthy. In mathematical texts, *rāśi* is usually "quantity, heap", while *aṅka* is "digit" and *saṅkhyā* is "number", but this terminology varies from author to author. Throughout his commentary, Siṃhatilakasūri uses *aṅka* in a pretty loose way: it sometimes denotes a "digit", sometimes a "number", sometimes a "quantity", for instance a fraction or a compound fraction composed of a fraction plus an integer.[21] In lines 14–15, Siṃhatilakasūri clarifies that, with a respect to a number composed of a "set of digits" (*aṅkarāśi*), a "digit" (*aṅka*) can be either *pūrva* or *agretana*. These are considered as such from a left to

19 It is worth noting that in the first line the commentary defines a number as a row, a string of digits.

20 Siṃhatilakasūri uses *aṅkarāśi*, for instance, while commenting upon the procedure of multiplication (line 15) and it denotes the number (which in the procedure is the multiplicand) 21586; in the square-root of integers (line 3), *aṅkarāśi* denotes the new remainder obtained by joining the digit 1 and 6, thus the number 16.

21 See, for instance, SGT 28–29, 23; SGT 39, 22, and SGT 39, 15–16.

right direction: with regard to the *aṅkarāśi* 317, the digit 3 is the *pūrvāṅka*, while 1 and 7 are the *agretanāṅkas*.[22]

I shall now investigate further the term *pakṣa* first as presented in the GT and then in the SGT, in order to clarify how it should be understood in relation to the procedure of addition. In the GT, *pakṣa* is found only twice: i) here in the *karaṇasūtra* of addition, and ii) in the *karaṇasūtra* of the "rule of five quantities" (*pañcarāśika*) (GT 107), where *pakṣa* denotes one of the "sides" in which a notational layout is subdivided to arrange numbers and perform the algorithm. With respect to the SGT, after his passage on the procedure of addition Siṃhatilakasūri uses again *pakṣa* in SGT 42, 11 while explaining the division of fractions, and in SGT 51, 7 while elucidating the cube-root of fractions, as well as various times after the "rule of three quantities" (*trairāśika*). In all these occurrences, *pakṣa* refers to the "side" of a notational layout.

Coming back to the commentator's explanation, the term *pakṣa* acquires a specific meaning in the context of the rule of addition; *pakṣa* denotes each "side" of an *aṅkarāśi*, that is, each row of the notational places (thus each digit) which compose a number and which numbers vertically arranged in order to be added have in common. Furthermore, the commentator explains that digits of numbers can be added either by the "regular order" or by the "reverse order", which is from the top to the bottom and from the bottom to the top respectively.

The SGT on GT 14 (ex.): In line 26, the row of numbers presenting the first set of numerical data is horizontal. It must be recognised that in the manuscript, a long vertical row would have occupied much space; nevertheless, it is possible to infer from the language used that the commentator refers to a vertical arrangement of numbers, and thus to a vertical layout. In lines 26–27, Siṃhatilakasūri clarifies how to place numbers in the working surface; to better elucidate the commentator's explanation, I shall use the same data given by Śrīpati in the sample problem, a layout similar to those used by the SGT, and employ the SGT's vocabulary:

7
8
9
16
93
60
76
50

22 This is specified by the commentator while explaining the procedure of cube, where this very number is found (SGT 28–29, 23).

This *ūrdhvāṅkaśreṇi* is composed of digits placed each below the other, the digits composing the given numbers are in their own "side" (*pakṣa*) according to their notational place. With respect to the "number" (*aṅkarāśi*) 16, the "digit" (*aṅka*) 1 is the *pūrvāṅka,* while the digit 6 is the *agretanāṅka*. An interesting feature is that when performing the sum of the numbers according to the "regular order" (*krama*), which the SGT clarifies to be from top to bottom, in line 28 the commentator explains: "Eight added to seven becomes fifteen". In the same way, he explains the "reverse method" (*utkrama*), which he clarifies to be from below to the top, by saying that the 3 of 93 is added to the 6 of 76 (line 29) and the obtained 9 is added to the 6 of the 16 above.

Subtraction (vyavakalita) *[GT $\overline{15}$]*

GT $\overline{15}$: As in Śrīpati's previous passage the two terms *saṅkalita* and *yuti* denote two different features of the rule of addition, a similar pattern occurs in relation to the operation of subtraction, and it regards the terms *vyavakalita* on the one hand and *viśodhana/viśuddhi/ pātana* on the other hand.[23] The term *vyavakalita* represents the process of finding the subtraction, while *viśodhana* and *viśuddhi* indicate the deduction itself, the result of the process of subtracting values from certain quantities.[24] On a semantic level, it is interesting to observe that the root *viśudh-* is lit. "to purify". Other terms often found in both the GT and the SGT related to the deduction are:

- forms of the verbs *hā-* (lit. "to abandon"), *apanī-* (lit. "to remove, take away"), and *rah-* (lit. "to separate, abandon")
- the nouns *vyutkalita* and *vyutkalana* (lit. "expelling"), *śodhana* (lit. "clearing"), *viyoga* (lit. "separation")
- the adjective *ūna* (lit. "deficient") used in composition with numbers which are to be subtracted
- the remainder called *śeṣa*
- *antara, agra,* and *viśeṣa,* used to indicate the residue, the difference from a deduction.

One should bear in mind that in the original manuscript this half verse on the rule of subtraction must have been coupled together into one stanza with the previous half verse, which is the *karaṇasūtra* of addition.[25] *Prima facie,* it is tempting to read *amunā krameṇa* as "by that regular order", since in the rule of addition Śrīpati uses the word *krama* to denote one of the methods – the

23 While explaining the "subtraction of fractions" (*bhinnavyavakalita*), the commentator uses the terms *viśleṣa* and *pāta* to mean "subtraction, deduction" (SGT $\overline{38}$,8). Another term used by the commentator with the same connotation is *vivara* (see for instance SGT 70, 15).

24 In this context, "result" does mean the "remainder" of a subtraction, which is instead denoted by *śeṣalavaśeṣa*.

25 See Hayashi (2013, 62).

regular order – of this operation. In lines 11–14, the commentator clarifies his understanding and thereby brings to light the idea implied by *amunā krameṇa*.

The SGT on GT 15: In mathematical literature, the term *krama* means "manner" as well as "regular order". Textual analysis helps to clarify that for Siṃhatilakasūri, *apy amuṇā krameṇa*, which I render as "also in that manner", denotes "that same manner" of computing digits, which consists of the *krama* and *utkrama* methods just mentioned in the half verse of the *karaṇasūtra* of *saṅkalita*. The fact that Siṃhatilakasūri reads *amuṇā krameṇa* as "in that manner", or else "according to that manner", is supported by a close investigation of his vocabulary in this very passage. He glosses *krama* with *rīti*, a term often occurring in the commentary and which means "method".[26] That the commentator here interprets *krama* as "method" is confirmed by the structure of his passage: the two sentences included between the two *daṇḍa*s, whose first unit is *apīti* (line 12) and the last is *nītam aṅkasthānam* (line 14), run parallel. After having glossed *krameṇa* with *prāguktarītyā,* the SGT uses two clauses where one finds two absolutives (in both cases *nyasya* or "having placed"), two finite verbs (*pātyate* and *niṣkāśyate*, both meaning "it is subtracted, removed") and the same object, which is *pātya* or "the subtrahend". In the first clause, the subtrahend is specified as *svalpāṅka* or the "smaller number"; the second clause starts with *tathā,* which emphasises a comparison and expresses the correlation between the two clauses. Furthermore, the commentator expands *apy amunā krameṇa* by repeating the term *krama* twice: the expression *iti krameṇa* is found at the end of each of the two clauses aforementioned, so that the construction *iti krameṇa*...*iti krameṇa* marks the explanation of the two methods of subtraction.

The commentator's argument turns around the arrangement of numbers in the layout drawn on the calculating board. This is confirmed by the use of *nyasya* ("having placed"), *ūrdhva* ("top, above"), and *adhas* ("below, underneath"). The first line of his passage focuses on "subtraction" (*viyojana*) as the mathematical procedure in which a smaller digit is removed from a larger. Then, he specifies the meaning of *apy amunā krameṇa* and states that such a procedure can be executed in two ways: by placing the subtrahend at the bottom or by placing it at the top; in the first case, the *krama* method (from top to bottom, as explained by the SGT) is applied, while in the second case, the *utkrama* method takes place (from the bottom to the top). The first method is explained in the first clause which begins with *bṛhadaṅkād adhaḥ* ... *nyasya* or "having placed...below from the larger number", while the second method is mentioned in the clause which immediately follows the first *ity krameṇa*; it is introduced by *tathā* or "in the same way" and continues with *bṛhadaṅkopari*...*nyasya* or "having placed...above the larger number". At

26 It is found, for instance, at the beginning of his explanation of the methods of multiplication: *atrāṅkaguṇanavidhau rīticatuṣṭayam uktam* is "here four methods (*rīticatuṣṭayam*) regarding the procedure of multiplication of numbers (*aṅkaguṇanavidhau*) are explained".

the end of this second method, *ity krameṇa* is again found. The adjectives *alpa*/*svalpa* (here meaning "smaller") and *bṛhat* (here "larger"), and the prepositions *adhas* ("below") and *upari* ("above") expands upon *apy amunā krameṇa*. Lastly, Siṃhatilakasūri illustrates what the "remainder" (*avaśeṣa*) is and briefly comments upon the sample problem given by Śrīpati.

The SGT on GT 16 (ex.): The commentator's explanation of this sample problem is concise. However, I would like to draw attention to the expression *ūrdhvādhorītyā* (line 22),[27] which refers to the subtraction of numbers by means of two distinct methods: (a) the subtrahend is placed below the minuend, or (b) the subtrahend is placed above it.[28] It obviously does not refer to a commutative principle but rather to two conceivable ways of arranging the two operands in the layout to carry out the procedure of deduction. The reference to the layout is an aspect which significantly characterises Siṃhatilakasūri's literary style and mathematical pedagogy.

Multiplication (guṇakāra) *[GT 17–$\overline{18}$]*

GT 17–$\overline{\textbf{18}}$: The methods of "multiplication" (*guṇakāra*) mentioned by Śrīpati are:

- the *kapāṭasandhi* (lit. "door-junction") method, which can be carried out in the "regular order" (*anuloma*) or in the "reverse order" (*viloma*)[29]
- the *tatstha* or "as it stands" method
- the *khaṇḍa* or "portion" method, which is sub-divided into: a) *sthāna*, and b) *rūpa*.

With respect to the vocabulary used, one can observe:

- a certain preoccupation with the way the two operands, which are the multiplicand-*guṇya* and the multiplier-*guṇaka*, are arranged in the layout and computed: see, for instance, the use of the verbs *vinyas*- ("to place") and *utsṛ*- ("to shift"), and prepositions and adverbs such as *adhas*, *kramaśa*, *anuloma*, and *pratiloma*
- the occurrence of technical terms – *sthāna*, *rūpa*, *khaṇḍa* – and forms semantically related to the verb "to kill, to strike" as the root *han*- and the noun *santāḍana,* which in mathematics denote "to multiply" and

27 The syntagm *ūrdhvādhas* is also found in SGT 90, 7 and in SGT 132–$\overline{133}$, 2. In the first case, it denotes the exchange ("above and below") of the two operands of the fraction-divisor, in the latter refers to the multiplication of the number "below by [the number] above".

28 In his English introduction, Kāpadīā (1937, LI) observes: "Siṃhatilaka mentions two methods for subtraction viz. of placing subtrahend below the minuend and *viceversa.*"

29 These are from right to left and from left to right respectively. Note that the terms *anuloma* and *viloma* are also found in the rule of multiplication in PG 18–29, TŚ 5–8, GSS 2.1, and L 14–16.

"multiplication" respectively. In this regard, other terms occurring in the GT and SGT denoting the multiplication are, for instance, *āhati, vadha, hanana,* and *ghāta* – all having the literal meaning of "the act of killing, destroying".[30]

The SGT on GT 17–18: In the introductory passage, Siṃhatilakasūri mentions for the first time Śrīdhara's TŚ and states that the purpose of multiplication is the increase of numbers by means of summation.[31] This recalls Bhāskara's I remark at the very beginning of his commentary on the *Āryabhaṭīya*, where he states that "mathematics as a whole" (*aśeṣagaṇita*) consists of "increase" (*vṛddhi*) and "decrease" (*apacaya*).[32]

Siṃhatilakasūri's explanation of the *karaṇasūtra* includes a vocabulary which probably belongs to local traditions. Although the commentator explains the *modus operandi* of each multiplication method, he does not go into detail; here Siṃhatilakasūri is visibly concerned with providing clear verbal explanations rather than showing the graphic manipulations of the algorithms. The SGT does not illustrate the various sets of calculations characterising each algorithm mentioned by Śrīpati, and intermediate passages are not discussed at all. This makes it difficult for us to fully understand the procedure or to reconstruct the main steps.[33] As a matter of fact, the overall treatment of the multiplication methods, as they are formulated in the GT and elucidated in the SGT, requires some preliminary knowledge of multiplication, demonstrating thus that the multiplication of numbers smaller than ten was a prerequisite.

Siṃhatilakasūri begins his explanatory passage by observing that, in the GT, the methods of multiplication explained are four. First of all, the commentator clarifies the operands involved in the operation (i.e., the multiplicand and the multiplier) and thereby he uses the numbers which are mentioned by Śrīpati in the sample problem on this same rule. Siṃhatilakasūri elucidates that the "multiplicand" (*guṇya* in Śrīpati's rule) is in this case – meaning in the sample problem given by Śrīpati – is the number 21586, while the "multiplier" (*guṇaka*) is the number 96. Let us look in detail at the way the commentary analyses the various multiplication methods:

30 Datta and Singh ([1935] 1962, 134) claim that the term *guṇana*, among the terms which stand for multiplication, "appears to be the oldest as it occurs in Vedic literature. The terms *hanana, vadha, kṣaya*, which mean 'killing' or 'destroying', have also been used for multiplication. These terms came into use after the invention of the new method of multiplication with the decimal place-value numerals; for in the new method the figures of the multiplicand were successively rubbed out and in their places were written the figures of the product".

31 In GSS 2.1, the multiplication is also called *pratyutpanna*.

32 Bhāskara I also quotes an unnamed source describing multiplication as a form of increase, and division and root extractions as forms of decrease.

33 A study on various methods of multiplication is by Datta and Singh ([1935] 1962). A recent paper on procedures of multiplication is by Keller and Morice-Singh (2014).

lines 2–8: the first method elucidated by the SGT is the *kapāṭasandhi*. In the *karaṇasūtra*, Śrīpati states that the "junction of two doors" method can be executed either according to the "regular order" or according to the "reverse order". Siṃhatilakasūri explains that, having placed the multiplicand 21586 below the multiplier 96, according to the regular order one should first multiply 86 by 96 and then shift the multiplier above the multiplicand 96 and multiply 15 by 96. He uses the preposition *upari* ("above") to clarify the position of the multiplier 96 in each step, since this is shifted above the multiplicand. According to the reverse order, first multiply 21 by 96, and then shift the multiplier 96 above the multiplicand 21586. To better elucidate the SGT's argument, I show below the reconstruction of the initial steps of this method:

$$\left\{\begin{matrix} 96 \text{ is the multiplier} \\ 21586 \text{ is the multiplicand} \end{matrix}\right\}$$

According to the regular order (*anuloma*/*anukūla*):

$$1^{st}: \longleftarrow 96\times 86$$

$$\begin{matrix} 96 \\ 21586 \end{matrix}$$

$$2^{nd}: \longleftarrow 96\times 15$$

$$\begin{matrix} 96 \\ 21586 \end{matrix}$$

Unfortunately, techniques concerning carry-overs are not spelled out in the commentary and this much affects our understanding and reconstruction of the mathematical procedures. According to the reverse order (*pratiloma*/*pratikūla*), the multiplier 96 shifts above the multiplicand in the following way:

$$1^{st}: \longrightarrow 96\times 21$$

$$\begin{matrix} 96 \\ 21586 \end{matrix}$$

$$2^{nd}: \longrightarrow 96\times 58$$

$$\begin{matrix} 96 \\ 21586 \end{matrix}$$

lines 8–10: Simhatilaka devotes a rather brief description to the method called *tatstha*. He explains that this method too can be executed in the regular order or in the reverse order; each digit of the multiplicand is multiplied by the multiplier, which stands above. While in the *kapāṭasandhi* method the multiplier is placed above the multiplicand and gradually shifted, in the *tatstha* method the multiplier stands where it is, thus the very name of this procedure.[34] Taking as example the same numbers, according to the *anuloma* and *viloma* order, the multiplier 96 stands above the number 21 (in the *anuloma* order) or above the number 86 (in the *viloma* order) of the multiplicand 21586, and the digits of the multiplicand are thus multiplied by the digits of the multiplier

lines 11–17: introduced by *yadvā* ("alternatively"), the commentator explains the third method of multiplication. He analyses the method *khaṇḍa* (lit. "portion") as a procedure which can present the following two cases: the multiplier-*sthāna* or the multiplicand-*rūpa* is divided into *khaṇḍa*s (here *khaṇḍa* denote an aliquot, a divisor). In line 11, Siṃhatilakasūri specifies that by the word *sthāna* (lit. "place"), which is mentioned in Śrīpati's rule,[35] the "multiplier" is meant; in this method, the multiplier-*sthāna* is divided into two or three aliquots. The multiplicand should be multiplied by the two or three quotients obtained and the results summed. In brief, the steps of this procedure are:

$$
\begin{array}{l}
21586 \text{ (multiplicand)} \\
\quad 96 \text{ (multiplier)} \\
\quad\; 96 \div 3 = 32 \\
\qquad\quad \text{or} \\
\quad\; 96 \div 2 = 48
\end{array}
$$

$$(21586 \times 32) + (21586 \times 32) + (21586 \times 32) = 2072256$$
$$\text{or } (21586 \times 48) + (21586 \times 48) = 2072256$$

lines 14–17: the commentator explains the same method but concerning the multiplicand this time. He states that by the word *rūpa,* the "multiplicand" is meant; Siṃhatilakasūri shows that the multiplicand-*rūpa* 21586 is divided into two equal parts and, having written the quotients in two places, these are multiplied by the multiplier 96:

$$
\begin{array}{l}
21586 \text{ (multiplicand)} \\
\quad 96 \text{ (multiplier)} \\
21586 \div 2 = 10793
\end{array}
$$

$$(10793 \times 96) + (10793 \times 96) = 2072256$$

34 The commentator uses the verbal adjectives *anutsārita* and *acālita* to underline that, in this method, the multiplier is not shifted.

35 Kāpadīā (1937, LI, f1) observes that: "Siṃhatilaka's interpretation of *sthāna* as multiplier and *rūpa* as multiplicand (p.5) seems to be rather peculiar".

lines 17–24: in line 17, by *iti caturthaṃ* the commentator introduces the fourth method, which involves two procedures with the latter being twofold. In the first procedure, the multiplier is divided into as many *sthāna*s as these are, and the multiplicand is multiplied by each; this time *sthāna* denotes the notational "place" and no longer the "multiplicand". The commentator explains that the multiplicand 21586 is multiplied once by the digit 9 and once by the digit 6 of the multiplier 96 and the products are added having arranged the lower number by one notational place next with respect to the number above. The two *sthāna*s of the multiplier 96 are 9 and 6; in line 11 one finds the only layout on the multiplication methods and this shows how to place the two products obtained. In this layout, the abbreviation *gu* for *guṇita* or "multiplied" is found:

21586 *gu* 9 = 194274

21586 *gu* 6 = 129516

|194274 |
| 129516|

lines 24–28: from line 24 until the end of the passage, Siṃhatilakasūri illustrates a different twofold *khaṇḍa*-method. In the first method (lines 24–26), the multiplier (he uses again the term *sthāna*)[36] is added to a chosen number. Then, the multiplicand is multiplied once by the sum of the multiplier plus the chosen number, once by the chosen number itself, and the smaller product is subtracted from the larger. He specifies that in this procedure the multiplier is made larger. In order to illustrate this method, the commentator supplies a practical example and mentions again the same operands 21586 and 96, while the chosen number is 2:

$$\begin{array}{r} 21586\,(\text{multiplicand}) \\ 96\,(\text{multiplier}) \end{array}$$

$$96 + 2 = 98$$
$$(21586 \times 98) - (21586 \times 2) = 2072256$$

lines 26–28: the SGT explains a *khaṇḍa*-procedure which presents the reverse case of the preceding. The multiplier-*sthāna* is divided into "portions" which make the multiplier itself smaller. The multiplicand is multiplied by both *khaṇḍa*s, and the products are added. According to the example given by the commentator, the reconstruction is as follows:

36 This means that the commentator interprets *sthāna*, in Śrīpati's *sthānaṃ ca rūpaṃ ca*, as denoting more than one procedure of multiplication.

21586 (multiplicand)
96 (multiplier)

the *sthāna* 96 is subtracted by 4, it gives 92

*khaṇḍa*s of 96 are, in this case, 92 and 4

(21586 × 92) + (21586 × 4) = 2072256

To sum up, the multiplication methods explained by the commentator are:

1) the twofold *kapāṭasandhi*
2) the twofold *tatstha*
3) the *khaṇḍa*-method according to which the multiplier-*sthāna* is divided into aliquots
4) the *khaṇḍa*-method according to which the multiplicand-*rūpa* is divided into aliquots
5) the *khaṇḍa*-method according to which the multiplicand is multiplied by each notational place of the multiplier (place division method)
6) the *khaṇḍa*-method according to which the multiplier-*sthāna* is made larger (this procedure is called *guṇakādhikakārikhaṇḍasañjñaka*)
7) the *khaṇḍa*-method according to which the multiplier-*sthāna* is made smaller (this procedure is called *guṇakalīnatākārikhaṇḍasañjñaka*).

The SGT on GT 19–20: Having explained the multiplication methods while commenting upon the *karaṇasūtra*, Siṃhatilakasūri does not show how to solve this sample problem. Notably, he clarifies the numerical and symbolical association for each term of the *bhūtasaṃkhyā* notation used by Śrīpati.

Division (*bhāgahāra*) [GT 21]

GT 21: Śrīpati's rule on the operation of "division" (*bhāgahāra*) resembles TŚ 9. The GT, as well as L 18, uses the absolute locative *sati sambhave* to denote that one should reduce, "whenever possible", the two operands of the division by a same number. Śrīpati specifies that the division should be carried out according to the reverse order but he does not specify how the two operands should be arranged. Other significant terms are:

- the "dividend" (*bhājya*) and the "divisor" (*hara*)
- the verbal root denoting "to divide", *vibhaj-*
- the absolutive *apavartya* or "having reduced", which indicates the reduction of the two operands by a same quantity.

In mathematical texts, the terms used for "division" are *bhāga* (from the verbal root *bhaj-* lit. "to remove"), *bhāgahāra*, and *bhājana*. Interestingly, there is a group of terms relating to the division, such as *hara/hāra, hṛta, hārita* (from the root *hṛ-* "take away, remove"), that semantically highlight the division as a form of decrease.

The SGT on GT 21: In line 25, the commentator glosses *apavartya* with *khaṇḍitvā* and hence explains that "to reduce" means "to divide", as in fact the operation of reducing two or more quantities by a common factor implies a division. The SGT then elucidates the operands of the division by providing synonyms as well as by supplying new terms to the definitions enunciated by Śrīpati; having specified the role played by the divisor (*hara*), the commentary states that this is also termed *bhāgagrāhaka*. The commentator uses various forms of the root *grah-* "to grasp, to take away" in relation to the operation of division; at the end of this passage he glosses, in fact, *vibhajet* or "one should divide" with *bhāgaṃ grāhayet*. In this regard, it is noteworthy that, throughout his work, Siṃhatilakasūri often uses the term *bhāga* to denote the "division". Having defined the two operands, the commentator explains how to arrange these quantities in the working surface: they should be placed one below the other (the wording suggests that the divisor is below)[37] and, if they are both compatible,[38] before performing the division they should be reduced by the same quantity. Unfortunately, no layouts are given to show how the procedure was carried out.

The SGT on GT 22 (ex.): The commentator emphasises again that, whenever possible, one should carry out the reduction of both operands "by a same quantity". Siṃhatilakasūri then elucidates what the sample problem must determine and by *atra karaṇaghaṭanā* (line 9) he highlights that, by means of this exercise, it is possible to perform the "verification" (*ghaṭanā*) of the "execution" (*karaṇa*). The underlying idea is that via the sample problem on division so formulated by Śrīpati, one can verify (hence *ghaṭanā*) the previously obtained results on the operation of multiplication. In brief, Śrīpati's pedagogical aim is to bring attention to the relationship between multiplication and division as reciprocal operations.[39] The commentator clarifies that the first product obtained in the previous sample problem is 2072256; this should be divided by the initial multiplier 96; that which is obtained is the *mūlaprakṛti* ("primary base") 21586. Therefore, the multiplicand and *mūlaprakṛti* 21586 is the obtained quotient; by means of this reverse process one can verify the results previously calculated. In fact, in the sample problem on multiplication (GT 19):

$A \times B = C$, where A is the *guṇya* 21586
B is the *guṇaka* 96
C is the *guṇita* 2073356

37 Cf. GSS 2.19.

38 To express this idea, he uses the verb *sah-* ("to be capable of, able of").

39 Śrīpati uses this educational aid other times in the GT to simplify the learning process for young students and show the connection between the fundamental mathematical operations: addition/subtraction, multiplication/division, square/square-root, and cube/cube-root. See the same pattern in the treatment of the eight arithmetical operations with fractions.

In the sample problem on division (GT 22):

$C \div B = A$
C is the *prāgguṇita* ("previous product") with respect to $A \times B = C$
B is the *prāgguṇaka* ("previous multiplier") in $A \times B = C$

the quotient A is the *mūlaprakṛti* ("primary base"), which verifies the accuracy of the result of the previous multiplication.

The quotients obtained in solving the sample problem on division given by Śrīpati are hence the *mūlaprakṛtis* and *guṇyas*[40] 93685, 98510, and 12987013.

Square (varga) *[GT 23–$\overline{24}$]*

GT 23–$\overline{\textbf{24}}$: Śrīpati provides three methods of the procedure of square. According to the first method (GT 23), one should perform the square of the *antyapada* or "last place" and multiply then the "last place" by two and by each of the "remaining places" (*śeṣapada*), which are gradually shifted.[41]

The SGT on GT 23: Siṃhatilakasūri's explanation contributes to our understanding of the way numbers were shaped for the execution of the operation of square. In lines 26–27, the commentator clarifies the meaning of *antyapada* ("last place") and states that this denotes the "last digit" (*antyāṅka*) of a given number according to the regular direction (hence from right to left). He also emphasises that, at different stages of the computation, each digit "gradually" (see *iti krameṇa*) becomes the "last digit". The SGT then shows the execution of the square of the number 163, which is the highest number of a set of numbers given by Śrīpati in the next sample problem. The commentator observes that, in the number 163, the digit 1 represents the *antyapada*. The GT's procedure requires one to first determine the square of this last digit, but as Śrīpati has not yet explained what a square is, Siṃhatilakasūri supplies this information[42] and, in line 28, he states that to square a number means to perform the multiplication of two equal numbers. Having clarified what a square is, the commentator elucidates the algorithm meant in the procedural rule by showing its execution.

40 These numbers are the multiplicands mentioned in the previous sample problem (GT 19–20).
41 This is an algorithm for calculating the square by means of notational places.
42 This observation on the structure of the text is made by the commentator in his passage on the third method of square.

The SGT provides eight layouts[43] illustrating various steps of the algorithm for squaring the number 163:[44]

- the first *antyapada* 1 is squared and placed below the given number:

$$1^2 = 1\left|\begin{array}{l}163\\1\end{array}\right|$$

- by the expression *krameṇeti śeṣaḥ* (line 2), the commentator provides important information and specifies that, once doubled, the last digit should be multiplied by the remaining places one after the other: this means that the last digit should be multiplied by the "remaining place" 6 and by the "remaining place" 3. The *antyapada* 1 is doubled and the product 2 is multiplied by the "remaining places"; the results are 12 and 6 respectively. The SGT explains that the two digits of the number 12 are written below the 1 and the 6 of the original number 163 and that the 1 previously obtained (the result of the square of the first *antyapada* 1) is moved below:

$$(1\times 2)\times 6 = 12\left|\begin{array}{l}163\\\mathbf{12}\\1\end{array}\right|$$

$$(1\times 2)\times 3 = 6\left|\begin{array}{l}163\\12\mathbf{6}\\1\end{array}\right|$$

$$(1\times 2)\times 3 = 6\left|\begin{array}{r}163\\126\\1\end{array}\right|$$

- the commentator clarifies that "having caused to shift" (*samutsārya*) means to move the remainder above, which is composed of the two

43 The layouts given in the published edition are partly mistaken. In his English introduction, Kāpadīā (1937, LII, f4) gives an improved version of the layouts and observes: "It appears that out of these 11 steps, the steps (vi) to (ix) have been wrongly given in the original MS. by the scribe, and through oversight this mistake has not been corrected by me on p.9, and that the steps (v), (vi) and (xi) have been omitted in the original MS. by the scribe". The editor has exactly the same view in relation to the procedure of cube. Kāpadīā (1937, LIII-LIV f2) shows improved layouts for the execution of the cube and observes: "In this case, the steps (i) and (v) have been misprinted on pp.11 and 12, and the 8th (last) has been omitted in the original MS. by the scribe." This consideration seems to confirm that: i) in the edition, misprints and mistaken layouts occur frequently, and that ii) Kāpadīā believes that steps are missing because omitted by the scribe. It should be kept in mind, however, that in mathematical works, only the essential steps were often sketched, while details were explained verbally.

44 For the sake of clarity, I highlight in bold the new results placed in the layouts.

"remaining places", by one notational place more in respect to the lower number obtained.[45] Therefore the *antyapada* 1 is erased and the *śeṣapada* 63 is shifted: $\left|\begin{array}{l} \mathbf{63} \\ 126 \\ 1 \end{array}\right|$

- the procedure starts again: the next "last digit" is now the 6, which is squared and placed in the layout: 6^2 $\left|\begin{array}{l} 63 \\ 126 \\ \mathbf{136} \end{array}\right|$
- the *antyapada* 6 is multiplied by 2 and by the *śeṣapada* 3: thus $(6 \times 2) \times 3 = 36$ $\left|\begin{array}{l} 63 \\ 126\mathbf{6} \\ 136 \\ \mathbf{3} \end{array}\right|$
- the *antyapada* 6, being computed, is erased and the *śeṣapada* 3 shifted: $\left|\begin{array}{l} \mathbf{3} \\ 1266 \\ 136 \\ 3 \end{array}\right|$
- the last *antyapada* 3 is squared. Siṃhatilakasūri explains that its square 9 is placed below the 3, in the line next to the previous results: $\left|\begin{array}{l} 3 \\ 1266\mathbf{9} \\ 136 \\ 3 \end{array}\right|$.

  Because there are no other remaining places, the procedure ends here. The number 3 should be erased, although the commentator does not say so explicitly and the numbers arranged are summed; the square of 163 is thus 26569.

Finally, it is noteworthy that the commentator uses two ways to refer to the digits of a given number, as he switches between i) the *anulomagati* or "regular course" (hence from right to left), which is the method to be used in order to mark the digits of the given number, and ii) a left to right direction, hence from the larger towards the smaller notational place, which is the way numbers are read according to their power of ten.

Regarding the second procedure (GT $\overline{24}$), the commentator's explanation is brief and clear, as the rule is fairly simple. Let the given number be 5; one should opt for a "chosen" (*iṣṭa*) number which is "convenient" (*abhirucita*). This remark implies that the chosen number is not arbitrarily assumed but must fit certain conditions; for instance, one understands that the *iṣṭa*-number must be smaller than the given.

45 The remainder above is moved according to a left to right direction.

Therefore:

the given number is 5
the chosen number is 2

one should write the 5 in two places and execute the following steps:

$$(5+2)\times(5-2)=21$$
$$2^2=4$$
$$21+2^2=25$$

Śrīpati gives a third procedure (GT $\overline{24}$), which is actually the definition of a square; it states that the square is the product of two identical numbers. After having explained that the equal multiplication denotes the multiplication of two same numbers such as 12 by 12, the commentator disapproves the order of presentation of the methods of square presented by the GT. In line 29–2, Siṃhatilakasūri argues that in the GT, the order of the methods of square is not convenient; he quotes a passage from the L to demonstrate a better arrangement of the various methods of square. In this way, he argues that, as in the L, it would have been better to first mention what a square is and only then to explain how to carry out it. In the GT, the opposite case in fact occurs and the commentator has first to explain what a square is (SGT 23, 28) in order to elucidate the first method of square.

Furthermore, Siṃhatilakasūri himself provides a fourth rule by quoting a passage from the L. He illustrates the rule by the following example: let the given number be 5, this should be broken into *khaṇḍa*s ("parts"), such as the numbers 2 and 3. These two *khaṇḍa*s are multiplied and the product obtained is then doubled. The two *khaṇḍa*s are squared, their squares summed, and this result is added to the resulting product which has been doubled. Below I show the reconstruction of this procedure:

5 is the given number
the numbers 2 and 3 are the two *khaṇḍas*

$$2\times3=6$$
$$2^2=4$$
$$3^2=9$$
$$4+9=13$$
$$13+(6\times2)=25$$

Finally, Siṃhatilakasūri quotes a rule from TŚ 11 which concerns arithmetical series.

Let 5 be the given number, the method explains that:

$$5^2 = 1 + 3 + 5 + 7 + 9 = 25.$$

The SGT on GT 25 (ex.): The SGT provides only the results in figures and in verbal numerical expressions, having already elucidated the methods of square in detail.

Square-root (*vargamūla*) [GT 26]

GT 26: Śrīpati's *karaṇasūtra* of square-root resembles the rule found in Śrīdhara's TŚ.[46] In mathematical texts, the square-root is *vargamūlā* or *kṛtipada*; in the context of square-root or cube-root procedures, one often finds the brief form *mūla* or *pada* meaning "root". For instance, 3 is the *mūla* or *pada* of its *varga* or *kṛti* 27.

In GT 26, the most significant elements mentioned are:

- the last "odd place" (*viṣamapada*) of a given number, from which one should subtract (see the verbal root *viśudh-*) the highest square the given number itself contains
- the "remainder" (*śeṣa*) and the "root" (*pada*), the latter being the divisor of the remainder which should be moved from its place and multiplied by two
- in the layout, the "result" (*phala*) should be placed in the "line" (*paṅkti*)
- having subtracted the square, the square-root will be in the half of the doubled number.

The SGT on GT 26: In the introductory passage, Siṃhatilakasūri underlines that the author gives the rule to arrive at the "root" (*mūla*) of the squares of the numbers previously (i.e., in the previous sample problem) obtained. The commentator explains the procedure by taking as an example the number 26569,[47] which is the square of 163 obtained by means of the execution shown in the SGT while commenting upon GT 24. As before in the operations of subtraction and division, here too the sample problem asks students to work with the numbers obtained in the previous sample problem, which regards the square, and to determine the "roots" of the squares previously obtained.

The commentator observes that, with respect to a given number, one should designate the digits one after the other as "odd" (*viṣama*) and "even" (*sama*). This way of marking numbers, which follows a right from left direction, is positional and not numerical; for instance, with respect to the number 26569,

46 See TŚ 12–13 and PG 25–26. This is an algorithm for calculating the square-root by notational places. Plofker (2009, 124), Datta and Singh ([1935] 1962), and SaKHYa (2009, 108) explains the way this procedure is expounded in other Sanskrit texts.

47 This is one of the numbers given by Śrīpati in the next sample problem.

the digit 9 is the first "odd place" (*viṣamapada*), the 6 is "even", and so forth. In line 29, Siṃhatilakasūri points out that, according to this calculation, there is a "resting place" (*viśrāma*), denoting the alternating sequence of odd and even places by which the digits of a number are subdivided.

In order to better understand the SGT's execution of this procedure, below I show the layouts found in the commentary, accompanied by a reconstruction and explanation of the intermediate passages:

- line 3: the given number is 26569
- lines 3–7: before arriving at the layout found in line 7, there are intermediate passages which are not shown in the SGT. The first layout should thus be:

26569
1

  , where 1 is the result of the subtraction of the square of its *pada* 1 ($2{-}1^2$) from the *viṣamapada* 2. Then, one should divide the "remainder" (*śeṣa*) by the *pada* ("root") moved from its place and multiplied by 2. By the expression "moved from its place", one understands that, during the algorithm, the *pada* is first written in another place of the calculating board, as a partial result. Importantly, the remainder is composed of the result of the subtraction of its highest square from the *antyapada* 2, in this case 1, and of the next place 6 (the 6 of 26569), which is found next to the *antyapada* 2. Thus, the first remainder is the number 16. The layout should be:

16569
2
6

  , where the remainder above is 1 (from the 2 of 26 minus the *pada* 1 squared and placed below, see the previous layout) and the 6 of [2]6569. The 2 below is the *pada* 1 moved from its place and multiplied by 2 ($1 \times 2 = 2$) and the 6 below 2 is the *yogyāṅka* or the "suitable number" (see line 5), here denoting the quotient. This quotient arises from the following steps: the remainder above, which is 16, is divided by the *pada* 1 multiplied by 2, thus: $16 \div 2$ In this case, however, the resulting quotient 8 from $16 \div 2$ is not a suitable number, since in this way the procedure cannot continue; this step requires, in fact, some readjustments. In fact, according to the next steps of this algorithm, from $16{-}(8 \times 2)$ there is no remainder and the square of the quotient (8^2), which is 64, cannot be subtracted from the 5 above (the 5 of 26569). One should try with a lower quotient, and the suitable number comes to be 6; therefore, $16 \div 2 = 6$, which is the chosen quotient and it is placed in the last line of the previous layout. This quotient becomes the next *pada*. Above, the number 16 is replaced by 4, which is the result of $16{-}12$, where 16 is the remainder and 12 is the result of the chosen quotient 6 multiplied by 2. In line 7, the SGT provides the following layout:

4569
2
6

- lines 7–8: the 6 below is placed in the same "line" (*paṅkti*) of the 2

- line 9: the quotient 6 is the next *pada*, which is to be squared. The result 36 is subtracted from the remainder 45 above, thus: 45–36= 9
- line 10: the "remainder" above is 96, since the 45 above has been replaced by 9; hence the layout should look like this:

969
26

- lines 10–12: between intermediate steps, digits are erased and results replaced. The remainder 96 is divided by 32; this 32 is the result of the union (see *yojanā*, line 11) of the two *pada*s 1 and 6, hence 16, multiplied by 2, thus 96÷32 = 3. The commentator observes that this 3 is placed first below the 32 and then in the same line next to it. The layouts should look like the following:

969	969
32	323
3	

- lines 13–14: the remainder above, which is 96, is erased. The quotient 3 becomes the next *pada*, its square is 9, which is subtracted from the remainder above, which is now the digit 9, hence 9–9 = 0. Thus, above there is no remainder
- lines 15–19: the SGT explains that the *pada* 16 multiplied by 2 is 32, halved is 16; the last *pada* 3 is joined to 16.[48] In this way, the final result is 163.

The SGT on GT 27: The commentator paraphrases the root-text, elucidates the meaning of the word-numerals used and, having explained the algorithm while commenting upon the *karaṇasūtra*, gives the results without any additional details.

Cube (*ghana*) [GT 28–30]

GT 28–30: Śrīpati gives four procedures to calculate the "cube" (*ghana*) of a number; since the method which corresponds to the definition of a cube occurs only at the very end (GT 30), Siṃhatilakasūri disagrees with Śrīpati's expository structure.[49] At the end of the explanation of the third method of cube, the commentator states, in fact, that the author should have first mentioned what a cube is and only afterwards enunciate the methods for calculating it.[50] To supply this information, before showing the execution of the first procedure, Siṃhatilakasūri defines a cube by quoting the last line from Śrīpati's passage, where the definition of the cube is found.

In the first method of cube (GT 28, 3–6),[51] the main elements are:

48 The commentator explains that the number 3 is not doubled and not halved but it is placed (see *sthāpanīyam*) as it is (see *tadavastham eva*).

49 See also the SGT on GT 23, which is the procedure of square.

50 For this reason, the commentator quotes L 24; in L 24–26, the definition of a cube corresponds, in fact, to the first method expounded.

51 This is an algorithm for calculating the cube by means of notational places.

- the "last" digit (*antya*)
- the "preceding" digit (*ādi*)
- and the way partial results are placed in the layout. In this regard, it is important to note that, in the *sūtra*, the expression *sthānādhikatva* denotes that way results should be placed in the layout before being finally added.

The second method (GT 29, 7–9) concerns a "series" (*pracaya*) of integers of which the number "one is the first" (*ekādi*). The third procedure (GT 29, 10) is the definition of a cube and the last (GT 30) involves two "parts" (*khaṇḍa*) of a given "number" (*rāśi*).

The SGT on GT 28–29: In order to explain the first procedure enunciated by Śrīpati, Siṃhatilakasūri performs the cube of the number 317:[52]

- the commentator clarifies that, in 317, the *antya* or "last" digit is 3
- the cube of the *antya* 3 is placed below the given number and moved over one notational place, thus $3^3 = 27$;

317
27

- lines 13–16: the square of the *antya* 3 is multiplied by 3 and by the "preceding" digit 1 (of 317), thus: $3^2 \times 3 \times 1 = 27$, this result is placed in two different lines, which are the line of units (above) and the line of tens (below) respectively:

317
277
2

- lines 16–19: the square of this preceding digit 1 is multiplied by the last digit 3 and by 3, thus: $1^2 \times 3 \times 3 = 9$;

317
277**9**
2

- lines 19–20: the cube of the preceding digit 1 is placed in the layout: $1^3 = 1$;

317
2779**1**
2

- lines 20–1: the commentator explains that, at this point, the procedure starts again from the beginning but he points out that not every step should be performed. Between the preceding and the following layouts (line 20 and line 1), the SGT provides various explanations and shows partial results in figures. In lines 25–29, the commentator observes that 31 is the next *antya*; the square of 31 is multiplied by 3 and by the preceding

52 The SGT explains the four methods of cube by means of the numbers 317, 3, 4, and 5, which are among the numbers given by Śrīpati in the next sample problem (GT 31).

digit 7, thus: $31^2 \times 3 \times 7 = 20181$. This is written below and moved over one notational place:[53] $\begin{array}{|l|} 317 \\ 27791 \\ \;\;2 \\ \;\;\mathbf{20181} \end{array}$

- lines 4–7: the square of the "preceding" digit 7 is now multiplied by the *antya* 31 and by 3, thus: $7^2 \times 31 \times 3 = 4557$. This is placed below and moved over one notational place (line 4): $\begin{array}{|l|} \;\;317 \\ 27791 \\ \;\;2 \\ \;\;20181 \\ \;\;\;\;\;\;\mathbf{4557} \end{array}$. The last step concerns the cube of the "preceding" digit 7. Hence, 343 is placed one additional place more with respect to the last result and then all the numbers are added together (the *mūla* 317 should be first erased): $\begin{array}{|l|} 317 \\ 27791 \\ \;\;2 \\ \;\;20181 \\ \;\;\;\;\;\;4557 \\ \;\;\;\;\;\;\;\;\mathbf{343} \end{array}$. Therefore, the cube of 317 is 31855013.

The expressions *anyatra sthāpayitvā* and *anyatra lekhya* (line 14, 26, and 2) indicate that intermediate results were written "elsewhere" (*anyatra*) in the calculating board and not in the main layout were the algorithm was carried out. Also, it is noteworthy the commentator's simultaneous use of two ways to denote digits: a) one emphasises the method to be applied in order to perform the algorithm, and b) the other corresponds to the way digits are read according to their power of ten. The first is from right to left, the other is from left to right. For instance, in order to perform the procedure of cube, in the given number 317 the 3 is the "last" (*antya*) and the 1 is the "preceding" (*ādi*); however, according to the way numbers are read by means of the decimal place value notation, the 1 is the digit which occurs next (see *agretanāṅka*) to the 3.

The second method (GT 29) of cube given by Śrīpati involves a series of numbers in arithmetical progression. This procedure is also found in GSS 2.44 and TŚ 15. Siṃhatilakasūri observes that the arithmetical series should be placed in a *ūrdhvāṅkaśreṇi* or in "a vertical row of numbers", although he does not show it. The SGT explains that the cube of the number 3 is equal to the cube of the "preceding" number (hence 2) plus 1, the result is added to the last term multiplied by 3 and by the preceding. Hence, with respect to 3,

53 It can be noticed that a change takes place and now "one [notational] place more" is from left to right; this change corresponds to the moment during which the procedure starts again from the beginning.

the "preceding" is the number 2 and the "last" is the number 1. The cube thus should be:

$3^3 = (2^3 + 1) + (3 \times 3 \times 2) = 27.$

The third method (GT 29) represents the definition of a cube: the cube is the multiplication of a collection of three equal numbers, which the commentator illustrates by means of the number 4.

In lines 26–3, Siṃhatilakasūri explains the last method (GT 30) of square by means of the number 5. He shows that that this number is first divided into two "parts" (*khaṇḍa*) and that these are written in two places (line 26). The *khaṇḍa* 2 is multiplied by the given number 5, and the result multiplied by the *khaṇḍa* 3 and then again by 3. This result is added to the cubes of the two *khaṇḍas*. Thus: $5^3 = [(2 \times 5) \times 3] \times 3 = 90$

$90 + 2^3 + 3^3 = 125$, which is the cube of 5.

The SGT on GT 31 (ex.): The commentator presents the *nyāsa* of the numerical data and the results in both digits and verbal expressions and provides no further explanations.

Cube-root (*ghanamūla*) [GT 32–33]

GT 32–33: In Śrīpati's *karaṇasūtra* on *ghanamūla* or "cube-root", the most important elements to note are:

- the subdivision of the digits of the given number into the following positional scheme: one "cubic place" (*ghana*) and two "non-cubic places" (*aghana*)
- the highest cube included in the last cubic place is subtracted and its "root" (*mūla*) is brought below the third place
- the remainder is divided by the square of the root multiplied by three
- the quotient is placed in the line, its square multiplied by the last and by three is then subtracted from the remainder above
- the cube of the quotient is subtracted from the result.

The SGT on GT 32–33: The commentator shows how to perform the algorithm of the cube-root of a given number by means of the cube obtained while solving the previous sample problem, which is the cube of 317, namely 31855013. This is, in fact, one of the "cubes" of which Śrīpati requires students to ascertain the "root" in the next sample problem.

The procedure of cube-root involves first of all the subdivision of the digits of the given number 31855013 into "one cubic place" and "two non-cubic places". Below I provide the explanation of the various steps according to the SGT:

31855013
3

  (line 29): the number above, which is the given number 31855013, is the *ghana* or "cube" of which the *mūla* or "root" is to be determined. The first step is: subtract the highest possible cube contained in the "last cubic place" (*antyaghana*), which is here 1 (from right to left). However, although Siṃhatilakasūri does not directly mention it, in such a case in order to carry out the calculation one should consider a larger number, and 31 becomes thus the *antyaghana*. In the layout shown above, the 3 below is the *mūla*, as it is the "root" of the highest possible cube (27) contained in 31. The commentator explains that this 3 should be placed below the 1 of 31
4855013
3

  (line 4): the *antyaghana* 31 has been lessened by the cube of 3 which is 27 and above the remainder 4 has been placed instead. As the *karaṇasūtra* states that "one should bring the root below the third place" from the last cubic place, Siṃhatilakasūri explains that the *mūla* 3 is moved below the digit 5, since this is the third *pada*
4855013
273

  (line 6): the commentator explains (lines 4–6) the step from the *karaṇasūtra*: "one should divide the remainder by its square multiplied by three". He observes that *asya* ("its") in *trinighnakṛtyāsya* refers to the "root", which is in this case 3, whose square multiplied by 3 is the number by which the remainder is divided. Thus, $3^2 \times 3 \times = 27$. Siṃhatilakasūri notes that this result should be placed one notational place less (see *ekasthānonatayā*) and behind (from right to left) the *mūla* 3
- between the preceding (line 6) and the following layout (line 12), there are intermediate passages. The next layout is

2155013
31

  , it can be noticed that 27 has been subtracted from 48 which was the remainder above; the 1, which was placed below,[54] is brought in the line in front of the 1, and the 27 has been erased
- then: $1^2 \times 3 \times 3 = 9$, which subtracted from the remainder above (215) becomes 206:

2065013
31
- at this point, the algorithm goes back to the first step. The cube of the quotient, which is the number 1 from the division aforementioned (the remainder 48 divided by 27), should be subtracted from the 5 above. In the next layout, this is replaced by the number 4 ($5-1^2$). In the next step,

54 See line 7. The preceding layout should be as follows:

4855013
273
1

.

as earlier, one should bring the root 31 below the third place, from the last cubic place which here is 4: $\left|\begin{array}{r}2064013\\ \mathbf{31}\end{array}\right|$

- 31 is squared and multiplied by 3, thus: $31^2 \times 3 = 2883$, which is placed below by one notational place less in respect to the 31: $\left|\begin{array}{l}2064013\\ 31\\ \mathbf{2883}\end{array}\right|$
- the remainder above 20640 is divided by the result 2883, and the quotient 7 is placed next to 3 and 1: $\left|\begin{array}{r}2064013\\ 317\\ 2883\end{array}\right|$. The numbers 28, 8, and 3 of 2883 are multiplied by the quotient 7 and each result is subtracted from the remainder above, which changes accordingly, and below the divisor 2883 is erased.

In order to better understand the procedure, I reconstruct the last steps and layouts not shown in the SGT:

- $20640 \div (31^2 \times 3) = 20640 \div 2883 = 7$
- the numbers 28, 8, and 3 of 2883 are multiplied by the quotient 7; hence, $28 \times 7 = 196$
- $206 - 196 = 10$. This 10 is the new remainder, so erase 206 and write 10 instead. The layout should look like this: $\left|\begin{array}{l}\mathbf{10}4013\\ 317\\ 2883\end{array}\right|$
- $8 \times 7 = 56$; $104 - 56 = 48$. This 48 is the new remainder, so erase 104 and place 48. The layout is: $\left|\begin{array}{r}\mathbf{48}013\\ 317\\ 2883\end{array}\right|$
- $3 \times 7 = 21$; from the remainder above, $80 - 21 = 59$. The layout is: $\left|\begin{array}{r}\mathbf{459}13\\ 317\\ 2883\end{array}\right|$

  Next, because it has been manipulated, the divisor 2883 is erased.

In line 28, the SGT shows the following layout: –. This is the result of the step of the *karaṇasūtra* paraphrased by the SGT in line 24: *tatkṛtim antyanighnīṃ trisaṅgunāṃ cāpanayed* or "one should subtract its square multiplied by the last and by three". In fact, in this layout one finds the number 4557, which is the result of the square of the quotient 7 multiplied by 31 and by 3. The commentator explains that the 7 of 57 should be placed below the 1 of 31. Then, one should carry out the subtraction of this result from the remainder above. Thus:

$45 - 45 = 0$
$91 - 57 = 34$

The layout as this point should be the following:
$$\begin{array}{|r|} \mathbf{343} \\ 31 \\ 7 \end{array}$$

The commentator explains (lines 30–1) that from the remainder above the cube of 7 should be subtracted. Thus, $343 - 7^3 = 0$; the remainder above is erased and the procedure is concluded: the cube-root of 31855013 is 317.

Lines 2–10: Siṃhatilakasūri himself supplies a method and introduces it by the expression *āmnāyena* or "according to a traditional usage". This passage is unfortunately corrupt and a further challenge is that there are two terms which are not found elsewhere in the SGT, namely *bhāgāpahāra* and *bhāgāṅka*. I render these terms as "multiplication by means of the division" and "divisor" respectively. The procedure presented is a kind of verification by which the result of a cube, the cube-root itself, is involved.

The commentator first states that the cube 31855013 is divided by 100489, which is the square of the root that just obtained above, namely 317. From now onwards, the procedure involves the "multiplication" (*apahāra*) of the "divisor" (*bhāgāṅka*) 100489 by each of the three digits composing the obtained quotient and cube-root[55] 317. Such as multiplication is based on the quotient resulting from this division and this is, in my opinion, the meaning of the term *bhāgāpahāra*. The result of $31855013 \div (317^2)$ is 317, but in the first layout (line 5):
$$\begin{array}{|l|} 31855013 \\ 100489 \\ \;\;3 \end{array}$$
only the 3 is shown, and this is below 100489 which is the square of 317. The 17 of 317 is placed, in fact, somewhere in the calculating board. The SGT specifies that the multiplication of the divisor 100489 by the quotient 3 should be performed. Then, the product obtained, which is 301467, should be placed in a new layout with the cube standing above. The product 301467 is subtracted from the cube 31855013 but this subtraction is performed with the two numbers placed exactly one below the other from the left; the digits are not placed not according to their decimal value. The layout should look like the following:
$$\begin{array}{|l|} 31855013 \\ 301467 \\ \;\;1708313 \end{array}$$
, with 1708313 being the result of the subtraction of the two numbers above.[56] This 1708313 replaces the original cube 31855013 and the divisor is again written below:
$$\begin{array}{|l|} \;\;\;\;1708313 \\ 100489 \end{array}$$
. Next, the number above is moved and aligned at the far left so as to be perfectly above the divisor 100489:
$$\begin{array}{|l|} 1708313 \\ 100489 \end{array}$$
. Then, the divisor 100489 is multiplied by the other digit of the quotient 317, which is 1, and this is placed

55 The three digits are 3, 1, and 7 of 317.

56 Thus, $31855013 - (100489 \times 3)$, with the numbers placed as explained above.

below: $\begin{array}{|c|}1708313\\100489\\1\end{array}$. When there is the subtraction of the product obtained by multiplying 100489 by 1 from the number above,[57] the result is 703423. The two quotients 3 and 1 (hence 31) are placed elsewhere in the calculating board, and the other quantity, the 7 of the quotient 317, is placed below: $\begin{array}{|c|}703423\\100489\\7\end{array}$.

After having aligned the number above with the number below $\begin{array}{|c|}703423\\100489\\7\end{array}$, the divisor 100489 is multiplied by the digit 7 and the product, which is 703423, is subtracted from the number standing above. The layout is: $\begin{array}{|c|}703423\\703423\end{array}$. The number below is subtracted from the number above and the result, which is 0, proves that 317 is indeed the cube-root of 31855013.

The SGT on GT 34 (ex.): The SGT provides only the numerical presentation of the initial data and the roots required to be determined, and ends the passage by *etat samāptyā pūrvāṇy aṣṭau parikarmāṇi samāptāni* or "With the completion of this [subject], the first eight arithmetical operations are accomplished".

4.5 The eight arithmetical operations with fractions [GT $\overline{35}$–51]

Some remarks

In the GT, the eight fundamental operations with integers (GT $\overline{13}$–34) are followed by the treatment of the same arithmetical operations applied to fractions. In Sanskrit mathematical treatises, a term used to denote a fraction in general is *bhinna* (lit. "broken, split"); in the SGT, the word more frequently used to denote a fraction is "*bhāga*"; Siṃhatilakasūri's vocabulary testifies the polysemic usage of *bhāga*: this is frequently "division", "fraction", and occasionally "numerator". In both the GT and the SGT, the terms used for "numerator" are *aṃśa* and *lava*,[58] the "denominator" is *cheda*/*chedaka,* and *hara*/*hāra.* Fractions are denoted by the use of a cardinal number compounded with the terms *bhāga, lava,* and *aṃśa* meaning "part" or "portion";[59] for instance, *tribhāga* as well as *trilava* mean "one-third". With respect to the arrangement of the numerator and of the denominator, in Sanskrit mathematical texts the numerator is placed above and the denominator below. Fractions are written

57 Note that the numbers are placed below each other as previously explained.

58 The term *aṃśa* is also "fraction" (see GT 57) and "denominator" (SGT 49, 29).

59 In the SGT, one also finds the use of ordinal numbers followed by *bhāga* to express fractions.

without the dividing line and inside rectangular boxes such as $\left|\begin{matrix}1\\2\end{matrix}\right|$ which is, for instance, the way they are also found in the Bakhshālī Manuscript.

The expression *bhinnaparikarman* denotes an "arithmetical operation with fractions" (*bhinna*). In the GT, the eight arithmetical operations with fractions are treated just before the five varieties or "classes" (*jāti*) for the simplification of fractions. It is important to note that the order of presentation of Śrīpati's work requires the commentator to modify his own presentation; in order to explain and carry out the sample problems on the operations with fractions given by Śrīpati, the commentator has to anticipate some of rules of the *jāti*s. While showing the execution of the sample problems on the operations with fractions, Siṃhatilakasūri quotes passages from the rule of the "class of simple fractions" (*bhāgajāti*), the "class of fractional increase" (*bhāgānubandhajāti*), and the "class of fractional decrease" (*bhāgāpavāhajāti*);[60] in this way, he provides the explanation of some mathematical methods in advance, since in the various computations these are required earlier than Śrīpati includes them in the root-text. Despite this feature, in the last two lines of his exposition of the cube-root of fractions (*bhinnaghanamūla*) Siṃhatilakasūri points out that, in the root-text, the topic of arithmetical operations with integers is rightly followed by the arithmetical operations with fractions. Most of the treatises apply this same structure,[61] while in the L the classes of simplification of fractions are explained before the operations with fractions.

Addition of fractions (bhinnasaṅkalita) *[GT $\overline{35}$]*

GT $\overline{35}$: In the *karaṇasūtra* of *bhinnasaṅkalita* or the "addition of fractions", Śrīpati states that in order to perform addition one should reduce fractions to the "same denominator" (*sadṛśahara*). Also, he clarifies that in the case of a "quantity devoid of denominator", which denotes a whole number, the number one is to be arranged as its denominator. The algorithm, which is elucidated by the commentary, can be represented as[62]:

$$\left|\begin{matrix}a\\b\end{matrix}\right| + \left|\begin{matrix}c\\d\end{matrix}\right| = \left|\begin{matrix}a\\b\\d\end{matrix}\right| + \left|\begin{matrix}c\\d\\b\end{matrix}\right| = \left|\begin{matrix}ad\\bd\\d\end{matrix}\right| + \left|\begin{matrix}cb\\db\\b\end{matrix}\right| = \left|\begin{matrix}ad+cb\\bd \text{ or } db\end{matrix}\right|$$

The SGT on GT $\overline{35}$: Commenting upon this passage, Siṃhatilakasūri uses a wide range of technical terms. In line 22, it is interesting to note the contrast between the expression *pūrṇarūpa,* which denotes a "whole number",

60 GT $\overline{53}$, 57, and 60.
61 See the PG, TŚ, GSS, and GK.
62 Note that the formulae *bd* and *db* represent the same result.

and *khaṇḍitarūpa*, which is lit. a "broken number", used by the commentator as a synonym for *bhinna* or "fraction". The SGT also employs *pūrṇa* and *rūpa* as substantives to denote an "integer". While commenting upon the sample problem of GT 36, Siṃhatilakasūri uses the expression *rūpakhaṇḍa* to gloss *bhinna*; in this way, fractions are denoted as "parts of integers". Regarding the term *rūpa*, in Siṃhatilakasūri's passage this denotes a "whole number, an integer" as well as the number "one, unity".

The SGT on GT 36 (ex.): Siṃhatilakasūri presents the *nyāsa* of the four fractions given in the sample problem: $\begin{array}{|c||c||c||c|}1&1&1&1\\2&3&9&18\end{array}$. In order to calculate the same denominator and add then the numerators, the commentator has to quote and explain in advance the class of simple fractions enunciated in GT 53. The commentator manipulates the first pair of fractions from left to right and adds the resulting fraction to the next one. Starting from the first on the left, in the SGT each next fraction is denoted by the expression *agretanāṅka* or the "quantity which occurs next"; a fraction is so denominated according to the space it occupies in the layout with respect to the fraction just performed. The SGT explains how to carry out the sum of these fractions and illustrates the execution by means of the following steps:

- Siṃhatilakasūri quotes steps and paraphrases passages from the rule of *bhāgajāti*. He observes that the "exchange" (*vinimaya*) of the two denominators of the first two fractions should be made. They are placed below the original denominators:[63] $\begin{array}{|c||c|}1&1\\2&3\\3&2\end{array}$
- the numerators and denominators are multiplied by the exchanged denominator: $\begin{array}{|c||c|}3&2\\6&6\\3&2\end{array}$
- after that the fractions have been reduced to the same denominator, the sum of their numerators is calculated and the original denominators are erased:[64] $\begin{array}{|c|}5\\6\end{array}$

63 In this regard, the commentator uses the correlatives *ekatra/aparatra* to denote the exchange and setting down of the two denominators.

64 In the SGT (lines 7), another layout is given before this, where the denominator of the *agretanāṅka* is still retained although the numerators have been added and the first denominator is the obtained common denominator. In lines 7–8, this is then said to be erased and the new layout is shown. The SGT first gives $\begin{array}{|c|}5\\6\\2\end{array}$ and then $\begin{array}{|c|}5\\6\end{array}$.

- in line 8, the *agretanāṅka* $\left|\begin{array}{c}1\\9\end{array}\right|$ is shown; there is the exchange of the denominators: $\left|\begin{array}{c}5\\6\\9\end{array}\right|\left|\begin{array}{c}1\\9\\6\end{array}\right|$
- the result of the multiplication of the original terms of the first fraction by the new denominator 9 is $\left|\begin{array}{r}45\\54\\9\end{array}\right|$, while the second computed fraction becomes[65] $\left|\begin{array}{r}6\\54\\6\end{array}\right|$
- the result of the sum of the two fractions is $\left|\begin{array}{c}51\\54\end{array}\right|$
- introduced by *athāgretanāṅka*, the next fraction $\left|\begin{array}{r}1\\18\end{array}\right|$ is shown and the operation proceeds as before until the final result $\left|\begin{array}{c}972\\972\end{array}\right|$, which corresponds to 1, is obtained.

In these last lines, Siṃhatilakasūri elucidates the expression *haravirahita* ("devoid of denominator") mentioned in the *karaṇasūtra* and states that it denotes a whole number. When an integer such as the number 6 – the number mentioned by Śrīpati in the next sample problem – is added to a fraction, the number one is to be placed as its own denominator. In this context, the use of the term *rūpa* in *ṣaḍ rūpaṇi* highlights that the number 6 is an "integer". The commentator shows that the sum of two fractions having different denominators is calculated only after they have been reduced to the same denominator.

The SGT on GT 37 (ex.): The *karaṇasūtra* formulated in GT 35 does not explain how to execute the addition or subtraction of an integer plus or minus a fraction; therefore, this sample problem assumes the knowledge of mathematical procedures which, in fact, have not yet been explained in the root-text. For this reason, the commentary has to quote and explain passages from the *bhāgānubandhajāti* and the *bhāgāpavāhajāti* which are mentioned later in the GT.[66] These two rules on fractions involve the "class of fractional increase" (*bh āgānubandhajāti*)[67] and the "class of fractional decrease" (*bhāgāpavāhajāti*).

65 In line 12, this layout is missing and the sentence ends abruptly with *yathā*.
66 GT 57 and GT 60.
67 In GT 57, it is denoted by Śrīpati as *aṃśānubandhajāti*.

The SGT gives the following *nyāsa*: $\begin{array}{|c|}3\\1\\2\end{array}\begin{array}{|c|}6\\1\\ \end{array}\begin{array}{|c|}9\\01\\4\end{array}\begin{array}{|c|}7\\1\\3\end{array}$, which is read as $3+\frac{1}{2}$, 6, $9-\frac{1}{4}$, $7+\frac{1}{3}$. The integer 6 has 1 as denominator and there is a zero in front of the numerator 1 of the fraction $\frac{1}{4}$. This is an operational sign and it denotes that the quantity is "negative" because it has to be subtracted. In order to solve the sample problem and to perform the first step of the procedure, Siṃhatilakasūri observes that first the *bhāganubandhajāti* should be carried out and quotes a passage (line 4) of the rule enunciated in TŚ 24. The first procedure of the rule of the class of fractional increase involves the sum between an integer and a fraction, which in modern notation is expressed as $(a+\frac{b}{c})$, while in the SGT is given as $\begin{array}{|c|}a\\b\\c\end{array}$ and performed as $\begin{array}{|c|}ac+b\\c\\ \end{array}$. Similarly, the first procedure of the class of fractional decrease involves the deduction of a fraction from an integer, which in modern notation is expressed as $(a-\frac{b}{c})$, while in the SGT is given as $\begin{array}{|c|}a\\0b\\c\end{array}$ and manipulated as $\begin{array}{|c|}ac-b\\c\\ \end{array}$.

Let us look now at Siṃhatilakasūri's execution of the sample problem:

- the commentator emphasises that, in the first quantity, the number 3 is a whole number; he similarly characterises as such the other integers of this sample problem, namely 6, 9, and the 7. In brief, students should first distinguish the given integers from the fractions. Regarding the first quantity $\begin{array}{|c|}3\\1\\2\end{array}$, by applying the *bhāganubandhajāti* one obtains the fraction $\begin{array}{|c|}7\\2\end{array}$, which is added to the "quantity which occurs next" (*agretanāṅka*):[68] $\begin{array}{|c|}6\\1\\ \end{array}$

68 The intermediate passages are: $\begin{array}{|c|}7\\2\end{array}\begin{array}{|c|}6\\1\end{array}$, $\begin{array}{|c|}7\\2\\1\end{array}\begin{array}{|c|}6\\1\\2\end{array}$, $\begin{array}{|c|}7\\2\\1\end{array}\begin{array}{|c|}12\\2\\2\end{array}$. Every quantity which in the *nyāsa* (line 4) occurs next to the first on the left is denoted by the expression *agretanāṅka*. See lines 7, 9, and 21.

- these two simple fractions are manipulated by the method of the "class of simple fractions" (*bhāgajāti*); one arrives at the same denominator and the two numerators are summed. The resulting $\begin{array}{|c|}19\\2\end{array}$ is to be added to the *agretanāṅka* $\begin{array}{|c|}9\\01\\4\end{array}$
- the commentator observes that first the procedure of the "class of fractional decrease" (*bhāgāpavāhajāti*) should be carried out in order to compute $\begin{array}{|c|}9\\01\\4\end{array}$,and quotes a passage from the *karaṇasūtra* concerned (GT 60). This procedure involves the "subtraction" (*apavāha*) of a "fraction" (*bhāga*) from an integer; the resulting $\begin{array}{|c|}35\\4\end{array}$ is then added to the previous fraction $\begin{array}{|c|}19\\2\end{array}$
- the SGT explains that, to perform the procedure, one should carry out the "reduction" (*apavartana*)[69] of the denominators to their lowest terms by the common factor 2 and place the resulting quotients by exchanging their place below the original denominators: $\begin{array}{|c||c|}19&35\\2&4\\2&1\end{array}$. The numerators obtained can be thus added: $\begin{array}{|c||c|}38&35\\4&4\end{array}=\begin{array}{|c|}73\\4\end{array}$
- by means of the *bhāgajāti,* $\begin{array}{|c|}73\\4\end{array}$ is added to the next quantity $\begin{array}{|c|}7\\1\\3\end{array}$. This, however, has to be first simplified by means of the *bhāgānubandhajāti* and the resulting fraction $\begin{array}{|c|}22\\3\end{array}$ is then added to $\begin{array}{|c|}73\\4\end{array}$, it gives $\begin{array}{|c|}307\\12\end{array}$
- the final fraction $\begin{array}{|c|}307\\12\end{array}$ is an improper fraction, since the numerator is larger than the denominator. In such a case, Siṃhatilakasūri always transforms this type of result into a compound fraction, a mixed number composed of an integer plus a proper fraction. The method followed by the SGT

69 This term is found in GT 42, which is the *karaṇasūtra* of the division of fractions.

is the following: the numerator of the improper fraction is divided by the denominator and the quotient obtained represents the integer of the compound fraction; the fraction is composed of the difference between the initial numerator and the product of the multiplication of the quotient by the given denominator, while the denominator remains the same denominator. Taking into the account the numbers of this last fraction, one obtains:

- $307 \div 12 = 25$; $307 - (25 \times 12) = 7$; hence $25 + \dfrac{7}{12}$, which is given as $\begin{vmatrix} 25 \\ 7 \\ 12 \end{vmatrix}$.

Subtraction of fractions (bhinnavyavakalita) *[GT $\overline{38}$]*

GT $\overline{38}$: This *karaṇasūtra* repeats the same principle of the preceding but with a different operational sign: determine the same denominator, and carry out the "deduction of the numerators" (*aṃśaviśleṣa*).

The SGT on GT $\overline{38}$: Siṃhatilakasūri glosses *viśleṣa* with *pāta* (both meaning "deduction") and refers again to "fractions" by both the term *bhinna* and the compound *rūpakhaṇḍa*. As often in the SGT, in order to effectively comment upon this rule he links the technical terms mentioned by Śrīpati to the numbers mentioned in the next sample problem. This involves two mathematical exercises in which the student is asked to subtract different amounts of money from one *dramma*, a coin denomination mentioned in GT4. In line 6, the commentator highlights that, in this case, there is the "quantity-capital" (*āyarāśi*) and the "quantity-expenditure" (*vyayarāśi*); the "deduction" (*viśleṣa*/*pāta*) is the subtraction of the latter from the former.

The SGT on GT 39–40 (ex.): The commentator clarifies the quantities mentioned in the first half of the sample problem and presents then the *nyāsa*: $\begin{vmatrix} 1 & 1 & 1 & 1 \\ 1 & 6 & 2 & 3 \end{vmatrix}$. The first quantity represents the unity, the integer one which is the *āyarāśi* or "quantity capital" one *dramma*, and the other quantities are each a *vyayarāśi* to be subtracted from it. The commentator first adds all the fractions to then subtract the resulting fraction $\begin{vmatrix} 36 \\ 36 \end{vmatrix}$ from the *āyarāśi* one *dramma*. In line 27, he quotes a passage from the class of simple fractions which should be applied and provides the layout: $\begin{vmatrix} 1 & 36 \\ 1 & 36 \\ 36 & 1 \end{vmatrix}$. In this case, there has been the "exchange" (*vinimaya*) of the denominators 1 of the *āyarāśi* one *dramma* and the 36 of the obtained sum of the quantities-expenditure. When each numerator is multiplied by the exchanged denominator, the two alike fractions $\begin{array}{|c|}36\\36\end{array}$ arise. The "subtraction" (*apanayana*) of the obtained numerators 36 is carried out and the remainder obtained is zero".

Keeping in mind that the *bhāgajāti* is to be first applied in order to arrive at the same denominator, it can be observed that the procedure has been performed by the SGT as:

$$\begin{array}{|c|}a\\b\end{array} - \begin{array}{|c|}c\\d\end{array} = \begin{array}{|c|}a\\b\\d\end{array} - \begin{array}{|c|}c\\d\\b\end{array} = \begin{array}{|c|}ad\\bd\\d\end{array} - \begin{array}{|c|}cb\\db\\b\end{array} = \begin{array}{|c|}ad-cb\\bd \text{ or } db\end{array}$$

The second exercise involves the sum and the deduction of an integer and a fraction. In order to perform these mixed quantities, the SGT has to quote again the class of fractional increase and to quote and explain in advance the class of fractional decrease (*bhāgāpavāhajāti*) too. Having presented the *nyāsa* $\begin{array}{|cccc|}6 & 1 & 1 & 1\\ 1 & |1 & |01 & |1\\ 3 & 2 & 4 & 8\end{array}$, Siṃhatilakasūri clarifies that (lines 7–8) the "first quantity is the quantity-capital, the remaining are the quantities-expenditure". The SGT explains that one should first calculate the quantities-expenditure and determine the final amount which represents the total quantity expenditure, and then to subtract it from the quantity-capital. In order to perform the compound fraction one minus one-fourth, the commentator quotes and explains (line 11) the *bhāgāpavāhajāti* (see GT 60). Therefore, $\begin{array}{|c|}19\\3\end{array}$ is the *āyarāśi* $\begin{array}{|c|}6\\1\\3\end{array}$ once simplified by means of the *bhāgānubandhajāti,* $\begin{array}{|c|}27\\8\end{array}$ is the total quantity expenditure. The final steps mentioned are: $\begin{array}{|c|}19\\3\end{array}\begin{array}{|c|}027\\8\end{array}$; $\begin{array}{|c|}19\\3\\8\end{array}\begin{array}{|c|}027\\8\\3\end{array}$; $\begin{array}{|c|}152\\24\end{array}\begin{array}{|c|}081\\24\end{array}$; $\begin{array}{|c|}71\\24\end{array}$ is the final result which expressed as a compound fraction is $\begin{array}{|c|}2\\23\\24\end{array}$.

Multiplication of fractions (bhinnapratyutpanna) *[GT $\overline{40}$]*

GT $\overline{40}$: This *karaṇasūtra* states that in the multiplication of fractions, one should multiply the numerators and the denominators and then, in case the result is an improper fraction, divide the two products. The algorithm can be represented as: $\begin{array}{|c|}a\\b\end{array} \times \begin{array}{|c|}c\\d\end{array} = ac \div bd$.

The SGT on GT 40: The SGT comments upon this *karaṇasūtra* rather briefly by supplying synonyms and clarifying some terms; in fact, it explains the algorithm involved during the execution of the exercises given next by Śrīpati.

The SGT on GT 41 (ex.): The first *nyāsa* is: $\begin{array}{|c|}3\\1\\2\end{array}\begin{array}{|c|}9\\1\\3\end{array}$. The commentator distinguishes between the procedure of the class of fractional decrease and the procedure of the class of fractional increase, he then quotes from the rule of fractional increase and, while carrying out the procedure, uses the expressions *prāgaṅka* ("the first quantity") and *paratra* ("the other place") to specify the quantity manipulated. The final product $\begin{array}{|c|}196\\6\end{array}$ is an improper fraction, the numerator is thus divided by the denominator to obtain the compound fraction $\begin{array}{|c|}32\\2\\3\end{array}$.

In the second example, the given quantities are: $\begin{array}{|c|}1\\4\end{array}\begin{array}{|c|}1\\2\end{array}$. Regarding the obtained product $\begin{array}{|c|}1\\8\end{array}$, in line 24 the SGT observes that since this, being a unit fraction, is not further divided and it stands as it is.

Division of fractions (*bhinnabhāgahāra*) *[GT 42]*

GT 42: In the rule of "division of fractions" (*bhinnabhāgahāra*), Śrīpati states that first the "interchange" (*parīvartana*) of the numerator and denominator of the fraction-divisor should be performed and, whenever possible, one should carry out the "cross-reduction" (*kuliśāpavartana*). Finally, multiply the numerators and the denominators.

The algorithm is computed as: $\begin{array}{|c|}a\\b\end{array} \div \begin{array}{|c|}c\\d\end{array} = \begin{array}{|c|}a\\b\end{array} \times \begin{array}{|c|}d\\c\end{array} = \begin{array}{|c|}ad\\bc\end{array}$

The SGT on GT 42: The SGT first clarifies that the "interchange" involves two numbers, which are the numerator and the denominator, of the "divisor", and this is the quantity found in the layout next to the quantity-dividend. The commentary also explains that before the division one should simplify the quantities if these involve compound fractions. Having transformed the compound fractions into simple ones, one should perform the "cross-reduction" whenever possible, and then apply the rule of multiplication of fraction.[70]

The SGT on GT 43 (ex.): The first exercise asks students to divide ten plus one-fourth by six plus one-third. The *nyāsa* given by the commentator is: $\begin{array}{|c|c|}10 & 6\\ 1 & 1\\ 4 & 3\end{array}$. Siṃhatilakasūri quotes a passage from the rule of fractional increase to denote that here the class of fractional increase should be applied and:

- the two compound fractions are "simplified" (*savarṇita*) to simple ones: $\begin{array}{|c|c|}41 & 19\\ 4 & 3\end{array}$
- one should perform the interchange of the operands in the fraction "divisor" (*hara*)
- the commentator quotes a passage from the previous *karaṇasūtra* (see the rule of multiplication of fractions) to illustrate that, having inverted the operands of the fraction-divisor, the multiplication should be carried out, the result is: $\begin{array}{|c|}123\\ 76\end{array}$
- since this an improper fraction, in order to indicate that the fraction obtained should be transformed into a proper fraction and the numerator divided by its denominator, Siṃhatilakasūri quotes (see *haratāḍeti*) the last part of the previous rule; the final result is: $\begin{array}{|c|}1\\ 47\\ 76\end{array}$.

The execution of the second exercise follows the same procedure but with the difference that while the fraction-dividend is a quantity involved in a fractional increase, the fraction-divisor involves a fractional decrease.[71] In this case, the rules of the *bhāgānubandhajāti* and *bhāgāpavāhajāti* are respectively applied and the resulting improper fraction is transformed into a quantity composed of a proper fraction plus an integer.

In the third example, the two unit fractions $\begin{array}{|c|c|}1 & 1\\ 2 & 6\end{array}$ are divided and the final result is the integer 3. In the last example, a unit fraction is to be divided by

70 One should bear in mind the difference between *apavarta/apavartana* and *savarṇa/savarṇana*: while *apavartana* is the "reduction" of a fraction to its lowest terms, *savarnana* is the "homogeneization, simplification" of a mixed number, a compound fraction into a simple one. The term *savarṇa* is lit. "having the same colour or appearance"; it involves the procedures concerning the *jāti*s of fractions (GT 53–62).

71 In this regard, see the SGT's remark while commenting upon the *karaṇasūtra* (SGT 42, 9–10).

the integer 3. The presentation of the data is: $\begin{vmatrix}1 & 3\\4 & 1\end{vmatrix}$. Having performed the exchange of the numerator and denominator of the fraction-divisor, the final result is the unit fraction $\begin{vmatrix}1\\12\end{vmatrix}$.

Square of fractions (*bhinnavarga*) [GT $\overline{44}$]

GT $\overline{44}$: This rule specifies that in order to perform *bhinnavarga* or the "square of fractions", the square of the numerator is divided by the square of the denominator; the rule clarifies the procedure to execute when improper fractions occur. The algorithm can be represented as follows:

$\begin{vmatrix}a\\b\end{vmatrix}^2$ which is performed as: $\begin{vmatrix}a\\b\end{vmatrix} \times \begin{vmatrix}a\\b\end{vmatrix} = a^2 \div b^2$

The SGT on GT $\overline{44}$: Siṃhatilakasūri first observes that the numerator is the number above and the denominator is the number underneath, and he then explains that the fraction should be written in two places, i.e., twice (see *sthānadvaye*). The SGT repeats that the square consists of the multiplication of two equal quantities (*sadṛśarāśighāta*).

The SGT on GT 45 (ex.): This sample problem includes four exercises. The commentator does not show every step of the procedure for squaring fractions as this is just the application, on the numerator and denominator, of the procedure for squaring integers enunciated in GT 23. It can be observed that the *nyāsa* of the fraction to be squared is written in two different places such as: $\begin{vmatrix}1 & 1\\2 & 2\end{vmatrix}$. The rule of fractional increase and fractional decrease are applied to the numerical data given in order to simplify the compound fractions into simple fractions and to calculate then the square. In the case of improper fractions, the last part of the procedural rule is not carried out (see the third and the fourth exercises where unit fractions are obtained).

Square-root of fractions (*bhinnavargamūla*) [GT $\overline{46}$]

GT $\overline{46}$: The GT explains that, in case of an improper fraction, the square-root of the numerator is divided by the square-root of the denominator. The formula is:

$$\begin{vmatrix}a\\a\end{vmatrix}^{\begin{vmatrix}1\\2\end{vmatrix}} = \sqrt{a} \div \sqrt{b}$$

The SGT on GT $\overline{46}$: The commentator explains in detail this procedure while solving the next sample problem. Since the square-root of fractions implies the knowledge and understanding of the square-root of integers, the SGT quotes relevant passages from the rule of square-root of integers (GT 26–27). It is interesting to observe that the commentator glosses the word *mūla* ("root") with *bīja* (lit. "seed, primary element"), which here denotes the "root, primary base" of a square-root, hence the number-root which is squared.

The SGT on GT 47 (ex.): Student should ascertain the square-root of the square of the fractions obtained in the previous sample problem on square of fractions.[72] In lines 7–15, the SGT calculates the first mixed number $\begin{array}{|c|}22\\9\\16\end{array}$; since this involves a fractional increase, the *bhāgānubandhajāti* is applied and it becomes $\begin{array}{|c|}361\\16\end{array}$. The commentator quotes a passage (see *samaviṣemety*) from his own explanation of square-root of integers to demonstrate that this procedure is to be applied to the operands of this fraction. The SGT does not show all the steps required to arrive at the root but mentions only the final result $\begin{array}{|c|}19\\4\end{array}$ which, being an improper fraction, is transformed into the compound fractions $\begin{array}{|c|}4\\3\\4\end{array}$.Below I shall demonstrate how to ascertain the root of the numerator 361 by performing the same procedure explained by the commentator in SGT 26, which illustrates the execution of the square-root of integers. This is also in order to verify the number twenty-nine mentioned by the commentator in line 10:

- the last odd place of 361 is the digit 3, the highest square to be subtracted from it is 1, which becomes the first *pada*
- the square of this first *pada* is multiplied by 2 and placed below; the *pada* 1 is squared and the result is subtracted from the last odd place 3. The remainder of the subtraction, which is 2, is written above instead of the 3 $\begin{array}{|l|}\mathbf{261}\\\mathbf{2}\end{array}$
- above, the remainder 26 is divided by the *pada* 1 multiplied by 2. In order to carry out the procedure, some readjustments are needed and the chosen quotient comes to be 9. This is placed below 2 $\begin{array}{|l|}261\\2\\\mathbf{9}\end{array}$

72 A similar pattern which highlights the relation between opposite operations is found in the sample problems given by the GT on multiplication, division, square/square-root, and cube/cube-root of integers.

- the result from [26– (9× 2)] is 8, which replaces the 26 above, and the new *pada* and quotient 9 is moved into the line next to the 2:[73] $\begin{array}{|c|}\mathbf{81}\\ 29\end{array}$. The *pada* 9 is squared and subtracted from the remainder above, which is 81, giving 0, and the procedure is hence completed. The two *padas* 1 and 9 give 19, this is multiplied by 2 and divided by 2 becoming 19 which is the root of the number 361.

In lines 16–29, the commentator presents a second method which, he observes, has not been explained by the author of the root-text. According to this procedure, the square of the two terms of a fraction is divided by the root-fraction. The algorithm can be represented as:

$$\begin{array}{|c|}a\\ b\end{array}^{2} \div \begin{array}{|c|}a\\ b\end{array} = \begin{array}{|c|}a^2\\ b^2\end{array} \times \begin{array}{|c|}b\\ a\end{array}$$

Siṃhatilakasūri applies this procedure in the following manner:

i) the original quantity is the fraction $\begin{array}{|c|}5\\ 01\\ 4\end{array}$, which is found in the first example of the previous sample problem (GT 45)
ii) the first step is to simplify it,[74] hento transform the compound fraction into a simple fraction by applying, in this case, the rule of the class of fractional increase; $\begin{array}{|c|}19\\ 4\end{array}$ is obtained
iii) this is the fraction-divisor, hence there is the "interchange" (*parīvartana/viparyaya*) of its numerator and denominator: $\begin{array}{|c|}4\\ 19\end{array}$
iv) at this point, the commentator explains that one should place the following fractions: $\begin{array}{|c|c|}361 & 4\\ 16 & 19\end{array}$, where the first is the square of the given fraction
v) the SGT mentions the reduction to lowest terms and quotes (see lines 21–24) the term *apavartana* from the expression *kuliśāpavartanam* or "cross-reduction" occurring in GT 42. The commentator executes the "reduction" (*apavartana*) and explains that one should divide 16 by 4 and 361 by 19. The resulting fractions are $\begin{array}{|c|c|}19 & 1\\ 4 & 1\end{array}$

73 Note that the number twenty-nine mentioned by the SGT in line 10 refers to the 29 which is underneath in this layout.

74 See the use of the term *savarṇita* (line 17) to denote this process.

vi) finally, the numerators and denominators are multiplied and the result, being an improper fraction, is transformed into a compound fraction: $\begin{array}{|c|}19\\4\end{array} = \begin{array}{|c|}4\\3\\4\end{array}$.

The second example (lines 1–6) concerns the quantity $\begin{array}{|c|}72\\1\\4\end{array}$. Siṃhatilakasūri explains that the *bhāgānubandhajāti* is to be applied, and one obtains $\begin{array}{|c|}289\\4\end{array}$. By the expression *samaviṣemety,* the commentator quotes again his own passage on square-root of integers and in doing so he emphasises that the root of the numerator 289 is determined by first marking every other digit as odd or even. One should thus determine the root by applying the procedure explained in SGT 26. The commentator does not show all the steps of this execution but mentions the final result.

I shall demonstrate how to ascertain the root of the numerator 289 by performing the same procedure shown by the commentator while explaining the square-root of integers:

- the last odd place of 289 is the digit 2, the highest square to be subtracted from it is 1, which becomes the first *pada*
- the square of this first *pada* is multiplied by 2 and placed below; the *pada* is squared and the result is subtracted from the last odd place 2. The result of the subtraction, which is 1, is written above instead of the 2: $\begin{array}{|l|}\mathbf{189}\\\mathbf{2}\end{array}$
- the remainder 18 is divided by the *pada* 1 multiplied by 2. In order to carry out the procedure, neither the quotients 9 nor 8 can be taken, hence the chosen quotient is 7. This is placed below 2: $\begin{array}{|l|}189\\2\\7\end{array}$
- the result from [18 – (7 × 2)] is 4, which replaces the remainder 18 above, and the new *pada* and quotient 7 is moved into the line next to the 2, thus:[75] $\begin{array}{|l|}\mathbf{49}\\27\end{array}$. The *pada* 7 is squared and subtracted from the remainder 49 giving 0, and the procedure is completed. The two *pada*s 1 and 7 give 17, this is multiplied by 2 and divided by 2 giving 17 which is the root of the number 289.

75 The number twenty-seven (see the following layout) is mentioned by the SGT in line 3.

The third and fourth examples are shortly explained by the SGT, since they involve a straightforward procedure concerning the unit fractions $\left|\begin{array}{c}1\\9\end{array}\right|$ and $\left|\begin{array}{c}1\\4\end{array}\right|$.

Cube of fractions (*bhinnaghana*) [GT $\overline{48}$]

GT $\overline{48}$: Śrīpati states that the cube of the numerator should be divided by the cube of the denominator. This rule, which specifies the case of improper fractions, can be represented as:

$$\left|\begin{array}{c}a\\b\end{array}\right|^3 = a^3 \div b^3$$

The SGT on GT $\overline{48}$: Commenting upon this *karaṇasūtra*, Siṃhatilakasūri briefly clarifies the meaning of terms such as "numerator", "denominator", and "cube" but shows the various steps of the procedure while solving the sample problem given next by Śrīpati.

The SGT on GT 49 (ex.): In the introductory line, the commentator observes that the exercise requires students to determine the cube of four quantities which involve different procedures. The first quantity concerns the *bhāgānubandhajāti*, the next the *bhāgāpavāhajāti*, and the last two the *bhāgajāti*. While calculating the cube of quantity given in the second example, Siṃhatilakasūri shows how to determine the cube of its numerator step by step.

The given quantity to manipulate is sixth less by one-third: $\left|\begin{array}{r}6\\01\\3\end{array}\right|$. By applying the rule of fractional decrease, this is reduced to $\left|\begin{array}{c}17\\3\end{array}\right|$. At this point, the commentator executes the cube of the numerator. The passages explained are:

- (lines 5–7): Siṃhatilakasūri quotes steps from the cube of integers formulated in GT 28 to emphasise that this rule is to be here applied; the cube of the "last" digit 1 is placed below by being one notational more in respect to the given number: $\left|\begin{array}{l}\ \ 17\\\mathbf{1}\end{array}\right|$
- the square of the "last" digit 1 is multiplied by 3 and by the next digit 7. The resulting 21 is placed in the layout in the following way: $\left|\begin{array}{l}\ \ 17\\\mathbf{11}\\2\end{array}\right|$. Here, the unit 1 of the product 21 is placed next to the cube of the last digit 1, as this is the line of the units, while the 2 of 21 is placed in the line below, which is the line of the tens

- the square of the "preceding" digit (*ādi*) 7 of the initial number 17 is multiplied by the "last" digit 1 and by 3. The resulting 147 is placed in the layout: $\left|\begin{array}{l} \;\;17 \\ 117 \\ 24 \\ 1 \end{array}\right|$. One can observe that the 1 of 147 is placed below the line of the tens, hence in the line of the hundreds, while the 4 stands in the line of the tens next to the 2, and the 7 in the lines of the units next to the 1
- the cube of the preceding number 7, which is 343, is placed: $\left|\begin{array}{l} \;\;\;17 \\ 1173 \\ 244 \\ 13 \end{array}\right|$
- all the digits are summed according to the place they occupy in the layout and the initial root 17 is erased; the obtained sum 4913 is the cube of the numerator 17. Finally, as the cubes of the operands give the improper fraction $\left|\begin{array}{r} 4913 \\ 27 \end{array}\right|$, according to Śrīpati's instruction, the cube of the denominator 4913 is to be divided by the cube of the denominator 27. The mixed number obtained is $\left|\begin{array}{r} 181 \\ 26 \\ 27 \end{array}\right|$.

Cube-root of fractions (*bhinnaghanamūla*) [GT $\overline{50}$]

GT $\overline{50}$: Śrīpati enunciates the *karaṇasūtra* of *bhinnaghanamūla* or the "cube-root of fractions", and states that this is arrived at when the cube-root of the numerator is divided by the cube-root of the denominator.[76] The algorithm is:

$$\left|\begin{array}{c} a \\ b \end{array}\right|^{\left|\begin{array}{c} 1 \\ 3 \end{array}\right|} = \sqrt[3]{a} \div \sqrt[3]{b}$$

The SGT on GT 50: Siṃhatilakasūri clarifies that, according to the sample problem given next by the root-text, this rule involves the cubes obtained in the previous sample problem and that their "roots" are to be ascertained. He paraphrases Śrīpati's passage but prefers to elucidate the rule while showing the execution of the sample problem.

The SGT on GT 51 (ex.): In this sample problem, Śrīpati requires to determine the "roots" of the cubes found in the previous sample problem, which

76 The division refers to the case of improper fractions.

are respectively: $\begin{array}{|c|}791\\29\\64\end{array}\begin{array}{|c|}181\\26\\27\end{array}\begin{array}{|c|}1\\216\end{array}\begin{array}{|c|}1\\27\end{array}$. Siṃhatilakasūri transforms the first mixed number back into an improper fraction, and $\begin{array}{|c|}791\\29\\64\end{array}$ becomes[77] $\begin{array}{|r|}50653\\64\end{array}$. He then quotes passages from the cube-root of integers (GT 32–33) and determines the cube-root of the numerator but, unfortunately, these passages are not fully explained by the commentator. Let us look at the layouts which accompany Siṃhatilakasūri's explanations:

- the commentator observes that one should subtract the cube of 3 from the number 50.[78] Although details are missing one understands that, according to the procedure of the cube of integers shown by the commentator while commenting upon GT 32, the last *ghana*[79] or "cubic" place of 50653 is 50. One should subtract its highest possible cube, which is the cube of 3. The remainder 23 (from $50-3^3$) replaces the manipulated *ghana* 50, while the root 3 is placed below: $\begin{array}{|c|}23653\\\mathbf{3}\end{array}$ (line 16)
- the root 3 is squared and multiplied by 3, the resulting 27 is placed by one notational place less with respect to the 3:[80] $\begin{array}{|c|}23653\\\mathbf{273}\end{array}$ (line 18)
- above, the remainder 236 is divided by the product 27:[81] as the quotient 8 is not suitable,[82] a lesser number is to be chosen, which is in this case 7. This step is not mentioned by the commentator, although the procedure

77 See the numbers mentioned in SGT 49, 26–27.

78 The first layout should be the following: $\begin{array}{|r|}50653\\3\end{array}$.

79 The digits of the number should be marked from right to left as one cubic and two not cubic. See GT 32–33. In lines 13–14, the commentary provides interesting information by explaining that the cube-root of the last cubic place be placed below the zero "belonging to" (see *satka*) 50. Siṃhatilakasūri implies that, in order to perform the procedure, the last cubic place is 50.

80 In GT 32–33, Śrīpati explains that the "root" is brought below the third place from the last cubic place. The last cubic place is the 3 of 23, hence the root 3 is brought below the 5 of 53.

81 In line 17, see the expression *pūrvanyastayā* or "by the [seventeen] previously written".

82 The reason is the following: $236 \div 27 = 8$, the remainder 20 (from $[236-(27\times 8)]$) is placed above instead of the dividend 236, and the quotient 8 stands below next to the 3: $\begin{array}{|r|}2053\\38\end{array}$. The quotient 8 is squared, multiplied by 3, and by the 3 (the digit 3 next to the 8 in the 38 below).

is carried out in this very way; in fact, the next layout is $\begin{array}{|r|}4753\\3\end{array}$ (line 19). Here the 47 above is the remainder of the division of 236 by 27 when taking 7 as quotient, so that $236 \div 27 = 7$, with 47 as remainder

- Siṃhatilakasūri quotes the step *paṅktyāṃ niyojya* from GT 32–33 to denote that the quotient 7 is placed in the line in front of three (see *trikāgrato*): $\begin{array}{|r|}4753\\37\end{array}$ (line 20)
- the last layout shown is $\begin{array}{|r|}343\\37\end{array}$ (line 22). Unfortunately, the SGT does not explain how one arrives at this result.[83]

Below I am going to provide a reconstruction of the intermediate steps before arriving at the last layout $\begin{array}{|r|}343\\37\end{array}$: the chosen quotient 7 has been squared and multiplied by 3 and by the last digit 3 (the 3 next to the 7 in the layout);[84] the result 441 is subtracted from the remainder above 475 of 4753; the remainder is 34 since $475 - (7^2 \times 3 \times 3) = 34$. This is placed above instead of 475 and next to the remaining 3

- lastly, one should calculate the cube of the quotient 7, which is 343, and this subtracted from the remainder above 343 gives 0. In this way, the number underneath, which is 37, represents the cube-root of 50656. This 37 is given in line 26 as $\begin{array}{|c|}9\\1\\4\end{array}$, which is the root-fraction[85] given in GT 49.

The SGT ends the explanation of the eight arithmetical operations with fractions by positively commenting upon the structure of the root-text. I have pointed out that in order to carry out the sample problems given by Śrīpati, Siṃhatilakasūri had to quote and to illustrate beforehand the execution of passages from the *karaṇasūtras* on the classes of simplification of fractions. Nevertheless, the commentator underlines that the classes of simplification of fractions require first the understanding of the operations with

The product is 576, and as this should be subtracted from the remainder 205 above, the procedure will abruptly end.

83 The result of the cube of the quotient 7 is 343; when this is subtracted from the remainder above which is 343, no remainder is left and the result of the cube of the initial numerator 50653 is the number below 37.

84 In GT 32–33, see *tatkṛtim antyanighnīṃ trisaṅguṇāṃ* [*cāpanayed ghanaṃ ca.*

85 The commentator explains that the integer 9 is placed above the remainder one-fourth (see, in line 26, *labdham copari niyojyam*)

fractions; the order of presentation chosen by the Śrīpati is therefore appropriate: operations with fractions first, and *jāti*s next.

Arithmetic of zero [GT 52]

GT 52: Arithmetic of zero is a topic commonly found in texts on *pāṭīgaṇita*.[86] Śrīpati's rule occupies overall a kind of conclusive position in the overall structure of his work: it ends the part on the arithmetical operations and it precedes the treatment of the *jāti*s of fractions.

Śrīpati, as well as other mathematicians, uses the term *śūnya* to denote "zero"; synonyms of words for "sky", such as *gagana, vyad, antarikṣa,* and *vyoman* are also widely employed.[87]

The operations with zero explicitly mentioned by the GT are:

$0 + n = n$
$n \pm 0 = n$
$n \times 0 = 0$
$n \div 0 = 0$
$0 \div 0 = 0$
$0^2 = 0$
$0^3 = 0.$

It is noteworthy that while earlier texts share approximately the same formulae on the operations with zero, Bhāskarācārya seems to be the first to provide a different formula. The GSS[88] and the GT state that: $n \div 0 = 0$; in L 45–46, Bhāskarācārya observes that the same formula gives $\begin{vmatrix} n \\ 0 \end{vmatrix}$ instead, which means that a quantity divided by zero has zero as divisor (*khahara*).[89] Moreover, Bhāskarācārya is the first to mention the following: $n \div 0 \times 0 = n$, which is pointed out by the SGT in lines 12–13.

The SGT on GT 52: In the introductory line,[90] the commentator observes that:

> Because it occurs together with numbers [...], he now enunciates [the procedural rule concerning] the relationship (*vyāpti*) of zero [and numbers].

86 Some details on the arithmetic of zero in Sanskrit mathematical works are discussed by Plofker et al. (2017).
87 In GSS 1.53–63, Mahāvīrācārya provides a list of terms used to denote "zero" according to the *bhūtasaṃkhyā* notation.
88 See GSS 1.49, where it is said that: $n \times 0 = 0$; $n \pm 0 = n$; $n \div 0 = n$; $0 + n = n$; $0 \times 0 = 0$; $0 \div 0 = 0$; and so forth.
89 That a quantity divided by zero remains "zero-divided" was earlier stated in BSS 18.30–35.
90 The SGT uses the expression *śunyasvarūpa* (the "proper form of zero").

The term *vyāpti* occurs only once in the SGT and it refers to the way zero affects numbers in computation.[91] Nevertheless, philosophical speculations on zero are found neither in the GT nor in the SGT; both texts provide purely mathematical content.

Having commented upon the first three operations with zero enunciated by Śrīpati, Siṃhatilakasūri mentions the L and later fully quotes the *karaṇasūtra* on zero (L 45–46) in order to supply the specific case mentioned in Bhāskarācārya's work, which is $n \times 0 \div 0 = n$. Also, it is significant that the SGT does not seem to clearly object that while the L states that a quantity divided by zero is "zero-divided" (*khahara*), according to the GT it gives zero. In lines 16–17, the commentary refers to the commutative principle of multiplication and observes that, as the root-text mentions the formula: $n \times 0 = 0$, the same result is arrived at when there is: $0 \times n$. In lines 13–14, Siṃhatilakasūri refers to the three cases mentioned by Śrīpati in which a quantity remains unchanged when computed with zero. In mathematical terms, these are: $0 + n = n$ and $n \pm 0 = n$.

In the whole passage, the SGT's intention is to emphasise that Śrīpati has rightly mentioned all the eight arithmetical operations with zero. As the GT does not provide an exercise related to this *karaṇasūtra*, the commentator supplies it by quoting a sample problem which corresponds to L 47. The layouts regarding the multiplication by zero, the division by zero, and the final *nyāsa* are noteworthy. In fact, the multiplication of the number 5 by zero (hence zero as multiplier) is represented as: $\begin{array}{|c|c|} 0 & 0 \\ 5 & \end{array}$; the division of the number 10 by zero (hence zero as divisor) is: $\begin{array}{|c|c|} 10 & 0 \\ 0 & \end{array}$, which seems to show that the result is zero. Zero as multiplier and divisor of 63 is represented as: $\begin{array}{|c|} 0 \\ 63 \\ 0 \end{array}$, which is clearly a reference to the case specified in the L: $n \times 0 \div 0 = n$.

4.6 Classes of simplification of fractions [GT 53–63]

Preliminary observations

The *jātis* or "classes" of fractions formulated by Śrīpati in the GT are five. Verses 53–63 present reduction techniques for the simplification of fractions which involve five kinds of manipulations.[92] In texts on *pāṭīgaṇita*, this topic

91 It is interesting to observe that in the Nyāya philosophical tradition, *vyāpti* represents a category of thought; it denotes the logical relation of one notion to another in a proposition, and hence a law of concomitance such as "where there is the smoke, there is fire".

92 Procedures for the simplification of fractions are investigated in Kusuba (2004).

is usually found under the name *kalāsavarṇana* or "assimilation, simplification (*savarṇana*) of fractions". Brahmagupta gives rules concerning five *jāti*s, Mahāvīrācārya and Śrīdhara expound six *jāti*s, while Bhāskarācārya presents four *jāti*s.[93]

The five *jāti*s for the reduction of fractions expounded by Śrīpati are:

i) the class of simple fractions (*bhāgajāti*)
ii) the multi-part class (*prabhāgajāti*)
iii) the class of fractional increase (*bhāgānubandhajāti*)
iv) the class of fractional decrease (*bhāgāpavāhajāti*)
v) the chain-simplification class (*vallīsavarṇanajāti*)

With respect to this scheme, one should take into account the rule of *bhāgajāti* quoted by Siṃhatilakasūri[94] from PG 37.

The class of simple fractions (*bhāgajāti*) [GT $\overline{53}$]

GT $\overline{53}$: The first procedure on the assimilation of fractions is the *bhāgajāti* or the "class of simple fractions"; this method involves the addition or subtraction of fractions having different denominators. The formula of the algorithm is:

$$\left|\begin{matrix} a \\ b \end{matrix}\right| + \left|\begin{matrix} c \\ d \end{matrix}\right| = \left|\begin{matrix} a \\ b \\ d \end{matrix}\right| + \left|\begin{matrix} c \\ d \\ b \end{matrix}\right| = \left|\begin{matrix} ad \\ bd \\ \, \end{matrix}\right| + \left|\begin{matrix} cb \\ db \\ \, \end{matrix}\right| = \left|\begin{matrix} ad + cb \\ bd \end{matrix}\right|$$

The SGT on GT $\overline{53}$: This rule has already been quoted by Siṃhatilakasūri while commenting upon GT $\overline{35}$, which is the rule on addition of fractions. After having restated that the numerator is the number above and the denominator is the number underneath, the commentator observes that this rule involves two fractions. He then uses the expressions *prācyāṅka* and *agretanāṅka* to indicate each of the two fractions respectively and explains that the numerator and denominator of the "first quantity" (*prācyāṅka*) is multiplied by the denominator of the "next quantity" (*agretanāṅka* or *parāṅka*). The two numerators obtained as products are either added or subtracted, depending upon the sign of the operation. At the end of his brief comment Siṃhatilakasūri emphasises that this procedure yields the same denominator.

93 I disagree with Kāpadīā (1937, LIV) who observes that the GT gives four *jāti*s; in this way, the editor does not include the *vallīsavarṇana* which, as the name itself suggests, also involves "simplification" of fractions.

94 This is found after GT $\overline{53}$.

The SGT on GT 54 (ex.): Siṃhatilakasūri presents the *nyāsa*: $\begin{array}{|c|c|c|c|c|c|} 1 & 1 & 1 & 1 & 1 & 1 \\ 2 & 3 & 4 & 6 & 5 & 7 \end{array}$.

He shows that, in order to arrive at the same denominator, starting from the first pair these quantities are summed two by two, and the result is added to the next fraction. The first step is the exchange of the two denominators which are brought below the original denominators, next is the multiplication of the original numerator and denominator by the exchanged denominator. Once the same denominator has been obtained, the numerators are added and the exchanged numerators are "erased" (*bhañjanīya*). The result is then summed to the next fraction and the procedure continues in this way until all the quantities have been added. In brief, the SGT shows the following steps:

- the first two fractions are $\begin{array}{|c|} 1 \\ 2 \end{array}$ and $\begin{array}{|c|} 1 \\ 3 \end{array}$. By applying the *bhāgajāti* rule, the SGT demonstrates that: $\begin{array}{|c||c|} 1 & 1 \\ 2 & 3 \\ 3 & 2 \end{array} = \begin{array}{|c||c|} 3 & 2 \\ 6 & 6 \\ 3 & 2 \end{array} = \begin{array}{|c|} 5 \\ 6 \end{array}$
- the obtained fraction $\begin{array}{|c|} 5 \\ 6 \end{array}$ is added to the next quantity, which is $\begin{array}{|c|} 1 \\ 4 \end{array}$, thus: $\begin{array}{|c||c|} 5 & 1 \\ 6 & 4 \\ 4 & 6 \end{array}$ $= \begin{array}{|c|c|} 5 & 1 \\ 6 & 4 \\ 2 & 3 \end{array}$ (the two denominators have been reduced by half) $= \begin{array}{|c|c|} 10 & 3 \\ 12 & 12 \\ 2 & 3 \end{array} = \begin{array}{|c|} 13 \\ 12 \end{array}$
- $\begin{array}{|c|} 13 \\ 12 \end{array}$ is then added to $\begin{array}{|c|} 1 \\ 6 \end{array}$, thus: $\begin{array}{|c||c|} 13 & 1 \\ 12 & 6 \\ 1 & 2 \end{array} = \begin{array}{|c|c|} 13 & 2 \\ 12 & 12 \\ 1 & 2 \end{array} = \begin{array}{|c|} 15 \\ 12 \end{array}$
- the obtained fraction $\begin{array}{|c|} 15 \\ 12 \end{array}$ added to $\begin{array}{|c|} 1 \\ 5 \end{array}$ becomes: $\begin{array}{|c|c|} 15 & 1 \\ 12 & 5 \\ 5 & 12 \end{array} = \begin{array}{|c|c|} 75 & 12 \\ 60 & 60 \\ 5 & 12 \end{array} = \begin{array}{|c|} 87 \\ 60 \end{array}$
- the obtained fraction $\begin{array}{|c|} 87 \\ 60 \end{array}$ added to $\begin{array}{|c|} 1 \\ 7 \end{array}$ becomes: $\begin{array}{|c||c|} 87 & 1 \\ 60 & 7 \\ 7 & 60 \end{array} = \begin{array}{|c||c|} 609 & 60 \\ 420 & 420 \\ 7 & 60 \end{array} = \begin{array}{|c|} 669 \\ 420 \end{array}$
- this final result is reduced by 3; the fraction arrived at is $\begin{array}{|c|} 223 \\ 140 \end{array}$.

Regarding the terminology used by the SGT, when there is the exchange of the denominator, *adho neya* (see lines 3, 9, and 13) denotes that the denominator of the *agretanāṅka* is "brought below" the original denominator of the *prācyāṅka*. The commentator uses the expressions *śeṣaṃ bhañjanīyaṃ* (line 2) and *śeṣaṃ gatam* (lines 7 and 16) to indicate that the two original denominators are erased once the same denominator is obtained by multiplying the two operands by the exchanged denominator.

Rule corresponding to PG 37

The last part of the explanatory passage on the class of simple fractions is corrupt; an *āryā* stanza occurs after the commentator's explanation of GT 53 which is a quotation from PG 37.[95] This rule presents another procedure on *bhāgajāti*, the algorithm can be represented as follows:

$$\left|\begin{array}{c} a \\ b \\ c \\ d \end{array}\right| = \left|\begin{array}{c} ad+bc \\ bd \end{array}\right|$$

This means that:

$$\left|\begin{array}{c} a \\ b \end{array}\right| + \left|\begin{array}{c} c \\ d \end{array}\right| = \left|\begin{array}{c} ad+bc \\ bd \end{array}\right|$$

The commentary elaborates the terms used by the PG and pays particular attention to specify the operands involved in the various steps of the algorithm by means of qualifiers (adjectives, adverbs, and prepositions).[96] The SGT reframes the expressions adopted by Śrīdhara to establish a correspondence between these and his terminology with respect to the notational layout. In order for the students to compare the two methods of *bhāgajāti,* the commentator applies Śrīdhara's rule while solving the previous sample problem (GT 54). The fractions to be calculated are the six unit fractions: $\begin{array}{|c|c|c|c|c|c|} 1 & 1 & 1 & 1 & 1 & 1 \\ 2 & 3 & 4 & 6 & 5 & 7 \end{array}$.
The first two fractions are placed each below the other (see *ūrdhvagatyā aṅkaśreṇeḥ*): $\left|\begin{array}{c} 1 \\ 2 \\ 1 \\ 3 \end{array}\right|$. Then, the *ūrdhvāṃśa* ("upper numerator") 1 is multiplied by the *adharahara* ("lower denominator") 3; the *adharahara* 3 is then multiplied by the *ūrdhvahara* ("upper denominator") 2. The product of the multiplication of these two denominators becomes the new denominator of the resulting "simplified" (*savarṇita*) fraction. The commentator expands upon the expression *madhyāṃśahara* and explains that the "intermediate numerator

95 K 31, 25–26. Notably, this verse is not found in the TŚ by the same author Śrīdhara; this means that Siṃhatilakasūri had both texts available to him (see Hayashi 2913, 60) while commenting upon the GT.

96 See *adhara* ("lower"), *ūrdhva* ("upper"), *madhya* ("intermediate"), and *uparima* ("upper").

and denominator" are the denominator of the fraction above and the numerator of the fraction below respectively and that these compose another fraction, which is the middle fraction. The *madhyāṃśa* 2 is multiplied by the *madhyahara* 1 and this product 2 is added to the product 3 obtained from the multiplication of the *ūrdhvāṃśa* 1 by the *adharahara* 3. The resulting fraction $\left|\begin{array}{c}5\\6\end{array}\right|$ is placed in a vertical layout with the next fraction $\left|\begin{array}{c}1\\4\end{array}\right|$ standing underneath: $\left|\begin{array}{c}5\\6\\1\\4\end{array}\right|$. The same process is repeated: the *ūrdhvāṃśa* 5 is multiplied by the *adharahara* 4; the *adhahara* 4 is then multiplied by the *ūrdhvahara* 6 giving the new denominator 24. The *madhyāṃśa* 6 is multiplied by the *madhyahara* 1 and the product 6 is added to the product 20. The resulting fraction $\left|\begin{array}{c}26\\24\end{array}\right|$ is placed in a vertical layout with the next fraction $\left|\begin{array}{c}1\\6\end{array}\right|$ standing underneath: $\left|\begin{array}{c}26\\24\\1\\6\end{array}\right|$. The procedure continues in this very way until the last quantity is added. As the resulting fraction $\left|\begin{array}{c}8028\\5040\end{array}\right|$ is an improper fraction, it is transformed into a compound fraction and, when reduced, it becomes 1 plus $\left|\begin{array}{c}83\\140\end{array}\right|$.

The multi-part class (*prabhāgajāti*)[97] [GT $\overline{55}$]

GT $\overline{55}$: Kāpadīā (1937, LXVII) observes that in the manuscript folio 64 is lacking. As I have mentioned in the translation, folio 64 must have included the *karaṇasūtra* on the *prabhāgajāti* or the "multi-part class",[98] the commentator's explanation, and the sample problem provided by the root-text.[99] The algorithm enunciated in the rule can be represented as:

$$\left|\begin{array}{c}a\\b\end{array}\right| \times \left|\begin{array}{c}c\\d\end{array}\right| = \left[\frac{a\times c}{b\times d}\right]$$

97 The verse is lost because it was on the missing folio 64.

98 Cf. PG 38a and TŚ 23b.

99 The sample problem has been numbered by Hayashi as GT 56.

The SGT on GT 56 (ex.): The commentator executes the numerical data given by Śrīpati in the stanza of the sample problem, which must have been written in four *padas* or "quarters". Folio 65 begins with the result of the first *pada*,[100] which is $\begin{array}{|c|}1\\24\end{array}$, and the remaining part of the procedure explained by the SGT is extant. The sample problem concerns three men and the amount of money they give to a beggar; the first part of the execution of this sample problem is missing; however, according to the reconstruction provided, the *nyāsa* of this first *pada* must have been the following: $\begin{array}{|c|c|c|}1&1&1\\2&3&4\end{array}$. By applying the rule of multi-part class enunciated in GT 55, one should multiply the numerators and denominators of the first two fractions and the resulting fraction by the last one: $\begin{array}{|c|c|}1&1\\6&4\end{array} = \begin{array}{|c|}1\\24\end{array}$. This is, in fact, the result of the first *pada* and the fraction given in line 28. Then, the commentary presents the *nyāsa* of the data of the second *pada*: $\begin{array}{|c|c|c|c|c|}1&1&1&1&1\\8&5&3&2&6\end{array}$. By means of the same procedure, he shows that: $\begin{array}{|c|c|}1&1\\40&3\end{array} = \begin{array}{|c|}1\\120\end{array}$; $\begin{array}{|c|c|}1&1\\120&2\end{array} = \begin{array}{|c|}1\\240\end{array}$; $\begin{array}{|c|c|}1&1\\240&6\end{array} = \begin{array}{|c|}1\\1440\end{array}$. This is the result of the second *pada*. The *nyāsa* of the third *pada* is: $\begin{array}{|c|c|c|c|}1&1&1&1\\7&8&4&10\end{array}$. By applying the rule of multi-part class, one obtains: $\begin{array}{|c|c|}1&1\\56&4\end{array} = \begin{array}{|c|}1\\224\end{array}$; $\begin{array}{|c|c|}1&1\\224&10\end{array} = \begin{array}{|c|}1\\2240\end{array}$. The results of the three *padas* are: $\begin{array}{|c|c|c|}1&1&1\\24&1440&2440\end{array}$. The SGT adds the first two fractions and in doing so it first reduces the two quantities to the same denominator by applying the rule of simple fractions so that:[101] $\begin{array}{|c|c|}1&1\\24&1440\\60&1\end{array} = \begin{array}{|c|}61\\1440\end{array}$. The commentator explains that this fraction is now added to the next one, which is $\begin{array}{|c|}1\\24\end{array}$. By the expression *śeṣaṃ vinaṣṭam* (lines 22 and 1), the commentary points out that the previous quantities are erased because they have been computed and a new result has arisen. Next there is the "assimilation" (*savarṇana*) of the two fractions, which implies the exchange of the denominators and, in this case,

100 See K 32, 28.

101 The *bhāgajāti* is quoted by the SGT in line 16.

the reduction of these by the common factor 160. The two quotients obtained are exchanged: $\left|\begin{matrix}61\\1440\\14\end{matrix}\right|\left|\begin{matrix}1\\2240\\9\end{matrix}\right| = \left|\begin{matrix}863\\20160\end{matrix}\right|$ of one *dramma*.

The commentator transforms this result into the sub-multiples of a *dramma*, according to the conversion ratios given by Śrīpati in the section on technical terms (*paribhāṣā*): 1 *dramma* is equal to 16 *paṇa*s; 1 *paṇa* is equal to 4 *kākiṇī*s, and 1 *kākiṇī* is equal to 20 *kapardaka*s. The first step is to multiply the numerator 863 by 16 which becomes 13808 and since this numerator is still smaller than the denominator 20160, it is not possible to divide the two in order to obtain the number of *paṇa*s, which means that the number of *paṇa*s in $\left|\begin{matrix}863\\20160\end{matrix}\right|$ are 0. The monetary unit *kākiṇī* should be calculated by multiplying the obtained numerator 13808 by 4. The resulting fraction is $\left|\begin{matrix}55232\\20160\end{matrix}\right|$, now it is possible to divide the larger numerator by the denominator in order to obtain the number of *kākiṇī*s contained in $\left|\begin{matrix}863\\20160\end{matrix}\right|$ of one *dramma*. The result is 2 plus $\left|\begin{matrix}14912\\20160\end{matrix}\right|$, the quotient integer 2 represents the number of *kākiṇī*s occurring in $\left|\begin{matrix}863\\20160\end{matrix}\right|$ of one *dramma*. Then, in order to obtain the number of *kapardaka*s, the numerator 14912 is multiplied by 20; being larger, the resulting numerator 298240 is divided by the denominator 20160, thus: 298240 ÷ 20160 = 14 *kapardaka*s, the remainder is $\left|\begin{matrix}16000\\20160\end{matrix}\right|$, which reduced to the common factor 320 gives $\left|\begin{matrix}50\\63\end{matrix}\right|$.

Therefore, the three men gave altogether 2 *kākiṇī*s and 14 $\left|\begin{matrix}50\\63\end{matrix}\right|$ *kapardaka*s to the beggar.

The class of fractional increase (*bhāgānubandhajāti*) [GT 57]

GT 57: In Śrīpati's formulation, one can notice the polysemic use of some technical terms, for example *aṃśa* means "fraction" in *aṃśānubandha* while is "numerator" in *sāṃśaka,* and *bhāga*is "fraction" in *bhāgānubandha* (used by the commentator) but is "numerator" in *bhāgaṃ kṣipet.* The terms for "denominator" are three: *hara, cheda,* and *chedana*. The word *rūpa* denotes the "integer", which in the first method of *bhāgānubandhajāti* is added to the fraction.

The *bhāgānubandhajāti* or the "class of fractional increase" involves two different procedures of which the first is given in the first quarter of the stanza; this is emphasised by Siṃhatilakasūri while introducing the sample problem given by Śrīpati in GT 58.[102] The first type of *bhāgānubandhajāti* involves the association, the sum of an integer and a fraction. GT 57 states that one should add the numerator to the integer multiplied by the denominator; it can be represented by the following algorithm:

$n + \frac{a}{b}$, which in the SGT is given as:[103] $\begin{array}{|c|}n\\a\\b\end{array}$ and read as: $\frac{(n \times b) + a}{b}$

The second type concerns fractions in association with fractions, that is, the sum of a fraction and a fraction of itself. The SGT shows, during the execution of the sample problem on this type of *bhāgānubandhajāti* in GT 59, that the two fractions are placed in the layout one below the other: $\begin{array}{|c|}a\\b\\c\\d\end{array}$. The rule explains that:

- the upper denominator is multiplied by the lower denominator: $b \times d$
- the first numerator is multiplied by the denominator below added to its own numerator, which is thus: $a \times (c+d)$.

The SGT on GT 57: At the very beginning of its explanation, the SGT clarifies that the rule concerns two procedures. The first procedure is said to be dependent on the class of simple fractions (see line 19).[104] The second method of *bhāgānubandhajāti* (see lines 19–24) is described by the commentator in the introductory line to the second sample problem (GT 59); Siṃhatilakasūri refers to this procedure by the expression *bhāgānubandhabhāga* or "fractions in association with fractions". This method is twofold since the initial fraction can be either attached to an initial integer or not.

The SGT on GT 58 (ex.): This sample problem concerns the sum of a fraction and an integer, and it thus requires to simplify the given quantities by means of the first method of *bhāgānubandhajāti*. In this way, the quantities are simplified and then the resulting fractions are added together by means of the *bhāgajāti*. The three given quantities, composed by an integer

102 See line 25.
103 In line 18, the commentator specifies that the "integer" (*rūpa*) is "situated above" (*uparistha*). SGT 58 shows the execution of the method of *bhāgānubandhajāti*.
104 This is quoted, in fact, while solving the next sample problem (see SGT 58, 4 and 8).

plus a fraction, are placed as follows: $\begin{array}{|c|c|c|} 10 & 1 & 2 \\ 1 & 1 & 1 \\ 4 & 2 & 3 \end{array}$. In order to simplify the first quantity, the first method of fractional increase is applied: the numerator 10 is multiplied by the denominator 4, and the result added to the numerator 1. It gives $\begin{array}{|c|} 41 \\ 4 \end{array}$, the second quantity is simplified in the same way and the previously obtained fraction is added to it becoming $\begin{array}{|c|} 47 \\ 4 \end{array}$. The last quantity is also first simplified and then added to $\begin{array}{|c|} 47 \\ 4 \end{array}$; it becomes $\begin{array}{|c|} 169 \\ 12 \end{array}$ which, being an improper fraction, is reduced to the compound fraction $\begin{array}{|c|} 14 \\ 1 \\ 12 \end{array}$.

The SGT on GT 59 (ex.): In the introductory line to this sample problem, the SGT gives these three definitions to illustrate the relationship between the quantities involved: fractions in association with fractions,[105] the association of fractions devoid of integer,[106] and an integer joined in a fractional increase.[107]

Before showing the execution, the commentator clarifies (lines 19–24) the connection between the given quantities and specifies that the first line of the sample problem regards fractions attached to an integer,[108] the second line concerns the association of fractions with fractions.[109] The sample problem is, in fact, divided into two parts: in the first part (lines 25–10), both methods of *bhāgānubadhajāti* are carried out, since the given quantities include first the case of an integer joined in fractional increase and then fractions in association with fractions, from the fraction obtained as the result from the integer added to a fraction, a fraction of itself is attached to it, and to it another fraction, and so forth in a "chain-like" arrangement.[110] The second part of the sample problem (lines 11–26) involve fractions in association with fractions and but because no integer is attached to the fractions, only the second method of *bhāgānubadhajāti* is applied.[111]

105 *bhāgānubandhabhāga.*
106 *rūparahitabhāgānubandha.*
107 *bhāgānubandhasahitāṅka.*
108 *rūpapratibaddhabhāga.*
109 *bhāgānubandhabhāga.*
110 See the commentator's remark in line 25.
111 This is pointed out by the SGT in line 11.

Having provided important background information, in line 25 the commentator gives the first *nyāsa*: $\begin{array}{|c|}1\\1\\4\\1\\2\\1\\3\\1\\6\end{array}$

- the first step is to apply the first method of the class of fractional increase. Hence, the first quantity one plus one-fourth becomes $\begin{array}{|c|}5\\4\end{array}$
- this result is added to its half, the SGT explains that the second method of fractional increase should thus be applied and the two fractions are placed one below the other: $\begin{array}{|c|}5\\4\\1\\2\end{array}$. The denominator above, which is 4, is multiplied by the denominator underneath, which is 2; the denominator underneath is added to its own numerator 1 and the result is multiplied by the numerator above. It becomes $\begin{array}{|c|}15\\8\end{array}$
- its one-third is then joined: $\begin{array}{|c|}15\\8\\1\\3\end{array}$. The same procedure is applied, thus: 3 + 1 = 4; 8 × 3; 15 × 4 = 60
- to the resulting $\begin{array}{|c|}60\\24\end{array}$, its one-sixth is joined: $\begin{array}{|c|}60\\24\\1\\6\end{array}$. It gives $\begin{array}{|c|}420\\144\end{array}$
- in line 10, the commentator explains that this quantity should be placed aside.

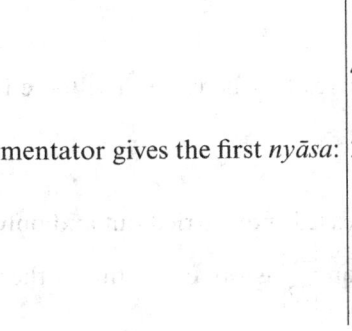

The second part of the execution of this sample problem begins with the following layout (line 11): $\begin{array}{|c|}1\\3\\1\\6\\1\\4\end{array}$. The SGT observes that here, because there is no integer, the first method of fractional increase is not carried out and only the second method is performed. The fraction $\begin{array}{|c|}35\\72\end{array}$ is obtained, this is then added to the previous result $\begin{array}{|c|}420\\144\end{array}$, which has been placed aside. By applying the *bhāgajāti* and by reducing the improper fraction obtained into a compound fraction, the final result is $\begin{array}{|c|}3\\29\\72\end{array}$.

The class of fractional decrease (*bhāgāpavāhajāti*) [GT 60]

GT 60: The passage on the *bhāgāpavāhajāti* or the "class of fractional decrease" follows a similar structure of the preceding *karaṇasūtra*; the first part presents the first method of this class of fractions where an integer is diminished by a fraction, while the second part gives a method on subtraction of fractions of fractions.

Regarding the first type of *bhāgāpavāhajāti,* GT 60 states that one should subtract the numerator from the integer multiplied by the denominator. This can be represented by the following algorithm:

$n - \frac{a}{b}$, which in the SGT is given as: $\begin{array}{|cc|} & n\\ 0 & a\\ & b\end{array}$ and read as: $\frac{(n \times b) - a}{b}$

The second procedure of *bhāgāpavāhajāti* involves the deduction from a fraction of a fraction of itself and, as shown in the SGT while commenting upon the next sample problem, the two fractions should be placed one below the other as follows: $\begin{array}{|cc|} & a\\ & b\\ 0 & c\\ & d\end{array}$. The *karaṇasūtra* explains that:

i) one should multiply the [lower] denominator by the [upper] denominator, which is thus: $b \times d$

ii) and one should multiply the foremost numerator by the [lower] denominator decreased by the lower numerator, which is thus: $a \times (c-d)$. It can be observed thus that the only difference between the *bhāgāpavāhajāti* and the *bhāgānubandhajāti* is the operational sign.

The SGT on GT 60: In the first three lines, the commentator clarifies three situations in which the two methods of *bhāgāpavāhajāti* are applied: i) when there is the deduction of a fraction from an integer; ii) when the difference of an integer diminished by a fraction is further diminished by a fraction of itself (i.e., of the difference), and iii) when there is the deduction of a fraction and a fraction of itself. These three situations recall the three cases which Siṃhatilakasūri has mentioned in the introductory line to the previous sample problem (GT 59). Having clarified the meaning of some technical terms relating to the first method of fractional decrease, the last three lines are dedicated to a brief explanation of the second method. Notably, in line 9 the SGT observes that a "zero" (*śūnya*) stands "behind" (*paścāt*) the numerator of the fraction which is subtracted.

The SGT on GT 61 (ex.): This sample problem asks students to manipulate the same quantities of the previous two sample problems,[112] relating to the class of fractional increase, by changing this time their operational sign. This is further clarified by the commentator in the first two lines of his explanation of this sample problem. It is a pedagogical stratagem use by Śrīpati to demonstrate that the two classes of fractions are opposite procedures.

The execution of this sample problem is divided into two as it involves two examples. Since it manipulates the quantities mentioned in GT 58, which concerns the first method of fractional increase, the first part (lines 19–6) involves the subtraction of a fraction from an integer; the second part (lines 7–4) can be subdivided into three parts. In the first section (lines 7–18) the quantities are initially manipulated, as in the first part of SGT 59 but with a different operational sign, by means of the first method of fractional decrease and then by means of the second method of fractional decrease as the resulting fraction is decreased by a fraction of itself and so forth. In the second section, fractions of fractions are involved and thus only the second method is applied. In the last section (lines 1–4), the results obtained in the two previous sections are added to arrive at the final result.

In the first part, the SGT applies the first procedure of *bhāgāpavāhajāti*; hence the integers 10, 1, and 2 (see GT 58) are decreased, instead of being increased, by the same fractions previously mentioned. The three results are then added by means of the *bhāgajāti*. The final result is $\left|\begin{array}{c}11\\11\\12\end{array}\right|$.

112 GT 58 and GT 59.

The second part (line 7) involves first a fraction subtracted from an integer – and thus the first method of fractional decrease – and to this resulting fraction, a fraction of itself is removed and successive fractions of the last result are removed by means of the second method of fractional decrease. The result $\begin{vmatrix} 5 \\ 24 \end{vmatrix}$ is placed aside[113] as the second section of this second part of the sample problem begins. This concerns removing fractions of fractions, by applying the second method of fractional decrease the result obtained is $\begin{vmatrix} 15 \\ 72 \end{vmatrix}$, which reduced by the common factor 3, becomes $\begin{vmatrix} 5 \\ 24 \end{vmatrix}$. In the last section, the two alike fractions $\begin{vmatrix} 5 \\ 24 \end{vmatrix}$ and $\begin{vmatrix} 5 \\ 24 \end{vmatrix}$ are added giving $\begin{vmatrix} 10 \\ 24 \end{vmatrix}$, which reduced by 2, is $\begin{vmatrix} 5 \\ 12 \end{vmatrix}$.

The chain-simplification class (*vallīsavarṇanajāti*)[114] [GT 62]

GT 62: The procedure called *vallīsavarṇana* regards a chain of measures and involves the sum of different denominations of a same measuring unit. These, which the SGT observes to be related to each other like in a "chain" (*vallī*), are finally reduced to the same unit. In Śrīpati's formulation, the most interesting detail is the use of the terms *ṛṇa* and *dhana*, which mean lit. "debt" and "wealth" respectively. Siṃhatilakasūri clarifies that they denote a negative and positive numerator, meaning that a fraction is considered negative when subtracted from another quantity, while it is positive when it is added.

The SGT on GT 62: The commentator observes that this procedure concerns a pair of quantities. The chain-simplification class involves, in fact, the relationship between the values of two denominations of the same measuring unit (in the given example, it is a unit of money), which are placed one below the other and manipulated two by two, since one is a part of the other. The term *vallī* highlights that in this procedure the measures are in relation to each other as in a chain. However, the main aim of the whole passage is to elucidate the meaning of the generic *vivadhyāt* or "one should perform" used by Śrīpati. The commentator relates this verb to the terms *ṛṇa* and *dhana*, and shows how its meaning is affected according to the context. The SGT observes that the numerator is called "negative" (*ṛṇa*) "in relation to the first numerator" when it is subtracted from it; behind a negative numerator, a zero is written. The numerator is called "positive" (*dhana*) "in relation to the first numerator", when it is added; behind a positive numerator, no zero occurs. Śrīpati has opted to use the verbal root *vidh-* to embrace both situations, and

113 See line18.
114 K 39, 7–10.

to leave its meaning open depending upon whether a subtraction or an addition takes place. In line 15, the commentator observes that when the numerator is positive, then the *bhāgānubandhajāti* needs to be carried out, but in the case of a "negative" numerator, one should perform the *bhāgāpavāhajāti*. Since the procedure involves the sum or the deduction of a fraction and a fraction of itself, the second method of the class of fractional increase and the class of fractional decrease is performed.[115]

The SGT on GT 63 (ex.): Lines 23–29 clarify the values and the relationship between the units of money mentioned by Śrīpati. In line 29, the commentator presents the *nyāsa*: $\begin{array}{|r|} 2 \\ 1 \\ 5 \\ 16 \\ 1 \\ 4 \\ 01 \\ 20 \\ 01 \\ 4 \end{array}$; it can be observed that below each unit of money a denominator is placed, this represents the conversion ratio between the unit and its preceding (larger) unit. In the GT, the largest unit is the *dramma*, hence its denominator is 1. The last fraction represents the smallest unit of money, namely the *kaparda* which, since it is to be subtracted, has a zero in front of its numerator.

The main steps of the procedure executed by the commentator are:

- in lines 1–4, Siṃhatilakasūri relates the larger unit *dramma* and its submultiple *paṇa*. The first numerator 2 is multiplied by the denominator underneath: $2 \times 16 = 32$; the first denominator 1 is multiplied by 16, the denominator underneath: $1 \times 16 = 16$. The fraction $\begin{array}{|r|} 32 \\ 16 \end{array}$ is obtained; at this point, the first layout shown is: $\begin{array}{|r|} 32 \\ 16 \\ 5 \\ 16 \end{array}$. Next, the step enunciated in the second part of the *karaṇasūtra* is applied: the first numerator (*pūrvalava*) 32 is manipulated (see *vidadhyāt*), which here implies that it is added, to

115 In the given example, the first two quantities concerns an integer and a fraction and are added by means of the first method of *bhāgānubandhajāti*.

5 and the resulting fraction is $\begin{array}{|c|}37\\16\end{array}$ is obtained. In this way, the reduction between the first two measures *dramma* and *paṇa* has been executed. In fact, the fraction $\begin{array}{|c|}37\\16\end{array}$ is the result (expressed in *paṇa*s) of the sum between 2 *dramma*s and 5 *paṇa*s

- in lines 5–8, the *paṇa* and the *kākiṇī* are compared. The fraction $\begin{array}{|c|}37\\16\end{array}$ is hence added to 1 *kākiṇī*, which is said to be one-fourth of a *paṇa*: $\begin{array}{|r|}37\\16\\1\\4\end{array}$.
The first denominator 16 is multiplied by the lower denominator 4 becoming the denominator 64. In the same way, one should multiply 37, which is the numerator above, by the denominator 4, it gives: $\begin{array}{|r|}148\\64\end{array}$.
One should then add the denominator 148 and 1; the fraction $\begin{array}{|r|}149\\64\end{array}$ is the result of the sum of 2 *dramma*s, 5 *paṇa*s, and 1 *kākiṇī*. This result is given in *kākiṇī*s and has been arrived at by adding the previously obtained fraction, given in *paṇa*s, plus 1 *kākiṇī*
- this result compared with the next unit, i.e., the *kaparda*s, so that the previous resulting quantity of *kākiṇī*s is turned into *kaparda*s. The sample problem requires one to subtract one *kaparda*; in line 10, the SGT shows this layout: $\begin{array}{|r|}149\\64\\01\\20\end{array}$.
- The procedure continues as before. During the second step of this calculation, in line 13 the commentator points out that, this time, 1, which is the numerator below, is "negative", hence in this case *vidadhyāt* means that "one should subtract" (*apanayet*) it from the first numerator obtained, which is 2980 (from 149 × 20). The fraction obtained is $\begin{array}{|r|}2979\\1280\end{array}$ *kaparda*s. In the same manner, this result is decreased by one-fourth of itself:[116] $\begin{array}{|r|}2979\\1280\\01\\4\end{array}$.

116 The fraction is already expressed in the same unit of money, which is the *kapardaka*.

The result obtained is $\left|\begin{matrix}11915\\5120\end{matrix}\right|$, which is the quantity of *kaparda*s arrived at; having reduced the chain of measures into the lowest unit the commentator performs the reduction of these by the common factor 5, so it becomes: $\left|\begin{matrix}2383\\1024\end{matrix}\right|$

- having summed up all the quantities and having reduced them into the same lowest unit, the SGT determines the number of the various units of money contained in the fraction $\left|\begin{matrix}2383\\1024\end{matrix}\right|$. Since this an improper fraction, when the numerator is divided by the denominator, the obtained quotient 2 is the number of *dramma*s contained in this quantity of *kaparda*s. The remainder is 325 (the denominator is 1024)
- in order to arrive at the number of *paṇa*s, the remainder 325 is multiplied by 16 (1 *dramma* is equal to 16 *paṇa*s). The numerator of the resulting improper fraction $\left|\begin{matrix}5360\\1024\end{matrix}\right|$ is divided by its denominator; the quotient 5 represents the *paṇa*s and the remainder is 240
- in order to calculate the number of *kākiṇī*s, the remainder is multiplied by 4 (1 *paṇa* is equal to 4 *kākiṇī*s); the resulting fraction $\left|\begin{matrix}960\\1024\end{matrix}\right|$ is a proper fraction and hence, in line 25, the SGT observes that there is no division and with respect to the number of *kākiṇī*s, a zero is placed instead
- in order to arrive at the number of *kaparda*s, the numerator 960 is multiplied by 20 (1 *kākiṇī* is equal to 20 *kaparda*s). The fraction obtained is $\left|\begin{matrix}19200\\1024\end{matrix}\right|$; when the numerator is divided by the denominator, it gives 18 *kaparda*s, and $\left|\begin{matrix}768\\1024\end{matrix}\right|$ is the remainder
- in order to arrive at the fraction of a *kaparda* contained in this result, the numerator is multiplied by 4 giving 3072, and when this is divided by the denominator it becomes $\left|\begin{matrix}3\\4\end{matrix}\right|$.

Hence, the final sum of the "simplified" units of money is 2 and $\left|\begin{matrix}335\\1024\end{matrix}\right|$ *dramma*s or 2 *dramma*s, 5 *paṇa*s, and 18 plus three-fourths *kaparda*s.

4.7 Type-problems of fractions [GT $\overline{64}$–92]

Preliminary remarks

In the GT, verses $\overline{64}$–92 are dedicated to presenting procedures that we could classify as solving equations with one unknown; these are divided into categories depending on the characteristics of the given quantities. The GT and the GSS use the word *jāti* to denote these procedures, while in Śrīdhara's PG one finds *uddeśa*. I have explained that in GT $\overline{53}$–63, the same term *jāti* indicates the "classes" of homogenization of fractions.[117] SaKHYa (2009, xxvi and 208–214) emphasises that some mathematicians categorise the classes of reduction of fractions as well as several types of linear and quadratic equations as *jāti*s: "Thus the word *jāti* ('birth, family, caste, class') is used by them side by side for the two different categories, the mixed fractions and the algebraic equations, each comprising several 'classes.'" For the sake of a clear translation, I have rendered the term *jāti* as "class" to denote the simplification of fractions, and as "type-problem" to denote the procedures with fractions which involve linear and quadratic equations.

Type-problems of fractions are differently classified by authors.[118] In GSS 4.1–72, Mahāvīrācārya presents ten *jāti*s, while in the GT these are nine. Significantly, in the introductory lines Siṃhatilakasūri specifies the title of each of the nine *jāti*s.

The visible quantity type-problem (*dṛśyajāti*) [GT $\overline{64}$]

GT $\overline{64}$: The first type-problem with fractions given in the GT is the *dṛśyajāti* or the "visible quantity type-problem". According to the rule, one should divide the "visible quantity" (*dṛśya*) by the number "one" (*rūpa*) lessened by the "sum of the fractions" (*bhāgaikya*). Algebraically, this rule can be expressed as:

$x = \left|\begin{matrix} d \\ 1-f \end{matrix}\right|$ where d is the *dṛśya* and f is the sum of the fractions.

The SGT on GT $\overline{64}$: The commentator explains this procedure by means of the data given by the GT in the next sample problem, which concerns a "pillar" (*stambha*).[119] In order to clarify the meaning of the term *dṛśya*, he explains that there is a visible part and an invisible part of the pillar.

117 In the thirteenth chapter of SŚ, Śrīpati presents only the *jāti*s or "classes" for the simplification of fractions.

118 See table 2.5 in SaKHYa (2009, xl–xli).

119 The sample problem given by Śrīpati in GT 65, which corresponds to TŚ 25, regards a pillar. The "pillar type-problem" (*stambhoddeśa*) is the name of a procedural rule found in the TŚ; as in the GT, in the TŚ too this follows the rule on *vallīsavarṇana*. The "pillar type-problem" is also given in PG 74, TŚ 27, and GK 2.14. A similar sample problem is found in GSS 4.5 and GK 2.15.

Siṃhatilakasūri clarifies that the *dṛśya* is defined as such by direct perception (see *lokapratyakṣa*), and that it is one and a half cubits of the pillar. There is, then, an "invisible" part (*adṛśya*) which is represented by the various fractions of the pillar being under water and so forth; the *bhāgaikya* mentioned by Śrīpati denotes the sum of these invisible parts which should be subtracted from the number one. The result of this subtraction becomes the divisor of the "visible quantity" (*dṛśya*). Finally, the commentator adds that the class of simple fractions should be applied to calculate the sum of the fractions and that the number one is the "positive quantity" (*āyarāśi*), from which the "deduction" (*viśleṣa*) of the sum of the fractions is carried out; the term *āyarāśi* denotes the number one in contrast to the fractions, which are "negative" because subtracted.

The SGT on GT 65 (ex.): This sample problem asks students to determine the length of a pillar. Siṃhatilakasūri gives the following *nyāsa*: $\left|\begin{array}{c|c|c|c} 1 & 1 & 1 & 1 \\ 2 & 12 & 6 & 1 \\ & & & 2 \end{array}\right|$. By applying the class of simple fractions to the first two fractions on the left, and then by applying it to the obtained result and the next fraction and so forth, the sum of fractions arrived at is $\left|\begin{array}{c} 9 \\ 12 \end{array}\right|$. Then, the commentator explains that (lines 23–28), this, which is the result of the sum of the fractions, should be subtracted from the number one; it gives the "quantity-divisor" (*hararāśi*) $\left|\begin{array}{c} 3 \\ 12 \end{array}\right|$. At this point, the *dṛśya* $\left|\begin{array}{c} 1 \\ 1 \\ 2 \end{array}\right|$ (see the *nyāsa*) is to be divided by the quantity-divisor obtained. In order to perform this step, it is first simplified by means of the *bhāgānubandhajāti,* and the *dṛśya* becomes $\left|\begin{array}{c} 3 \\ 2 \end{array}\right|$. The division of fractions requires that the quantity-divisor $\left|\begin{array}{c} 3 \\ 12 \end{array}\right|$ becomes the multiplier, and hence its quantities are interchanged. The final product of the multiplication by the *dṛśya* $\left|\begin{array}{c} 3 \\ 2 \end{array}\right|$ gives six cubits: $\left|\begin{array}{c} 6 \\ 1 \end{array}\right|$

The SGT quotes L 51, which is the *karaṇasūtra* on *iṣṭakarman* or the "supposition operation".[120] L 51 formulates a rule to find an unknown quantity from a given result. The reason for such a quotation is that Bhāskarācārya

120 This has been already mentioned by the commentator at the very end of his passage on zero (see SGT ad L 47, 11). This rule explains the "optional number operation", which in medieval Europe was known as *regula falsi* or "false position method", a procedure for solving an equation with one unknown.

applies the *iṣṭakarman* procedure to solve a sample problem (L 53) which involves the *dṛśyajāti*.

The SGT on GT 66 (ex.): This sample problem deals with a herd of elephants. It is interesting to observe that the commentator solves this exercise by two methods: first by applying the supposition operation quoted from L 51, and then by applying the rule of the visible quantity type-problem given by the root-text.

The SGT presents the *nyāsa*: $\left|\begin{array}{c|c|c} 1 & 1 & 1 \\ 2 & 6 & 8 \\ 1 & 1 & 1 \\ 3 & 7 & 9 \end{array}\right|$ | dṛśya 4 |.The given quantities represent fractions of a herd of elephants, and the last quantity on the right is the "visible quantity". The fractions are: one-half plus its one-third which are the elephants that went to a mountain, one-sixth plus its one-seventh are drinking water in a river, and one-eighth plus its one-ninth are playing among lotuses. Siṃhatilakasūri observes that these quantities constitute fractions of fractions, and one should simplify them by applying the second method of the class of fractional increase. Therefore, the SGT simplifies the three groups of quantities and, by means of the class of simple fractions, the three simplified fractions are added; the result is $\left|\begin{array}{c} 502 \\ 504 \end{array}\right|$. This fraction represents the sum of the various fractions of the herd of elephants. The commentator explains that this fraction corresponds to what the L terms *uddiṣṭa* or the "specified quantity" (see L 51). The SGT then clarifies that the *iṣṭarāśi* or "assumed quantity" is the number 4. The two quantities $\left|\begin{array}{c} 502 \\ 504 \end{array}\right|\left|\begin{array}{c} 4 \\ 1 \end{array}\right|$ are added by applying the quoted class of simple fractions; it gives[121] $\left|\begin{array}{c} 2016 \\ 504 \end{array}\right|$. The quantity-*uddiṣṭa* $\left|\begin{array}{c} 502 \\ 504 \end{array}\right|$ is multiplied by the *iṣṭarāśi* $\left|\begin{array}{c} 4 \\ 1 \end{array}\right|$, and the resulting $\left|\begin{array}{c} 2008 \\ 504 \end{array}\right|$ is subtracted from $\left|\begin{array}{c} 2016 \\ 504 \end{array}\right|$; it becomes the divisor $\left|\begin{array}{c} 8 \\ 504 \end{array}\right|$. Afterwards, the *dṛśya* or "visible quantity" 4, corresponding to the elephants seen playing together among lotuses, is multiplied by 4 and the result $\left|\begin{array}{c} 16 \\ 1 \end{array}\right|$ becomes the dividend of $\left|\begin{array}{c} 8 \\ 504 \end{array}\right|$. The result

121 As I have mentioned in the translation, the result of the sum should be $\left|\begin{array}{c} 2518 \\ 504 \end{array}\right|$.

of the division between $\begin{vmatrix}16\\1\end{vmatrix}$ and $\begin{vmatrix}8\\504\end{vmatrix}$ is $\begin{vmatrix}1008\\1\end{vmatrix}$, which corresponds to the size of elephants.

In the verification, starting from the number obtained the commentator calculates the various fractions given in the sample problem; the result corresponds to the number of the groups of elephants occupied in different activities, which are: 504, 168, 168, 24, 126, 14, and 4 which, when summed, become 1008.

The procedure followed by the commentator can be represented as:

$$x = (i \times d) \div [(f + d) - (f \times d)]$$

where d is the *dṛśya*, f is the sum of the fractions, and i is the *iṣṭa*,

From line 27 onwards, the commentator illustrates how to solve this sample problem by means of the rule of the visible quantity type-problem enunciated by Śrīpati. By applying the aforementioned formula: $x = \begin{vmatrix}d\\1-f\end{vmatrix}$, Siṃhatilakasūri clarifies that, in this case, $\begin{vmatrix}502\\504\end{vmatrix}$ is the *bhāgaikya* or "sum of the fractions". Then, one should subtract the *bhāgaikya* from the number 1, and one arrives at $\begin{vmatrix}2\\504\end{vmatrix}$. The commentator observes that this is the *hararāśi* or the "quantity-divisor". When the *dṛśya* 4 is divided by it, the result obtained is 1008.

The remainder type-problem (*śeṣajāti*) [GT $\overline{67}$]

GT $\overline{67}$: The *śeṣajāti* represents a type-problem with fractions where the unknown amount is composed of a given quantity and fractional parts of successive "remainders" (*śeṣa*). The GT states that the product of the denominators lessened by their own numerator should be divided by the product of the denominators. The "known "(*prakaṭa*) quantity is then divided by this quotient.

The rule can be represented as:

$$x = \frac{d}{\dfrac{p(a-b)}{pa}}$$

where d is the *dṛśya*, which in Śrīpati's verse is termed *prakaṭa*, a represents the denominators, b denotes the numerators, and p is the product.

The SGT on GT 67: The commentator observes that the denominators should be written twice and specifies that, in the next sample problem (GT 68), the remainder type-problem concerns three quantities. These are the successive remainders of the first fraction one-half, which is the quantity of elephants that are playing together. The SGT clarifies that the *śeṣajāti* consists, in fact, of "remainders" (*śeṣa*). In the case presented in the next sample problem, the successive remainders of the initial fraction one-half are three: one-third, one-fourth, and one-fifth. Siṃhatilakasūri then paraphrases the root-text, expands the concepts formulated in the *sūtra*, and observes that here, referring to the following sample problem, the *prakaṭarāśi* or "known quantity" is 60 (this is the quantity of elephants which are "seen"). An important point is that the commentary makes a distinction between the "denominators written first" and the "denominators written a second time"; these expressions point out that there are two sets of denominators in the same layout. The procedure is, however, better elucidated by the SGT while solving the next exercise.

The SGT on GT 68 (ex.): This sample problem concerns a herd of elephants. The commentator presents the *nyāsa*: $\begin{array}{|c|c|c|c|} 1 & 1 & 1 & 1 \\ 2 & 3 & 4 & 5 \\ 2 & 3 & 4 & 5 \end{array}$. The numbers at the bottom are those "denominators written a second time" mentioned by the SGT in line 15. The first step involves the subtraction of the numerators from "the denominators written first". In fact, in line 25 the following *nyāsa* occurs: $\begin{array}{|c|}1\\2\end{array}\begin{array}{|c|}2\\3\end{array}\begin{array}{|c|}3\\4\end{array}\begin{array}{|c|}4\\5\end{array}$ where, as the commentator explains, the results of this subtraction are above, while below are "the denominators written a second time". Then, the quantity-divisor is to be determined: this is represented by the product of the denominators of the fractions obtained above:[122] $(2 \times 3 \times 4 \times 5) = 120$. The dividend (see *bhājya,* line 3) is composed of the result of the reciprocal multiplication of the numbers above, which is the result of "the denominators written first" minus their own numerator: $(1 \times 2 \times 3 \times 4) = 24$. Then,[123] they should be divided (see *bhāgaṃ dātum*, line 28): $\begin{array}{|c|}1\\120\end{array}\begin{array}{|c|}24\\1\end{array}$, the result is $\begin{array}{|c|}1\\5\end{array}$. This becomes the divisor of the *prakaṭarāśi* 60; the final result obtained is 300, which is the number of elephants in the herd.

122 The number 120 is the product of the "denominators written a second time" (see the SGT in line 15). This becomes the quantity-divisor.

123 The commentator has obtained the two products: the product of the denominators minus their numerators is 24, and the product of the denominators is 120.

The execution of the sample problem is followed by its proof. The number of the elephants playing together is 150, which is one-half of 300. One-third of the remainder, which is 150, is 50; this represents thus the elephants which disappeared on the mountain. The next remainder is 100 [(300−150)−50], of which one-fourth is 25, which are the elephants that are removing an itching sensation from the cheek. Then, one-fifth of the remainder 75 (100 −25) is 15, which represents the number of elephants that entered the river to drink water. Their sum, together with the known quantity 60, are added, it gives 300: $\begin{vmatrix} 150 \\ 50 \\ 25 \\ 15 \\ 60 \end{vmatrix}$.

The SGT on GT 69 (ex.): This exercise requires students to determine the total number of a herd of swans. The commentator gives the *nyāsa*: $\begin{array}{|c|c|c|c|} 1 & 2 & 3 & 4 \\ 2 & 3 & 4 & 5 \\ 2 & 3 & 4 & 5 \end{array}$; the SGT first subtracts the numerators from their denominators; according to the results obtained, the layout should be: $\begin{vmatrix} 1 \\ 2 \end{vmatrix}\begin{vmatrix} 1 \\ 3 \end{vmatrix}\begin{vmatrix} 1 \\ 4 \end{vmatrix}\begin{vmatrix} 1 \\ 5 \end{vmatrix}$. The initial numerators are then replaced with the results of this subtraction. Then, the product of the denominators lessened by the numerator is: [(2 − 1) × (3 − 1) × (4 − 1) × (5 − 1)] = 1, while the product of the denominators below is 120 (2 × 3 × 4 × 5). This last product becomes the divisor of the first product 1, which is the product of the denominators lessened by the numerators. The resulting fraction $\begin{vmatrix} 1 \\ 120 \end{vmatrix}$ becomes the divisor of the *prakaṭarāśi* or "known quantity" 3. The result of the division gives 360, which is the total number of swans.

Lines 1–6 provides the verification, where the SGT ascertains the various remainders: one-half of the result 360 is 180, two-thirds of the remainder 180 is 120, three-fourths of the remainder 60 is 45, and four-fifths of the remainder 15 is 12: $\begin{vmatrix} 180 \\ 120 \\ 45 \\ 12 \\ 3 \end{vmatrix}$.

The difference type-problem (*viśleṣajāti*) [GT 70]

GT 70: In the first line (line 10), it is noteworthy that Śrīpati explicitly refers to the name of this *jāti.* This procedure involves the quantity called "visible quantity" and the difference of fractions of successive remainders.

The SGT on GT 70: The SGT expands the *karaṇasūtra* by glossing terms and providing synonyms; it also clarifies that the class of simple fractions should be applied (see the next sample problem) and that the last part of this procedure is the same as the visible quantity type-problem.

The SGT on GT 71–72 (ex.): Students have to determine the total number of a swarm of bees landing on different flowers and trees. The data mentioned in this sample problem can be represented in the following manner:

- $(A-B)\times 2 = C$; where A is the number of bees staying on the mango tree and B are those on the lotus
- $C+\dfrac{C}{2}=D$, the number of bees on the oleander
- $6\dfrac{(D-A)}{2}=E$
- $E+(\frac{1}{3}E\times 3)-\frac{1}{3}(\frac{1}{3}E\times 3)=F$ is the number of bees on the jasmine flower
- G is the *dṛśya* or "visible quantity", the 10 bees sitting on the tilak tree
- $A+B+D+F+G=S$, which is the sum total of the swarm of bees.

In the SGT, the first lines (up to line 9) show the calculation of what I have indicated above as D, which represents the number of bees landing on the oleander. By means of the class of simple fractions, by subtracting the smaller (*vihīna*) fraction $\left|\begin{matrix}1\\8\end{matrix}\right|$, which is the quantity of bees staying on the lotus (see above the letter B), from the larger fraction $\left|\begin{matrix}1\\5\end{matrix}\right|$, which is the quantity of bees landing on the mango tree (letter A), one obtains $\left|\begin{matrix}3\\40\end{matrix}\right|$. When this is multiplied by 2 and, by means of the second procedure of the class of fractional increase mentioned by the commentator in lines 5–6, added to its half, it gives $\left|\begin{matrix}9\\40\end{matrix}\right|$, which is the number of bees on the oleander (letter D). In line 10, the fractions of bees onthe mango tree (A), on the lotus (B), and on the oleander (D) are shown one next to the other: $\left|\begin{matrix}1\\5\end{matrix}\right|\left|\begin{matrix}1\\8\end{matrix}\right|\left|\begin{matrix}9\\40\end{matrix}\right|$. At this point, the commentator clarifies that the denominator 40 represents the whole size of the swarm (line 10) and next determines the number of bees on the jasmine flower (F). In order to arrive at this, one should first subtract the fraction of bees on the mango tree from the fraction of bees on the oleander and multiply their difference by 6; in this way, one arrives at the quantity E. By applying again the class of simple fractions, the SGT shows that: $\begin{array}{|c|c|}1&9\\5&40\\8&1\end{array}$; $\begin{array}{|c|c|}8&9\\40&40\\8&1\end{array}$; $\begin{array}{|c|}1\\40\end{array}$. Half this remainder is multiplied by 6 (the result is what I have denoted by the letter *E*) and, by means of the class of fractional increase partly quoted by the SGT in line 18, is added to its one-third, the number of bees on the jasmine flower (*F*). Thus: $\begin{array}{|c|c|}6&1\\80&3\end{array}$, $\begin{array}{|c|}6\\80\\1\\3\end{array}$, $\begin{array}{|c|}24\\240\end{array}$ which is then multiplied by 3 and lessened by its one-third: $\begin{array}{|c|}72\\240\\01\\3\end{array}$; the result $\begin{array}{|c|}1\\5\end{array}$ is the number of bees on the jasmine flower (*F*).

In line 28, the commentator shows the fractions of the swarm of bees thus far obtained: $\begin{array}{|c|c|c|c|}1&1&9&1\\5&8&40&5\end{array}$, which are the fractions of bees on the mango, lotus, oleander, and jasmine respectively. One should carry out the step *apāsya bhāgaikyam athaikataś* or "having subtracted the sum of the fractions from one": by applying the *bhāgajāti,* the first two fractions on the left are added, and the result $\begin{array}{|c|}13\\40\end{array}$ is added to the fraction of the next fraction, which represents the bees on the oleander. The result $\begin{array}{|c|}22\\40\end{array}$ (= A + *B* + *D*) is added to the fractions of bees on the jasmine flower, the result is "the sum of the fractions" (*bhāgaikya*). In this way, $A + B + D + F$ is $\begin{array}{|c|}30\\40\end{array}$. The commentator then explains (lines 9–17) that this should be subtracted from the number one by means of the class of simple fractions: $\begin{array}{|c|c|}30&1\\40&1\\1&40\end{array}$. It becomes $\begin{array}{|c|}10\\40\end{array}$, one should divide the "visible quantity" ten by the result of one lessened by the sum of the fractions, which is the remainder $\begin{array}{|c|}10\\40\end{array}$. The quotient obtained is $\begin{array}{|c|}40\\1\end{array}$, which represents the size of the swarm of bees.

In the last four lines, the SGT provides the *vāsanā* or "verification", the bees landing on the various flowers and trees are respectively: $\begin{array}{|c|}8\\5\\9\\8\\10\end{array}$. Their sum gives 40.

The remainder-root type-problem (*śeṣamūlajāti*) [GT 73]

GT 73: This *karaṇasūtra* enunciates the rule of the remainder-root type-problem.[124] The name of this procedure is specified in the first line of the commentary's passage upon the following sample problem (GT 74) and in the last line of the second sample problem: the commentator uses the syntagm *śeṣamūla* to denote the "root of the remainder", and observes that this type-problem is so called because the result is derived from the "visible quantity near the root-remainder". In the layout of the sample problem, the number of the "visible quantity" is, in fact, next to the root of the remainder".

As shown in the following exercises given by Śrīpati, this procedure involves the square-root of successive remainders. The steps of the rule can be represented as:

$$\sqrt{(d \times 4)+r^2} = \mathrm{D}$$

$$\frac{(D+r)^2}{2} = \mathrm{G}$$

$$\frac{G}{1-f} = \mathrm{x}$$

where d is the *dṛśya,* r is the root, and f is the fraction.

The SGT on GT 73: The commentator first clarifies some of the technical terms used by the root-text. In lines 27–28, Siṃhatilakasūri provides important information while explaining the meaning of *svaguṇamūlayute* or "added to the root multiplied by itself" found in the *karaṇasūtra*; he observes that in some cases,[125] such as in the next sample problem where the "visible quantity" is 2, the "root" is the number 1, and consequently *svaguṇa* (the root "multiplied by itself") is *ekaguṇa* ("multiplied by one"). In lines 3–5, the SGT supplies a significant detail concerning *punar vidhi* which, in GT 73, is the final

124 Referring to GT 73, SaKHYa (2009, 210) translates *śeṣamūlajāti* as the "root-remainder type".

125 Whenever a "number" (*saṅkhyā*) does not have a "root" (*mūla*), in brief, whenever no perfect squares occur.

expression occurring: the commentator explains that the procedure should be performed again but without the "the fraction which is in the middle". This expression refers to the "fraction" (*lava*), which in the layout is placed in the middle: in fact, in the layout found in SGT 74, 11, the fraction is denoted by *śela* (and it should be the same in SGT 75, 6, where the last syllable *la* is, however, missing). The abbreviation *śela* stands for *śeṣalava* or the "fraction of the remainder" and it occurs, in fact, in the middle.

The SGT on GT 74 (ex.): This exercise asks students to determine the total number of a collection of leaves. In the first line (line 11), the commentator provides the following *nyāsa*: $\begin{array}{|r|r|r|}\text{mū } 1 & \text{śela } 2 & \text{śemū } 1 \\ 1 & 3 & 1\end{array}$ $|\text{dṛśya } 2|$. While explaining the first two mathematical steps from the *karaṇasūtra*, which are *padasamīpacaturguṇadṛśyake* and *svaguṇamūlayute*, Siṃhatilakasūri clarifies that the *dṛśya* is 2, that *samīpa* ("near") qualifies *pada* or the "root" and denotes the *śeṣamūla* or the "remainder of the root" (see the layout), and that in this case the *mūla* is the number 1. Thus, in mathematical terms *padasamīpacaturguṇadṛśyake svaguṇamūlayute* means that:

- the *dṛśya* 2 is "multiplied by four" (*caturguṇa*), thus $2 \times 4 = 8$
- the *mūla* 1 is "multiplied by itself" (*svaguṇa*), thus: $1 \times 1 = 1$
- thus, $8 + 1 = 9$

The next steps (lines 14–17), denoted by the expressions *kṛtamūlike, padayute, dalite,* and *nijatāḍite*, are:

- $\sqrt{9} = 3$ (see *kṛtamūlike*)
- $3 + 1 = 4$ (see *padayute*)
- $4 \div 2 = 2$ (see *dalite*)
- $2 \times 2 = 4$ (see *nijatāḍite*)
- then, "[when] divided by one lessened by the fraction" (*vilavarūpahṛte*) is:

rūpa denotes the number 1 and *vilava* means "lessened by the fraction" which is, in this case, the fraction $\begin{array}{|c|}2\\3\end{array}$, thus: $\begin{array}{|c|c|}2&1\\3&1\end{array}$; $\begin{array}{|c|c|}2&1\\3&1\\1&3\end{array}$; $\begin{array}{|c|}1\\3\end{array}$.

The previous result 4 is divided by it and gives 12, which represents the new *dṛśya*. At this point, the procedure is performed again (see Śrīpati's *punar vidhi*) but without the fraction, which has been worked out and thus is erased (see lines 20–21). Therefore the new layout, although this is not provided by the SGT, should be $\begin{array}{|c|c|}\text{mū } 1 & \text{śemū } 1\\ 1 & 1\end{array}$ *dṛśya* 12.

The *mūla* of the *dṛśya* 12 is 1 and, as mentioned above by the commentator, the *antarālalava* $\begin{array}{|c|}2\\3\end{array}$ is erased because has been calculated. The procedure continues with:

- *padasamīpacaturguṇadṛśyake svaguṇamūlayute*:
- the *dṛśya* $12 \times 4 = 48$
- its *mūla* 1 is "multiplied by one" (*svaguṇa*), thus: $1 \times 1 = 1$
- $48 + 1 = 49$

With respect to the *karaṇasūtra*, the next steps (lines 22–25) are *kṛtamūlike, padayute, dalite,* and *nijatāḍite*:

- $\sqrt{49} = 7$ (see *kṛtamūlike*)
- $7 + 1 = 8$ (see *padayute*)
- $8 \div 2 = 4$ (see *dalite*)
- $4 \times 4 = 16$ (see *nijatāḍite*)

This 16 is the final result, the total sum of the leaves.

Lines 26–28 are dedicated to verifying the result obtained: the number of leaves fallen in the bed is the square-root of 16, hence 4; two-thirds of the remainder 12 (from $16 - 2$) is 8, the number of leaves fallen on the ground; the remainder is 4 (from $12 - 8 = 4$), its square-root is 2, the leaves which are also fallen on the ground. When these quantities are added to the *dṛśya* 2, it is 16: $\begin{array}{|c|}4\\8\\2\\2\end{array}$.

The SGT on GT 75 (ex.): This sample problem involves determining the total number of a flock of parrots. The *nyāsa* is: $\begin{array}{|c|c|c|c|}\text{mū } 3 & \text{śe } 1 & \text{mū } 3 & \text{dṛ } 0\\ 1 & 10 & 1 & \end{array}$. The SGT observes that, in respect to the previous exercise, here the "visible quantity near the root" of the remainder is zero because it has not been explicitly specified in the formulation of the sample problem. Regarding the first mathematical step of the procedural rule, which is *padasamīpacaturguṇadṛśyake*, the commentator points out that the *dṛśya* has been not mentioned and hence the "visible quantity" (*dṛśyaka*) which stands "near the root" (*padasamīpa*) is 0. Therefore, *caturguṇadṛśyake* means: 0×4

- the next mathematical step is *svaguṇamūlayute*: the *mūla* is 3 but, as Siṃhatilakasūri has explained in SGT 73, 27, when the number does not have a root, this is taken to be the number 1, hence: 3 is multiplied by its *mūla* 1 (see, in the sample problem, *triguṇaṃ padaṃ*) = 3

- the step *svaguṇamūlayute* is: 0 + (3 × 3) = 9; this 9 becomes the quantity "derived from the visible quantity" (see *dṛśyottha,* line 13). In fact, the commentator explains (see line 8) that this result replaces the 0 which is the initial *dṛśya* and, in the next steps, is manipulated as the *dṛśya*
- the next steps (lines 9–10) are *kṛtamūlike, padayute, dalite,* and *nijatāḍite*, and involve the derived *dṛśya* 9
- $\sqrt{9} = 3$ (see *kṛtamūlike*)
- 3+ the "root" of the *dṛśya* 9 which is 3, it gives 6 (see *padayute*)
- 6 ÷ 2 = 3 (see *dalite*)
- 3× 3 = 9 (see *nijatāḍite*)
- then, *vilavarūpahṛte* means that 1 (*rūpa*) is lessened by the *śeṣamūla* one-tenth; it becomes $\begin{array}{|r|}9\\10\end{array}$, which is the divisor (see *hṛte* in *vilavarūpahṛte*) "derived from the visible quantity" (in line 13, see *dṛśyottha*) 9
- the commentator quotes from the division of fractions of GT 42 and the operands of the divisor are inverted: $\begin{array}{|r|}10\\9\end{array}\begin{array}{|r|}9\\1\end{array}$; by means of the reciprocal multiplication first and by the reduction by 9 then, it becomes: $\begin{array}{|r|}10\\1\end{array}\begin{array}{|r|}1\\1\end{array}$, and hence the final result is $\begin{array}{|r|}10\\1\end{array}$. This represents the new *dṛśya* (line 17)
- then, *punar vidhi*: the procedure starts again but without the "fraction" (*lava*) which, because it has been calculated, is erased. At this point, the *nyāsa* should appear as: $\begin{array}{|lr|}\text{mū} & 1\\ & 1\end{array}\begin{array}{|lr|}\text{śemū} & 1\\ & 1\end{array}\begin{array}{|lr|} & 10\\ \text{dṛ} & \\ & 1\end{array}$. Then,
- line 17: *daśadṛśyaṃ caturguṇaṃ* = 10 × 4 = 40
- line 17: *triguṇapadam traya eva* = 1 × 3 = 3
- line 17–18: *svaguṇamūlayute* is 40 + 9 (the *mūla* 3 × 3) = 49
- the next steps (lines 18–20) are *kṛtamūlike, padayute, dalite,* and *nijatāḍite*:
- $\sqrt{49} = 7$ (see *kṛtamūlike*)
- 7 plus the *pada* 3 = 10 (see *padayute*)
- 10 ÷ 2 = 5 (see *dalite*)
- 5 × 5 = 25 (see *nijatāḍite*)

This is the final result, the total number of parrots.

The type-problem of the fraction [executed] with the root and the visible quantity (*mūlāgrabhāgajāti*) [GT 76]

GT 76: The quantities mentioned in this rule are: the "visible quantity" (*dṛśya*), the number "one" (*rūpa*), the "root" (*mūla*), and the "fraction" (*bhāga*). The steps of the algorithm can be represented as:

$$\left|\frac{d}{1-f}\right| = A; \left|\frac{r}{1-f}\right| = B$$

$$\sqrt{\left(A + \left|\frac{f}{2}\right|\right)^2} = C$$

$$\left(C + \left|\frac{B}{2}\right|\right)^2 = x$$

where d is the visible quantity, f is the fraction, and r is the root.

Notably, in SGT 79, 18–19 at the end of the third sample problem on this type-problem, the commentator elucidates the expression *mūlāgrabhāgajāti*. The name of this procedure could be so explained: the "fraction" (*bhāga*) is computed first with the *agra*/*dṛśya* and then with the "root" (*mūla*). The remaining part of the procedure is carried out by means of the results of these calculations. Regarding the terminology used by the GT, it can be observed that: *dṛśya* is the "visible quantity"; both *mūla* and *pada* mean "root", which is represented as $\left|\begin{matrix}1\\1\end{matrix}\right|$; in the locative *pade*, *pada* is "square-root"; *bhāga* is "fraction" in the expression *bhāgonarūpa* while compounded with *dvi* in *dvibhāga* it denotes the fraction $\left|\begin{matrix}1\\2\end{matrix}\right|$, and the terms *karaṇī* and *kṛti* denote the "square".

The SGT on GT 76: In the introductory line, the commentary refers to the procedure formulated by Śrīpati by the expression *mūlāgrabhāgajāti*. Here, the meaning of *agra*, (lit. "forepart, tip") is ambiguous; nevertheless, according to the commentator (see SGT 79, 18; 80, 25–26), the term *agra* indicates the "visible quantity" (*dṛśya*). This is, in fact, found at the "end, at the tip" of the *nyāsa*s given by the SGT while setting down the initial data of the next sample problems (see SGT 77, 15; 78, 20, and 79, 21). It can be observed that in the next type-problem of fractions, which Siṃhatilakasūri terms *ubhayāgradṛśyajāti, agra* specifies again the *dṛśya* which is, in fact, found at the tip of the *nyāsa*s occurring in SGT 81, 12; 82, 12, and 83, 23. In this procedure, there is a first *dṛśya* at the beginning and a second at the end (see *ubhayāgradṛśya* as the "visible quantity at both tips").[126]

The first passage is dedicated to observations on grammar. Line 5 clarifies that, whenever it is not possible to calculate "half" (*ardha*) the "root" (*mūla*/*pada*), one should then multiply the denominator by the number two. Also, since *karaṇī* ("square") is used by Śrīpati for the first time, the commentator

126 Interestingly, *agra* is used with the same connotation in GSS 4.47 in relation to the problem of fractions termed *śeṣabhāgajāti* by Mahāvīrācārya.

provides its synonym *varga*. The remaining part of Siṃhatilakasūri's passage paraphrases and expands the last part of the *karaṇasūtra.*

The SGT on GT 77 (ex.): Students are asked to determine the total number of deer in a herd. The *nyāsa* is: $\begin{array}{|r|r|r|r|} \text{rū } 1 & \text{bhā } 1 & \text{mū } 1 & \text{dṛ } 2 \\ 1 & 3 & 1 & \\ & 1 & & \\ & 3 & & \end{array}$. In lines 15–20, the first mathematical step, which is *bhāgonarūpa* or "one lessened by the fraction", is carried out.[127] The commentator calculates the sum of the fractions, which are one-third and its one-third, by means of the second method of the class of fractional increase; the result $\begin{array}{|c|}4\\9\end{array}$ is then, by applying the class of simple fractions, subtracted from the number 1. The fraction $\begin{array}{|c|}5\\9\end{array}$ is the obtained result (see *etad bhāgonarūpam,* line 20) and the *dṛśya* and the *mūla* are divided by it (see *anena vihṛte dṛśyamūle,* line 21). In order to perform the step *bhāgonarūpavihṛte dṛśyamūle* or "when the visible quantity and the root are divided by one lessened by the fraction", the commentator quotes the rule of division of fractions and the operands of the divisor are hence interchanged. When the *dṛśya* $\begin{array}{|c|}2\\1\end{array}$ is divided by the obtained *bhāgonarūpa* $\begin{array}{|c|}5\\9\end{array}$, the result is $\begin{array}{|c|}18\\5\end{array}$, which becomes the new *dṛśya.*[128] The *mūla* $\begin{array}{|c|}1\\1\end{array}$ divided by $\begin{array}{|c|}5\\9\end{array}$ becomes $\begin{array}{|c|}9\\5\end{array}$, which represents the new *mūla.*[129] In the next step, the obtained *mūla* is halved (*padārdha*) and that result is added to the obtained *dṛśya*; the square-root of the result is then extracted. The obtained fraction[130] $\begin{array}{|c|}21\\10\end{array}$ is added to the *mūladvibhāga* (in the GT, meant by *padārdha* too) $\begin{array}{|c|}9\\10\end{array}$, which represents half of the *mūla* $\begin{array}{|c|}9\\5\end{array}$. The result of this sum is $\begin{array}{|c|}30\\10\end{array}$, this is squared and the number obtained is 9, which is the number of deer in the herd.

The verification provided shows that one-third of 9 is 3, this added to its one-third, which is 1, becomes 4. Hence, 4 is the number of deer which

127 In the algorithm I have provided above, this is denoted by the formula: $1 - b$.
128 In my formulation, it is denoted by A.
129 In my formulation, it is denoted by B.
130 In my formulation, it is denoted by C.

disappeared. The root of 9 is 3, and this represents the number of deer longing for *gīta*, plus the number of deer in the forest, which is 2, thus: $\begin{array}{|c|}3\\1\\3\\2\end{array}$. When these are added, the result is 9.

The SGT on GT 78 (ex.): This sample problem concerns a troop of monkeys. The *nyāsa* $\begin{array}{|r|r|r|c|}\text{rū } 1 & \text{bhā } 5 & \text{mū } 1 & \\ & & & \text{dṛ } 10\\ 1 & 9 & 1 & \end{array}$ is followed by these calculations:

- (lines 20–23): the step "one lessened by the fraction" is the first to be carried out; the commentator quotes the class of simple fractions and shows the following layout where *rūpa* or "one", which is the *prācyāṅka* ("the first quantity"), and the fraction five-ninths, which represents the *parāṅka* ("next quantity"), exchange their respective denominators: $\begin{array}{|c||c|}1&5\\1&9\\9&1\end{array}$.

  The difference is $\begin{array}{|c|}4\\9\end{array}$
- the "visible quantity" 10 and the root 1 are to be divided by this result. Lines 23–28 are devoted to carrying out these divisions; the SGT quotes the *sūtra* of division of fractions according to which the numerator and the denominator of the obtained difference, since it is the divisor, are interchanged and the multiplication is executed. The new "visible quantity" becomes $\begin{array}{|r|}90\\4\end{array}$, the new "root" (*mūla*) is $\begin{array}{|c|}9\\4\end{array}$. In this manner, the step *bhāgonarūpavihṛte dṛśyamūle* is accomplished
- the commentator halves the obtained root (see the step *padārdha*) in order to sum its square and the obtained "visible quantity"; because it is not possible to divide the numerator 9 (see line 28) the *mūla* $\begin{array}{|c|}9\\4\end{array}$ is multiplied by $\begin{array}{|c|}1\\2\end{array}$; it becomes $\begin{array}{|c|}9\\8\end{array}$
- this obtained "half the root" is squared and then added to the "visible quantity" $\begin{array}{|r|}90\\4\end{array}$ (see the step *dṛśyāt padārdhakaraṇīsahitāt*). Thus,

line 4 shows the following layout: $\begin{array}{|r|r|} 81 & 90 \\ 64 & 4 \\ 1 & 16 \end{array}$. The two denominators have been reduced by the common factor 4 and the quotients, being the new denominators, are placed below. By means of the *bhāgajāti*, the result becomes $\begin{array}{|r|} 1521 \\ 64 \end{array}$

- its square-root is extracted and the obtained fraction $\begin{array}{|r|} 39 \\ 8 \end{array}$ is then added to the result of the step *mūladvibhāga* (in the GT also referred to as *padārdha*) which is $\begin{array}{|r|} 9 \\ 8 \end{array}$ the resulting 6 is squared. The number of monkeys is thus 36.

The verification provided by the SGT shows that: five-ninths of 36 is 20, the square-root of 36 is 6, and 20 plus 6 are the monkeys swinging in a jack-fruit three. Ten are the monkeys "seen" (the initial *dṛśya*) playing, thus: $\begin{array}{|r|} 20 \\ 6 \\ 10 \end{array}$.

The SGT on GT 79 (ex.): In this third exercise, students are asked to determine the total number of a herd of pigs. The numerical data are: $\begin{array}{|lr|} \text{rū} & 1 \\ & 1 \end{array}$ $\begin{array}{|lr|} \text{bhā} & 1 \\ & 8 \end{array}$ $\begin{array}{|lr|} \text{mū} & 1 \\ & 1 \\ & 2 \end{array}$ $\begin{array}{|lr|} \text{dṛ} & 8 \\ & 1 \end{array}$. The SGT performs the following steps:

- after having calculated the "root", which comes to be $\begin{array}{|r|} 3 \\ 2 \end{array}$, the first step concerns the subtraction of the fraction from "one" (*rūpa*). Thus the layout is (line 24): $\begin{array}{|r|r|} 1 & 1 \\ 1 & 8 \\ 8 & 1 \end{array}$. The result $\begin{array}{|r|} 7 \\ 8 \end{array}$, which represents the difference, becomes the divisor of the "visible quantity" and the "root"
- the commentator quotes the *sūtra* of division of fractions; the operands of the divisor are exchanged and the multiplication is carried out: $\begin{array}{|r|r|} 8 & 8 \\ 7 & 1 \end{array}$ giving the new *dṛśya* $\begin{array}{|r|} 64 \\ 7 \end{array}$. The "root" $\begin{array}{|r|} 3 \\ 2 \end{array}$ divided by the obtained difference $\begin{array}{|r|} 7 \\ 8 \end{array}$ becomes $\begin{array}{|r|} 24 \\ 14 \end{array}$

- in this way, the step *bhāgonarūpavihṛte dṛśyamūle* is accomplished, and line 2 shows the layout with the obtained quantities: $\begin{array}{|c|c|} \hline \text{mū } 24 & \text{dṛ } 64 \\ 14 & 7 \\ \hline \end{array}$
- the commentator then halves the obtained "root", which is squared and new "visible quantity" is added to it by means of the quoted class of simple fractions: $\begin{array}{|c|c|} 144 & 64 \\ 196 & 7 \\ 1 & 28 \end{array}$. The two denominators above are the results of the reduction of the original denominators by the common divisor 7; when these are added, the result is $\begin{array}{|c|} 1936 \\ 196 \end{array}$
- its square-root is extracted and the obtained fraction $\begin{array}{|c|} 44 \\ 14 \end{array}$ is added to "half the root" (*mūladvibhāga*) $\begin{array}{|c|} 12 \\ 24 \end{array}$ (which is also defines as *padārdha,* see lines 2–3). The result $\begin{array}{|c|} 56 \\ 14 \end{array}$ is reduced by the common factor 14 becoming: $\begin{array}{|c|} 4 \\ 1 \end{array}$
- this is squared, and the result 16 represents the total number of pigs.

The verification provided by the commentary demonstrates that one-eighth of the total number 16 is 2, which is the number of hogs playing in the pond; the square-root of 16 is 4 which together with its half 2 becomes 6, and this is the number of hogs digging up the *mustā*-grass. The "visible quantity" is 8, thus: $\begin{array}{|c|} 2 \\ 4 \\ 2 \\ 8 \end{array}$. Their sum gives the final result 16.

The commentator ends the treatment of this *jāti* by observing (lines 18–19) that, as the solution has been obtained from the fraction manipulated with the root and the visible quantity at the "tip", the type-problem of the fraction executed with the root and the visible quantity is so termed.

The type-problem of the visible quantity at both tips (*ubhayāgradṛśyajāti*) [GT 80]

GT 80: The steps of the algorithm enunciated by Śrīpati can be represented in the following manner:

$$\frac{ad}{1-f} = Ad; \frac{r}{1-f} = R$$

$$Ad + \left(\frac{R}{2}\right)^2 = C$$

$$C + \bar{a}d = D$$
$$\sqrt{D} = E$$
$$\left(\frac{R}{2}+E\right)^2 = x$$

where *ad* is the *antadṛśya* (the "last visible quantity"), *f* is the product of the result of 1 lessened by each fraction, *Ad* is the obtained *antadṛśya*, *r* is the root, *R* is the obtained root, and *ād* is the *ādyadṛśya* (the "first visible quantity").

The SGT on GT 80: As in GT 76, here too *agra* ("tip") denotes the "visible quantity" (*dṛśya*). In the first lines of his explanation, Siṃhatilakasūri clarifies that this *jāti* concerns two *agra*s: the *ādyāgra* and the *antāgra*. In the sample problems concerning this procedure, Śrīpati mentions, in fact, two *dṛśya*s: one at the beginning of the sample problem and one at the end. With respect to the first mathematical step, which is *niraṃśarūpāhati* or the "multiplication of one devoid of the fractions", the commentary explains that *rūpa* is to be deducted by as many fractions as they are, and that the "multiplication" (*āhati*) denotes the multiplication of the fractions so obtained. This is to be carried out by means of the multi-part class, which is then quoted by the SGT. In line 28, the commentator specifies that the visible quantity and the root are divided by this product, and notes that the *dṛśya* which is involved in this step is the "last visible quantity" (*paryantadṛśya*). The next passage explains that the "sum" (*aikya*) to be carried out concerns both visible quantities, i.e., the *ādyadṛśya* ("first visible quantity") and the newly obtained *antadṛśya* ("last visible quantity").[131] The last lines of the commentator's passage elucidate, by paraphrasing the *karaṇasūtra*, that the *antadṛśya* is added to the square of half the root, the *ādyadṛśya* is then added to this result, and the square-root of this sum is extracted. Finally, the result is added to half the root, and the result multiplied by itself.

The SGT on GT 81 (ex.): Students have to ascertain the total number of a herd of elephants. In line 12, the following *nyāsa* occurs:

dṛ 1	śe 1	1	1	mū 1	dṛ 7
1	6	5	4	1	1

The elephant "standing near a pillar" represents the "first visible quantity" (*ādyadṛśya*), while the seven elephants which are "seen" represent the "last visible quantity" (*antadṛśya*). The steps executed are:

- lines 13–20 are dedicated to carrying out the first mathematical step, which is the multiplication of the fractions obtained by subtracting one from the given fractions. The commentator quotes the *karaṇasūtra* of the class of simple fractions and performs the subtraction of each "fraction"

131 This has been obtained by carrying out the first step (*niraṃśarūpāhatibhaktadṛśyamūle*) where the original *antadṛśya* is involved.

(*bhāga*) from "one, unity" (*rūpa*). The calculations performed are: $\begin{array}{|c|c|}1&1\\1&6\end{array}$;

$\begin{array}{|c|c|}1&1\\1&6\\6&1\end{array}$; $\begin{array}{|c|c|}6&1\\6&6\end{array}$; $\begin{array}{|c|}5\\6\end{array}$, and so forth the obtained fractions are: $\begin{array}{|c|c|c|}5&4&3\\6&5&4\end{array}$. The multiplication is accomplished by means of the multi-art class, and the result is $\begin{array}{|c|}1\\2\end{array}$

- this result becomes hence the divisor of the "last visible quantity" $\begin{array}{|c|}7\\1\end{array}$ and the "root" $\begin{array}{|c|}1\\1\end{array}$. The results are the new "last visible quantity" $\begin{array}{|c|}14\\1\end{array}$ and the new "root" $\begin{array}{|c|}2\\1\end{array}$
- the next step is *padadvyaṃśakavargayuktāt dṛśyaikato mūlam*: half the new root obtained, which is 2, is 1; its "square" (*varga*) is 1 and when the new "last visible quantity" is added to it, the result is $\begin{array}{|c|}15\\1\end{array}$. The *ādyadṛśya* $\begin{array}{|c|}1\\1\end{array}$ added to $\begin{array}{|c|}15\\1\end{array}$ becomes $\begin{array}{|c|}16\\1\end{array}$; its *mūla* or "square-root" is 4
- the new root w is halved giving one; four added to it becomes 5
- the last step is *svanighnaṃ*: 5 "multiplied by itself" is 25, which is the total number of elephants in the herd.

The verification shows that one-sixth of the remainder 24 is 4, which is the number of elephants on the shores of a lake. One-fifth of the remainder 20 is 4, which is the number of elephants being on the top of a mountain. One-fourth of the remainder 16 is 4, which is the number of elephants frightened by a lion. The square-root of the total sum 25 is 5, the visible quantities 1 and 7 are added too, thus: $\begin{array}{|c|}1\\4\\4\\4\\5\\7\end{array}$. Their sum gives 25.

The SGT on GT 82 (ex.): The total of a swarm of bees is to be calculated. The numerical data are: $\begin{array}{|cc|c|c|}\text{dṛ}&\begin{array}{c}2\\1\end{array}&\text{śe}\ \begin{array}{c}1\\2\\1\\7\end{array}&\text{mū}\ \begin{array}{c}1\\1\end{array}\ \text{dṛ}\ \begin{array}{c}2\\1\end{array}\end{array}$. Lines 12–19 are dedicated to carrying out the first mathematical step, which is the multiplication of one devoid of the fractions. The fractions of the remainder are calculated by means of the second method of the class of fractional increase, the result is: $\left|\begin{matrix}8\\14\end{matrix}\right|$. By means of the class of simple fractions, this is subtracted from one and it becomes $\left|\begin{matrix}3\\7\end{matrix}\right|$. At this point, the commentator explains that the "multiplication" (*āhati*) is not possible because of the absence of other fractions. In order to perform the step "the [last] visible quantity and the root are divided by [the result of] the multiplication of one devoid of the fractions", the fraction $\left|\begin{matrix}3\\7\end{matrix}\right|$ becomes the divisor of the "last visible quantity" (*antadṛśya*) 2 and of the "root" (*mūla*) 1; the fractions obtained are the new *antadṛśya* $\left|\begin{matrix}14\\3\end{matrix}\right|$ and the new *mūla* $\left|\begin{matrix}7\\3\end{matrix}\right|$. Afterwards, the new "root" $\left|\begin{matrix}7\\3\end{matrix}\right|$ is halved by being multiplied by $\left|\begin{matrix}1\\2\end{matrix}\right|$; the square of the resulting $\left|\begin{matrix}7\\6\end{matrix}\right|$ is $\left|\begin{matrix}49\\36\end{matrix}\right|$, and the new *antadṛśya* is then added to it, the result is $\left|\begin{matrix}217\\36\end{matrix}\right|$. The first visible quantity is added to it: hence, the "first visible quantity" $\left|\begin{matrix}2\\1\end{matrix}\right|$ added to $\left|\begin{matrix}217\\36\end{matrix}\right|$ becomes $\left|\begin{matrix}289\\36\end{matrix}\right|$. Its "square-root" is $\left|\begin{matrix}17\\6\end{matrix}\right|$; $\left|\begin{matrix}7\\6\end{matrix}\right|$, which is half the new "root", is added to it giving $\left|\begin{matrix}4\\1\end{matrix}\right|$ that multiplied by itself (the last step is *svanighnaṃ*) is $\left|\begin{matrix}16\\1\end{matrix}\right|$, the measure of bees in the swarm.

The verification is as follows: 2 bees (which represent the "first visible quantity") are seen on a lotus, half of the remainder 14 is 7. One-seventh of 7 is 1, which added to it becomes 8; this is the number of bees resting on the temples of an elephant. The square-root of the total sum 16 is 4, which is the number of bees humming on a jasmine flower, plus the *antadṛśya* 2: $\left|\begin{matrix}2\\7\\1\\4\\2\end{matrix}\right|$. Their total sum gives, in fact, 16.

The SGT on GT 83 (ex.): This exercise concerns the amount of money given by someone to a twice-born. The presentation of the numerical data is: $\left|\begin{matrix}\text{dṛ } 1\\4\end{matrix}\right|\begin{matrix}\text{śe } 1\\3\end{matrix}\left|\begin{matrix}\text{śe } 1\\4\end{matrix}\right|\begin{matrix}\text{mū } 1\\1\end{matrix}\left|\begin{matrix}\text{dṛ } 1\\2\end{matrix}\right|$. The commentator points out that, in

Śrīpati's sample problem, the two *dṛśya*s are not mentioned. In performing the first mathematical step, which is *niraṃśarūpāhatibhaktadṛśyamūle,* the two quantities-remainder $\begin{array}{|c|}1\\3\end{array}\begin{array}{|c|}1\\4\end{array}$ are each subtracted from one and the resulting fractions are multiplied; the result is $\begin{array}{|c|}1\\2\end{array}$. This fraction becomes the divisor of the "last visible quantity" $\begin{array}{|c|}1\\4\end{array}$ and the "root" $\begin{array}{|c|}1\\1\end{array}$; the fractions obtained are the new *antyadṛśya* $\begin{array}{|c|}2\\2\end{array}$ and the new *mūla* $\begin{array}{|c|}2\\1\end{array}$. The next step is *padadvyaṃśakavargayuktāt dṛśyaikato mūlam*: the new "root" $\begin{array}{|c|}2\\1\end{array}$ is halved, it gives $\begin{array}{|c|}1\\1\end{array}$ and its square, which is again 1, is added to the new "first visible quantity" $\begin{array}{|c|}2\\2\end{array}$, and the result is $\begin{array}{|c|}4\\2\end{array}$. The new *antyadṛśya* is added to it: hence, the *ādyadṛśya* $\begin{array}{|c|}1\\4\end{array}$ added to $\begin{array}{|c|}4\\2\end{array}$ gives $\begin{array}{|c|}9\\4\end{array}$ (*etad dṛśyaikyam,* line 12). Its square-root is $\begin{array}{|c|}3\\2\end{array}$, the step *padārdhayuktaṃ* means here that the previous found $\begin{array}{|c|}1\\1\end{array}$ is added to the last result becoming $\begin{array}{|c|}5\\2\end{array}$, which multiplied by itself (the last step is *svanighnaṃ*) is $\begin{array}{|c|}25\\4\end{array}$. This improper fraction is transformed into the compound fraction $\begin{array}{|c|}6\\1\\4\end{array}$. Therefore, the amount of money given to the twice-born was 6 plus one-fourth *drammas*.

The verification demonstrates that, with respect to this final sum, one-fourth of a *dramma* represents the "first visible quantity". Regarding the two fractions one-third and one-fourth: one-third of 6 is 2; the remainder is 4, its one-fourth is 1. Then, the root of the final fraction $\begin{array}{|c|}25\\4\end{array}$ is $\begin{array}{|c|}5\\2\end{array}$. In line 24, the commentator presents the various quantities: $\begin{array}{|c|}1\\4\end{array}\begin{array}{|c|}2\\1\end{array}\begin{array}{|c|}1\\1\end{array}\begin{array}{|c|}2\\1\\2\end{array}\begin{array}{|c|}1\\2\end{array}$, where the first two fractions on both sides represent the "first" and the "last visible quantity" respectively.

The type-problem of the fractions and the visible quantity which is a fraction (*bhinnabhāgadṛśyajāti*) [GT 84]

GT 84: This rule formulates an algorithm which can be represented as follows:

$$\frac{1-d}{p} = \mathrm{x}$$

where d is the *dṛśya* (which is here a fraction) and p is the product of the parts of the pillar which are not visible.

The SGT on GT 84: At the end of the explanation of the second sample problem (SGT 86, 14), the SGT terms this procedure *bhinnabhāgadṛśyajāti*. In this expression, two different terms mean "fraction": *bhinna* denote the fractions which represent the invisible parts (see the pillar and bamboo mentioned in the two sample problems) and *bhāga* is the fraction which is the "visible quantity" (*dṛśya*). As emphasised by the commentator in the first two lines of his explanation, the main characteristic of this rule is, in fact, the relationship between the fraction representing the visible quantity and the fractions representing the "invisible" (*adṛśya*) parts.

The SGT on GT 85 (ex.): This sample problem asks students to ascertain the measure of a pillar. In line 8, the SGT presents the layout: $\begin{array}{|c|c|c|} 1 & 1 & \text{dṛ} \; 1 \\ 7 & 10 & 2 \end{array}$. The first step carried out by the commentator is to subtract the fraction of the visible quantity from "one", the result is $\begin{array}{|c|} 1 \\ 2 \end{array}$. In order to carry out its division by the product of the multiplication of the parts of the pillar, the multi-part class is performed; it gives $\begin{array}{|c|} 1 \\ 70 \end{array}$. This is the divisor by which the result $\begin{array}{|c|} 1 \\ 2 \end{array}$ is divided, the quotient is $\begin{array}{|c|} 35 \\ 1 \end{array}$ *hastas*, which is the measure of the pillar.

The verification provided by the SGT shows that one-seventh of 35 is 5; one-tenth of 35 is 3 and one-half, which multiplied by 5 becomes 17 and one-half *hastas*. This is the part of the pillar which is invisible, which is the first half. The other half, 17 and one-half *hastas*, is the part of the pillar which is visible.

The SGT on GT 86 (ex.): This sample problem asks students to ascertain the measure of a bamboo. In line 25, one finds the following layout: $\begin{array}{|c|c|c|} 3 & 5 & 13 \\ 50 & 25 & 25 \end{array}$, where the last fraction on the right represents the visible quantity and the two fractions on the left are the parts of the bamboo which are invisible. By means of the class of simple fractions, "one" (*rūpa*) is lessened by the "visible quantity"; the result is $\left|\begin{matrix}12\\25\end{matrix}\right|$. Then, by means of the multi-part class, the product of the two fractions representing the invisible part is calculated giving $\left|\begin{matrix}15\\1250\end{matrix}\right|$, this result corresponds to the step "the [product of the] multiplication of the parts of the pillar" but this is also the divisor (see *hara*, line 2) of $\left|\begin{matrix}12\\25\end{matrix}\right|$. The final result of the division of the two fractions is 40 *hasta*s, which is the measure of the bamboo. Keeping in mind that according to GT $\overline{8}$, 1 *hasta* is equal to 24 *aṅgula*s, the verification shows that:

$$40\ ha \times \frac{3}{50} = \frac{40 \times 24}{50} \times 3 = 19\frac{1}{5}\ a\dot{n} \times 3 = 2\ ha\ 9\frac{3}{5}\ a\dot{n}$$

$$40\ ha \times \frac{5}{25} = 8\ ha;\ 2\ ha\ 9\frac{3}{5}\ a\dot{n} \times 8\ ha = 19\ ha\ 4\frac{4}{5}\ a\dot{n}$$

$$40\ ha \times \frac{13}{25} = 20\ ha\ 19\frac{1}{5}\ a\dot{n}$$

$$19\ ha\ 4\frac{4}{5}\ a\dot{n} + 20\ ha\ 19\frac{1}{5}\ a\dot{n} = 40\ ha.$$

where *ha* is *hasta* and *aṅ* is *aṅgula*.

The type-problem of the [coefficient of the] [square] root of the fraction (*bhāgamūlajāti*) [GT 87]

GT 87: This rule involves three quantities: the "visible quantity", the given "fraction", and the "coefficient of the square-root" of the given fraction from the which the name of this type-problem derives. Notably, the commentary elucidates the name of this procedure at the end of the execution of the third sample problem (SGT 89, 18–19); it explains that this type-problem is so called because the visible quantity is manipulated with the "root" (*mūla*) of the "fraction" (*bhāga*). The steps of the algorithm are:

$$\frac{d \times 4}{f} + r^2 = D;$$

$$\sqrt{D} = \mathrm{E}$$

$$\left(\frac{E+r}{2}\right)^2 \times f = x$$

where d is the *dṛśya*, f is the fraction, and r is the given coefficient of the square-root.

The SGT on GT 87: The SGT clarifies the terms used by Śrīpati and specifies the two ambiguous passages *svamūlasaṃvargayutāt* or "added to the square of its own [coefficient of the] [square] root" and *samūlam* or [that result is] united with the [coefficient of the] [square] root [...]". The ambivalence of these steps lies on the use of the term *mūla*, which here does not denote the operation of "square-root" – as it does *mūla* which is the subject of the first two lines of the *karaṇasūtra* – but instead the coefficient, that "times" the square-root as specified in a given sample problem. It is significant that in line 23, the commentator uses the term *aṅka* ("number") to indicate that, with respect to the expressions mentioned above, it is the number, the coefficient of the square-root to be concerned. In the last lines the SGT explains that whenever there are two or three visible quantities, the fractions and the given coefficient of the square-roots are also two or three; this is a clear reference to the last sample problem on this *jāti* given by the root-text where such a case occurs.

The SGT on GT 88 (ex.): This sample problem involves determining the total number of a herd of elephants. The SGT presents the following *nyāsa*: $\begin{array}{|c|r|r|} 1 & \text{mū } 18 & \text{dṛśya } 18 \\ 8 & 1 & 1 \end{array}$. The first box on the left presents the *lava* or "fraction", which in this sample problem is one-eighth; the second box represents the *mūla* or the "coefficient of the root" of the fraction, which is 18, and the "visible quantity" is 18. The steps carried out are:

- lines 3–10 explain the first mathematical step. The visible quantity is multiplied by 4, divided by the given fraction one-eight, and added to the square of the coefficient of the square-root of the fraction. The square-root of this result is then extracted. The calculations are: $18 \times 4 = 72$; $72 \times 8 = 576$; $18^2 = 324$; $324 + 576 = 900$; $\sqrt{900} = 30$
- in lines 10–14, the resulting 30 is added to the *mūla* 18, the result is halved, squared, and multiplied by the given part one-eighth. The calculations are: $30 + 18 = 48$; $48 \div 2 = 24$; $24^2 = 576$; $576 \times \begin{array}{|c|} 1 \\ 8 \end{array} = \begin{array}{|r|} 576 \\ 8 \end{array} = 72$, which is the number of elephants in the herd.

In the verification, the commentary demonstrates that one-eighth of 72 is 9. Its square-root is 3, multiplied by 18 it becomes 54, which is the number of elephants that moved to the top of a mountain. The visible quantity is 18, and 54 plus this 18 produces the final result 72.

The SGT on GT 89 (ex.): Students have to ascertain the total number of a flock of swans. In the introductory line, Siṃhatilakasūri emphasises that this sample problem involves a double visible quantity accompanied by a double root; the second *dṛśya*, however, is not given but it is obtained during the computation. The two visible quantities are each computed with their respective fractions and their roots. The *nyāsa* is: $\left|\begin{array}{c|c|c|c|c} 2 & \text{mū } 9 & \text{śeṣa } 3 & \text{mū } 6 & \text{dṛ } 24 \\ 3 & 1 & 5 & & \end{array}\right|$. The steps performed by the commentator are the following:

- lines 24–3 explain the first mathematical step; the visible quantity is multiplied by 4, divided by the fraction three-fifths, and added to the coefficient 6 of the square-root; the square-root of this result is extracted. The calculations are hence the following: $24 \times 4 = 96$; $96 \div \left|\begin{array}{c} 3 \\ 5 \end{array}\right| = 160$; $6^2 = 36$; $160 + 36 = 196$; $\sqrt{196} = 14$
- lines 3–7: in order to carry out the final step, the resulting 14 is added to the root 6, the result is halved, squared, and multiplied by the fraction three-fifths. The calculations are: $14 + 6 = 20$; $20 \div 2 = 10$; $10^2 = 100$; $100 \times \left|\begin{array}{c} 3 \\ 5 \end{array}\right| = 60$
- in line 7, the process starts again from the beginning and 60, which is the previous result, is the new *dṛśya* computed with the fraction two-thirds and the *mūla* 9
- having obtained the new visible quantity 60, the layout should thus be: $\left|\begin{array}{c||c||c} 2 & \text{mū } 9 & \text{dṛ } 60 \\ 3 & 1 & \end{array}\right|$. These represent the given fraction, the coefficient of its square-root, and the new visible quantity respectively. Hence in this sample problem, the first set of data are $\left|\begin{array}{c} 3 \\ 5 \end{array}\right|$, the *mūla* 6, and the *dṛśya* 24. Now the visible quantity 60 and its related *lava* $\left|\begin{array}{c} 2 \\ 3 \end{array}\right|$ and *mūla* $\left|\begin{array}{c} 9 \\ 1 \end{array}\right|$ are computed (lines 7–18). The calculations performed are: $60 \times 4 = 240$; $240 \div \left|\begin{array}{c} 2 \\ 3 \end{array}\right| = 360$; $9^2 = 81$; $360 + 81 = 441$; $\sqrt{441} = 21$; $21 + 9 = 30$; $30 \div 2 = 15$; $15^2 = 225$; $225 \times \left|\begin{array}{c} 2 \\ 3 \end{array}\right| = 150$, which is the final number of swans in the flock.

The verification presented by the commentator shows that two-thirds of 150 is 100, its square-root is 10 which multiplied by 9 is 90, the number of swans which flew away. The remainder is 60, its three-fifths is 36, six times its square-root gives 36, which is the number of swans that perished. Twenty-four is the initial *dṛśya* and when these are added, the final result is one hundred fifty: $\left|\begin{array}{c} 90 \\ 36 \\ 24 \end{array}\right|$.

The type-problem of the subtractive square (hīnavargajāti) *[GT 90]*

GT 90: In the introductory line, the commentator refers to this procedure by the expression *hīnavargajāti*.[132] Here the quantities involved are three: the "visible quantity", the "subtractive quantity" (*ūna*), and the given "fraction". The operands of the given fraction are interchanged and this is then written twice: one is manipulated with the subtractive quantity and the other with the square of its half. As specified by the commentary, the name of this procedure derives from the fact that the square of the given fraction (see the sample problems) is lessened by the quantity which is the "subtractive quantity", thus the square is "subtractive".[133]

The various phases of the algorithm are:

$$\frac{a}{b} \rightarrow \frac{b}{a}$$

$$\left[\left(\frac{b}{a} \times \bar{u}\right) + \left(\frac{b}{a}\Big/2\right)^2\right] - d = D$$

$$\bar{u} + \left(\frac{b}{a}\Big/2\right) = E$$

$$E + \sqrt{D} = \mathrm{H}$$

$$\mathrm{H} \div \frac{a}{b} = \mathrm{x}$$

where $\frac{a}{b}$ is the given fraction, $\frac{b}{a}$ represents this same fraction with its operands interchanged, $\bar{u}$ is the "subtractive quantity" (*ūna*), and d is the "visible quantity" (*dṛśya*).

The SGT on GT 90: The SGT first clarifies that the given fraction should be written twice having interchanged its numerator and denominator; thereafter, in order to clarify the rule, the commentator provides synonyms (see *āhata/guṇita, kṛti/varga, pada/mūla, ūna/hīna)* and specifies that the term *anya* or "other" refers to the "other" fraction, the second of the two equal fractions which have been placed and among which the first is multiplied by the *ūna* or "subtractive quantity". In lines 4–5, the commentator explains the meaning of the expression *tatpadam* or "the [square] root of that", and

132 The *hīnavargajāti* is the last type-problem with fractions expounded by the GT.
133 See SGT 90, 1.

observes that *tat* ("of that") refers to the number obtained by carrying out the step "[and the result is] lessened by the visible quantity".[134] The last passage is dedicated to clarifying another ambiguous passage, which is *tatpadam ūnam anyadalānvitaṃ bhāgavibhaktam āptam* or "the subtractive quantity is added to half of the other [and the result is added to] the [square] root of that. [The result is] divided by the fraction, [and the final result is] obtained." The commentary shows that the verbal adjective *vibhakta* ("divided") does not qualify *ūnam anyadalānvitaṃ,* as it would be naturally to interpret, but it refers to the whole expression *tatpadam ūnam anyadalānvitaṃ,* and thus to the result obtained when carrying out this step.[135] This will be the dividend of the fraction (see *bhāga* in *bhāgavibhaktam*).

The SGT on GT 91 (ex.): This exercise asks students to ascertain the total number of a muster of peacocks. The SGT presents the following *nyāsa*: $\begin{array}{|c|c|c|}3 & \text{ū } 6 & \text{dṛ } 6\\ 5 & 1 & 1\end{array}$. The first step carried out by the commentator is to write twice the fraction $\begin{array}{|c|}3\\5\end{array}$ with its operands inverted: $\begin{array}{|c|c|}5 & 5\\ 3 & 3\end{array}$. Next, in lines 16–20 the first fraction on the left[136] is multiplied by the "subtractive quantity" 6, and it becomes $\begin{array}{|c|}30\\3\end{array}$. The "other" fraction $\begin{array}{|c|}5\\3\end{array}$ is halved and it bec omes $\begin{array}{|c|}5\\6\end{array}$; by means of the quoted class of simple fractions, its "square" (*kṛti*) is added to the previous result $\begin{array}{|c|}30\\3\end{array}$ giving $\begin{array}{|c|}385\\36\end{array}$. In lines 23–26, this result is lessened by the "visible quantity" 6: $\begin{array}{|c|c|}385 & 6\\ 36 & 1\\ 1 & 36\end{array} = \begin{array}{|c|}169\\36\end{array}$, and in line 28, the commentator explains that *tatpadam* or "the root of that" from the *karaṇasūtra*, refers to this resulting fraction which is the "remainder" (*śeṣa*) of the subtraction; its square-root becomes $\begin{array}{|c|}13\\6\end{array}$. The commentator subtracts $\begin{array}{|c|}5\\6\end{array}$ from $\begin{array}{|c|}6\\1\end{array}$, it becomes $\begin{array}{|c|}41\\6\end{array}$ which added to $\begin{array}{|c|}13\\6\end{array}$ gives $\begin{array}{|c|}54\\6\end{array}$, and when this is reduced to its lowest terms is $\begin{array}{|c|}9\\1\end{array}$. In lines 5–10, the last step is carried out: the last result is divided by the initial given fraction $\begin{array}{|c|}3\\5\end{array}$, the result becomes $\begin{array}{|c|}15\\1\end{array}$.

134 In my formulation above, "the [square] root of that" is the square-root of D.
135 With respect to the two following sample problems, it refers to the number 9 and 24, respectively.
136 See the commentary in line 17: *asau eko'ṅkaḥ* [...].

In the verification, the GST shows that 15 divided by 5 is 3, multiplied by 3 is 9, which lessened by 6 becomes 3, its square is 9, and the known quantity is 6: 9 plus 6 gives the final result 15, the total number of peacocks.

The SGT on GT 92 (ex.): This exercise requires students to determine the total number of a herd of elephants. In the first lines of the explanation (lines 18–21), the commentator carries out some initial calculations, the data are then ready to be placed in the *nyāsa* and be manipulated. The sample problem begins with "three times the eighth part of a herd [of *viṣkas*] is halved": the commentator hence multiplies one-eighth by three and then halves it, giving $\left|\begin{matrix}3\\16\end{matrix}\right|$. In the sample problem, sixteen is the "subtractive quantity" (*ūna*), and this multiplied by four gives the "visible quantity" (*dṛśya*). These numbers are then placed in the *nyāsa* given in line 21: $\left|\begin{matrix}3\\16\end{matrix}\right|$ |ū 16||dṛ 64|. At this point, the data are ready and the *hīnavargajāti* can be applied:

- line 21: the operands of the given fraction are interchanged and written in two places: $\left|\begin{matrix}16\\3\end{matrix}\right|\left|\begin{matrix}16\\3\end{matrix}\right|$
- lines 22–23: the first quantity is multiplied by the "subtractive quantity" 16 giving $\left|\begin{matrix}256\\3\end{matrix}\right|$
- lines 23–28: the commentator shows that half of the "other" (*anya*) fraction $\left|\begin{matrix}16\\3\end{matrix}\right|$ is $\left|\begin{matrix}8\\3\end{matrix}\right|$, its "square" (*kṛti*) is $\left|\begin{matrix}64\\9\end{matrix}\right|$. By means of the class of simple fractions quoted in line 24, this is added to $\left|\begin{matrix}256\\3\end{matrix}\right|$ giving $\left|\begin{matrix}832\\9\end{matrix}\right|$
- lines 28–5: this resulting fraction is lessened by the visible quantity 64. When the *dṛśya* 64 is deducted by means of the quoted *bhāgajāti*, it gives $\left|\begin{matrix}256\\9\end{matrix}\right|$
- lines 5–6: "the root of that" denotes the square-root of this fraction, which is $\left|\begin{matrix}16\\3\end{matrix}\right|$
- lines 6–12: the *ūna* 16 is added to half of the fraction, and it becomes $\left|\begin{matrix}56\\3\end{matrix}\right|$. This result is added to "the root of that": $\left|\begin{matrix}56\\3\end{matrix}\right|\left|\begin{matrix}16\\3\end{matrix}\right|$; $\left|\begin{matrix}24\\1\end{matrix}\right|$

- lines 13–16: the final step is *bhāgavibhaktam* or "divided by the fraction", hence $\begin{vmatrix}24\\1\end{vmatrix}$ divided by $\begin{vmatrix}16\\3\end{vmatrix}$ gives $\begin{vmatrix}128\\1\end{vmatrix}$, which is the size of the herd of elephants".

The verification shows that one-eighth of 128 is 16, multiplied by 3 is 48; halved is 24, diminished by 16 is 8. This is multiplied by itself, and 64 is the number of *viṣka*s playing in the cavity of a mountain. When this is added to the "visible quantity" 64, which is the number of elephants wandering in the forest, it becomes 128.

4.8 Inverse operation [GT 93–94]

***Inverse operation* (viparītoddeśaka)**

GT 93: In the GT, this rule occupies an awkward position as it stands between the end of the treatment of the type-problems with fractions and the rules on proportion which follow. In the edition of the GSS, it is part of the chapter on miscellaneous methods (GSS 6.286), while in Śrīpati's SŚ it is found after the classes of fractions and before the rule of three. Śrīpati's formulation of the "inverse operation" method is fairly standard and does not differ from those in other works.

The SGT on GT 93: In the introductory line, the commentator refers to this procedure by the expression *viparītoddeśaka*. He begins by pointing out the *modus operandi* characterising this method: all the quantities mentioned in the sample problem should be manipulated with their operational sign reversed. The commentator paraphrases the *karaṇasūtra* and provides some synonyms such as *bhāgadāyin*, which is a term peculiar to Siṃhatilakasūri's idiom, for *hara* or "divisor"; *hīna* for *kṣaya* or "subtractive, negative number"; *madhyakṣepa* for *dhana* or "additive, positive number".

The SGT on GT 94 (ex.): The commentator presents the *nyāsa*: |gu5|dha 9|mū 1|ū 2|kṛti 1|hīna 1|bhā 8|dṛśyarūpa 3|, where *gu* stands for *guṇa* or "multiplier", *dha* for *dhana* or "additive number", *mū* for *mūla* or "square-root, *ū* for *ūna* or "subtractive quantity", *kṛti* for "square", *hīna* for "subtractive number", *bhā* for "divisor", and *dṛśyarūpa* for the "known number". Siṃhatilakasūri explains that numbers are to be calculated by inverting their signs by following a right to left direction. The known number 3 is multiplied by 8, the result added to 1, its square-root extracted, added to 2, squared, lessened by 9, and divided by 5. The number obtained is 8, which is the final result. The commentator provides also the *ghaṭanā* or "verification" to show the correctness of the procedure.

L 48–49: After this sample problem, the SGT quotes L 48–49 which provide a rule for the inverse procedure concerning fractions; in order to supply a sample problem, the SGT quotes a verse which corresponds to L 50. The rule formulated in L 48 is the same as GT 93, while L 49 concerns fractions.

The SGT ad L 49: The SGT explains this rule by means of the fraction $\left|\begin{array}{c}3\\4\end{array}\right|$, which is given in the next sample problem quoted by the commentator from the L. The commentary elucidates both cases mentioned in Bhāskarācārya's formulation: the case of a fraction of an integer which is added to it (i.e., to the integer), and the case of a fraction of an integer which is subtracted from it. They both refer to quantities involved in the reverse operation. It is noteworthy that the SGT expands the rule by providing important details, such as the qualifier *ūrdhva* ("upper") to denote which denominator is computed.

The SGT ad L 50 (ex.): In line 6, the SGT gives the *nyāsa*:

gu3	svaca $\left|\begin{array}{c}3\\4\end{array}\right|$	bhā 7	svatryaṃśa $\left|\begin{array}{c}1\\3\end{array}\right|$	svagu 1	hīna 52	mū 1	dha 8	bhā 10	dṛśya 2

Siṃhatilakasūri first determines the number to be added to its own fraction which in the sample problem is $\left|\begin{array}{c}3\\4\end{array}\right|$. He then shows that 2 is to be multiplied by 10, lessened by 8, and squared becoming 144, which added to 52 gives 196, whose root is 14. This number is to be decreased by its own one-third: $\left|\begin{array}{c}14\\1\\01\\3\end{array}\right|$.

Then:

- the lower denominator 3 is lessened by its numerator 1; it becomes 2
- the upper denominator 1 is multiplied by the result, and becomes 2, which is the new lower denominator; the numerator 1 remains unchanged
- the SGT underlines that the fraction becomes an additive quantity. At this point, the layout should be like the following: $\left|\begin{array}{c}14\\1\\1\\2\end{array}\right|$
- the lower numerator 1 is added to the its denominator 2 giving 3 which is the multiplier of the upper denominator 14; it becomes 42 and the denominator is 2: $\left|\begin{array}{c}42\\2\end{array}\right|$, which reduced by the common term 2 is $\left|\begin{array}{c}21\\1\end{array}\right|$. As mentioned in the sample problem, this number is then multiplied by 7 becoming 147, which is then added to its three quarters: $\left|\begin{array}{c}147\\1\\3\\4\end{array}\right|$
- the lower denominator 4 is added to its numerator 3

- the upper denominator 1 is multiplied by the resulting 7; it gives 7, which is the new lower denominator. Because of the reverse procedure, the fraction becomes a subtractive quantity. The layout should appear like the following: $\begin{array}{|r|}147\\1\\03\\7\end{array}$. The lower denominator 7 is lessened by its numerator 3, it gives 4 which is the multiplier of the upper numerator 147. It gives: $\begin{array}{|r|}588\\7\end{array}$
- this fraction is then divided by 3 (instead of being multiplied by 3), becoming $\begin{array}{|r|}588\\21\end{array}$ which reduced by 21 gives the integer 28.

4.9 Rules on proportion [GT 95–117]

Preliminary observations

GT 95–117 are dedicated to presenting rules and sample problems on proportion: the rule of three and more terms, barter, and the rule of the sale of living beings. *Trairāśīka*[137] or the "rule of three quantities" (commonly referred to as "the rule of three") involves a procedure for solving linear problems of the type:[138]

> If n corresponds to z, to what will p correspond?
> The rule states that the answer is: $(p \times z) / n$

The GT comes up with a significant number of sample problems dealing with the rule of three, many of which draw directly from business and commercial situations; these provide interesting insights into actual commercial practices. In mathematical works, the most common problem-type involving the rule of three concerns prices and quantities of a variety of products. Typically, the following business presentations occur: i) knowing the price of a certain quantity of a given product, one needs to find out the price of another quantity of that same product, or ii) knowing the price of a certain quantity of a given product, one needs to determine the quantity which can be bought by a given amount of money. In the GT, sample problems mention weight measuring units and monetary units which must have been prevalent in Śrīpati's geographical area and time. There are also recreational problems (see, for instance, GT 98 and 102). Śrīpati gives solutions for direct proportions by

137 *trairāśika* – lit. the "[computation] related to three quantities".
138 On the rule of three in the history of mathematics, see Høyrup (2012) and Sarma (2003).

the rule of three (GT 95), where he also briefly mentions the inverse rule of three, and the rule of five (GT 107). Between GT 95 and GT 107 are sample problems on the rule of three, the rule of five, and quotations provided by the commentator from the TŚ and the L on the rules of seven, nine, and eleven with their corresponding sample problems.

The rules of five, seven, nine, and eleven depend on the rule of three; the rule of three can be, in fact, expanded by adding additional terms. Some works, such as the GSS, the TŚ, and the L dedicate a separate verse to enunciate the rules of seven, nine, and eleven. Moreover, these works mention explicitly the inverse rule of three and L 77–78 describes some of the situations in which it should be applied. The rule of five can be used to solve sample problems on the rules of seven, nine, and eleven as the procedure is exactly the same but with more terms;[139] nevertheless, Siṃhatilakasūri quotes from both the TŚ and the L and provides *karaṇasūtra*s and sample problems on these procedures.[140]

The rule of three quantities [GT 95]

GT 95: Śrīpati's passage on the rule of three is fairly standard; students have to pay attention to the characteristics of the three terms in order to distinguish among them and carry out the procedure correctly. The author underlines, in fact, that the first and the last term should be of the same type,[141] the second is then multiplied by the last, and the result divided by the first.

The three quantities are termed:

pramāṇa or "measure"
abhīpsā/*samicchā* or "demand, requisition"
phala or "fruit".

Therefore, the rule of three involves these three quantities: the measure, the requisition, and the fruit. With respect to the way quantities are written down on the working surface, GT 95 states that the "measure" and the "requisition" should be written at the beginning and at the end respectively,[142] while the fruit is in the middle. This emphasises the spatial relations among the terms and appears to point to a particular layout in which the given quantities can be manipulated easily. The GT states that the quantity *phala* is of a "different kind" than the other two quantities; this is most likely the reason why the fruit is the middle term, so as to distinguish it from the other two which are of the

139 In this regard, the commentator's remark in the introductory line to L 52 is significant.
140 Hayashi (2013, 57) observes that, since in GT 108 Śrīpati refers to this rule by the expression *pañcarāśikavidhi*, it may be that GT 107 was specifically meant only for the rule of five.
141 See the use of the term *jāti*. Here the same "type" means that both quantities should be either the "cost" or the "commodity".
142 The formulation clearly denotes a left to right direction.

same kind. GT 95 explains that one should multiply the fruit by the requisition, and the divide the result by the measure:

$$(\textit{fruit} \times \textit{requisition}) \div \text{measure}.$$

Śrīpati finally adds that, in the case of the inverse rule of three, the procedure is reversed:

$$(\text{measure} \times \text{fruit}) \div \text{requisition}.$$

The SGT on GT 95: The commentator explains that "measure" (*pramāṇa*) indicates either the "number which is the commodity" (*vastusaṅkhyā*) or the "number which is the cost" (*mūlyasaṅkhyā*), and that it is written at the beginning of the layout. Similarly, the "requisition" (*abhīpsā* is glossed with *icchā*) is either the number which is the commodity or the quantity which is the cost and it is found at the end. Siṃhatilakasūri clarifies that the quantities placed at the beginning and at the end of the layout should be of the same kind (see *sadṛśajātir eva*, line 10): if the first quantity, which is the "measure", represents the commodity, then the other quantity representing the commodity should be written at the end of the layout; in the same way, if the first quantity represents the cost, the other quantity representing the cost should be written down at the end of the layout. Finally, the commentator expands Śrīpati's brief reference to the inverse rule of three by quoting L 78. This explains, in fact, that when the three quantities are inversely proportional, so that an increase of the requisition produces a decrease of the fruit or else a decrease of the requisition produces an increase of the fruit, the inverse rule of three should be applied.

L 78 states that the inverse rule of three concerns the following situations: (i) the selling of living beings – such as slaves and cattle – where their value is determined by their age, which means that the price of the older is less; (ii) procedures with gold in which the weight in *varṇa*s determines the price of the piece of gold; and (iii) the subdivision of heaps of grain.[143]

To sum up, while commenting upon GT 95, the commentary provides important additional information:

- it explains what the three terms concerning the rule stand for and the place they occupy in the layout
- it clarifies the relation between the three terms
- it supplies the inverse rule of three by quoting passages from the L.

The SGT on GT 96 (ex.): This sample problem requires students to ascertain the price at which seven and one-third *pala*s of musk can be obtained. The

143 Here the L mentions mathematical procedures which are commonly found in texts on *pāṭīgaṇita*; see GT 106 and 115.

SGT provides the presentation: $\begin{array}{|r|r|r|}\text{va } 1 & \text{mū } 12 & \text{va } 7\\ 1 & 1 & 1\\ 2 & 4 & 3\end{array}$, where the abbreviation *va* stands for *vastu* or "commodity" and *mū* for *mūlya* or "cost". The first quantity is the "measure" and represents the commodity, in the middle there is the "fruit", which is the quantity-cost, and the last term is of the same kind as the first – thus a *vastusaṅkhyā* – and it is the "requisition". At this stage, these quantities are compound fractions and need to be simplified. The commentator treats the first quantity by means of the *bhāgānubandhajāti* which he partially quotes, and it becomes $\begin{array}{|r|}3\\2\end{array}$; the second quantity becomes $\begin{array}{|r|}49\\4\end{array}$ and the third $\begin{array}{|r|}22\\3\end{array}$. In line 5, he clarifies that the quantity in the middle, which is $\begin{array}{|r|}49\\4\end{array}$, is the "fruit", which in this case is a *mūlyasaṅkhyā* and should be multiplied by the "requisition", which is $\begin{array}{|r|}22\\3\end{array}$. The result $\begin{array}{|r|}1078\\12\end{array}$ is the dividend (see line 8). When the "measure", which is the divisor and which the SGT emphasises to be the first quantity, is divided by the dividend, the quotient is 59 *dramma*s, and the remainder is $\begin{array}{|r|}8\\9\end{array}$. This is the price at which seven and one-third *pala*s of musk can be obtained. As this point, the commentator determines all the lower monetary units contained in $\begin{array}{|r|}8\\9\end{array}$ *dramma*s, and lines 16–19 are dedicated to carrying out these conversions. As in the *paribhāṣā* GT 4 states that there are 16 *paṇa*s in 1 *dramma*, the commentator multiplies the remainder $\begin{array}{|r|}8\\9\end{array}$ by 16, it becomes the proper fraction $\begin{array}{|r|}128\\9\end{array}$ which transformed into a mixed number is 14 $\begin{array}{|r|}2\\9\end{array}$. This means that there are 14 *paṇa*s in $\begin{array}{|r|}8\\9\end{array}$ *dramma*s. As GT 4 states that there are 4 *kākiṇī*s in 1 *paṇa*, the remainder 2 is then multiplied by 4 in order to calculate the number of *kākiṇī*s, which are the sub-multiple of the *paṇa*; it becomes the fraction $\begin{array}{|r|}8\\9\end{array}$. Since this is a proper fraction, the division of the numerator by its denominator cannot be performed, which means that there are no *kākiṇī*s. The numerator 8 is then multiplied by 20 to calculate the number of *kaparda*s, since GT 4 states that there are 20 *kaparda*s in 1 *kākiṇī*. It becomes $\begin{array}{|r|}160\\9\end{array}$, which transformed into a compound fraction is 17 $\begin{array}{|r|}7\\9\end{array}$. The final result, the *icchāphala* or "fruit of the requisition", is 59 *dramma*s, 14 *paṇa*s, 0 *kākiṇī*, and 17 $\begin{array}{|c|}7\\9\end{array}$ *kaparda*s.

The SGT on GT 97 (ex.): The *nyāsa* is: $\begin{array}{|r|r|r|}\text{mū } 16 & \text{va } 1 & \text{mū } 100\\ 1 & 1 & 1\\ 3 & 2 & \end{array}$, where *mū* stands for *mūlya* ("cost") and *va* for *vastu* ("commodity"). The first quantity is simplified by means of the *bhāgānubandhajāti,* and it becomes $\begin{array}{|r|}49\\3\end{array}$; similarly, the second quantity is reduced, and it becomes $\begin{array}{|r|}3\\2\end{array}$. The commentator specifies that the *mūlya* $\begin{array}{|r|}100\\1\end{array}$ is the *samicchā* or "requisition" by which the *vastu*-quantity is multiplied giving $\begin{array}{|r|}300\\2\end{array}$. The first *vastu*-quantity is the *pramāṇa* or "measure" and hence the divisor of this obtained product. The division of fractions involves the interchange of the operands of the fraction-divisor and the reciprocal multiplication; when the numerator of the obtained fraction $\begin{array}{|r|}900\\98\end{array}$ is divided by its numerator, it gives the number of *pala*s: 9 plus $\begin{array}{|r|}18\\98\end{array}$. This quantity is further reduced to $\begin{array}{|r|}9\\49\end{array}$, which Siṃhatilakasūri specifies to be the fraction of *pala*s, and he next converts this quantity into lower weight measuring units. Since GT 6 states that 1 *pala* is equal to 10 *dhaṭaka*s, the SGT multiplies the numerator 9 by 10 which gives 90, and the fraction is $\begin{array}{|r|}90\\49\end{array}$. Since this is an improper fraction, it is reduced and becomes 1, which represents the number of *dhaṭaka*s; the remainder is $\begin{array}{|r|}41\\49\end{array}$. To arrive at the next unit, which is the *niṣpāva* (called *valla* by the SGT), 49 is multiplied by 14, since GT 6 specifies that there are 14 *niṣpāvaka*s in 1 *dhaṭaka;* it becomes $\begin{array}{|r|}574\\49\end{array}$, which expressed as a mixed number is 11 plus $\begin{array}{|r|}35\\49\end{array}$. Hence, 9 *pala*s, 1 *dhaṭaka,* and 11 $\begin{array}{|r|}35\\49\end{array}$ *valla*s of camphor are worth 100 *dramma*s.

The SGT on GT 98 (ex.) (TŚ 37): Siṃhatilakasūri explains that one should first add 6 to 100, and gives the *nyāsa*: $\begin{array}{|r|r|r|}106 & 6 & 1000\\ 1 & 1 & 1\end{array}$. He then multiplies the middle term 6 by the last term 1000, it becomes 6000. The first term, which is the "measure", is the divisor and its operands are thus interchanged in order to carry out the multiplication by 6000, it becomes $\begin{array}{|c|}6000\\106\end{array}$. The SGT expresses this improper fraction as a compound fraction which is 56 plus $\begin{array}{|c|}32\\53\end{array}$, having halved the previous remainder $\begin{array}{|c|}64\\106\end{array}$.

The SGT on GT 99 (ex.): The *nyāsa* is: $\begin{array}{|c|c|c|}1&5&1\\1&1&1\\2&4&3\end{array}$, where the first quantity represents the commodity one and a half *dhaṭaka*s (the weight) of saffron, in the middle is its *mūlya* or "cost", and the last is the quantity of saffron of which the student has to determine the price, namely one plus one-third *pala*s. The commentator quotes and applies the class of fractional increase, the three quantities become respectively: $\begin{array}{|c|c|c|}3&21&4\\2&4&3\end{array}$. The middle term and the last term are multiplied and the result $\begin{array}{|c|}84\\12\end{array}$ is then divided by the "measure" $\begin{array}{|c|}3\\2\end{array}$. With respect to the "measure", since it is the divisor (see *haratvāt*), there is the interchange of its operands: $\begin{array}{|c|}2\\3\end{array}$. When $\begin{array}{|c|}84\\12\end{array}$ is multiplied by it, having reduced the numbers to lowest terms, the result is $\begin{array}{|c|}28\\6\end{array}$. The commentator then applies the conversion ratio of the units of money mentioned in GT 4:

- in $\begin{array}{|c|}28\\6\end{array}$ *paṇa*s, there are 4 (from 28 ÷ 4) and $\begin{array}{|c|}4\\6\end{array}$ *dramma*s
- to calculate the number of *paṇa*s in $\begin{array}{|c|}4\\6\end{array}$ *dramma*s, the numerator 4 is multiplied by 16, the result $\begin{array}{|c|}64\\6\end{array}$ represents 10 *paṇa*s (from 64÷ 6) and the remainder is $\begin{array}{|c|}4\\6\end{array}$
- to calculate the number of *kākiṇī*s, the numerator 4 is multiplied by 4 becoming $\begin{array}{|c|}16\\6\end{array}$
- when there is the division of the numerator by the denominator, the quotient is 2 and the remainder is $\begin{array}{|c|}4\\6\end{array}$

- to calculate the number of *varāṭaka*s, the numerator 4 is multiplied by 20 giving $\left|\begin{array}{r}80\\6\end{array}\right|$
- when these operands are divided, the quotient is 13 *kapardaka*s and the remainder $\left|\begin{array}{r}2\\6\end{array}\right|$ is reduced to $\left|\begin{array}{r}1\\3\end{array}\right|$.

Hence, the final result is 4 *dramma*s, 10 *paṇa*s, 2 *kākiṇī*s and 13 $\left|\begin{array}{r}1\\3\end{array}\right|$ *kapardaka*s.

The SGT on GT 100 (ex.): In the introductory line, Siṃhatilakasūri specifies that this sample problem concerns grain. The *nyāsa* is: $\left|\begin{array}{r|r|r}2 & 8 & 100\\1 & 01 & 1\\2 & 8 & 3\end{array}\right|$. Since these are compound fractions, the procedure of simplification of fractions is first performed. The commentator applies the class of fractional decrease to the "first" and "last" quantities (*ādyāṅka* and *antyāṅka*) – the *pramāṇa* and the *phala* – and the obtained fractions are $\left|\begin{array}{r}5\\2\end{array}\right|$ and $\left|\begin{array}{r}301\\3\end{array}\right|$. The quantity in the middle (see *madhyāṅka*) represents the price of two and a half *mānikā*s of grain, and since it is mixed number, the SGT quotes and carries out the class of fractional decrease by means of which $\left|\begin{array}{r}63\\8\end{array}\right|$ is obtained. Line 11 shows the three obtained fractions $\left|\begin{array}{r|r|r}5 & 63 & 301\\2 & 8 & 3\end{array}\right|$; the middle quantity, which is the "fruit", is multiplied by the last quantity, which is the "requisition", and it becomes $\left|\begin{array}{r}18963\\24\end{array}\right|$. The first fraction, which is the "measure", is the divisor of the obtained product $\left|\begin{array}{r}18963\\24\end{array}\right|$; the commentator quotes the rule of division of fractions (GT 42) and the operands are interchanged in order to transform the divisor into the multiplier and carry out the multiplication. The result is $\left|\begin{array}{r}18963\\60\end{array}\right|$, which represents the number of *paṇa*s one hundred and one-third *mānikā*s of grain cost. In order to determine the various units of money contained in $\left|\begin{array}{r}18963\\60\end{array}\right|$ *paṇa*s, the SGT carries out the following conversion ratio:

- the denominator 60 is multiplied by 16, since 16 *paṇa*s make one *dramma*; it becomes $\begin{array}{|r|}18963\\960\end{array}$ which is an improper fraction, hence it is reduced to 19 plus $\begin{array}{|r|}723\\960\end{array}$
- in order to calculate the number of *paṇa*s, the numerator 723 is multiplied by 16; it becomes $\begin{array}{|r|}11568\\960\end{array}$, which expressed as a mixed number is 12 whose remainder is $\begin{array}{|r|}48\\960\end{array}$
- to arrive at the *kākiṇī*s, the numerator 48 is multiplied by 4; it becomes the improper fraction $\begin{array}{|r|}192\\960\end{array}$, the SGT observes that the division of the numerator by the denominator cannot take place and the number of *kākiṇī*s is thus 0
- the fraction $\begin{array}{|r|}192\\960\end{array}$ is reduced by the common factor 96, it becomes $\begin{array}{|r|}2\\10\end{array}$
- to arrive at the *kapardaka*s, the numerator 2 is multiplied by 20, it is forty; when it is divided by the denominator 10, the quotient is four *kapardaka*s.

The final result is thus 19 *dramma*s, 12 *paṇa*s, and 4 *kapardaka*s, which is the cost of one hundred and one-third *mānikā*s of grain.

The SGT on GT 101 (ex.): The *nyāsa* is: $\begin{array}{|r|r|r|}6&2&80\\1&1&1\\3&4&2\end{array}$. The SGT quotes and applies the *bhāgānubandhajāti* in order to transform the given quantities into simple fractions, and these become respectively $\begin{array}{|c|c|c|}19&9&161\\3&4&2\end{array}$. The middle quantity, which is the "fruit", is multiplied by the last quantity, which is the "requisition", and it becomes $\begin{array}{|r|}1449\\8\end{array}$. In the first quantity, which is the "measure" and hence the divisor, the operands are interchanged in order to transform it into the multiplier and carry out the multiplication by the product obtained. Therefore $\begin{array}{|r|}1449\\8\end{array}$ is multiplied by $\begin{array}{|r|}3\\19\end{array}$ giving $\begin{array}{|r|}4347\\152\end{array}$ *mānikā*s. Since this is an improper fraction, it is transformed into the mixed number 28 plus $\begin{array}{|r|}91\\152\end{array}$ *mānikā*s. Siṃhatilakasūri next applies the conversion ratios concerning units of capacity mentioned in GT 7. In order to arrive at the number of *hārikā*s contained in $\begin{array}{|c|}91\\152\end{array}$ *mānikā*s, the numerator 91 is multiplied by 4; it becomes 364. This is divided by the denominator 152, the quotient is 2 *hārikā*s, the remainder is $\begin{array}{|c|}60\\152\end{array}$. This fraction reduced by the common factor 4 becomes $\begin{array}{|c|}15\\38\end{array}$. Hence, the final result is 28 *mānikā*s and 2 $\begin{array}{|c|}15\\38\end{array}$ *hārikā*s.

The SGT on GT 102 (ex.): This sample problem concerns the time it takes an elephant to cover a certain distance. The commentator explains that first the *savarṇana* or "simplification" of one-third of one-half *yojana* should be applied, and gives the following layout: $\begin{array}{|c|c|c|}1&1&1\\1&2&3\end{array}$. The first quantity represents the whole, 1 *yojana*; the SGT shows that, by means of the *prabhāgajāti* or the "multi-part class", it becomes $\begin{array}{|c|}1\\6\end{array}$. Having reduced the first quantity one-third of one-half *yojana*, now the rule of three is to be performed. The three quantities are: $\begin{array}{|c|c|c|}1&\text{di }2&\text{yo }70\\6&01&1\\&3&\end{array}$, where the abbreviation *di* stands for *dina* or "day", and *yo* for the unit of length *yojana*. The second quantity is a compound fraction and it is reduced by means of the class of fractional decrease, it becomes $\begin{array}{|c|}3\\2\end{array}$ which represents the "fruit". This is then multiplied by the "requisition", which is the last quantity, and it becomes $\begin{array}{|c|}210\\2\end{array}$. The first quantity $\begin{array}{|c|}1\\6\end{array}$ is the "measure" and, since it is the divisor of the previous product, there is the interchanges of its operands, and it becomes $\begin{array}{|c|}6\\1\end{array}$. The product is $\begin{array}{|c|}630\\1\end{array}$, which represents the total sum of the days which the elephant would need to travel seventy *yojana*s. Lastly, the commentator divided the product obtained by 360, which represents the length of one *saṃvatsara* or "year". The final result comes to be 1 year and 9 months.

The SGT on GT 103 (ex.): This is also a sample problem on time and distance; one is asked to determine the time a snake will fully enter a hole. The SGT presents the following *nyāsa*: $\begin{array}{|c|c|c|}1&1&84\\1&3&1\\2&&\end{array}$, where the first quantity is the depth of the hole the snake enters, the second is the time it takes the snake to travel that distance, and the third is the length of the snake. The commentator first transforms into the same unit of length (which is the *aṅgula* or "finger") the two quantities representing the length of the snake and the depth of the hole: in GT $\overline{8}$, it is said that 1 *kara*/*hasta*= 24 *aṅgula*s; three and a half *kara*s or "cubits" are equal to $\left|\begin{array}{c}7\\2\end{array}\right|$ *kara*s which multiplied by 24 becomes 84 *aṅgula*s. By means of the class of fractional increase, the first quantity, which denotes the depth of the hole, becomes $\left|\begin{array}{c}3\\2\end{array}\right|$. Then, the middle quantity, which is the "fruit", is multiplied by the "requisition", which is the last quantity: it becomes $\left|\begin{array}{r}84\\3\end{array}\right|$. The first quantity $\left|\begin{array}{c}3\\2\end{array}\right|$ is the divisor of this product and hence its operands are interchanged in order to perform the multiplication. When $\left|\begin{array}{r}84\\3\end{array}\right|$ is multiplied by $\left|\begin{array}{c}2\\3\end{array}\right|$, it gives 18 plus $\left|\begin{array}{c}2\\3\end{array}\right|$ *ghaṭikā*s. This fraction is transformed into the sub-multiple of the unit *ghaṭikā*, which is the *pala* and it is equal to 60 *ghaṭikā*s. It becomes $\left|\begin{array}{r}120\\3\end{array}\right|$ *pala*s, and thus 40 *pala*s. Therefore, the snake will fully enter the hole in 18 *ghaṭikā*s and 40 *pala*s.

The SGT on GT 104 (ex.): The SGT gives the following presentation: $\left|\begin{array}{r|r|r}\text{dra } 14 & 1 & 90\\ 1 & 1 & 01\\ 2 & 2 & 3\end{array}\right|$, where the first and last quantities represent the cost, while the quantity in the middle is the weight of the piece of gold. The commentator illustrates that since the first two quantities involves a fractional increase, the *bhāgānubandhajāti* is performed; the last quantity is expressed by means of a fractional decrease and hence the SGT quotes and applies the *bhāgāpavāhanajāti* in order to simplify it. In line 22, the three simplified fractions are given: $\left|\begin{array}{r|r|r}29 & 3 & 269\\ 2 & 2 & 3\end{array}\right|$.The middle number, which is the "fruit" (and the weight of the piece of gold), is then multiplied by the last number, which is the "requisition": it becomes $\left|\begin{array}{r}807\\6\end{array}\right|$. Then, in the first number, which is the "measure" and hence the divisor, there is the interchange of the operands: $\left|\begin{array}{r}2\\29\end{array}\right|$. This is the multiplier of $\left|\begin{array}{r}807\\6\end{array}\right|$, and the product becomes $\left|\begin{array}{r}807\\87\end{array}\right|$. The SGT transforms this improper fraction into a compound fraction: it is 9 plus $\left|\begin{array}{r}24\\87\end{array}\right|$

*gadyāṇaka*s, which reduced gives $\begin{vmatrix}8\\29\end{vmatrix}$ *gadyāṇaka*s. This remainder is converted into the submultiples of a *gadyāṇaka*; the first is the unit *dharaṇa*: GT 5, which presents measuring units of gold, states that 1 *gadyāṇaka* is equal to 2 *dharaṇa*s. Hence, the numerator 8 is multiplied by 2 and the remainder becomes $\begin{vmatrix}16\\29\end{vmatrix}$. The SGT explains that, in this case, the division cannot occur, meaning that the fraction is a proper fraction and therefore the number of *dharaṇa*s is $\begin{vmatrix}0\end{vmatrix}$. Then, in order to arrive at the number of *niṣpāva*s, the remainder $\begin{vmatrix}16\\29\end{vmatrix}$ is multiplied by 8 since a *dharaṇa* equals 8 *niṣpāva*s, and it becomes $\begin{vmatrix}128\\29\end{vmatrix}$, which transformed into a compound fraction is 4 *niṣpāva*s, the remainder is $\begin{vmatrix}12\\29\end{vmatrix}$. In order to arrive at the next sub-multiple, which is the *yava*, the numerator is multiplied by 6 and becomes $\begin{vmatrix}72\\29\end{vmatrix}$, which is expressed as the mixed number 2 plus $\begin{vmatrix}14\\29\end{vmatrix}$ *yava*s. Thus, the quantity of gold purchased with ninety minus one-third *dramma*s is 9 *gadyāṇakā*s, 4 *niṣpāva*s, and 2 plus $\begin{vmatrix}14\\29\end{vmatrix}$ *yava*s.

The SGT on GT $\overline{105}$ (ex.): The SGT provides the *nyāsa*: $\begin{vmatrix}8 & 16 & 6\\1 & 1 & 1\end{vmatrix}$. In this case, the inverse rule of three is applied; as explained by the SGT while commenting upon L 78, this implies that the "measure" is multiplied by the "fruit" and the result divided by the "requisition". The commentator shows that the quantity in the middle, which is the "fruit" and the quantity of strings of pearls, is multiplied by the measure: 16× 8 gives the numerator 128, the denominator is one: $\begin{vmatrix}128\\1\end{vmatrix}$. Because of the inverse rule of three, one should divide this product by the "requisition", which is the last number: by applying the rule of division of fractions, the "requisition" $\begin{vmatrix}6\\1\end{vmatrix}$ becomes the multiplier and hence its operands are interchanged: $\begin{vmatrix}1\\6\end{vmatrix}$. The product $\begin{vmatrix}128\\1\end{vmatrix}$ is multiplied by $\begin{vmatrix}1\\6\end{vmatrix}$; it becomes $\begin{vmatrix}128\\6\end{vmatrix}$. This improper fraction is transformed into a compound fraction: the final result is thus 21 plus $\begin{vmatrix}1\\3\end{vmatrix}$ string of pearls each weighing 6 *setikā*s.

The SGT on GT $\overline{106}$ (ex.): In the introductory line, the SGT underlines that this latter half of the metrical stanza concerns the inverse rule of three.

The SGT presents the *nyāsa*: $\begin{array}{|c|c|c|} 16 & 90 & 11 \\ 1 & 1 & 1 \end{array}$. The quantity in the middle is the weight of the initial piece of gold in *gadyāṇaka*s, the first and the last quantities represent the *varṇika*s of the two pieces of gold. The commentator specifies that, in this case, because the inverse rule of three occurs, the middle number, which is the "fruit", is multiplied by the first number, which is the "measure": hence, $\begin{array}{|c|} 16 \\ 1 \end{array}$ multiplied by $\begin{array}{|c|} 90 \\ 1 \end{array}$ becomes $\begin{array}{|c|} 1440 \\ 1 \end{array}$. This is then divided by the "requisition", which is the last quantity $\begin{array}{|c|} 11 \\ 1 \end{array}$. In order to carry out the division, the "requisition", which is the divisor, is transformed into a multiplier and its operands are interchanged; therefore $\begin{array}{|c|} 1440 \\ 1 \end{array}$ is multiplied by $\begin{array}{|c|} 1 \\ 11 \end{array}$ and the result is $\begin{array}{|c|} 1440 \\ 11 \end{array}$. This improper fraction is transformed into a compound fraction, which is 130 plus $\begin{array}{|c|} 10 \\ 11 \end{array}$. This is the final result; 130 plus $\begin{array}{|c|} 10 \\ 11 \end{array}$ is thus the weight, in *gadyāṇaka*s, of a piece of gold of eleven *varṇika*s of pure gold which can be obtained when a piece of gold of sixteen *varṇika*s and weighing ninety *gadyāṇaka*s is exchanged with.

4.10 The rule of five (*pañcarāśika*) [GT 107]

GT 107: In the introductory line, Siṃhatilakasūri clarifies that verse 107 is the *karaṇasūtra* of the "rule of five quantities" (*pañcarāśika*), which is a derivative of the rule of three. Regarding the terminology, it can be observed that:

- *pakṣa* is the "side"
- *ānīya* or "having moved" denotes that the "fruit" is moved to the other side of the layout[144]
- *anyarāśipakṣa* is the "side of the other quantities"
- *viparyayavidhi* is the "procedure of the exchange"
- *nijapakṣa* is "each side"
- *nijapakṣarāśighāta* denotes the "multiplication" (*ghāta*) of the "quantities" (*rāśi*) of "each "side (*nijapakṣa*).

Śrīpati employs different expressions to indicate the position of the quantities on the working surface. It clearly implies, as the commentator notes, a layout divided into two "sides" (*pakṣa*); one finds, in fact, the "other side"

144 The second side becomes larger; see, in this regard, the quotation from the TŚ in SGT 108, 24.

(see *ānīya pakṣam aparaṃ*), "each side" (*nijapakṣa*), "one side" (see *pakṣam aparaṃ vibhajet*), and the "side of the other quantities" (see *anyarāśipakṣeṇa*).

The SGT on GT 107: By glossing *phala* with *vyāja*, the commentator underlines that the "fruit" denotes the "interest-rate" (*vyāja*), and in doing so he refers to the sample problem which follows GT 107, where the "fruit" is the interest-rate 5%. He then clarifies that moving the "fruit" to the other side means to place it in the second side. Lastly, Siṃhatilakasūri elucidates the meaning of the expression "mutual multiplication", and observes that "one side" refers to the second side where there is the larger number, while the "side of the other quantity" refers to the first side.

The SGT on GT 108 (ex.): In this sample problem, the author gives some data involving a financial transaction and asks students to determine the "capital" (*mūladhana*), the "accrued interest" (*phala*), and the "time" (*kāla*). With respect to the capital, the SGT shows the calculation of: i) the "given capital" (*prathamadattadhana*),[145] which is 76, and ii) the "measure-value" (*pramāṇadhana*), which is 100.[146] With respect to the interest, the SGT ascertains: i) the "interest which accrues" (*kalāntara*) in a year, which comes to be 45 plus $\begin{array}{|c|}3\\5\end{array}$ *drammas*, and ii) the monthly "interest on the measure-value" 100 (*pramāṇaphala*), which is 5.[147] Regarding the time, the SGT calculates: i) the time concerning the accrued interest on the borrowed capital (76 *drammas*), which is 12 months, and ii) the time regarding the interest-rate on the measure-value 100 (*pramāṇakāla*), which is 1 month.

The commentator presents the following initial *nyāsa*: $\begin{array}{|lr|c|}\text{mā} & 1 & 12\\ \text{dra} & 100 & 76\\ \text{vyā} & 5 & \end{array}$, where *mā* stands for *māsa* ("month"), *dra* for *drammas* (the unit of money), and *vyā* for *vyāja* ("interest-rate"). The number 5, which is the "fruit", is moved to the other side: $\begin{array}{|r|r|}1 & 12\\ 100 & 76\\ & 5\end{array}$. The product of "one side" is divided by the product of the "other side" 100: $\begin{array}{|r|}4560\\100\end{array}$. The quotient is 45, and the remainder $\begin{array}{|r|}60\\100\end{array}$ when reduced becomes $\begin{array}{|c|}3\\5\end{array}$. The quantity 45 and $\begin{array}{|c|}3\\5\end{array}$ *drammas* represents the accrued

145 It represents the loan amount which has been borrowed.

146 See lines 15–26 and 13–16 (next page) The syntagm *pramāṇarāśi* indicates the quantity-measure, the 100 taken as the whole in a percentage-rate.

147 See lines 10–14 and 5–12 (next page).

interest and this is written at the bottom of the next layout: $\begin{array}{|r||r|} 1 & 12 \\ 100 & 0 \\ 5 & 45 \\ & 3 \\ & 5 \end{array}$. It can be observed that the interest-rate 5 is back in the first side, and the 0 below 12 represents the "unknown", because it will be ascertained, initial capital which is, in fact, 76.

The next passage (lines 27–4) is dedicated to determining the "time" (*kāla*). The sample problem clarifies the monthly interest-rate and asks students to calculate the interest which will accrue in one year. In this procedure, the commentator demonstrates that the time of the above obtained interest accrued on the initial given capital 76 is 12 months. In the layout, above 76 there is a zero in place of 112: $\begin{array}{|r||r|} 1 & 0 \\ 100 & 76 \\ 5 & 45 \\ & 3 \\ & 5 \end{array}$, as this time the quantity to be determined is the "time". The compound fraction denoting the accrued interest is transformed back into an improper fraction and moved, together with its denominator 1, to the first side. Then, the products of the two sides are obtained and the larger quantity is divided by the smaller quantity; the time 12 months is obtained.

Afterwards, the "interest on the measure-value" (*pramāṇaphala*) is determined. The accrued interest 45 and $\begin{array}{|r|} 3 \\ 5 \end{array}$ is transformed back into the improper fraction $\begin{array}{|r|} 228 \\ 5 \end{array}$, and it is moved to the first side. Then, there is the exchange of the denominators but because there is no a second "fruit", 5 only is moved to the second side. As before, in the first side the multiplication gives 22800. The second side gives 4560; when these are divided, the quotient is the monthly interest 5 *dramma*s.

The next step shows the calculation of the "measure-value" (*prāmaṇadhana*) which is, in fact, 100. As before, the first fruit is moved to the first side and the exchange of the denominators takes place. When there is the division of the products of the two sides, the quotient is 100.

Finally, the "time regarding the measure-value" (*pramāṇakāla*) is determined. In this case too, as before, the accrued interest is given back as an improper fraction and it becomes $\begin{array}{|r|} 22800 \\ 5 \end{array}$. There is then the exchange of the denominators five by the one of the second interest-rate: $\begin{array}{|r|r|} 0 & 12 \\ 100 & 76 \\ 228 & 5 \\ 1 & 5 \end{array}$. There is the multiplication of the quantities of the first side, it becomes 22800. In the other side, the product of the reciprocal multiplication is also 22800. The division of these two products gives the quotient 1, which is the time regarding the measure-value and thus one month. The final result is 45 plus $\begin{array}{|r|} 3 \\ 5 \end{array}$ *drammas* (the accrued interest), 12 months (the time), and 100 (the measure-value).

Among the technical terms used by the commentator are: *pramāṇaphala,* which is the "interest on the measure-value" 100, and it is 5%; *pramāṇadhana,* the "measure-value" 100, and *pramāṇakāla,* which is the time (i.e., one month) concerning the measure-value. The synonyms *vyāja, phala,* and *kalāntara* are "interest-rate" or "accrued interest" according to the context. Moreover, it is noteworthy that the commentary always tries to clarify the position of each quantity in the layout: one finds, in fact, *prākpakṣa, dvitīyapakṣa, aparapakṣa, prācyāṅkapakṣa, pūrvāṅka*, and *parapakṣa.*

The SGT on GT 109 (ex.): As underlined by the commentator in the introductory line, Śrīpati provides a sample problem with fractions. The presentation given by the commentator is: $\begin{array}{|r|r|} 1 & 8 \\ 3 & 1 \\ 100 & 4 \\ 1 & 20 \\ 2 & 01 \\ 2 & 4 \\ 1 & \\ 2 & \end{array}$. The first side on the left gives the known data, while the second presents the set of data which is part of the question. In total, there are five quantities as this sample problem concerns, in fact, *pañcarāśika.* By means of the class of fractional increase, the SGT first transforms the mixed numbers of the first side into simple fractions, except the unit fraction one-third; the first side becomes $\begin{array}{|r|} 1 \\ 3 \\ 201 \\ 2 \\ 5 \\ 2 \end{array}$. In the second side, by applying both the rules of fractional increase and fractional decrease, the compound fractions are transformed into improper fractions; the second side becomes $\begin{array}{|r|}33\\4\\79\\4\\\\\\\end{array}$. Then, by applying the step *ānīya pakṣam aparaṃ phalam* formulated in GT 107, the known "fruit", which is the fraction $\begin{array}{|r|}5\\2\end{array}$ and which is at the very bottom of the first side, is moved to the other side, and it becomes $\begin{array}{|r|}33\\4\\79\\4\\5\\2\\\\\end{array}$.

All denominators of the fractions of both sides are interchanged: $\begin{array}{|r|r|}1&33\\4&3\\201&79\\4&2\\&5\\&2\end{array}$. The denominator 2 of the "fruit" moved to the second side is also interchanged and moved back to the first side: $\begin{array}{|r|}1\\4\\201\\4\\2\end{array}$. The mutual multiplication of the quantities of each side is performed and the products obtained are respectively 6432 and 78210, which when reduced by half, become 3216 and 39105. In lines 13 and 17, the commentator emphasises that these are the quantity-divisor and the quantity-dividend respectively. After the division, the quotient is 12 plus $\begin{array}{|r|}513\\3216\end{array}$ which reduced by 3 becomes $\begin{array}{|r|}171\\1072\end{array}$. The interest which will accrue on twenty less one-fourth *dramma*s in eight plus one-fourth months is therefore 12 plus $\begin{array}{|r|}171\\1072\end{array}$ *dramma*s.

The SGT on GT 110 (ex.): This sample problem requires students to determine the amount of money that two groups of workers earn during two different intervals. The SGT gives the *nyāsa*: $\begin{array}{|l||r|} \text{di } 2 & \text{di } 5 \\ \text{ka } 3 & \text{ka } 8 \\ \text{pa } 20 & 0 \end{array}$. Regarding the abbreviations found in this layout, *di* stands for *divasa* or "day" (it is interesting that the commentator does not use the same term used by Śrīpati, which is *vāsara*), *ka* stands for *karmakārā* or "worker", and *pa* for the unit of money *paṇa*. It can be observed that there are five quantities arranged, and the unknown quantity to be determined (which is the second fruit) is replaced by a zero.[148] The commentator shows that the number 20, which is the amount three workers earn in two days and hence the known "fruit", is moved to the other side: $\begin{array}{|r|} 5 \\ 8 \\ 20 \end{array}$. There is the mutual multiplication of the quantities of each side and the products obtained are 6 and 80 respectively. After the division of the larger quantity 80 by the smaller quantity 6, the result is 133 plus $\begin{array}{|c|} 2 \\ 6 \end{array}$ *paṇa*s, which reduced by 2 is $\begin{array}{|c|} 1 \\ 3 \end{array}$. This represents the second "fruit" and the amount the eight workers earn in five days.

The SGT on GT 111 (ex.): As underlined by Siṃhatilakasūri in the introductory line, this fourth sample problem on the rule of five concerns grain. The commentator presents the following *nyāsa*: $\begin{array}{|l||r|} \text{m } 8 & 63 \\ \text{kro } 4 & \text{kro } 18 \\ \text{pa } 6 & 0 \end{array}$, where *mā* stands for the unit of capacity *mānīka*, *kro* stands for the unit of length *krośa*, and *pa* for the unit of money *paṇa*. The known "fruit", which is represented by 6 *paṇa*s, is moved to the other side: $\begin{array}{|r|} 63 \\ 18 \\ 6 \end{array}$. There is the multiplication of the quantities of each side and the products obtained are the divisor 32 and the dividend 6804. When the division is performed, the quotient obtained is 212 plus the remainder $\begin{array}{|c|} 20 \\ 32 \end{array}$ which reduced by 4 becomes $\begin{array}{|c|} 5 \\ 8 \end{array}$. Hence, the second "fruit" is 212 plus $\begin{array}{|c|} 5 \\ 8 \end{array}$ *paṇa*s, which is the cost to carry sixty-three *māṇika*s of grain over 18 *krośa*s.

148 This feature does not occur in the layout found in SGT 108, 10.

The rules of seven, nine, and eleven

The GT does not provide *karaṇasūtra*s on the rules of seven, nine, and eleven; as I mentioned before, the rule of five can be applied to solve sample problems on the rules of seven, nine, and eleven, the difference is that these involve more terms. The rules of five, seven, nine, and eleven are also called "compound proportion"; Colebrooke ([1817]1973, 35) observes that the commentator Sūryadāsa explains the rule of five as comprising two proportions, the rule of seven comprising three proportions, the rule of nine four, and the rule of eleven five. The SGT supplies rules and sample problems on the rules of seven, nine, and eleven by quoting from the TŚ and the L.

L 82: The commentator quotes L 82, which gives the rules of five, seven, nine, and eleven; it is specified that in the rules of five, nine, eleven, or more terms, one should transpose the fruit and the divisor[149] to the other side and divide the larger quantity by the smaller. In this way, the quotient obtained becomes the fruit required. After this quotation, the SGT provides a sample problem on the rule of seven, which corresponds to TŚ 51. The sample problem on the rule of nine corresponds to TŚ 52, the sample problem on the rule of nine is the same as L 86, and the sample problem on the rule of eleven corresponds to L 87.

The SGT on TŚ 51: Having provided the *nyāsa* $\begin{array}{|c|c|} 1 & 2 \\ 2 & 3 \\ 8 & 9 \\ 10 & \end{array}$,the commentator emphasises that, in total, seven quantities should be arranged. In the first side, the number 1 represents the first piece of cloth, the numbers 2 and 8 are its measure, and 10 the cost; in the second side, 2 represents the two pieces of cloth measuring 3 and 9 respectively. Their cost (hence the second, the unknown "fruit") is to be determined. The first step is to move the known "fruit" (which is the cost of the piece of cloth) to the other side: $\begin{array}{|c|c|} 1 & 2 \\ 2 & 3 \\ 8 & 9 \\ & 10 \end{array}$.

The multiplication of the quantities of each side is carried out and the two products obtained are 16 and 540 respectively. When the larger product 540 is divided by 16, the quotient is 33 plus $\begin{array}{|c|} 12 \\ 16 \end{array}$ which reduced by 4 is $\begin{array}{|c|} 3 \\ 4 \end{array}$. The cost of the two pieces of cloth, the wished "fruit" is 33 plus $\begin{array}{|c|} 3 \\ 4 \end{array}$.

149 See the explanation by Colebrooke ([1817]1973, 35).

The SGT on TŚ 52: The *nyāsa* given by the commentator is $\begin{array}{|c|c|} 1 & 2 \\ 9 & 10 \\ 5 & 7 \\ 1 & 2 \\ 8 & \end{array}$, where the first side presents five quantities, and the second three. In the first side, the first three quantities denote the measures of the stone, then there is the quantity 1 stone, and at the bottom there is its cost. In the second side, the first three quantities are the measures, the number 2 at the bottom represents the quantity 2 stones, and the "fruit" is missing as it is to be ascertained. The first step is to move the "fruit" to the other side, and 8 is written below the 2 of the second side: $\begin{array}{|c|} 2 \\ 10 \\ 7 \\ 2 \\ 8 \end{array}$. Then, in the first side, the product of the mutual multiplication is 45; in the second side, it becomes 2240. After the division, the quotient is 33 plus $\begin{array}{|c|} 3 \\ 4 \end{array}$, this is the cost of the two stones which was asked to be ascertained and thus second "fruit".

The SGT on L 86: In order to supply another sample problem on the rule of nine, Siṃhatilakasūri quotes a sample problem, which corresponds to L 86. The SGT gives the *nyāsa*: $\begin{array}{|c|c|} 12 & 8 \\ 16 & 12 \\ 14 & 10 \\ 30 & 14 \\ 100 & \end{array}$. In the first side there are five quantities, while in the second side these are four; these are, in fact, 9 quantities in total. In the first side, the first three quantities represent the measures of the 30 pieces of cloth, and the number at the very bottom is the cost. The three different measures of the 14 pieces of cloth are arranged in the second side. The first step is to move the "fruit" and cost 100 *dramma*s to the second side: $\begin{array}{|c|} 8 \\ 12 \\ 10 \\ 14 \\ 100 \end{array}$. The mutual multiplication of the quantities of the first side gives the product 1344000, the second side becomes 80640. The division is carried out and the quotient obtained is the wished "fruit" 16 plus $\begin{array}{|c|}53760\\80640\end{array}$ which, when reduced by the common factor 26880, becomes $\begin{array}{|c|}2\\3\end{array}$. Hence, the cost of the 14 pieces of cloth is 16 plus $\begin{array}{|c|}2\\3\end{array}$ *drammas*.

The SGT on L 87: The SGT supplies this sample problem in order to provide an exercise on the "rule of eleven" (*ekādaśarāśika*). The presentation is: $\begin{array}{|r|r|}12&8\\16&12\\14&10\\30&14\\1&6\\8&\end{array}$, in the first side there are six quantities, while in the second side five. This sample problem is a modification of the previous, it involves the same pieces of cloth but in the first side the price is different and one more quantity that is the distance to cover occurs. In the second side, the quantities are five as the price is specified. The first step is to move the "fruit" and cost 8 *drammas* to the other side: $\begin{array}{|r|}8\\12\\10\\14\\6\\8\end{array}$. In the first side, when there is the mutual multiplication, the product is 80640, while in the side the product is 645120. When there is their division, the "fruit" 8 *drammas*, which is the final result, arises.

As a final remark, in lines 11–13 the SGT underlines that in the rules of five, seven, nine, and eleven which involve the discovery of the second, unknown fruit by first moving the known fruit to the second side, in the initial setting down of the data there is a quantity less, and this is, in fact, the wished "fruit".

4.11 Barter (*bhāṇḍapratibhāṇḍa*) [GT $\overline{112}$]

GT $\overline{112}$: The rule on barter enunciated by Śrīpati states that one should first perform the "exchange" (*vinimaya*) of the "prices" (*mūlya*) of the commodities and then carry out the rule of five.

The SGT on GT $\overline{112}$: In the introductory line, the commentator points out that the rule which follows depends upon the rule of five; this remark refers to Śrīpati's passage "the rule of five is to be performed" (*pañcarāśikavidhir* *vidhīyate*). Siṃhatilakasūri then provides a definition of *bhāṇḍapratibhāṇḍa* (lines 14–15). In commenting upon the rule, the commentator first underlines that each price of the two goods to be exchanged is written in the two sides (one in the first, the other in the second). Then, the exchange is performed, the "fruit" is moved to the other side and, as in the *pañcarāśika,* the quantities of the two sides are multiplied; the products obtained are then divided.

The SGT on GT 113 (ex.): By clarifying their price, Śrīpati asks students to determine the number of pomegranates which can be obtained when exchanging them with 12 mangoes. The SGT gives the *nyāsa*: $\begin{array}{|c|c|} 1 & 3 \\ 16 & 100 \\ 12 & \end{array}$. It can be observed that, as in the rule of five upon which the *bhāṇḍapratibhāṇḍa* depends, there are five quantities in total: three in the first side, and two in the second side. In the first row, the numbers 1 and 3 represent the cost of the two fruits to be exchanged; in the second row, 16 and 100 represent the quantities of mangoes and pomegranates respectively, and below is the quantity of mangoes to be exchanged, which is 12 (here representing the *phala*). The first step is to exchange the prices, and then to move the *phala* 12 to the other side: $\begin{array}{|c|c|} 3 & 1 \\ 16 & 100 \\ & 12 \end{array}$. As in the rule of five, there is the multiplication of the quantities of both sides, and the two products are finally divided. The quotient is the second *phala*, which is the quantity of pomegranates that can be obtained when they are exchanged with 12 mangoes. The final result is 1200÷48= 25.

The SGT on GT 114 (ex.): This sample problem concerns the barter of musk and agarwood. The SGT gives the presentation: $\begin{array}{|c|c|} 6 & 9 \\ 2 & 1 \\ 7 & \end{array}$. The commentator explains that one should first exchange the prices 6 and 9 and then move the "fruit" 7 to the other side, thus: $\begin{array}{|c|c|} 9 & 6 \\ 2 & 1 \\ & 7 \end{array}$. Then, the multiplication of the quantities of the two sides gives the products 42 and 18; after they are divided, the quotient is the final result and second fruit 2 *pala*s plus the remainder $\begin{array}{|c|} 6 \\ 18 \end{array}$ which, when reduced, is $\begin{array}{|c|} 1 \\ 3 \end{array}$. Thus, 2 plus $\begin{array}{|c|} 1 \\ 3 \end{array}$ *pala*s of musk can be obtained when they are exchanged with seven *pala*s of agarwood.

4.12 The rule regarding the sale of living beings (*jīvavikraya*) [GT $\overline{115}$]

GT $\overline{115}$: This rule concerns the sale of living beings, such as animals and slaves. With respect to the algorithm formulated, Śrīpati states that, having performed the "exchange" (*vyatyaya*) of the ages, the procedure is carried out "as before" (see *pūrvavat*). In his introductory line, Siṃhatilakasūri elucidates that this procedure involves the rule of five.

The SGT on GT $\overline{115}$: The commentator briefly comments upon this rule as his main preoccupation is to emphasise that this time all the quantities, and not just the "fruit" as in the rule of five, are involved in the step "having moved [the fruit] to the other side".

The SGT on GT 116 (ex.): The SGT gives the following presentation: $\begin{array}{|r||r|}1&1\\16&20\\70&\end{array}$.

The two ages16 and 20are exchanged and the "fruit" 70 is moved to the other side: $\begin{array}{|r||r|}1&1\\20&16\\&70\end{array}$. In the first side, when the quantities are multiplied, the product come to be 20; in the second side, it becomes 1120. When this is divided by 20, it becomes the quotient and price 56.

The SGT on GT 117 (ex.): This sample problem concerns two groups of 3 and 8 camels having different ages. Siṃhatilakasūri provides two methods to solve this sample problem. First, he applies the rule on the sale of living beings given in GT $\overline{115}$. The SGT gives the following presentation: $\begin{array}{|r||r|}3&8\\10&9\\108&\end{array}$; the age of the two groups of camels, which are 10 and 9, and the "fruit" and cost, which is 108, are exchanged: $\begin{array}{|r||r|}3&8\\9&10\\&108\end{array}$. In the first side, the multiplication produces 27; in the second side, it is 8640. These quantities are then divided and it becomes 320, which is the price eight camels nine years old cost.

The commentator then quotes a passage from L 78, which in the L occurs after the inverse rule of three. L 78 has been, in fact, fully quoted by the commentator while commenting upon the rule of three and its inverse (GT 95). It is noteworthy that Siṃhatilakasūri reflects upon the different presentation of this rule characterising the two texts, and emphasises that in the L the rule concerning the sale of living beings is mentioned in relation to the inverse rule of three. After having ascertained the result 320, Siṃhatilakasūri observes (lines 9–18) that this sample problem concerns the inverse rule of three and that in the L the sale of living beings is, in fact, mentioned straight after the inverse rule of three as, with respect to the sale of living beings, "when the age increases, the price conversely diminishes".[150] Therefore, the commentator reformulates the sample problem given by the GT and shows how to solve this sample problem by means of the inverse rule of three: (*measure* × *fruit*) ÷ *requisition*. The commentator provides an example which presents the same situation given by the root-text in the sample problem, but now dealing with two distinct camels and not with the two groups of 3 and 8 camels. The commentator gives the following presentation: |10|36|9|, with 3 quantities as in the rule of three. Thus, $(10 \times 36) \div 9 = 40$, which is the cost of one camel 9 years old. He then explains that this should be multiplied by its own "requisition" (*svecchā*) 8 resulting in 320, which is the final cost of the 8 camels, as it has been previously obtained (see line 9).

4.13 Practices [GT 118–133]

Preliminary observations

In mathematical texts, the term *vyavahāra* (lit. "practice, usage") connotes a group of mathematical procedures which deal with both practical and business-related computations. I have mentioned that the BSS is the first to present a subdivision of mathematical topics into twenty *parikarmāṇi* and eight *vyavahāras.* The GT is a fragmentary work which abruptly ends after the treatment of five rules on the first *vyavahāra*, which is *miśravyavahāra* (GT 118) or the "practice of mixture".[151] The expression *miśravyavahāra* given by the commentary in the introductory line to GT 118, is also found in TŚ 33, PG 47, GSS 5.2, L 90, SŚ 13.17, and GSK 3.1. Among standard topics of procedures on "mixture" (*miśra*) are interest, investments, and financial transactions. In the GT, the computations on proportional rules occurring in the section dedicated to the rule of three and its derivatives (GT 95–115 also treat business arithmetic. It has been shown that in that section some sample problems require students to calculate the rate of interest knowing the capital and time, or the period of time necessary to earn a certain amount of interest knowing the capital and the rate of interest. Computations on *miśravyavahāra* are, however, more elaborate and involve analysing combinations of quantities. The term *miśra*/*vimiśra* represents, in fact, the "mixed quantity" composing the combined sum[152] of the capital plus the accrued interest (see GT 118) or the total capital which has been lent and which should be divided into equal instalments (see GT 131). Another example of investment computations

150 See L 77 and SGT 116, 28.

151 Note that *miśravyavahāra* is both the name of the first of the eight *vyavahāras* and the name of the first procedure on *miśravyavahāra* enunciated in the GT and so termed by the SGT.

152 Colebrooke ([1817]1973, 39) points out that *miśrakavyavahāra* is "chiefly grounded on the rule of proportion".

is GT 124, which contains a rule for finding the time in which the given principal invested at a given rate will increase to a given multiple of itself.[153]

Practice of mixture (*miśravyavahāra*) [GT 118]

GT 118: Śrīpati's formulation is similar to TŚ 33. The procedure involves the computation of quantities – such as the accrued interest, the interest-rate, and the time – in order to separate the capital and the interest from their combined sum and determine their respective values. The steps of the algorithm formulated by the root-text can be represented as:

$$p \times t = A$$
$$T \times i = B$$
$$A \times V = D$$
$$B \times V = E$$
$$D \div (A + B) = C$$
$$E \div (A + B) = I$$

where p is the quantity-measure, t is its time, T is the other time, i is its interest-rate, and V is the mixed quantity. C is the capital and I the accrued interest.

The SGT on GT 118: The SGT explains this rule by linking the technical terms to the data mentioned in the sample problem which follows the *karaṇasūtra*. It is interesting to observe that the commentator clarifies the meaning of the expression *nijayogahṛtau,* points out that the result of the sum of the two products should be placed apart, and analyses the "mixed quantity" as *maulakyadrammavyāja*: the sum total of the original quantity of *drammas* (see *maulakyadramma*) and the "accrued interest" (*vyāja*).

The SGT on GT 119 (ex.): The commentator gives the *nyāsa*: |1||100||5||12||96|, where 1 is the time 1 month (*kāla*), 100 is the "quantity-measure" (*pramāṇarāśi*), 5 is the "interest-rate" (*phala*), 12 the "other time" (*parakāla*), and 96 is the "mixed quantity" (*miśraka*). The steps explained by the SGT are:

- the "quantity-measure" is multiplied by its time: $100 \times 1 = 100$
- the "other time" 12 is multiplied by 5, which is the "interest-rate": $12 \times 5 = 60$
- the two products are added:[154] $100 + 60 = 160$
- the first product is multiplied by the "mixed quantity": $100 \times 96 = 9600$
- the second product is multiplied by the "mixed quantity": $60 \times 96 = 5760$

153 This procedural rule corresponds to BSS 12.14.

154 In the SGT, in line 27 *pṛthak* underlines that this result should be placed aside. The result of this sum (see above $A + B$) is the divisor of the two dividends (see D and E) which are the two products mentioned in the *karaṇasūtra.*

- 9600 is divided by the previously obtained[155] 160 giving 60, which represents the *mūla* or initial "capital"
- 5760 is also divided by 160 giving 36, which becomes the "accrued interest", the interest that will build up and that the lender will earn in 12 months.

With respect to the terminology, the SGT uses the terms *phala* (line 8) and *kalāntara* (line 13) to denote the "interest-rate" and the "accrued interest" respectively, although these meanings are interchangeable according to the context.

4.14 Commission to the moneylender (*vyājopajīvivṛtti*) [GT 120]

GT 120: In the introductory line, the commentator specifies that this *karaṇasūtra* concerns "the commission (*vṛtti*) [paid] to the moneylender (*vyājopajīvin*)". Interestingly, GSK 3.2 gives a similar rule and terms it *bhāvyaka*; this work gives also a sample problem similar to GT 121. According to the way *bhāvyaka* is used in the SGT, I understand it to imply the commission, the fee which is paid when one borrows money from a creditor. In the documents of medieval Gujarat collected under the name *Lekhapaddati,* various types of pledges, sureties, and loan repayments are mentioned. These show that in medieval India, when an official contract to borrow money was made, different professionals, such as the accountant and the scribe writing the document, were involved. The *Lekhapaddhati* and the *Manusmṛti* provide examples of contracts and these testify that it was common for the creditor to take an adequate deposit, a surety before advancing a loan. Every detail was recorded: the date, the amount of money lent to the borrower, the interest-rate, and the names of the individuals involved. Furthermore, primary sources demonstrate that this type of financial transaction was recorded in the presence of a witness.

In this *karaṇasūtra* (GT 120), Śrīpati explains how to determine quantities such as the initial capital, the interest, and so forth concerning a certain amount of money lent to somebody, when the mixed quantity and the time are known. The quantities are:

- the "quantity-measure" (*pramāṇarāśi*)
- "its own time" (*nijakāla*)
- the "elapsed time" (*vyatītakāla*)
- the "interest and so forth" (*phalādi*)
- the "mixed quantity" (*miśra*)
- the "capital" (*mūla*)

155 While commenting upon the *karaṇasūtra,* the commentator refers to this sum by the expression *prāgudita* (line 1). In this regard, see also the expression *pṛthak sthāpyam* (line 9).

The SGT on GT 120: The commentator clarifies that "as before" in the previous procedural rule, here too there is the step "the quantity-measure is multiplied by its own time". In order to better elucidate the meaning of the expressions used by Śrīpati he refers to the data of the following sample problem. The SGT specifies the meaning of the expression *phalādi* or "the interest and so forth", and uses the term *bhāvyaka* which will be mentioned in the next sample problem. It can be understood that by *ādi* or "and so forth", Śrīpati denotes the fees to be paid to the various professionals involved at the time of stipulating a contract.[156]

The SGT on GT 121 (ex.): The SGT gives this presentation:

1	100	5	1	1	1	12	905
1	1	1	1	2	4	1	1

These quantities represent the "time" 1 month, the "quantity-measure" 100, the "interest" 5%, at this stage the fee for the "commission of surety" (*bhāvyaka*) is 1, half a *dramma* is the payment for another professional (who could be, for instance, the accountant) involved in the factual making of the contract, one-fourth is the payment for the scribe,[157] the "time" 12 months (thus one year), and the *miśraka* or "mixed quantity" is 905. The steps explained by the commentator are:

- lines 1–3: the "quantity-measure" 100 is multiplied by "its own time", which is one month (since the interest-rate 5% is a monthly rate) hence: $100 \times 1 = 100$. The interest-rate is then multiplied by the "elapsed time" (*vyatītakāla*), which the SGT clarifies to be 12 (i.e., it is a 12 months contract): $12 \times 5 = 60$
- the "elapsed time" 12 is multiplied also by the four sets of data shown in the *nyāsa* after the *pramāṇarāśi* and representing the various fees to be paid on the occasion of this contract. Line 4 shows the following layout:

100	60	12	12	12
1	1	1	2	4

- lines 5–13: their "own sum" (*svayuti*) is carried out by applying the *bhāgajāti*; the result is

724
4

- in lines 13–14, the commentator states that this sum is the "divisor" (*bhāgadāyin*) of the next product, which is the result of the multiplication of the mixed quantity by the quantity-measure
- lines 15–20: the *miśraka* 905 is multiplied by the quantity-measure 100 and it becomes 90500, which is specified by the SGT as the *bhājyarāśi*

156 In this regard, see the following sample problem.

157 Interestingly, GSK 3.4 gives exactly the same sample problem (the same data and the same situation, but differently formulated). The GSK mentions a fee to be paid to a *brāhmaṇa*; commissions paid to *brāhmaṇas*; and accountants are recorded in both the *Lekhapaddhati* and the *Manusmṛti*. However, these are mentioned neither by Śrīpati nor by Siṃhatilakasūri; in the sample problem, these authors specify only the fee to be paid to the scribe.

or "dividend". In order to perform the division, the divisor interchanges its operands and the multiplication is instead carried out. It gives $\begin{array}{|r|}90500\\181\end{array}$ which, when reduced, becomes the "capital" (*mūladhāna*) 500

- lines 20–22: the product 60, which is the result of the multiplication of the "elapsed time" 12 by the "interest" 5, multiplied by the "mixed quantity" 905 becomes 54300; this result is then divided by the divisor $\begin{array}{|r|}724\\4\end{array}$. The quotient 300 represents the interest (*kalāntara*) which will accrue in 12 months
- lines 23–25: the "mixed quantity" 905 is multiplied by the first (with respect to the layout occurring in line 4) 12 whose denominator is 1, and the result is divided by the divisor $\begin{array}{|r|}724\\4\end{array}$; the result 60 is the fee to be paid for the "commission" (*bhāvyaka*)
- lines 26–32: the second fraction $\begin{array}{|r|}12\\2\end{array}$ is also multiplied by the *miśraka* 905 and divided by the divisor $\begin{array}{|r|}724\\4\end{array}$; the result is 30, which is the "fee" (*vṛtti*) to be paid to the accountant
- lines 1–4: the third fraction $\begin{array}{|r|}12\\4\end{array}$ is multiplied by the *miśraka* 905 and divided by the divisor $\begin{array}{|r|}724\\4\end{array}$; the result is 15, which is the fee charged by the scribe lines 4–5: all the amounts to be paid are $\begin{array}{|r|}500\\300\\60\\30\\15\end{array}$. Their sum is, in fact, the "mixed quantity" 905.

Regarding the results: 500 *dramma*s are the capital which has been borrowed; the accrued interest, the money which the borrower has to give back in addition to the capital 500 is 300 *dramma*s; 60 is the fee paid for the commission, 30 is the fee paid to the other professional (the accountant?) involved, and 15 the fee charged by the scribe.

4.15 Rule on interest [GT 122]

GT 122: In the introductory line, the commentator observes that the following *karaṇasūtra* provides a rule to ascertain the interest alone. Regarding the GT's terminology, it can be observed that:

- *dravya* is the "wealth, capital", in the introductory line the SGT uses instead the term *mūladhana*
- *vṛddhi* is the "interest-rate"
- *phala* is the "interest which accrues", in the introductory line the commentator uses the term *kalāntara,* note that this is the same term used by the GT in the next sample problem where the *kalāntara* is the quantity to be determined

The algorithm enunciated in the rule can be represented as:

$$(c \times t \times i) \div 100 = I$$

where c is the capital, i is the interest rate, t is the time, and I is the accrued interest

The SGT on GT 122: In the introductory line, Siṃhatilakasūri underlines that the rule is stated in another work, he does not comment upon it.

The SGT on GT 123 (ex.): The commentator gives the following *nyāsa*: |1|100|5||12|, where 1 is one month, 100 is the wealth (the capital which has been borrowed), 5 is the *vṛddhi* or the monthly "interest rate", and 12 is the length of the loan agreement. The SGT shows that the first step is to multiply the wealth by the time one month: $100 \times 1 = 100$; the result is multiplied by 12 months, which is the "elapsed time" (*vyatīkāla,* see GT 120), and it becomes 1200. This is multiplied by the *vṛddhi* 5 becoming 6000; this result is finally divided by 100, and the quotient 60 represents the number of *dramma*s charged by the lender.

In 12 months, the borrower who has borrowed 100 *dramma*s has to pay back 160; the lender will therefore earn 60 *dramma*s.

4.16 Rule on time and double capital [GT 124]

GT 124: In the introductory line, the commentator specifies that this *karaṇasūtra* is a quotation from the BSS. This procedural rule corresponds, in fact, to BSS 12.14. The SGT emphasises that this rule concerns double capital and time. The main elements of this rule are:

- the "capital" (*pramāṇa*)
- the "accrued interest" (*phala*)
- the "time" (*kāla*)
- the "multiplier" (*guṇaka*)

The algorithm can be represented as:

$$(c \times t) \div I) \times m = T$$

where c is the capital, t is the time, I is the accrued interest, m is the multiplier, and T is the time by which the capital will increase.

The SGT on GT 124: The commentator illustrates the *karaṇasūtra* by referring to the numerical data given in next sample problem (GT 125). He clarifies that the "multiplier" represents the number times by which the capital will increase, and the quantity "time" refers to interest which accrues monthly.

The SGT on GT 125 (ex.): The SGT shows the following *nyāsa*: |1|200|6|gu3|, where 1 is 1 month (i.e., the interest which accrues monthly), 200 is the "capital", 6 is the monthly "accrued interest" in *dramma*s, and 3 is the "multiplier", denoted by the abbreviation *gu* (*guṇa*). Siṃhatilaka elucidates the following steps:

- the "capital" is multiplied by the time, hence $200 \times 1 = 200$
- the result is divided by the "interest": $200 \div 6 = 33$ plus $\left|\begin{array}{c}1\\3\end{array}\right|$
- this quotient is multiplied by the "multiplier" 3 lessened by 1, and it becomes $\left|\begin{array}{c}66\\2\\3\end{array}\right|$. This is the final time required, which the commentator explains to represent 5 years and six months (66 months are divided by 12), and when the fraction $\left|\begin{array}{c}2\\3\end{array}\right|$ is multiplied by 30, which represents the number of days in a month, it gives 20 days.

Hence, in 5 years 6 months and 20 days, the capital 200 *dramma*s will increase by three times.

The SGT on GT 126 (ex.): The SGT gives the *nyāsa*: $|2||20||5|\left|\begin{array}{c}\\ \text{gu}\\ \\ \end{array}\begin{array}{c}1\\1\\2\end{array}\right|$, where 2 is the "time" expressed in months, 20 is the "capital", 5 the "accrued interest", and the quantity one plus one-half is the "multiplier". The steps executed are:

- the "capital" 20 is multiplied by the "time": $20\times 2 = 40$
- this is divided by the "accrued interest" 5, and it becomes 8

- then, the "multiplier" $\begin{array}{|c|}1\\1\\2\end{array}$ should be subtracted by one, and the resulting 1 multiplied by 8; it gives 8 which divided by the denominator 2 is 4.

In 4 months, 20 *paṇa*s will become 30.

4.17 Conversion of several bonds into one (*ekapatrakaraṇa*) [GT 127]

GT 127: In the introductory line, the commentator underlines that this rule[158] concerns the procedure for the conversion of several bonds of various amounts and interest-rates into a single one. This procedure is termed *ekapatrīvidhāna* by the GT and *ekapatrakaraṇa* by the SGT.[159] In Śrīpati's formulation, the key-terms are the following:

- the "average time" (*gatakāla*): the single date to be determined on which the borrower will pay back the total capital by a single "bond" (*patra*)
- the "sum of the interest" (*phalaikya*), which denotes the total sum of the interest which will accrue on each given bond during each elapsed time; every bond has, in fact, a different interest-rate
- the "elapsed time" (*gatasamaya*), which is the time for the repayment of each bond
- the "sum of the monthly profits" (*māsalābhaikya*), which is the total accrued interest, thus the total profit obtained by the creditor
- the "sum of the [lent] capitals" (*draviṇayuti*), which represents the total amount lent by the creditor
- the "percentage-rate" (*śataphala*)
- and the expression *ekapatrīvidhāna,* which denotes the operation of converting various bonds into a single one.

The SGT on GT 127: As usual, the commentator explains this procedural rule by linking the technical terms used by Śrīpati to the data mentioned in the next sample problem (GT 128–129). In this regard, he elucidates that:

- 7, 8, 6, and 12 months denote the "elapsed time" (*gatasamaya*) of the four bonds[160]
- *phala* is the "interest-rate"; Siṃhatilakasūri explains that, by multiplying each interest-rate by the elapsed time of each bond, the numbers

158 A similar rule is found in the TŚ and in the GSK too.

159 In the *Lekhapaddhati*, *vyavahārapatra* and *ādhipatra* denote a debt deed; *patra* is lit. "document, letter". In the *Lekhapaddhati*, see also *paṭṭa*/*paṭṭaka*.

160 The next sample problem concerns four debt obligations; for each the capital is 100, 200, 300, and 400.

beginning with 14 arise (see line 6). The commentator specifies that their sum is 374

- the SGT then explains that in each month, there is an "increment" (*vṛddhi*, the "accrued interest"); for the creditor, this denotes the monthly profit that, with respect to each bond, comes to be 2, 6, 12 and 20. Their sum is 40
- Siṃhatilakasūri specifies that the "average time" – 9 months – is determined when 374, which is the previously obtained sum, is divided by the sum of the monthly profits[161]
- in lines 19 and 22, the commentator uses the expression *sarve māsā vilupya* ("having equalised all the [elapsed] months") to denote the process of arriving at a unique date on which the single bond will be paid
- the commentator also mentions that the *dravinayuti* or the "sum of the capitals" is 1000 and that when the product of the multiplication of 100 by the sum of the monthly profits, which is 40, is divided by the sum of the given bonds, one finds the equal percentage-rate for all the bonds.

The SGT on GT 128–129 (ex.): The commentator provides the following *nyāsa*:

1	7	1	8	1	6	1	12
100	100	100	200	100	300	100	400
2		3		4		5	

. The eight sides of this layout represent, two by two, the four *patra*s: each left side denotes the quantity-measure above and the monthly interest-rate (2% and so forth) underneath; in the right side, the number of the months (7 and so forth) is above, and the money which is lent is underneath. With respect to each bond, the capital is thus 100, 200, 300, and 400.

Siṃhatilakasūri first determines the *gatasamayaphalaikya* or "the sum of the interest which has accrued during the elapsed time", which will be then divided by the "sum of the monthly profits". The sum of the interest accumulated during the elapsed time is obtained in the following way: each interest-rate is multiplied by its elapsed time and by each capital according to its integer[162] (lines 3–7): $2 \times 7 \times 1 = 14$; $3 \times 8 \times 2 = 48$; $4 \times 6 \times 3 = 72$; $5 \times 12 \times 4 = 240$. In line 6, these products are shown:

14
48
72
240

and, when the sum is calculated, the result is 374 (it denotes the *gatasamayaphalaikya*). The commentator explains that every month, the monthly profits are 2, 6, 12, and 20, since each

161 In the next sample problem, the numbers shown in the layout (line 9) are obtained by dividing the interest (see the layout, line 6) by the elapsed time.

162 In the next sample problem, in line 25 the commentator clarifies that, in this computation, one should consider each amount of lent capital according to its integer.

resulting product is divided by its elapsed time: 14 ÷ 7 = 2; 48 ÷ 8 = 6; 72 ÷ 6 = 12; 240 ÷ 12 = 20. In line 9, the commentator shows the monthly profits earned by the creditor on each bond:

2
6
12
20

and states that their sum is 40; next, the SGT observes that the creditor gains 40 *dramma*s every month. At this point, the *gatasamayaphalaikya* 374 is divided by the *māsavṛddhyaikya* 40, and the result

9
7
20

is the *gatakāla* or the "average time" for the equivalent single bond, whose amount is yet to be determined. Then, the sum of the monthly profits is multiplied by 100 and divided by the "sum of the capitals" (*draviṇayuti*) of the various bonds: 40 × 100 = 4000, the sum of the lent capital is 1000, which is the sum of 100, 200, 300, and 400, thus: 4000 ÷ 1000 = 4, which is thus the "percentage-rate" (*śataphala*).[163]

Finally, the commentator specifies that in 9 and

7
20

months and at the interest-rate of 4%, the creditor will get back his money (the capital 1000 *dramma*s) plus an accrued interest of 374 *dramma*s in one only payment.

The SGT on GT 130 (ex.): The sample problem asks students to work out the same four *patra*s given in the previous sample problem but this time each interest-rate is increased by one-fourth and the months (the elapsed time) by one-third. The commentator provides the *nyāsa*:

1	7	1	8	1	6	1	12
100	1	100	1	100	1	100	1
2	3	3	3	4	3	5	3
1	100	1	200	1	300	1	400
4		4		4		4	

. This layout shows eight sides, two for each *patra.* In this regard, in lines 6, 13, and 19 the SGT uses the expressions *dvitīyapatre, tṛtīyapatre*, and *caturthapatre* to refer to the "bond" (*patra*) involved each time in the computation. The data concerning each bond are given in pair; the first side is denoted as *prathamāṅka* or the "first quantity" and the "second quantity" is *dvitīyāṅka*. In each pair, the side on the right represents the elapsed time expressed in months (see the top numbers 7, 8, 6, 12), and below the months, the fraction one-third denotes the increment. On the left side of each pair, at the very top the number 1 denotes 1 month,

163 Out of the four bonds, an equal interest-rate and period of time have been calculated in order for the borrower to pay back the money by a single payment.

the *pramāṇarāśi* 100 is below it, and the interest-rates 2, 3, 4, 5 are followed by the fraction one-fourth, which is the increment that Śrīpati requires the student to add to each interest-rate.

The method of the class of fractional increase is first carried out: the increments mentioned by Śrīpati, which are one-fourth and one-third, are added to the number of months and to the interest-rate of each bond respectively. The commentator then explains that that the numerators 198, 325 and so forth of the obtained fractions[164] are multiplied by the capital of each bond. The expression "its own integer" (*svarūpa*) means that, regarding the first capital 100, the "integer" (*rūpa*) is 1; regarding the second capital 200, the integer is 2, and so forth. Next, in order to accomplish the first step of this procedure (see, *gatasamayaphalaikye*, GT 127), each *gatasamayaphala* is obtained to determine the total sum. The calculations carried out by the SGT are the following:

- lines 25–5: in the first bond, the interest-rate 2 is increased by $\left|\begin{array}{r}1\\4\end{array}\right|$, while its time 7 months is increased by $\left|\begin{array}{r}1\\3\end{array}\right|$; the fractions $\left|\begin{array}{r}9\\4\end{array}\right|$ and $\left|\begin{array}{r}22\\3\end{array}\right|$ are obtained. These are multiplied giving the improper fraction $\left|\begin{array}{r}198\\12\end{array}\right|$. The numerator 198 is multiplied by the integer 1 which, with respect to the capital 100, is its *svarūpa*. This fraction is then transformed into the quantity $\left|\begin{array}{r}16\\1\\2\end{array}\right|$, which denotes the *gatasamayaphala* (the "interest which has accrued during the elapsed time").
- Lines 6–13: in the second bond, the first quantity becomes $\left|\begin{array}{r}13\\4\end{array}\right|$, while the second quantity becomes $\left|\begin{array}{r}25\\3\end{array}\right|$. These are multiplied and the improper fraction obtained is $\left|\begin{array}{r}325\\12\end{array}\right|$; the numerator 325 is then multiplied by the integer 2, which is the *svarūpa* of the capital 200 *dramma*s, and the improper fraction $\left|\begin{array}{r}650\\12\end{array}\right|$ is transformed into the quantity $\left|\begin{array}{r}54\\1\\6\end{array}\right|$. This is the *gatasamayaphala* of the second bond

164 See the procedures explained from line 25 to line 23. In regard to the improper fractions, see lines 2, 9, 15, and 21.

- lines 13–18: in the third bond, the first quantity becomes $\begin{array}{|c|}17\\4\end{array}$, while the second quantity becomes $\begin{array}{|c|}19\\3\end{array}$. These are multiplied and the improper fraction obtained is $\begin{array}{|c|}323\\12\end{array}$; the numerator 323 is then multiplied by the integer 3, which is the *svarūpa* of the capital 300, and the improper fraction $\begin{array}{|c|}969\\12\end{array}$ is transformed into the quantity $\begin{array}{|c|}80\\3\\4\end{array}$
- lines 19–23: in the fourth bond, the first quantity becomes $\begin{array}{|c|}21\\4\end{array}$, while the second quantity becomes $\begin{array}{|c|}37\\3\end{array}$. These are multiplied and the improper fraction obtained is $\begin{array}{|c|}777\\12\end{array}$; the numerator 777 is then multiplied by the integer 4, which is the *svarūpa* of the capital 400, and the improper fraction $\begin{array}{|c|}3108\\12\end{array}$ becomes $\begin{array}{|c|}259\\1\end{array}$
- then, the four obtained quantities are, with respect to each bond, the interest that has accrued during the elapsed time (i.e., the four *gatasamayaphalas*): $\begin{array}{|c|}16\\1\\2\end{array}\begin{array}{|c|}54\\1\\6\end{array}\begin{array}{|c|}80\\3\\4\end{array}\begin{array}{|c|}259\\1\\\ \end{array}$.
- In the next step (see *gatasamayaphalaikye māsavṛddhyaikyabhakte*), the sum of the interest that has accrued during the elapsed time should be divided by the sum of the monthly profits.[165] In order to carry out the sum, the interest accumulated during the elapsed time of each *patra*, is given back as an improper fraction (line 27): $\begin{array}{|c|}33\\2\end{array}\begin{array}{|c|}325\\6\end{array}\begin{array}{|c|}323\\4\end{array}\begin{array}{|c|}259\\1\end{array}$. By means of the method of the class of simple fractions, the first two fractions are added and the resulting fraction is added to the next one:

165 See the step *gatasamayaphalaikye māsavṛddhyaikyabhakte* (GT 127).

- the first two fractions $\begin{array}{|r|r|}33 & 325\\ 2 & 6\end{array}$, when their denominators are reduced and exchanged, become $\begin{array}{|r|r|}33 & 325\\ 2 & 6\\ 3 & 1\end{array}$, and when these are multiplied by the denominators 3 and 1 respectively, the results are $\begin{array}{|r|}99\\ 6\end{array}$ and $\begin{array}{|r|}325\\ 6\end{array}$. Their sum gives $\begin{array}{|r|}424\\ 6\end{array}$
- this is added to the next fraction $\begin{array}{|r|}323\\ 4\end{array}$ producing the fraction $\begin{array}{|r|}1817\\ 12\end{array}$ (line 7)
- this result is added to the last profit $\begin{array}{|r|}259\\ 1\end{array}$ becoming $\begin{array}{|r|}4925\\ 12\end{array}$ (line 11). With respect to the four bonds, this fraction represents the total sum of the interest which has accrued during the elapsed time (*gatasamayaphalaikya*).
- At this point, the four accrued interest $\begin{array}{|r|r|r|r|}33 & 325 & 323 & 259\\ 2 & 6 & 4 & 1\end{array}$ are each divided by their respective time[166] added to the increment $\begin{array}{|r|}1\\ 3\end{array}$, as mentioned in the sample problem. Hence, the four fractions $\begin{array}{|r|r|r|r|}33 & 325 & 323 & 259\\ 2 & 6 & 4 & 1\end{array}$ are divided by the obtained "increment of the months" which are $\begin{array}{|r|r|r|r|}22 & 25 & 10 & 37\\ 3 & 3 & 3 & 3\end{array}$. The steps executed by the commentator, although not fully shown,[167] are the following:
- with respect to the first bond, the interest which has accrued during the elapsed time is divided by the fraction representing its time multiplied by the increment one-third: when $\begin{array}{|r|}33\\ 2\end{array}$ is divided by $\begin{array}{|r|}22\\ 3\end{array}$, it means that it is multiplied by $\begin{array}{|r|}3\\ 22\end{array}$. The quotient $\begin{array}{|r|}13\\ 2\end{array}$, being an improper fraction, is transformed into the quantity $\begin{array}{|r|}2\\ 1\\ 4\end{array}$

166 The months are 7, 8, 6, and 12.
167 In line 12, the SGT gives the final quotients of these divisions.

- with respect to the second bond, when $\left|\begin{array}{r}325\\6\end{array}\right|$ is divided by $\left|\begin{array}{r}25\\3\end{array}\right|$, it is multiplied by $\left|\begin{array}{r}3\\25\end{array}\right|$. The quotient $\left|\begin{array}{r}51\\4\end{array}\right|$, being an improper fraction, is transformed into the quantity $\left|\begin{array}{r}6\\1\\2\end{array}\right|$
- with respect to the third bond, when $\left|\begin{array}{r}323\\4\end{array}\right|$ is divided by $\left|\begin{array}{r}10\\3\end{array}\right|$, it is multiplied by $\left|\begin{array}{r}3\\10\end{array}\right|$. The quotient obtained $\left|\begin{array}{r}51\\4\end{array}\right|$, being an improper fraction, is transformed into the quantity $\left|\begin{array}{r}12\\3\\4\end{array}\right|$
- with respect to the fourth bond, when $\left|\begin{array}{r}259\\1\end{array}\right|$ is divided by $\left|\begin{array}{r}37\\3\end{array}\right|$, it means that it is multiplied by $\left|\begin{array}{r}3\\37\end{array}\right|$. The quotient obtained is the integer $\left|\begin{array}{r}21\\1\end{array}\right|$
- in line 12, the four results are given: $\left|\begin{array}{r}2\\1\\4\end{array}\right|\left|\begin{array}{r}6\\1\\2\end{array}\right|\left|\begin{array}{r}12\\3\\4\end{array}\right|\left|\begin{array}{r}21\\1\\ \end{array}\right|$
- next, their sum is obtained and this is $\left|\begin{array}{r}170\\4\end{array}\right|$. This is the sum of the monthly profits (line 15, *etat māsavṛddhyaikyam*). This result becomes the divisor of $\left|\begin{array}{r}4925\\12\end{array}\right|$, which is the sum of the interest that has accrued during the elapsed time of the four *patra*s. In order to perform this division, the divisor becomes the multiplier and by means of the cross-reduction the two operands become $\left|\begin{array}{r}1\\170\end{array}\right|\left|\begin{array}{r}4925\\3\end{array}\right|$ respectively. The product $\left|\begin{array}{r}4925\\510\end{array}\right|$ is transformed into the quantity $\left|\begin{array}{r}9\\67\\102\end{array}\right|$. This is the *gatakāla* or the "average time"
- the *māsavṛddhyaikya* or "sum of the monthly profits", denoted also by the expression *māsalābhaikya* (line 22), is $\left|\begin{array}{r}170\\4\end{array}\right|$ (line 15). This is multiplied

by 100 and the product $\begin{array}{|r|}17000\\4\end{array}$ is divided by the *dravinayuti* or "sum of the capitals" which, in SGT 128–129, 16, the commentator has demonstrated to be 1000

- the result $\begin{array}{|r|}4\\1\\4\end{array}$ is the "percentage rate" (*śataphala*), this is multiplied by 10 and it becomes $\begin{array}{|r|}85\\2\end{array}$ (line 29)
- this is then multiplied by the average time $\begin{array}{|r|}9\\67\\102\end{array}$ and the result $\begin{array}{|r|}410\\5\\12\end{array}$ is the "total interest that will accrue in nine months".

In 9 plus $\begin{array}{|r|}67\\102\end{array}$ months and at the interest-rate of 4 plus $\begin{array}{|r|}1\\4\end{array}$%, the creditor will get back his money (the capital 1000 *drammas*) plus an accrued interest of $\begin{array}{|r|}410\\5\\12\end{array}$ in one only payment.

4.18 Equating instalments of capital (*samīkaraṇa*) [GT 131]

GT 131: As elucidated by the commentary in the introductory line, this rule explains how to determine, with respect to a given "lent quantity" (*prayuktarāśi*), equal instalments when they are initially unequal and have a different interest-rate. The steps are:

- the "time" one month is multiplied by the "quantity-measure" (this is specified by the commentator)[168]
- the "opposite numerator and denominator" (*vyastāṃśahāra*) are multiplied
- the denominators obtained become the "profits" (*lābha*), which are each separately multiplied by the "mixed quantity" (*vimiśrasva*), representing in this case the whole capital (the sum of the instalments)
- the products obtained are divided by the sum of the profits.

168 Note that "quantity-measure" (*pramāṇarāśi*, SGT 131, 15) and "wealth-value" (*pramāṇadhana*, SGT 132–133, 28) are synonyms.

Srīpati's passage presents a space-related vocabulary. As clarified by the commentator, terms such as "the opposite numerator and denominator" do not refer to the operands of a fraction, but rather to the way quantities are written one below the other in order to perform this procedure. Another consideration concerns the term *lābha* or "profit", which has the same meaning in the previous procedural rule (GT 127). In GT 132–133, the total profit (which represents the total accrued interest) earned by the creditor is obtained by multiplying the interest-rate by the elapsed time and the time one month by the quantity-measure 100.

The SGT on GT 131: The commentary first explains that in "each instalment", there are four elements: the "quantity-measure" (*pramāṇarāśi*), the "twofold time" (*kālasaṅkhyā*), and the "interest-rate" (*phaladhana*). The "twofold time" denotes the time of the monthly interest-rate (thus 1 month), and the elapsed time (see in the following sample problem: 2, 3, and 4 months). Each profit is determined by calculating separately each interest-rate, its time, the quantity-measure, and the elapsed time. The commentator then clarifies the meaning of the expressions "numerator" and "denominator". Siṃhatilakasūri glosses *vyasta* (here "opposite") with *viparyaya* or "conversely", supplies the verbal adjective *guṇair* or "multiplied" to *aṃśahāraiḥ* ("the numerator and the denominator"), and specifies that the term *pṛthak* denotes the way the "profits" (*lābha*), which represent the interest that accrues on one *dramma,* are arranged in the layout.

The SGT on GT 132–133 (ex.): The given *nyāsa* shows six sides:

1	3	1	2	1	4
100	1	100	1	100	1
2		3		4	

, two for each *khaṇḍa* ("instalment"). On the left-side, the lower numbers 2, 3, and 4 represent the elapsed time; the 1 at the top denotes 1 month, and 100 is the quantity-measure. In each *khaṇḍa,* the right-side denotes the interest-rate. In lines 27–5, the commentator carries out the first step of the procedure formulated by Śrīpati, which is *vyastāṃśahārair pṛthag ekarūpalābhair*:[169]

- with respect to the first "instalment": one month, which is the number at the top of the layout, is multiplied by one hundred, which is underneath and which is the "wealth-value" (*pramāṇadhana*): 1 × 100 = 100. The interest-rate 2%, which is the number below 100, is multiplied by the 3 at the top on the right side, which represents 3 months (the elapsed time): 3 × 2 = 6. In this way, the time has been multiplied by the "wealth-value" and the interest-rate by the "elapsed time"; it becomes

100
6

169 While performing the procedure, the commentator elucidates the meaning of the terms *aṃśa* and *hāra* found in *vyastāṃśahārair.*

- regarding the second instalment, the time one month is also multiplied by the 100; the opposite numerator 2 (i.e., the interest-rate) is multiplied by the elapsed time 3 months, which is the "denominator" underneath, it is: $\begin{array}{|r|}100\\6\end{array}$
- regarding the third instalment, 1 is multiplied by 100: $1 \times 100 = 100$; the denominator 4 is multiplied by the opposite numerator 4; the result is:[170] $\begin{array}{|r|}100\\16\end{array}$
- regarding each "unit" mentioned by Śrīpati in the rule, the denominators of these fractions, which are 6, 6 and 16, represent the "profits", i.e., the "accrued interest" (*lābha*)[171]
- the next step is *svakīyayogena* (lines 6–13): the commentator determines the sum of the obtained profits in order to ascertain the divisor (see in line 13 *bhādagāyī*) for the final step. By means of the method of the class of simple fractions, the three fractions are added obtaining $\begin{array}{|r|}1900\\48\end{array}$, which is the quantity-divisor
- the step carried out next is the multiplication of the three profits $\begin{array}{|r|}100\\6\end{array}\begin{array}{|r|}100\\6\end{array}\begin{array}{|r|}100\\16\end{array}$ by the mixed wealth 190, which produces $\begin{array}{|r|}19000\\6\end{array}\begin{array}{|r|}19000\\6\end{array}\begin{array}{|r|}19000\\16\end{array}$. In line 15, the commentator specifies that these are the dividends
- the dividends are each divided by the divisor $\begin{array}{|r|}1900\\48\end{array}$ (see line 12). The three quotientsobtained are $\begin{array}{|r|}80\\80\\30\end{array}$. From the mixed wealth 190, by means of the steps above, three equal "instalments" (*khaṇḍa*) are hence obtained. In line 26, the commentator emphasises that when these are added, the mixed wealth 190 occurs
- line 26: the next step is to determine the "equal profits" (*samavṛddhi*). The commentator explains that this is determined by applying the rule of five (*pañcarāśika*). The final results are the three instalments 80, 80, and 30 and the interest-rate is 4 plus $\begin{array}{|r|}4\\5\end{array}$%.

170 This layout is not given in the edition.

171 Regarding the three *khaṇḍa*s, the three *lābha*s or "profits" denoting the accrued interests are 6, 6, and 16 respectively. These are obtained by multiplying each elapsed time (in Śrīpati's formulation this is that termed "denominator") by its interest (in Śrīpati's formulation this is that termed "numerator").

L 92: On the same topic, the commentator supplies a rule from the L. It explains, in fact, how to determine equal instalments from a given mixed quantity. After this quotation, the commentator applies this rule to the sample problem given by the GT which he has just solved (GT 132–133). In this way, he illustrates the similarities and differences characterising these procedures. The *nyāsa* is: $\begin{array}{|r|r|r|r|r|r|} 1 & 3 & 1 & 2 & 1 & 4 \\ 100 & 1 & 100 & 1 & 100 & 1 \\ 2 & & 3 & & 4 & \end{array}$, the "mixed wealth" (*miśradhana*) is $\begin{array}{|r|} 190 \end{array}$. It can be observed that this *nyāsa* is exactly the same as the one shown by the commentator in SGT 132–133, 27. First, with respect to each instalment, the "time" (*kāla*) 1 month, which is the number at the very top and is the time of the monthly rate, is multiplied by the "quantity-measure" (*pramāṇa*) 100: $1 \times 100 = 100$. The "interest-rate" (*phala*) is 3, 2, and 4 respectively and represent what, in the *karaṇasūtra*, Śrīpati refers to as "numerators" (see *vyastāṃśahāraiḥ*); these are multiplied by the "elapsed time" (*vyatītakāla*), which are the months 2, 3, and 4 respectively and which represent what Śrīpati refers to as "denominators". In the first instalment, the interest-rate 3 is multiplied by the elapsed time 2 months, $3 \times 2 = 6$; this is placed below the obtained number 100: $\begin{array}{|r|} 100 \\ 6 \end{array}$. In the second instalment, the interest-rate 2 is multiplied by the 3 elapsed months, $2 \times 3 = 6$, and it becomes $\begin{array}{|r|} 100 \\ 6 \end{array}$. In the third instalment, the interest-rate 4 is multiplied by the elapsed time 4 months, $4 \times 4 = 16$, and it becomes $\begin{array}{|r|} 100 \\ 16 \end{array}$. The three obtained results are the three instalments: $\begin{array}{|r|r|r|} 100 & 100 & 100 \\ 6 & 6 & 16 \end{array}$.

The SGT underlines that the rest of the procedure is not shown since it is the same as the illustrated before, meaning while solving the sample problem given by Śrīpati. Interestingly, in lines 19–21, the commentator points out that he supplies this rule from the L in order to clarify the meaning of the expression "the numerator and the denominator" used by Śrīpati in his *karaṇasūtra*; this is exactly why Siṃhatilakasūri is interested in explaining only the first half of this sample problem. It is noteworthy that Bhāskarācārya uses the terms *vyatītakāla* and *phala* to denote what Śrīpati terms *hāra* or "denominator" (these numbers are, in fact, at the very bottom of the layout) and *aṃśa* or "numerator" (these numbers are, in fact, at the very top).

Appendix 1

List of mathematical rules and sample problems supplied by the SGT and found in other works[1]

L 19–20 (square of integers)
TŚ 11 (square of integers)
TŚ 15 (cube of integers)
L 24 (cube of integers)
TŚ 24 (class of fractional increase)
PG 37 (class of simple fractions)
L 45–46 (arithmetic of zero). Sample problem = L 47
L 51 (supposition operation)
L 48–49 (inverse operation). Sample problem = L 50
L 78 (inverse rule of three)
TŚ 31 (rules of five, seven, nine, and eleven)
L 82 (rules of five, seven, nine, and eleven). Sample problems which are the same as TŚ 51, TŚ 52, L 86, L 87
L 78 (use of the inverse rule of three)
L 88 (rule on barter)
L 92 (rule on calculating instalments from a given mixed quantity)

1 Passages from these rules are sometimes mentioned or paraphrased also in other parts of the commentary.

Appendix 2

Rules and sample problems of the GT occurring in other works

GT 65 (sample problem) = TŚ 25
GT 66 (sample problem) = TŚ 27
GT 98 = TŚ 37
GT 105 = TŚ 38
GT 122 = Rule on interest (unidentified source)
GT 124 = BSS 12.14

Appendix 3
Glossaries

3.1 Glossary of mathematical terms (Sanskrit–English)

abhirucita convenient
abhyāsa multiplication
adhas below
adhika plus, increased by
agra in front of
agretana occurring further, next
aikya sum
aṃśa fraction, portion, denominator
aṅka number, digit, quantity
antya last
anuloma regular course
apavarta/apavartana reduction
ardha half
āhati multiplication
āya income, positive quantity
bhagna broken, erased
√*bhañj* to break, remove
bhāga division, fraction, numerator
bhāgadāyin divisor
bhāgahāra division
√*bhāj* to divide
bhājya dividend
√*cal* to move, erase
cheda denominator
dhana capital, positive quantity
dṛśya visible quantity
dṛṣṭa found, obtained
gaṇita mathematics, calculation
gata erased
gati movement
ghana cube
guṇana/guṇanā multiplication

guṇa multiplier, multiplied, multiplication
guṇya multiplicand
ghaṭanā verification
hara divisor
√*hṛ* to divide
hṛti division
lava numerator
iṣṭa known quantity
jāta become, obtained
jāti class, type-problem
karaṇa procedure
karaṇasūtra procedural rule
krama regular order
kṛta performed, executed
kṛti square
kṣaya subtractive, negative quantity
kṣepa additive quantity
√*kṣip* to add
labdha quotient, obtained
√*lik* to write
√*maṇḍ* to arrange
miśraka mixed quantity
√*nikṣip* to add
√*niyoj* to add, unite
nīta brought, obtained
nyāsa numerical presentation, setting down
pada root, quarter
pakṣa side
paṅktī line
parasparam reciprocally
parikarman operation
paścāt afterwards, behind
√*pat* to subtract
pata/patana deduction
pātya subtrahend
phala fruit, accrued interest, interest-rate, result
pramāṇa measure, capital
prathama first
pratiloma reverse, irregular
pūrva original, previous
rāśi heap, quantity, number
rīti method
rūpa one, whole number
ṛṇa subtractive, negative quantity
sama even
samāpta completed, accomplished

saṃkhyā number, calculation
samuttha erased
saṃyoga sum
santāḍana multiplication
saṅkalita addition
√*sañc* to remove
savarṇana homogeneization, reduction of a compound fraction
savarṇita simplified
śeṣa remainder
√*śuddh* to subtract
śuddhi deduction
√*sthā* to place
sthāna place, notational place
sthita placed
udāharaṇa sample problem
ujjhita removed, subtracted
upari above
utpanna produced
ūna decreased by
ūrdhva upper, vertical
varga square
vargamūla square-root
varjita decreased by
vāsanā verification
√*vibhaj* to divide
vidhi procedure
√*vinyas* to put down, write
vṛddhi increment, interest-rate, accrued interest
viṣama odd
viśeṣa deduction
viyoga deduction
vyavahāra practice
vyavakalita subtraction
vyaya expenditure, negative
vyāja interest-rate, accrued interest
√*yā* to disappear, to be erased
yoga sum, union
yojana/*yojanā* sum, union
√*yuj* to add
yuti sum, union

3.2 Glossary of mathematical terms (English–Sanskrit)

Computation and other related terms

Algorithmic or procedural mathematics, arithmetic (including geometry) *pāṭī*
Computation *gaṇanā, gaṇita*
Numerical presentation, layout *nyāsa*
Procedural rule *karaṇasūtra*
Procedure, method *karaṇa, krama, rīti, vidhāna, vidhi*
Reasoning *yuktī*
Result *phala*
Sample problem *udāharana*
Side of the layout *pakṣa*
Verification *ghaṭanā, vāsanā*
To arrange √*maṇḍ*
To bring √*nī*
To erase √*gam*, √*vinaś*
To obtain √*ap*
To perform √*kṛ*
To write √*likh*, √*nyas*

Number and digit

Number/digit/quantity *aṅka, aṅkarāśi, rāśi, rūpa, saṅkhyā*
Positive number *āya, dhana*
Negative number *hīna, ṛṇa, vyaya*
Even number *sama*
Odd number *viṣama*
Notational place *sthāna*
Numerical place *aṅkapada, aṅkasthāna*
Zero *gagana, śūnya, vyad*

Arithmetical operations

Addition *saṅkalita*
Additive term *kṣepa*
Additional, plus *adhika*
Sum *aikya, anubandha, yoga, yojana/yojanā, yuti, samāsa, vṛddhi,*
To add √*yuj*
Subtraction *apavāha, viśuddhi, vyavakalita*
Deduction *patana, viśuddhi*
To subtract √*apanī*, √*hā*, √*rah*
Less by *ūna*

Subtrahend *pātya*
Remainder *agra, antara, śeṣa*
Multiplication *ghāta, guṇana, guṇakāra, hati, vadha*
To multiply √*guṇ*, √*han*, √*taḍ*
Multiplier *guṇa, guṇakāra*
Multiplicand *guṇya*
Division *bhāga, bhāgahāra*
To divide √*bhaj*, √*hṛ*
Divisor *bhāgadāyin, cheda, chedana, hara*
Dividend *bhājyarāśi*
Reduction by a common divisor *apavarta, apavartana*
To reduce √*apavarṭ*
Square *kṛti, varga*
Square-root *kṛtipada, vargamūla*
Cube *ghana*
Cube-root *ghanamūla*

Fractions

Fraction *aṃśa, bhāga, bhinna, kalā*
Numerator *aṃśa, lava*
Denominator *cheda, chedaka, hara, hāra*
Simplification of compound fractions *kalāsavarṇana, savarṇa/savarṇana*
Cross-reduction *kuliśāpavartana*
Exchange of the operands *parivartana, parivṛtti, vinimaya, viparyaya*
Class of simple fractions *bhāgajāti*
Multi-part class *prabhāgajāti*
Class of fractional increase *bhāgānubandhajāti*
Class of fractional decrease *bhāgāpavāhajāti*
Chain-reduction class *vallīsavarṇanajāti*
Visible quantity type-problem *dṛśyajāti*
Remainder type-problem *śeṣajāti*
Difference type-problem *viśleṣajāti*
Remainder-root type-problem *śeṣamūlajāti*
Type-problem of the fraction executed with the root and the visible quantity *mūlāgrabhāgajāti*
Type-problem of the visible quantity at both tips *ubhayāgradṛśyajāti*
Type-problem of the fractions and the visible quantity which is a fraction *bhinnabhāgadṛśyajāti*
Type-problem of the coefficient of the square-root of the fraction *bhāgamūlajāti*
Type-problem of the subtractive square *hīnavargajāti*
Inverse operation *viparītoddeśaka*

Proportion

Rule of three *trairāśika*
Inverse rule of three *vyastatrairāśika*
Rule of five *pañcarāśika*
Rule of seven *saptarāśika*
Rule of nine *navarāśika*
Rule of eleven *ekādaśarāśika*
Barter *bhāṇḍapratibhāṇḍa*
Sale of living beings *jīvavikrayavidhi*

Business-related terms

Expenditure *vyaya*
Wealth *āya, dhana, dravya*
Capital *pramāṇa, pramāṇarāśi*
Loan *mūladhana, pradattarāśi, prayuktarāśi*
Interest/interest-rate *kalāntara, phala, vṛddhi, vyāja,*
Instalment *khaṇḍa*
Bond *patra*
Percentage-rate *śataphala*
Creditor *dhanin, dhanika*
Borrower *grāhaka*
Practice of mixture *miśravyavahāra*
Commission to the moneylender *vyājopajīvivṛtti*
Conversion of several bonds into one *ekapatrakaraṇa*
Equating instalments of capital *samīkaraṇa*

3.3 Glossary of measuring units

Measuring units of money

20 *kapardas*/*varāṭakas* = 1 *kākiṇī*
4 *kākiṇīs* = 1 *paṇa*
16 *paṇas* = 1 *dramma*

Measuring units of gold

6 *yavas* = 1 *niṣpāva*
8 *niṣpāvas* = 1 *dharaṇa*
2 *dharaṇas* = 1 *gadyāṇaka*

Measuring units of length

24 *yava*s = 1 *aṅgula*
24 *aṅgula*s = 1 *hasta*
4 *hasta*s = 1 *daṇḍa*
20 *daṇḍa*s = 1 *rajju*
2 *rajju*s = 1 *nivartana*
2000 *daṇḍa*s = 1 *krośa*
4 *krośa*s = 1 *yojana*

Measuring units of capacity

4 *pādikā*s =1 *mānaka*/*mānikā*
4 *mānaka*s = 1 *setikā*
10 *setikā*s = 1 *hārī*/*hārikā*

Measuring units of weight

14 *niṣpāvaka*s = 1 *dhaṭaka*
10 *dhaṭaka*s = 1 *pala*

Measuring units of time

60 *praṇā*s = 1 *viṇāḍī*
60 *viṇāḍī*s = 1 *ghaṭikā*
60 *ghaṭikā*s = 1 *ahorātra*
30 *ahorātra*s = 1 *māsa*
12 *māsa*s = 1 *saṃvatsara*

Bibliography

Primary Sources

Gaṇitatilaka of Śrīpati, with the commentary of Siṃhatilakasūri, ed. H. R. Kāpadīā. Baroda: Oriental Institutes, 1937.

Gaṇitasārakaumudī of Ṭhakkura Pherū, ed. and transl. SaKHYa. Delhi: Manohar, 2009.

Ganitasārasangraha of Mahāvīrācārya, ed. and transl. M. Raṅgācārya. Madras: Superintendent Government Press, 1912.

Kāvyālaṅkārasūtrāṇi of Vāmana, ed. N. R. Acharya. Saṃskaraṇa: Bombay, 1953.

Kharataragaccha bṛhad gurvāvali of Jinapāla, ed. M. Jinavijaya. Bombay: Singhi Jain Shastra Shikshapith, 1956.

Līlāvatī of Bhāskarācārya, with Gaṇeśa's *Buddhivilāsinī* and Mahīdhara's *Līlāvatīvivaraṇa*, ed. D. Āpaṭe, 2 vols. Poona: Ānandāśrama Press, 1937.

Līlāvatī of Bhāskarācārya, with *Kriyā-kramakarī*, ed. K. V Sarma. Hośiāpuram: Viśveśvarānanda-vaidika-śodha-saṃsthānam, Panjab University, 1975.

Mantrarājarahasya of Siṃhatilakasūri, ed. M. Jinavijaya. Bambaī: Siṅghī Jainaśāstra śikṣāpīṭha, Bhāratīya Vidyā Bhavana, 1980.

Pāṭīgaṇita of Śrīdharācārya, with an ancient Sanskrit commentary, ed. and transl. K. S Shukla. Lucknow: Lucknow University, 1959.

Siddhāntaśekhara of Śrīpati, ed. B. Miśra. 2 vols. Calcutta: University of Calcutta, 1932, 1947.

Triśatīkā of Śrīdhara, ed. Dvivedī. Benares: Pandit Jagannātha Śarmā Mehtā, 1899.

Secondary Sources

Acharya, N. R. (ed.) 1953. *Kāvyālaṅkārasūtrāṇi* of Vāmana. Muṃbaī: Nirṇayasāgara Presa.

Āpaṭe, D. (ed.) 1937. *Līlāvatī* of Bhāskarācārya, with Gaṇeśa's *Buddhivilāsinī* and Mahīdhara's *Līlāvatīvivaraṇa*. 2 vols. Poona: Ānandāśrama Press.

Bag, A. K. 1979. *Mathematics in Ancient and Medieval India*.Varanasi: Vidya Vilas Press.

Balbir, N. 2003. "The A(ñ)calagaccha Viewed from Inside and from Outside". In: *Jainism and Early Buddhism: Essays in Honor of Padmanabh S. Jaini*, ed. O. Qvarnström, 47–77. Fremont: Asian Humanities Press.

—— et al. 2006. *Catalogue of the Jain Manuscripts of the British Library: Including the Holdings of the British Museum and the Victoria & Albert Museum.* 3 vols. London: British Library: Institute of Jainology.

——. 2012. "Genealogical Discourse and Jain Sectarian Promotion: Mantri-Karmacandra Vaṃśāvalī-Prabandha and the Kharataragaccha". In: *Jaina Studies: Proceedings of the DOT 2010 Panel in Marburg, Germany*, ed. Jayandra Soni, 33–63. Delhi: Aditya Prakashan.

Balcerowicz, P., and Potter, K. H. (eds.) 2013. *Jain Philosophy* (Part II). Delhi: Motilal Banarsidass.

Bronkhorst, J. 2006. "Commentaries and the History of Science in India". *Asiatische Studien/Études Asiatiques* 60 (4): 773–788.

——. 2007. "Science and Religion in Classical India". *Indologica Taurinensia* 33:183–196.

——. 2016. "Jaina versus Brahmanical Mathematicians". *International Journal of Jaina Studies* 12 (1): 1–10.

Caillat, C., and Kumar, R. 1981. *The Jain Cosmology*. Ravi Kumar: Basel.

Chemla, K. 2004. "History of Science, History of Text: An Introduction". In: *History of Science, History of Text*, ed. K. Chemla, vii–xxvii. Dordrecht: Springer.

Clark, W. E. 1930 (transl.) *The Āryabhaṭīya of Āryabhaṭa: An Ancient Indian Work on Mathematics and Astronomy*. Chicago: University of Chicago Press.

Colebrooke, H. T. 1973 [1817]. *Algebra with Arithmetic and Mensuration from the Sanscrit of Brahmegupta and Bhascara*. London: John Murray.

Cort, J. 1991. "The Śvetāmbar Mūrtipūjak Jain Mendicant". *Man* (N.S.) 24(4): 651-–671.

——. 2001. "The Intellectual Formation of a Jain Monk: A Śvetāmbara Monastic Curriculum". *Journal of Indian Philosophy* 29: 327–349.

——. 2009a. "Contemporary Jain Maṇḍala Rituals." In: *Victorious Ones: Jain Images of Perfection*, ed. P. Granoff, 140–157. New York: Rubin Museum of Art; Ahmedabad: Mapin Publishing.

——. 2009b. "An Epitome of Medieval Śvetāmbara Jain Literary Culture: A Review and Study of Jinaratnasūri's *Līlāvatīsāra*". *International Journal of Jaina Studies* (Online) 5(1): 1–33.

Datta, B. B. 1928–1929. "On Mahāvīra's Solution of Rational Triangles and Quadrilaterals." *Bulletin of Calcutta Mathematical Society* 20: 267–294.

——. 1929. "The Jaina School of Mathematics." *Bulletin of Calcutta Mathematical Society* 27: 115–145.

——. 1935. "Mathematics of Nemicandra". *The Jaina Antiquary* 1 (2): 25–44.

——. 1936. "A Lost Jaina Treatise on Arithmetic". *The Jaina Antiquary* 2 (2): 38–41.

Datta, B., and Singh, A. N. 1962 [1935]. *History of Hindu Mathematics: A Source Book*. 2 vols. Bombay: Asia Publishing House.

Delire, J. M. 2016. *Les mathématiques de l'autel védique: Le Baudhāyana Śulbasūtra et son commentaire Śulbadīpikā*. Édition critique, traduction et commentaire. Hautes Études orientales, 54. Genève: Librairie Droz.

Dundas, P. 1998. "Becoming Gautama". In: *Opening Boundaries: Jain Communities and Cultures in Indian History*, ed. J. E. Cort, 31–52. New York: SUNY Press.

——. 2000. "The Jain Monk Jinapati Sūri Gets the Better of a Nāth Yogī". In: *Tantra in Practice*, ed. D. G. White, 231–238. Princeton, NJ and Oxford: Princeton University Press.

——. 2002. *The Jains*. 2nd ed. London: Routledge.

——. 2007. *History, Scripture and Controversy in a Medieval Jain Sect*. London: Routledge.

——. 2009. "How Not to Install an Image of the Jina: An Early Anti-Paurṇamīyaka Diatribe." *International Journal of Jain Studies* (Online) 5 (3): 1–23.

Dvivedī, S. (ed.) 1899. *Triśatīkā* of Śrīdhara. Benares: Pandit Jagannātha Śarmā Mehtā.

Filliozat, P.S. 2004. "Ancient Sanskrit Mathematics: An Oral Tradition and a Written Literature". In: *History of Science, History of Text,* ed. K. Chemla, R. S. Cohen, J. Renn, et al., 137–157 (Boston Series in the Philosophy of Science), Dordrecht: Springer Netherlands.

Formigatti, C. A. 2011. *Sanskrit Annotated Manuscripts from Northern India and Nepal*. PhD Dissertation. University of Hamburg.

Goodall, D. 2004. *The Parākhyatantra, a scripture of the Śaiva Siddhānta. A critical edition and annotate translation*. Pondicherry: Institut Français de Pondichéry, École Française D'Extrême-Orient.

Gough, E. 2012. "Shades of enlightenment: a Jain tantric diagram and the colours of the Tīrthaṅkaras". *International Journal of Jaina Studies* (Online) 8 (1): 1–47.

——. 2017. "The Sūrimantra and the Tantricization of Jain Image". In: *Consecration Rituals in South Asia*, ed. I. Keul, 265–308. Leiden: Brill.

Gupta R.C. 1992a. "Jaina Cosmography and Perfect Numbers" *Arhat Vacana* 4: 69–94.

——. 1992b. "The First Innumerable Number in Jaina Mathematics" *Gaṇita Bhāratī* 14: 11–24.

Hayashi, T. 1991. "The Pañcaviṃśatikā in Its Two Recensions: A Study in the Reformation of a Medieval Sanskrit Mathematical Textbook". *Indian Journal of History of Science* 26, 399–48.

——. 1994. "Indian Mathematics". In: *Companion of the History and Philosophy of the Mathematical Sciences,* ed. I. Grattan-Guinness, vol. 1, 118–130. London and New York: Routledge.

——. 1995a. *The Bakhshālī Manuscript: An Ancient Indian Mathematical Treatise*. Groningen: Egbert Forsten.

——. 1995b. "Śrīdhara's Authorship of the Mathematical treatise *Gaṇitapañcaviṃśī*." *Historia Scientiarium* 4 (3): 233–250.

——. 2003. "Indian Mathematics". In: *The Blackwell Companion to Hinduism,* ed. G. Flood, 360–375. Oxford: Blackwell Publishing.

——. 2013. "Authenticity of the Verses in the Printed Edition of the *Gaṇitatilaka*". *Gaṇita Bhāratī* 35 (1–2): 55–74.

——. 2014. "Arithmetic in India: *Pāṭīgaṇita*". In: *Encyclopaedia of the History of Science, Technology, and Medicine in Non-Western Cultures*, ed. H. Selin, 239–240. Dordrecht and London: Springer. DOI: 10.1007/978-94-007-3934-5_9208-2.

——. 2017. "The Bālabodhāṅkavṛtti: Śambhudāsa's Old-Gujarātī Commentary on the Anonymous Sanskrit Arithmetical Work Pañcaviṃśatika". *SCIAMVS* 18: 1–132.

Høyrup, J. 2012. "Sanskrit-Prakrit Interaction in Elementary Mathematics as Reflected in Arabic and Italian Formulations of the Rule of Three – and Something More on the Rule Elsewhere". Contribution to the International Seminar on History of Mathematics, Ramjas College, Delhi University, November 19–20, 2012. Max-Planck-Institut für Wissenschaftsgeschichte. Preprint 435, Berlin.

Jain, K. C. 1972. *Ancient Cities and Towns of Rajasthan: A Study of Culture and Civilization*. Delhi: Motilal Banarsidass.

Janert, K. L., and Poti, N. 1982. *Indische Handschriften.* Wiesbaden: Steiner.

Jhaveri, M. B. 1944. *A Comparative and Critical Study of Mantraśāstra*. Ahmedabad: Sarabhai Maninal Nawals.

Jinavijaya, M. (ed.) 1956. *Kharataragaccha bṛhad gurvāvali* of Jinapāla. Bombay: Singhi Jain Shastra Shikshapith.

——. (ed.) 1980. *Mantrarājarahasya* of Siṃhatilakasūri. Bambaī: Siṅghī Jainaśāstra Śikṣāpīṭha, Bhāratīya Vidyā Bhavana.

Joseph, G. G. 2011. *The Crest of the Peacock: Non-European Roots of Mathematics*. 3rd ed. Princeton, NJ: Princeton University Press.

Kane, G. 1969. "Conjectural Emendation". In: *Medieval Literature and Civilization. Studies in Memory of G.N. Garmonsway*, ed. D. A. Pearsall and R. A. Waldron, 155–169. London.

Katre, S. M. 1954. *Introduction to Indian Textual Criticism*. 2nd ed. Deccan College Hand-Book Series, 5. Poona: Deccan College, Postgraduate and Research Institute.

Kaye, G. R. 1908. "Notes on Indian mathematics" *Journal of the Asiatic Society of Bengal* 4: 111–141.

——. 1915. *Indian Mathematics*. Calcutta, Simla: Thacker, Spink & Co.

——. 1927–1933. *The Bakhshālī Manuscript: A Study in Medieval Mathematics*. Calcutta: Government of India, Central Publication Branch.

Kāpadīā, H. R. 1934. "A Note on Jaina Hymns and Magic Squares". *Indian Historical Quarterly* 10: 140–153.

——. (ed.) 1937. *Gaṇitatilaka* of Śrīpati, with the Commentary of Siṃhatilakasūri. Baroda: Oriental Institutes.

——. 1941. *A History of the Canonical Literature of the Jainas*. Gopipura, Surat: Hiralal Rasikdas Kapadia.

Keller, A. 2006. *Expounding the Mathematical Seed: A Translation of Bhāskara I on the Mathematical Chapter of the Āryabhaṭīya*. Basel: Birkhäuser Verlag.

——. 2010. "On Sanskrit Commentaries Dealing with Mathematics (Fifth-Twelfth Century)". In: *Looking at It from Asia: Phe processes that Shaped the Sources of History of Science*, ed. F. Bretelle-Establet, 211–244. Dordrecht and New York: Springer.

——. 2014. "Mathematics Education in India". In: *Handbook on the History of Mathematics Education*, eds. Karp and Schubring, 55–83. New York: Springer Science+Business Media.

Keller, A., and Morrice-Singh, C. 2014. "Multiplying Integers: On the Diverse Practices of Medieval Sanskrit Authors". [PDF] available online at: halshs.archives-ouvertes.fr/halshs-01006135v2/document.

Kusuba, T. 1993. *Combinatorics and Magic Squares in India: A Study of Nārāyaṇa Paṇḍita's Gaṇitakaumudī*. Chapters 13–14. PhD dissertation. Providence, RI: Brown University.

——. 2004. "Indian Rules for the Decomposition of Fractions". In: *Studies in the History of the Exact Sciences in Honour of David Pingree*, eds. C. Burnett et al., 497–516. Leiden: Brill.

Kusuba, T., and Plofker, K. 2013. "Indian Combinatorics". In: *Combinatorics: Ancient and Modern*, eds. R. J. Wilson and J. J. Watkins, 41–64. Oxford: Oxford University Press.

Mehta, M. L., and Malvania, D. B. 1969. *Jaina sāhitya kā bṛrhad itihāsa*. vol. 5. Vārāṇasī: Pārśvanātha Vidyāśrama Śodha Saṃsthāna.

Miśra, B. (ed.) 1932, 1947. *Siddhāntaśekhara* of Śrīpati. 2 vols. Calcutta: University of Calcutta.

Morice-Singh, C. 2015. "Mathématiques et Cosmologie Jaina: Nombres et Algorithmes dans le Gaṇitasārasaṃgraha et la Tiloyapaṇṇattī". PhD dissertation. Paris 3: Universite Sorbonne Nouvelle.

Murthy, S. 1996. *Introduction to Manuscriptology*. Delhi: Sharada Publishing House.

Olivelle, P. 2009. *The Law Code of Viṣṇu: A Critical Edition and Annotated Translation of the Vaiṣṇava-Dharmaśāstra*. Cambridge, MA: Department of Sanskrit and Indian Studies, Harvard University: Distributed by Harvard University Press.

Patte, F. 2004. *Le Siddhāntaśiromaṇi: L'oeuvre mathématique et astronomique de Bhāskarācārya*. Geneva: Librarie Froz.

Petrocchi, A. 2015. "A new Theoretical Approach to Sample Problems and Deductive Reasoning in Sanskrit Mathematical Texts". *Journal of The British Society for the History of Mathematics* 30 (1): 2–19.

——. 2016a. "Early Jaina Cosmology, Soteriology, and Theory of Numbers in the Aṇuogaddārāiṃ: An Interpretation". *Journal of Indian Philosophy* 45 (2): 235–255.

——. 2016b. "The *Bhūtasaṃkhyā* Notation: Numbers, Culture, and Language in Sanskrit Mathematical Literature". In: *On Meaning and Mantras: Essays in Honor of Frits Staal*, ed. G. Thompson and R. K. Payne, 477–502. Berkeley: Institute of Buddhist Studies and BDK America.

——. 2017. "Philosophy of Space-Time in Early Jaina Thought: Quantification as a Means of Knowing". *Religions of South Asia* 10 (3) 237–256.

Pingree, D. 1970–1994. *Census of the Exact Sciences in Sanskrit*. Series A, vols. 1–5. Philadelphia: American Philosophical Society.

——. 1981. *Jyotiḥśastra: Astral and Mathematical Literature*. Weisbaden: Otto Harrassowitz.

Plofker, K. 2007. "Mathematics in India". In: *The Mathematics of Egypt, Mesopotamaia, China, India, and Islam: A Sourcebook*, ed. V. Katz, 385–514. Princeton, NJ and Oxford: Princeton University Press.

——. 2009. *Mathematics in India*. Princeton, NJ and Oxford: Princeton University Press.

——. 2009b. "Spoken Text and Written Symbol: The Use of Layout and Notation in Sanskrit Scientific Manuscripts". In: *Digital Proceedings of the Lawrence J. Schoenberg Symposium on Manuscript Studies in the Digital Age* 1 (1), available online at: repository.upenn.edu/cgi/viewcontent.cgi?referer=https://www.google.co.uk/&httpsredir=1&article=1013&context=ljsproceedings [Accessed on October 2016].

Plofker, K. et al. 2017. "The Bakhshālī Manuscript: A Response to the Bodleian Library's Radiocarbon Dating". *History of Science in South Asia* 5 (1) 134–150.

Prasad, P. 2007. *Lekhapaddhati: Documents of State and Everyday Life from Ancient and Early Medieval Gujarat, 9th to 15th Centuries*. New Delhi and Oxford: Oxford University Press.

Ramanujacharya and Kaye, G. R. 1912–1913. "The Triśatikā of Śrīdharācārya". *Bibliotheca mathematica* 3 (13): 203–2017.

Raṅgācārya, M. (ed. and transl.) 1912. *Ganitasārasangraha* of Mahāvīrācārya. Madras: Superintendent Government Press.

Rath, S. 2012. *Aspects of Manuscript Culture in Ancient India*. Leiden: Brill.

Roth, G. 1974. "Notes on Pamca-Namokkara Parama-Mangala in Jaina Literature". *Brahmavidya* (*Adyar Library Bullettin*) 38, 1–18.

Royse, J. R. 2008. *Scribal Habits in Early Greek New Testament Papyri*. Leiden: Brill.

SaKHYa (ed. and transl.) 2009. *Gaṇitasārakaumudī* of Ṭhakkura Pherū. Delhi: Manohar.

Sanderson, A. 2009. "The Śaiva Age – The Rise and Dominance of Śaivism during the Early Medieval Period." In: *Genesis and Development of Tantrism*, ed. S. Einoo, 41–349. Tokyo: University of Tokyo, Institute of Oriental Culture.

Sandesara, B. J. 1953. *Literary Circle of Mahāmātya Vastupāla and Its Contribution to Sanskrit Literature*. Bombay: Singhi Jain Shastra Sikshapith, Bharatiya Vidya Bhavan.

Sarasvati Amma, T. A. 1979. *Geometry in Ancient and Medieval India*. Delhi: Motital Banarsidass.

Sarma, K. V. (ed.) 1975. *Līlāvatī* of Bhāskarācārya, with *Kriyā-kramakarī*. Hoshiapur: VVBIS& IS, Panjab University.

Sarma, S. R. 1989. "Vedic Mathematics vs. Mathematics in the Veda". *Vāṇjyotiḥ* 4: 53–65.

—. 1997. "Some Medieval Arithmetical Tables". *Indian Journal of History of Science* 32: 119–198.

—. 2003. "Rule of Three and Its Variations in India". In: *From China to Paris: 2000 Years of Transmission of Mathematical Ideas*, ed. Y. Dold-Samplonius et al., 133–156. Stuttgart: Franz Steiner Verlag.

—. 2009. "On the Rationale of the Maxim Aṅkānāṃ Vāmato Gatiḥ" *GaṇitaBhāratī* 31 (1–2): 65–86.

Sen, S. N., and Bag, A. K. 1983. *The Śulbasūtras*. New Delhi: Indian National Science Academy.

Sheikh, S. 2010. *Forging a Region: Sultans, Traders, and Pilgrims in Gujarat, 1200–1500*. New Delhi and Oxford: Oxford University Press.

Shukla, K. S. (ed. and transl.) 1959. *Pāṭīgaṇita* of Śrīdharācārya, with an Ancient Sanskrit Commentary. Lucknow: Lucknow University.

Shukla, K. S., and Sarma, K. V. (ed. and transl.) 1976. *Āryabhaṭīya of Āryabhaṭa*. New Delhi: Indian National Science Academy.

Singh, A. N. 1959. "History of Mathematics in India from Jaina Sources". *The Jaina Antiquary* 15 (2): 46–53.

Sinha, K. N. 1982. "Śrīpati's Gaṇitatilaka: English Translation with Introduction". *Gaṇita Bhāratī* 4 (3–4): 114–133.

Speyer, J. S. 1886. *Sanskrit Syntax*. Leiden: Brill.

Srinivasiengar, C. N. 1967. *The History of Ancient Indian Mathematics*. Calcutta: World Press.

Staal, F. 1995. "T*he Sanskrit of Science*". *Journal of Indian Philosophy* 23: 73—127.

—. 1999. "Greek and Vedic geometry". *Journal of Indian Philosophy* 27: 105–127.

—. 2006. "Artificial Languages across Sciences and Civilizations". *Journal of Indian Philosophy* 34: 89–141.

Strauch, I. 2002. *Die Lekhapaddhati-Lekhapañcāśikā: Briefe und Urkunden im mittelalterlichen Gujarat: Text, Übersetzung, Kommentar.* Berlin: D. Reimer.

Tanselle, G. T. 1990. *Textual Criticism and Scholarly Editing*. Charlottesville: University Press of Virginia.

Thibaut, G. 1875. "On the Sulba-Sutra". *Journal of the Asiatic Society of Bengal* 44: 7–75.

Tov, E. 2012. *Textual Criticism of the Hebrew Bible*. 3rd ed. Minneapolis: Fortress Press.

Van Egmond, W. 1976. "The Commercial Revolution and the Beginnings of Western Mathematics in Renaissance Florence, 1300–1500." PhD dissertation. Indiana University.

Vinyasāgara, M. (ed.) 2005a. *Kharataragaccha pratiṣṭhā lekha saṃgraha*, vol. 1. Jaipur, Rajastham: Prakrit Bharati Academy.

—. 2005b. *Kharataragaccha kā Bṛhad Itihāsa.* Jaipur, Rajasthan: Prakrit Bharati Academy.

—. 2006. *Kharataragaccha Sāhitya Kośa*. Jayapura: Prākṛta Bhāratī Akādamī.

Vose, S. M. 2013. "The Making of a Medieval Jain Monk: Language, Power, and Authority in the Works of Jinaprabhasūri (ca. 1261–1333)". PhD Dissertation. University of Pennsylvania.

West, M. L. 1973. *Textual Criticism and Editorial Technique Applicable to Greek and Latin Texts*. Stuttgart: B.G. Teubner.

Whitney, W. D. 1924. *Sanskrit Gammar: Including Both the Classical Languages and the Older Dialects of Veda and the Brahmana.* 5th ed. New Delhi: D.K. Publishers Distributors (P) Ltd.

Yano, M. 2006. "Oral and Written Transmission of the Exact Sciences in Sanskrit". *Journal of Indian Philosophy* 34, 143–160.

Index

Note: Page numbers in **bold** refer to tables.

addition *(saṅkalita)* [GT 13–14] 51–54; of fractions *(bhinnasaṅkalita)* [GT 35–37] 82–90; of fractions *(bhinnasaṅkalita)* [GT 35–37] text analysis 314–319; text analysis of 288–291
arithmetic of zero [GT 52] 120–125; text analysis of 331–332
arithmetical operations with fractions [GT 35–51] 82–120; text analysis of 313–331
arithmetical operations with integers [GT 13–34] 51–82; text analysis of 287–313
Āryabhaṭīya by Āryabhaṭa (476 CE) 4

barter *(bhāṇḍapratibhāṇḍa)* [GT 112–114] 253–255; text analysis of 397–398
bees, sample problems with 172–177, 197–200, 354–356, 366–367
Benedictory verse *(maṅgalācaraṇa)* [GT 1] 47–48; text analysis of 285–286
Brahmagupta 4, 7–8, 16, 41, 263, 333
Brāhmasphuṭasiddhānta (BSS, Brahmagupta) 4, 9, 25, 41, 400, 405

classes of simplification of fractions [GT 53–63] 125–156; text analysis of 332–347
cube *(ghana)* [GT 28–31] 72–77; of fractions *(bhinnaghana)* [GT 48–49] 110–114; of fractions *(bhinnaghana)* [GT 48–49] text analysis 327–328; text analysis of 306–309
cube-root *(ghanamūla)* [GT 32–34] 77–82; of fractions *(bhinnaghanamūla)* [GT 50–51] 114–120; of fractions *(bhinnaghanamūla)* [GT 50–51] text analysis 328–331; text analysis of 309–313

deer, sample problems with 184–187, 361–362
division *(bhāgahāra)* [GT 21–22] 60–62; of fractions *(bhinnabhāgahāra)* [GT 42–43] 99–103; of fractions *(bhinnabhāgahāra)* [GT 42–43] text analysis 321–323; text analysis of 298–300
Dundas, P. 13, 15

elephants, sample problems with 194–197, 218–220, 230, 235–236, 350–353, 365–367, 371, 375–376

fractions addition of fractions *(bhinnasaṅkalita)* [GT 35–37] 82–90; addition of fractions *(bhinnasaṅkalita)* [GT 35–37] text analysis 314–319; arithmetical operations with fractions [GT 35–51] 82–120; arithmetical operations with fractions [GT 35–51] text analysis 313–331; classes of simplification of fractions [GT 53–63] 125–156; classes of simplification of fractions [GT 53–63] text analysis 332–347; cube of fractions *(bhinnaghana)* [GT48–49] 110–114; cube of fractions *(bhinnaghana)* [GT48–49] text analysis 327–328; cube-root of fractions *(bhinnaghanamūla)* [GT 50–51] 114–120; cube-root of fractions *(bhinnaghanamūla)* [GT 50–51] text analysis 328–331; division of fractions *(bhinnabhāgahāra)* [GT 42–43] 99–103; division of fractions *(bhinnabhāgahāra)* [GT 42–43] text analysis 321–323; multiplication of fractions *(bhinnapratyutpanna)* [GT 40–41] 96–99; multiplication of fractions *(bhinnapratyutpanna)* [GT 40–41] text analysis 320–321; square of fractions *(bhinnavarga)* [GT 44–45] 103–106; square of fractions *(bhinnavarga)* [GT 44–45] text analysis 323; square-root of fractions *(bhinnavargamūla)* [GT 46–47] 106–110; square-root of fractions *(bhinnavargamūla)* [GT 46–47] text analysis 323–327; subtraction of fractions *(bhinnavyavakalita)* [GT 38–39] 90–96; subtraction of fractions *(bhinnavyavakalita)* [GT 38–39] text analysis 319–320; type-problems of fractions [GT 64–92] 156–221; type-problems of fractions [GT 64–92] text analysis 348–376

Gaṇitasārasaṃgraha (GSS, Mahāvīrācārya) 5, 16, 286, 288, 308, 348, 376, 379, 400

Gaṇitatilaka (GT, Śrīpati) "variegated metrical stanzas" *(vicitravṛtta)* in 17, 286; authorship and content of 5–7; as fragmentary text 41; Kāpadīā's brief and vague description of MS of 27; Kāpadīā's edition's linguistic phenomena and oddities in 28–34; Kāpadīā's Sanskrit edition of 6, 26–28; lines of transmission and internal evidence in 34–36; methodological notes on philological perspective of study of 36–38; new verse numbering system for by Hayashi 26, 39–41, 40–41; pattern of exposition of 16–17; rules expounded in 39–41, 40–41; Siṃhatilakasūri as commentator on 12–16; textual criticism of 37–39; translation challenges in 26; as work of *pāṭīgaṇita* 7–12 *See also Ganitatilaka* translation; GT and SGT text analysis; Śrīpati

Gaṇitatilaka translation editorial conventions in 46; GT 1 Benedictory verse *(maṅgalācaraṇa)* 47–48; GT 112–114 barter *(bhāṇḍapratibhāṇḍa)* 253–255; GT 115–117 rule of the sale of living beings 255–257; GT 118–133 practices 257–281; GT 13–14 addition *(saṅkalita)* 51–54; GT 15–16 subtraction *(vyavakalita)* 54–55; GT 17–20 multiplication *(guṇakāra)* 56–60; GT 2–12 technical terms *(paribhāṣā)* 48–51; GT 21–22 division, *(bhāgahāra)* 60–62; GT 23–25, square *(varga)* 62–68; GT 26–27 square-root *(vargamūla)* 68–72; GT 28–31, cube *(ghana)* 72–77; GT 32–34, cube-root *(ghanamūla)* 77–82; GT 35–37 addition of fractions *(bhinnasaṅkalita)* 82–90; GT 38–39 subtraction of fractions *(bhinnavyavakalita)* 90–96; GT 40–41 multiplication of fractions *(bhinnapratyutpanna)* 96–99; GT 42–43 division of fractions *(bhinnabhāgahāra)* 99–103; GT 44–45 square of fractions *(bhinnavarga)* 103–106; GT 46–47 square-root of fractions *(bhinnavargamūla)* 106–110; GT 48–49 cube of fractions *(bhinnaghana)* 110–114; GT 50–51 cube-root of fractions *(bhinnaghanamūla)* 114–120; GT 52 arithmetic of zero 120–125; GT 53–63 classes of simplification of fractions 125–156; GT 64–92 type-problems of fractions 156–221; GT 93–94 inverse operation 221–226; GT 95–117 rules on proportion 226–257; Homage to the Jina 47; notes on English translation of 43–46 *See also* GT and SGT text analysis

GT. *See Gaṇitatilaka* (GT, Śrīpati)

GT and SGT text analysis 37–39; addition *(saṅkalita)* [GT 13–14] 288–291; addition of fractions *(bhinnasaṅkalita)* [GT 35–37] 314–319; arithmetic of zero [GT 52] 331–332; arithmetical operations with fractions [GT 35–51] 313–331; arithmetical operations with integers [GT 13–34] 287–313; barter *(bhāṇḍapratibhāṇḍa)* [GT 112–114] 397–398; Benedictory verse *(maṅgalācaraṇa)* [GT 1] 285–286; classes of simplification of fractions [GT 53–63] 332–347; cube *(ghana)* [GT 28–31] 306–309; cube of fractions *(bhinnaghana)* [GT48–49] 327–328; cube-root *(ghanamūla)*

[GT 32–34] 309–313; cube-root of fractions *(bhinnaghanamūla)* [GT 50–51] 328–331; division *(bhāgahāra)* [GT 21–22] 298–300; division of fractions *(bhinnabhāgahāra)* [GT 42–43] 321–323; inverse operation *(viparītoddeśaka)* [GT 93–94] 376–378; multiplication *(guṇakāra)* [GT 17–20] 293–298; multiplication of fractions *(bhinnapratyutpanna)* [GT 40–41] 320–321; practices [GT 118–133] 400–417; rule of the sale of living beings *(jīvavikraya)* [GT 115–117] 399–400; rules on proportion [GT 95–117] 378–400; square *(varga)* [GT 23–25] 300–304; square of fractions *(bhinnavarga)* [GT 44–45] 323; square-root *(vargamūla)* [GT 26–27] 304–306; square-root of fractions *(bhinnavargamūla)* [GT 46–47] 323–327; subtraction *(vyavakalita)* [GT 15–16] 291–293; subtraction of fractions *(bhinnavyavakalita)* [GT 38–39] 319–320; technical terms *(paribhāṣā)* [GT 2–12] 286–287; type-problems of fractions [GT 64–92] 348–376

Hayashi, T. 7, 10; new verse numbering system of for GT 26, 39–41, 40–41, 45
Homage to the Jina 47

integers arithmetical operations with integers [GT 13–34] 51–82; arithmetical operations with integers [GT 13–34] text analysis 287–313
inverse operation *(viparītoddeśaka)* [GT 93–94] 221–226; text analysis of 287–313

Kāpadīā. *See under Gaṇitatilaka* (GT, Śrīpati)

leaves, sample problems with 179–180, 357–358
Līlāvatī (Bhāskarācārya) 5, 8–11, 20, 41; arithmetic of zero in 121, 123; barter in 253; inverse operations with fractions in 223; mathematical topics covered in **10**; square in 66; visible quantity type-problem in 160

Malvania, D.B. 12
Mantrarājarahasya (Siṃhatilakasūri) 12–14, 33, 285
mathematics *śāstric* tradition of 4, 17, 36; topics in *Līlāvatī* and *Triśatī* **10**; topics in the GT **40–41** *See also* Sanskrit mathematical writings in medieval India
measuring and measurement 48–51
Mehta, M.L. 12
monkeys, sample problems with 189–191, 362–363
mūla-texts 18–19
multiplication *(guṇakāra)* [GT 17–20] 56–60; of fractions *(bhinnapratyutpanna)* [GT 40–41] 96–99; of fractions *(bhinnapratyutpanna)* [GT 40–41] text analysis 320–321; text analysis of 293–298

parrots, sample problems with 180–182, 213, 358–359
pāṭīgaṇita, Gaṇitatilaka as 7–12
peacocks, sample problems with 215–218, 374–375
philological studies 36–38
pigs, sample problems with 192–193, 363–364
Pingree, D. 12
practices [GT 118–133] 257–281; commission to the moneylender *(vyājopajīvivṛtti)* [GT 120–121] 259–262; conversion of several bonds into one *(ekapatrakaraṇa)* [GT 127–130] 265–275; equating instalments of capital *(samīkaraṇa)* [GT 131–133] 275–281; practice of mixture *(miśravyavahāra)* [GT 118–119] 257–259; rule on interest [GT 122–123] 263; rule on time and double capital [GT 124–126] 263–265; text analysis of 400–417
proportion, rules on [GT 95–117] 226–257; rule of eleven 251–253; rule of five *(pañcarāśika)* [GT 107–111] 240–249; rule of three *(trairāśika)* [GT 95–106] 226–239; text analysis of 378–400

rule of eleven 251–253
rule of five *(pañcarāśika)* [GT 107–111] 240–249; text analysis of 389–395
rule of the sale of living beings *(jīvavikraya)* [GT 115–117] 255–257; text analysis of 399–400
rule of three *(trairāśika)* [GT 95–106] 226–239; text analysis of 378–380
rule on interest [GT 122–123] 263; text analysis of 405

rules on proportion [GT 95–117] 226–257; rule of eleven 251–253; rule of five *(pañcarāśika)* [GT 107–111] 240–249; rule of three *(trairāśika)* [GT 95–106] 226–239; text analysis of 378–400

Sanskrit: Kāpadīā's Sanskrit edition of *Gaṇitatilaka* 6, 26–28
Sanskrit mathematical writings in medieval India 3–5, 36–37; dichotomy between orality and literacy in 17; verse-treatises and prose-commentaries genres in 16–19
śāstric tradition of mathematics 4, 17, 36
SGT (commentary on GT) 8; language and style in 19–22; *Līlāvatī* by Bhāskarācārya and 8, 10–11, 20; lines of transmission and internal evidence in 34–36; linguistic phenomena and oddities in 28–34; methodological notes on philological perspective of study of 36–38; notes on English translation of 43–46; pattern of exposition in 16–17; textual criticism of 36–39; translation challenges in 26 *See also* GT and SGT text analysis; Siṃhatilakasūri
siddhāntas 3–4
Siṃhatilakasūri (commentator on GT) 6, 11–12; Jainism of 13–16, 23–24, 285; *Mantrarājarahasya* by 12–14, 33, 285; Sanskrit works by 12–16 *See also* GT and SGT text analysis; SGT
square *(varga)* [GT 23–25] 62–68; of fractions *(bhinnavarga)* [GT 44–45] 103–106; of fractions *(bhinnavarga)* [GT 44–45] text analysis 323; text analysis of 300–304
square-root *(vargamūla)* [GT 26–27] 68–72; of fractions *(bhinnavargamūla)* [GT 46–47] 106–110; of fractions *(bhinnavargamūla)* [GT 46–47] text analysis 323–327; text analysis of 304–306
Śrīpati 4–5, 7–9, 11 *See also Gaṇitatilaka* (GT, Śrīpati)
subtraction *(vyavakalita)* [GT 15–16] 54–55; of fractions *(bhinnavyavakalita)* [GT 38–39] 90–96; of fractions *(bhinnavyavakalita)* [GT 38–39] text analysis 319–320; text analysis of 291–293
swans, sample problems with 169–170, 211–213, 353, 371–372

technical terms *(paribhāṣā)* [GT 2–12] text analysis 286–287; translation 48–51
textual criticism 36–37; theory and praxis of 38–39 *See also* GT and SGT text analysis
Triśatī (Śrīdhara) 5, 7–11, 20, 23, 41, 45, 47; mathematical topics covered in **10**; square in 68–69
type-problems of fractions [GT 64–92] 156–221; text analysis of 348–376

zero, arithmetic of [GT 52] 120–125; text analysis of 331–332

9781032338538